Principles and Practices of Chemistry

Principles and Practices of Chemistry

Edited by **Saul Rowen**

Willford Press

New York

Published by Willford Press,
118-35 Queens Blvd., Suite 400,
Forest Hills, NY 11375, USA
www.willfordpress.com

Principles and Practices of Chemistry
Edited by Saul Rowen

International Standard Book Number: 978-1-68285-269-9 (Hardback)

Printed in China.

Contents

Preface

Principles of chemistry are of significant relevance especially for interdisciplinary fields like biochemistry, biotechnology, etc. This book attempts to understand the multiple branches that fall under the discipline of chemistry. Supramolecular and macromolecular chemistry, organometallic compounds, chemical synthesis, quantum chemistry, etc. are some of the important topics that have been covered through the lucid chapters. A number of latest researches have been included to keep the readers up-to-date with the global concepts in this area of study. This text will prove to be immensely beneficial to students and researchers associated with the field of chemistry and allied disciplines.

This book is a result of research of several months to collate the most relevant data in the field.

When I was approached with the idea of this book and the proposal to edit it, I was overwhelmed. It gave me an opportunity to reach out to all those who share a common interest with me in this field. I had 3 main parameters for editing this text:

1. Accuracy - The data and information provided in this book should be up-to-date and valuable to the readers.

2. Structure - The data must be presented in a structured format for easy understanding and better grasping of the readers.

3. Universal Approach - This book not only targets students but also experts and innovators in the field, thus my aim was to present topics which are of use to all.

Thus, it took me a couple of months to finish the editing of this book.

I would like to make a special mention of my publisher who considered me worthy of this opportunity and also supported me throughout the editing process. I would also like to thank the editing team at the back-end who extended their help whenever required.

Editor

1

Studies on phytochemical screening, antimicrobial and toxicity effect of the shoot system of *Acalypha segetalis* Müell. Arg.

Aboaba, Sherifat A. and Yeye, Emmanuel

Department of Chemistry, University of Ibadan, Ibadan, Nigeria.

The dried and pulverized shoot system of *Acalypha segetalis* was extracted with hexane, ethyl acetate and methanol using cold extraction technique. The three extracts were subjected to phytochemical, toxicity and antibacterial screening. Phytochemical studies revealed the presence of alkaloids, flavonoids, saponins, tannins, phenols, glycosides, steroids and resins. The toxicity activity was investigated using brine shrimp (*Artemia salina*) lethality assay. The toxicity test showed that ethyl acetate and methanol extracts showed moderate toxicity with LC_{50} at 777.9711 and 221.8975 ppm respectively while hexane extract was non- toxic with LC_{50} 8252.5750 ppm. Antibacterial activity was tested employing the disk diffusion method revealed that the extract exhibited significant antibacterial activity against all the organisms used. Antimicrobial assay was concentration dependent; at higher concentrations such as 200, 100 and 50 mg/ml, the extracts exhibited strong inhibition and expectedly at lower concentrations such as 25 and 12.5 mg/ml, the activity of the plant extracts was low. The three extracts showed activity against *Pseudomonas aeruginosa* and *Klebsiella pneumoniae* at all concentrations.

Key words: *Acalypha segetalis*, Euphorbiaceae, phytochemical, antimicrobial activity, brine shrimp lethality assay.

INTRODUCTION

Acalypha is a plant genus of the family *Euphorbiaceae*. The common names are copper leaves or three-seeded mercuries. These plants are mostly tropical or sub-tropical (but not present in Hawai and several other Pacific Islands); with a few representative in temperate zones. America contains two thirds of the known species, distributed from Southern United State to Uruguay and northern Argentina.

Acalypha segetalis is a shrub and eaten as a vegetable in some part of Africa (Burkill, 1994). In spite of the uses of members of *Euphorbiaceae* family from literature, little has been done on the chemical composition, toxicity of some species of *Acalypha*. Some common species are; *A. segetalis, Acalypha wilkesiana, Acalypha ciliata, Acalypha ornata, Acalypha hispidia*, *Acalypha indica*, *Acalypha fimbriata.* They belong to same kingdom, division, class, family and genus but different species (Aworinde et al., 2009).

The essential oil of the leaves (Ogunwande et al., 2008) and whole plant (Aboaba et al., 2010) of *A. segetalis* have been studied. The brine shrimp toxicity assay of the essential oil revealed that it was toxic at LC_{50} 14.10 ppm. The secondary metabolites present in other species of *Acalypha* have also been the subject of discussion (Adesina et al., 2000; Bokshi et al., 2012; Rajaselvam et al., 2012). Flavonoids, alkaloids, phenolics, steroids, saponins, glycosides, phlobatanin were detected as some of the secondary metabolites in other *Acalypha* species. Different extracts of *A. wilkesiana*, *A. ornata*, *A. ciliata*, *A. indica*, *A. fimbriata* has been found to show moderate activities against different strains of bacteria and fungi (Gotep et al., 2010; Bokshi et al., 2012; Tehseen et al., 2012; Kasim et al., 2012) and also to *Artemia salina* and *anopheles gambiae* (Aboaba et al., 2012).

To the best of our knowledge, no record investigation has been reported on the phytochemical screening and isolation of the secondary metabolites in *A. segetalis*. The aim of this research is to determine the secondary metabolites in the extract of the shoot system of *A. segetalis,* the toxicity of the plant against brine shrimp larvae and to subject the extracts to antimicrobial screening.

MATERIALS AND METHODS

Plant collection and identification

Fresh plants of *A. segetalis* Muell. Arg. was collected from botanical garden of University of Ibadan, Ibadan. The plant was identified, authenticated and a copy was deposited at Forestry Research Institute of Nigeria (FRIN), herbarium section, Ibadan, Oyo State, Nigeria, with the herbarium number (109692) in November, 2013.

Plant extraction

The whole plant was separated from the root, air dried and pulverized the pulverized sample was weighed and kept for further analysis. The plant sample (600 g) was subjected to cold extraction by soaking in methanol for a week. The extract was concentrated using rotary evaporator at 35°C. The crude extract was then partitioned in n-hexane, ethyl acetate and methanol. Each extract; n-hexane, ethyl acetate and methanol were concentrated separately using rotary evaporator at 35°C.

Phytochemical screening

The extracts obtained above were screened for alkaloids, flavonoids, steroids, saponins, phenols, tannins, glycosides and resins using standard methods as described by Harborne (1998).

Toxicity analysis

Brine shrimp lethality test for toxicity

The brine shrimp lethality assay was used to test for the toxicity of the extracts. Brine shrimp eggs were hatched in sea water and allowed to stay for 48 h at room temperature. The nauplii (harvested shrimp) were attracted to one side of the vials with a light source. Solutions of the extracts were made in DMSO at varying concentration (1000, 100 and 10 µg/ml) and incubated in triplicate test tubes with brine shrimp larvae. Ten brine shrimp larvae were placed in each of the triplicate test tubes. Control brine shrimp larvae were placed in a mixture of sea water and DMSO only. After 24 h, the number of larvae that survived in each test was determined. The concentration at 50% mortality of the larvae (LC_{50}) was determined using the Finney computer programme (Meyer et al., 1982; McLaughlin et al., 1998).

Preparation of sample for antimicrobial analysis

1 g of each sample was weighed and dissolved in 5 ml of the solvent used for the extraction to give 200 mg/ml. This was serially diluted to obtain lower concentrations of 50, 25 and 12.5 mg/ml. Gentamycin (5 mg/ml) was used as the positive control for bacteria while griseolvuvin (70%) for fungi and the negative control contained the solvent used for dissolving each extract.

Agar diffusion: pour plate method for bacterial

Cultures of each of the organisms used were prepared. The organisms were *Staphylococcus aureus, Escherichia coli, Bacillus subtilis, Pseudomonas aeruginosa, Klebsiella Pneumonia* and *Salmonella typhi* were prepared. A sterile cork-borer was used to create wells (or holes) inside the set plates. The test solutions of oils (50 µl) at concentration of 40 mg/ml were then introduced into each of the designated cups on each plate ensuring that no spillage occurred. The same amount of the standard antimicrobial agent and solvents were introduced using syringes into the remaining cups on each plate to serve as positive and negative controls respectively. The plates were left at room temperature for 2 h, allowed to diffuse into the medium, turned upside-down and thereafter incubated at 37°C for 24 h in an incubator. Clear zones of inhibition were observed. Activity or inactivity of each extract was tested in triplicate and the diameters of zones of inhibition were measured in millimeter (mm) using a transparent well-calibrated ruler. The positive control for bacteria is gentamicin at the concentration of 5 mg/ml. The analysis was done in triplicates (Belboukhari and Cheriti, 2005).

Agar diffusion plate method for fungi

Sterile Sabouraud Dextrose Agar (SDA) was poured aseptically into the sterile plates and allowed to cool down for 45 min. 0.2 ml of 1:100 dilutions of the organisms *Candida albicans, Rhizopus stolonifer, Aspergillus niger* and *Penicillum nonatum* were spread on the surface using a sterile spreader. Then, a sterile cork-borer was used to create wells inside the plates. The same procedure described for anti-bacterial activity above was followed from this stage. The positive control for the fungi was 70% griseolvuvin. All the plates for the fungi were incubated at 28°C for 48 h. The clear zones of inhibition were observed and recorded using the same method as described in the case of bacteria (Andrew, 2001; Thongson et al., 2004).

RESULTS

The yields of the methanol, ethylacetate and hexane extracts were 2.25, 1.33 and 1.19% respectively. The

Table 1. Phytochemical screening of shoot system of *A. segetalis*.

Secondary metabolites	ASH	ASE	ASM
Alkaloids	+	+	+
Tannins	+	+	+
Saponins	+	+	+
Sterols	-	+	+
Flavonoid	-	+	+
Resins	+	+	+
Phenols	-	+	-
Glycosides	-	+	+

ASH =*Acalypha segetalis* hexane extract, ASE =*Acalypha segetalis* ethyl acetate extract, ASM = *A. segetalis* methanol extract + = positive, - = negative.

Table 2. Brine shrimp toxicity test.

Extract	LC_{50} (ppm)	G
Hexane	8252.58	0.5927
Ethyl acetate	221.90	0.2957
Methanol	777.997	0.2327

"LC_{50}", is the lethality concentration 50%, "G", is the degree of freedom.

percentage moisture content of the shoot system sample of *A. segetalis* was 13.57% this implies that the moisture content loss in the shrub is minimal. The result of phytochemical screening in Table 1 showed that methanol extract are rich in alkaloids, saponins, resins, tannins, sterols, glycosides and flavonoids while it gave a negative result for phenol. The ethylacetate extract contain alkaloids, tannins, saponins, sterols, flavonoids, resin, phenols, glycosides while the hexane extract gave positive results for alkaloids, tannins, saponins and resin but no sterols, flavonoids, resin, phenols and glycoside.

Toxicity of the extracts against brine shrimps show LC_{50} of 777.9711, 221.8975 and 8252.5750 ppm for methanol, ethylacetate and hexane extracts respectively (Table 2). The antimicrobial assay of the three extracts showed that at high concentrations; 200 and 100 mg/ml, the three extracts were active against the ten micro-organisms used but lower than the positive control. At 200 mg/mL, 100, 50 and 25 mg/ml, methanol extract was active against *S. aureus, B. subtilis, S. typhi, K. pneumonia, C. albican, A. niger, R. solonifer* and *P. notatum* but at 12.5 mg/ml, it displayed low activity against *E. coli, B. subtilis, C. albican* and *A. niger*. For ethyl acetate extract, it was active against all organisms tested for at 200, 100 and 50 mg/ml, but not active against *E. coli, S. typhi, K. pneumonia, C. albican and R. stolonifer.* However at 12.5 mg/L, no zone of inhibition was observed against *P. notatum*. For the hexane extract, at 200, 100, 50 and 25 mg/ml it was active against all organisms tested for, but at 12.5 mg/ml it was not active against *S. aureus, E. coli, B. subtilis, S. typhi, A. albican, A. niger* and *P. notatum*. Comparatively, at 12.5 mg/ml being the lowest concentration, there was no zone of inhibition shown against *E. coli, C. albican and A. niger* for the three extracts. Zone of inhibition decreases from the highest concentration 200 mg/ml to the lowest concentration 12.5 mg/ml. Ethyl acetate and methanol extracts have the highest zone of inhibition against *P. aeruginosa*, being 28 mm and 24 mm respectively. At the lowest concentration of 12.5 mg/ml, hexane extract was active against *P. aeruginosa, K. pneumonia, R. stolonifer* as shown in Table 3.

DISCUSSION

The yields of the three extracts under study that is; methanol, ethylacetate and hexane gave an expected range considering the polarity of the three solvents. Expectedly, methanol the most polar, ethylacetate moderately polar while hexane a non-polar solvent and as such methanol gave the highest yield followed by ethylacetate and then hexane. The phytochemical from this report is comparable with another species of *Acalypha* previously reported (Adesina et al., 2000). These classes of compounds present from the phytochemical screening are responsible for the curative activity against several pathogens and therefore could justify the ethno botanical uses of the plant for treatment of a wide range of illness. For example, glycosides,

Table 3. Antimicrobial Activity of *Acalypha segetalis* extracts.

Extract	Conc. (mg/ml)	*S. aureus*	*E. coli*	*B. subtilis*	*P. aeruginosa*	*S. typhi*	*K. pneumonia*	*C. albicans*	*A. niger*	*R. solonifer*	*P. notatum*
Methanol	200	20	20	18	28	18	20	18	16	20	20
	100	18	18	14	20	16	18	16	14	18	16
	50	16	14	12	18	14	14	14	12	16	14
	25	12	10	10	14	12	12	10	10	14	12
	12.5	10	-	-	10	10	10	-	-	10	16
Ethylacetate	200	24	18	20	28	18	18	16	14	18	16
	100	20	16	16	26	16	14	14	12	14	14
	50	18	14	14	20	12	12	12	10	12	10
	25	14	10	12	14	16	10	10	-	10	-
	12.5	10	-	10	10	-	-	-	-	-	-
Hexane	200	20	18	16	24	20	18	16	18	20	18
	100	18	14	14	20	16	16	14	14	16	14
	50	14	12	12	16	14	14	12	12	14	12
	25	12	16	16	12	10	12	10	10	12	10
	12.5	-	-	-	10	-	10	-	-	10	-
Controls	-ve	-	-	-	-	-	-	-	-	-	-
	+ve	38	36	34	40	36	34	24	26	28	26

S. aureus = Staphylococcus aureus. (Bacteria), *E. coli = Escherichia coli.* (Bacteria), *B. Subtilis = Bacillus subtilis.* (Bacteria), *P. Aeruginosa = Pseudomonas aeruginosa.* (Bacteria), *S. Typhi = Salmonella typhi.* (Bacteria), *K. pneumonia = Klebsiellae pneumonia.* (Bacteria), *C. albican = Candida albicans.* (Fungi), *A. niger = Aspergillus niger.* (Fungi), *R. stolonifer = Rhizopus stolonifer* (Fungi), *P. notatum = Penicillum notatum.* (Fungi), -ve: Negative controls; Methanol for methanol extract, ethylacetate for ethylacetate extract and hexane for hexane extract, +ve: Positive control; Gentamicin 5 g/ml; bacteria and Griseolvuvin 70%; fungi.

saponins, flavonoids, tannins and alkaloids have hypoglycemic activites, anti-inflammatory activities (Augusti and Cherian, 2008), sterols, triterpenoids and saponnins have also been reported to show analgesic properties and central nervous system activites (Argal and Pathak, 2006). Plants having alkaloids are used in medicines for reducing headaches and fever; these are attributed for antibacterial and analgesic properties (Pietta, 2000).

The results of the brine shrimp lethality test was interpreted according to Moshi et al. (2014); extracts with $LC_{50} \leq 1000$ ppm is toxic while extracts with $LC_{50} \geq 1000$ ppm. The LC_{50} of the methanol, ethyl acetate fraction and hexane fraction gave 777.9711, 221.8975 and 8252.5750 ppm respectively. This shows that hexane fraction is non toxic, while ethyl acetate and methanol fraction are regarded as moderate level of toxicity and very low toxicity respectively. In a related study, *A. hispida* showed displayed a very high level of toxicity (LC_{50} 7 ppm) even when compared to a standard drug chloramphenicol (Bokshi et al., 2012). Also the essential oils from *A. ciliata* and *A. ornata* plants revealed various larvicidal activities causing 100% mortality against *Artemia salina* and *A. gambiae* respectively. The LC_{50} values of *A. ornata* and *A.ciliata* oils estimated against *A. gambiae* s.l. were 77.59 and 73.96 ppm respectively (Aboaba et al., 2012).

Plant extracts contain different phytochemical with biological activity that can have valuable healing properties. The demonstration of broad spectrum of antimicrobial activity by *A. segetalis* may help to discover new antibiotic substances that could serve as agents for the control of infectious diseases. It was observed that methanol extracts gave the highest activity against the tested organisms compared to the ethylacetate and hexane extracts, the rational for this observation may be based on the ability of polar extracts permeability through the plasma membrane. The methanol extract showed inhibition against all the organisms used at all the tested concentrations except at the lowest which is 12.5 mg/ml where it displayed low activity against *E. coli, B. subtilis, C. albican and A. niger*. Ethyl acetate extract was active against all organisms tested for at 200, 100 and 50 mg/ml, but not active against *E. coli, S. typhi, K. pneumonia, C. albican and R. stolonifer* while the hexane extract was active against all organisms tested for at 200, 100, 50 and 25 mg/ml, but at 12.5 mg/ml it was not active against *S. aureus, E. coli, B. subtilis, S. typhi, A. albican, A. niger* and *P. notatum*.

Conclusion

Phytochemical analysis of plants is very important as it could explain the use of these plants traditionally for the treatment of some ailments. The extracts of *A. segetalis* from this study contain secondary metabolites like alkaloids, tannins, saponins, sterols, flavonoids, resin, phenols, glycosides which play a vital role in preventing various diseases. The antimicrobial assay results indicate that the plant possesses considerable antimicrobial activity that supports the traditional uses of this plant and other species of *Acalypha* in the treatment of various diseases. However, advance studies are required to identify and characterize the chemical compounds, that is, structures of the secondary metabolites responsible for these observed activities.

Conflict of Interest

The authors have not declared any conflict of interest.

ACKNOWLEDGEMENT

The authors are grateful to the Senate of the University of Ibadan, Ibadan, Nigeria for funding this work through the Senate Research Grant (SRG/FS/2010/26[A]).

REFERENCES

Aboaba S, Aiyelaagbe O, Ekundayo O (2010). Chemical Composition, Toxicity and Larvicidal Activity of the Essential Oil from the Whole Plant of *Acalypha segetalis* from South-West Nigeria. Nat. Prod. Comm. 5(3):481-483.

Aboaba S, Ibrahim K, Omotoso O (2012). Toxicity and mosquito larvicidal activities of the essential oils from the leaves of *Acalypha ornata* and *Acalypha ciliata* in southwest Nigeria. J. Vect. Borne Dis. 49:114–116.

Adesina SKO, Idowu AO, Ogundaina O, Olugbade TA, Onawunmi GO, Pais M (2000). Antimicrobial constituents of leaves of *Acalypha wikesiana and Acalypha hispida*. Phytother. 14:371 -374.

Argal A, Pathak A (2006). A KCNS activity of *Calotropis gigantean* roots. J. Ethnopharmacoogyl. 106:142-145. http://dx.doi.org/10.1016/j.jep.2005.12.024

Andrew JM (2001). Determination of minimum inhibitory concentration (MIC). J. Antimicrob. Chemother. 48(5):16.

Augusti KT, Cherian S (2008). Insulin sparing action of leucopelargonidin derivative isolated from *Ficus bengalosis* Linn. Indian J. Exp. Biol. 33:608-611.

Aworinde DO, Nwoye DU, Jayeola AA, Olagoke AO, Ogundele AA. (2009). Taxonomic significance of foliar Epidermis in some members of Euphorbiaceae family in Nigeria. J. Bot. 4:17-28.

Bokshi B, Sayeed MAS, Ahmed MI, Karmakar UK, Sadhu SK (2012). Assessment of Antimicrobial and Cytotoxic Activities of Ethanolic Extract of Leaves of *Acalypha hispida*. Int. J. Pure Appl. Sci. Res. 3(6):1705-1708.

Belboukhari N, Cheriti A (2005). Antimicrobial Activity of Aerial Part Crude Extracts from *Limoniastrum feei*. Asian J. Plt. Sci. 4(5):496-498. http://dx.doi.org/10.3923/ajps.2005.496.498

Burkill HM (1994). The Useful Plants of West Tropic Africa. Vol 2, Edition 2, Families E-I. Royal Botanical Gardens, KEW, U.K. P.24.

Gotep JG, Agada GOA, Gbise DS, Chollom S (2010). Antibacterial activity of ethanolic extract of *Acalypha wilkesiana* leaves growing in Jos, Plateau State, Nigeria. Malaysian J. Microbiol. 6(2):69-74.

Harborne JB (1998). Phytochemical Methods: A Guide to modern Techniques of plant Analysis 3rd Ed., Springer, London, ISBN: 9780412572609, P.302.

Kasim LSI, Ayodele O, Effedua HI, Adejumo OEI, Ekor M, Fajemirokun TO (2012). Antimicrobial activity of six selected plants against some strains of pathogenic organisms. J. Microbiol. Antimicrob. 4(3):54-59. http://dx.doi.org/10.5897/JMA11.013

McLaughlin JL, Rogers LL, Anderson JE (1998). The Use of Biological Assays to Evaluate Botanicals. Drug Info. J. 32:513- 524.

Meyer BN, Ferrigni NR, Putman JE, Jacobson LB, Nichols DE, McLaughlin JL (1982). Planta Med. 45:31-34.

Moshi MJ, Cosman JC, Mbwambo ZH, Kapingu M, Nkunya MHH (2004). Testing Beyond Ethnobotanical Claims: Brine Shrimp Lethality of Some Tanzania Plants. Pharm. Bio. 42:547-551. http://dx.doi.org/10.3109/13880200490897920

Ogunwande IA, Essien EE, Ogunbinu AO, Karioti A, Saroglou V, Skaltsa E, Adebayo MA (2008). Essential oil constituents of *Klainedoxa gabonensis* Pierre Ex Engl (Irvingiaceae), *Brachystegia nigerica* Hoyle & A. Jones (Caesalpinioideae) and *Acalypha segetalis* Muell Arg., (Euphorbiaceae). JEOR. 20(3):211-215. http://dx.doi.org/10.1080/10412905.2008.9699994

Pietta PG (2000). Flavonoids as antioxidants. J. Nat. Prod. 63:1035-1042. http://dx.doi.org/10.1021/np9904509

Rajaselvam J, Benila JM, Meena R (2012). A Study of Antimicrobial Activity of *Acalypha indica* Against Selected Microbial Species. Int. J. Pharm. Sci. Res. 3(9):473-476.

Tehseen Q, Salman A, Muhammad SA, Patricia AO, Iqbal A (2012). Antiemetic activity of *Acalypha fimbriata* Schumach. & Thonn. *Acalypha ornata* Hochs., *and Acalypha wilkesiana cv.* Godseffiana Muell Arg. J. Phytopharmacol. 3(2):335-340.

Thongson C, Davidson PM, Mahakarrchanakul W, Weiss J (2004). Antimicrobial activity of ultrasound associated solvent extracted species. Lett. Appl. Microbiol. 39:401-406. http://dx.doi.org/10.1111/j.1472-765X.2004.01605.x

2

Development of ordered and disordered macroporous silica from bagasse ash

S. Oluwagbemiga Alayande[1,2*], E. Olugbenga Dare[2], W. Babatunde Ayinde[2], Janet Bamigbose[2], P. A. Ayedun[3] and G. Ayodeji Osinkolu[1]

[1]Centre for Energy Research and Development, Obafemi Awolowo University, Ile-Ife, Osun State, Nigeria.
[2]Department of Chemistry, Federal University of Agriculture, Abeokuta, Ogun State, Nigeria.
[3]Department of Mechanical Engineering, Federal University of Agriculture, Abeokuta, Ogun State, Nigeria.

Silica from bagasse ash was nanostructured with templating concept, these resulted into ordered and disordered nano pores in the silica. The average pore widths were between 220.87 and 537 nm. These pores were classified as macro pores. The macro porous silica developed can be applicable in sensing, selective filtration, catalysis, drug delivery and membrane.

Key words: Silica, templating concept, macro pores.

INTRODUCTION

Nanoporous materials are advance materials which spur new technology for development. These materials are of numerous use in medicine, photonic materials, membrane Science, catalysis, sensing and dielectric coating (Randall et al., 2004; Polarz et al., 2010). According to Polarz et al. (2010) nanoporous materials can be classified into inorganic and organic materials. The group further divided the inorganic class: Inorganic oxide type, carbon materials, binary compound and single elements (Polarz et al., 2010). Nanoporous materials which are filled with pores are referred to as nanostructured materials (Polarz et al., 2010).

Nanoporous silica belongs to inorganic oxide group. Nanostructuring of silica had recorded several achievements ever since the first publication on MCM-41 in 1992 by Beck et al. (1992). Nanostructuring of silica had resulted to high degree of perfection in the pore design particularly in mesoporous silica (Polarz et al., 2010). The remarkable improvement in nanostructured silica is as a result of free energy of silica which is very close to that of its crystalline quartz state (Polarz et al., 2010). Also, silica can be produced by sol gel method and the precursors (alkoxysilane compounds) reactivity is low hence the control of the process is achievable (Sayari et al., 2001). The process is irreversible due to formation of strong Si-O bond (Sayari et al., 2001). Silica shape formed at room temperature is retained indefinitely because amorphous silica is stable over a very high temperature range and time. Silica can exist in anionic, neutral or cationic form depending on the pH. These make the synergy between silica and surfactant interesting chemistry (Polarz et al., 2010).

Nanoporous silica may be synthesized under any of these conditions: Acidic or basic medium, sub-ambient temperature to about 150°C, varied pH range, in the presence of surfactants and polymers (Affandi et al., 2009). Pore control and expansion from meso to macro dimension had been achieved in the material (Affandi et al., 2009). Ordered nanoporous silica had been reported to be much more interesting due to the control and well aligned pore size and shape (Polarz et al., 2010). Disordered formation has high polydisperity in pore size and irregular in shape. Generally, ordered porous materials are much more homogenous (Polarz et al., 2010). Ordered nanoporous silica is preferred for application as catalyst carrier, sensing devices, coatings, membrane and bioreactors (Polarz et al., 2010).

It is important to note that biosilification products contain amorphous silica and crystalline forms such as cristobalite, tridymite andquartz (Espindola-Gonzalez et al., 2010). The silica absorbed for terrestrial plants is

*Corresponding author. E-mail: gbengaalayande@gmail.com, salayande@cerd.gov.ng.

Table 1. The chemical composition of bagasse ash sintered at 900°C and silica after extraction.

Components (oxides)	Bagasse ash at 900°C (wt %)	After extraction (wt %)
SiO_2	56.50	98.86
Al_2O_3	0.02	0.00
Fe_2O_3	4.45	0.25
CaO	19.65	0.00
MgO	6.11	0.89
SO_3	0.01	0.00
Mn_3O_2	0.91	0.00
K_2O	12.35	0.00

around a fraction of 1% of the dry matter to several percent (Espindola-Gonzalez et al., 2010). Many plants sequester silica in biogenic phytoliths and soil later acquire reasonable amount of biogenic opal A (Espindola-Gonzalez et al., 2010). Several plants waste are noted to contain commercial quantity of silica; rice husk (Espindola-Gonzalez et al., 2010; Nittaya et al., 2008), sugar cane bagasse (Affandi et al., 2009; Espindola-Gonzalez et al., 2010), coffee husk (Espindola-Gonzalez et al., 2010). Further study on the morphology of these silica in rice husk and sugar cane bagasse showed mesoporous silica (Affandi et al., 2009; Kalapathy et al., 2000, 2002), with high purity. Mesoporous silica has pore size of 2 to 50 nm. Expansion of these pore size is attainable even to macro pore size dimension (above 50 nm) (Sayari et al., 2001) for the purpose of enhancing their applications. Expansion of pores in silica sourced from plants waste will reduce the burning of these wastes and increase dependence on this source for silica for application as catalyst carrier, sensing devices, coating, membrane and bioreactors.

This work is therefore aimed at developing macro pores in nanostructured silica sourced from bagasse ash.

EXPERIMENTAL

Silica production

Sugarcane bagasse from local dealers was used in this study. The bagasse was dried for 2 h and pre-ashed at 200°C. The pre-ashed sample was calcined to 900°C for 6 h to convert to ash. Silica was extracted from bagasse ash using cation exchange resin which had been reported for high purity silica (Affandi et al., 2009). All chemical and reagent used were of analytical grade (Aldrich and Fluka chemicals).

Nanostructuring of silica

Sayari et al. (2001) reported a recipe for nanostructuring of silica sourced from alkoxysilane compound, the recipe was adopted for the nanostructuring of extracted silica. 1.0: 0.17: x: 0.17: 17 SiO_2 extract: CTMAB: HMTA: NH_4OH: H_2O (Sayari et al., 2001) with a contact time and reaction temperature were kept constant at 6 h and 70°C, respectively. Cetyltrimethylammoniumbromide (CTMAB) is cationic surfactant and Hexamethylenetetraamine (HMTA) were used as templates. The samples were harvested and dried at 120°C for 2 h. All chemical and reagent used were of analytical grade without any further purification (Aldrich and Fluka chemicals).

Characterization

The chemical composition of the bagasse ash and silica extract were monitored X-ray powder diffraction (XRD), the measurements were taken with XRD Monochromatic CuKα radiation (wavelength = 1.5406) produced by Radicon limited. Phase identification and crystallographic structures of the silica was obtained. The quantitative analysis of chemical components was done using X-ray fluorescence (XRF).

Nanostructured silica samples were characterized using scanning electron microscope (SEM) for surface morphology examination.

RESULTS AND DISCUSSION

Table 1 shows the chemical composition of the bagasse ash, silica content of the ash is higher than that reported by Affandi et al. (2009) and Worathanakul et al. (2009) from Indonesia and Thailand respectively, this is attributed to different soil type that exist in these different geographical zones. The variety of the sugarcane used can also be responsible for the silica variation. After extraction, 98.86% of silica was obtained using cation exchange resin. The method was adopted due to its ability to lower Fe_2O_3 due to the better extraction of the oxide using acid resin (Affandi et al., 2009). There was reduction in concentration of other constituent oxides after extraction hence purity was enhanced by the method. The purity was lower than that reported by Affandi et al. (2009), this can be attributed to the variation in oxides composition. The X-ray diffractogram of the bagasse ash sintered at 900°C as shown in Figure 1, phase identification of the peak at 20.52° gave hexagonal symmetry of quartz silica [JCPDS value (47 to 1144)]. Annealing the ash below the melting point of silica will enhance the crystallinity of the silica content while other constituent oxides existing as impurity will be reduced before extraction. Annealing to this temperature will also aid the reduction of the constituents oxides therefore reducing impurity before extraction. Figure 2 shows

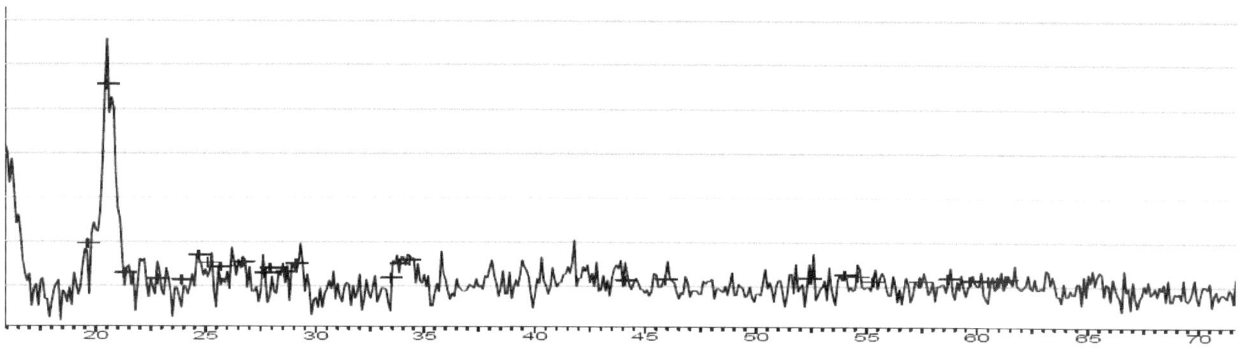

Figure 1. X-ray diffractogram of bagasse ash.

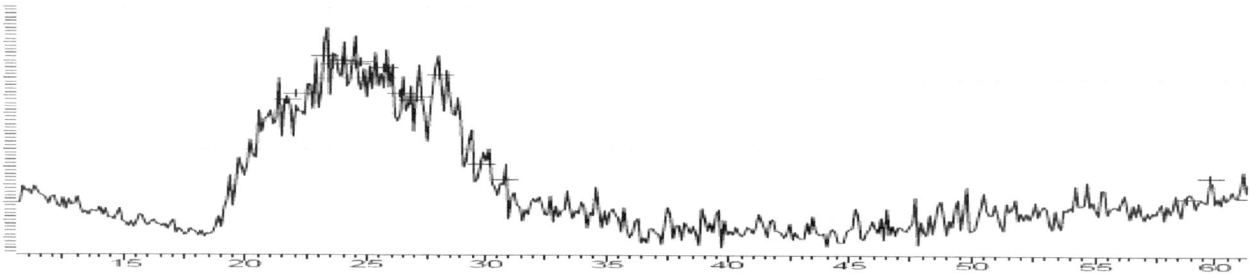

Figure 2. X-ray diffractogram of silica.

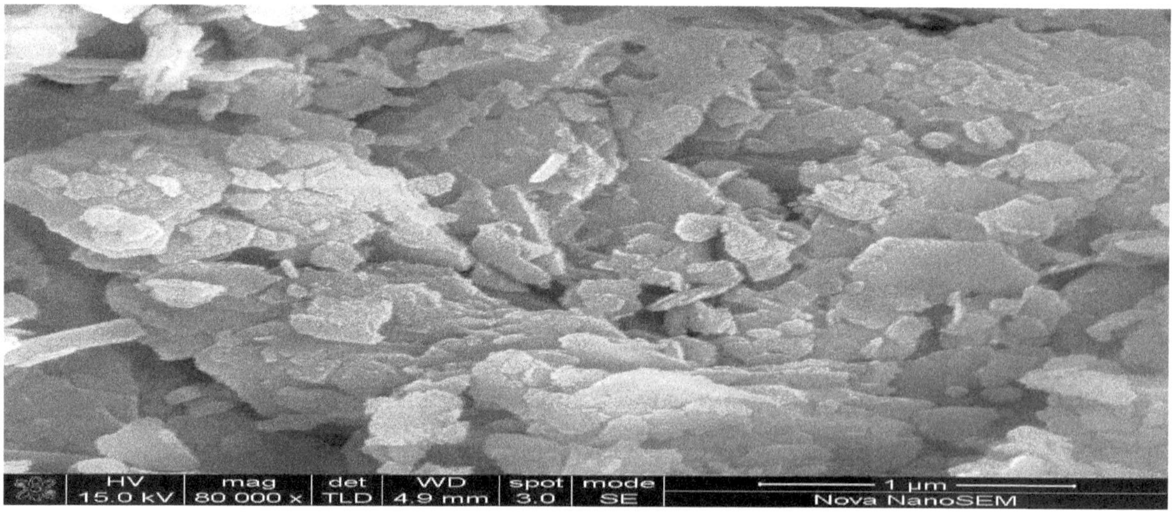

Figure 3. The surface morphology of silica extracted from bagasse.

amorphous silica obtained after the extraction which confirms the absence of any ordered crystalline structure. It also shows the purity of the silica (Kalapathy et al., 2000, 2002). Generally, amorphous silica is stable over a long period of time (Iler, 1968). To further characterize the physical property of the silica extracted, scanning electron microscope was used. The morphology shows irregular shape which varies in size and agglomerated as shown in Figure 3.

Following the recipe, the nanostructured silica in Figure 4 shows visible pores in the silica. The templating system had successfully transformed the silica into a porous material. However, the pores developed were highly polydispersed with irregular shape. The irregularity in shape of the pores can be attributed to the irregular shape of the silica extracted. Classification of the resultant pores formed gave open, closed and enclosed pores (Sing et al., 1994). Average open pore width is

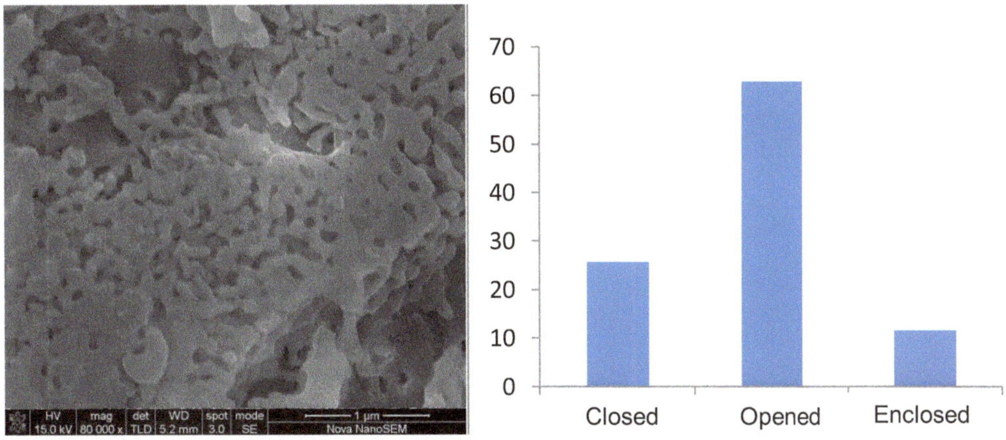

Figure 4. The surface morphology of porous silica with percentage frequency distribution chart.

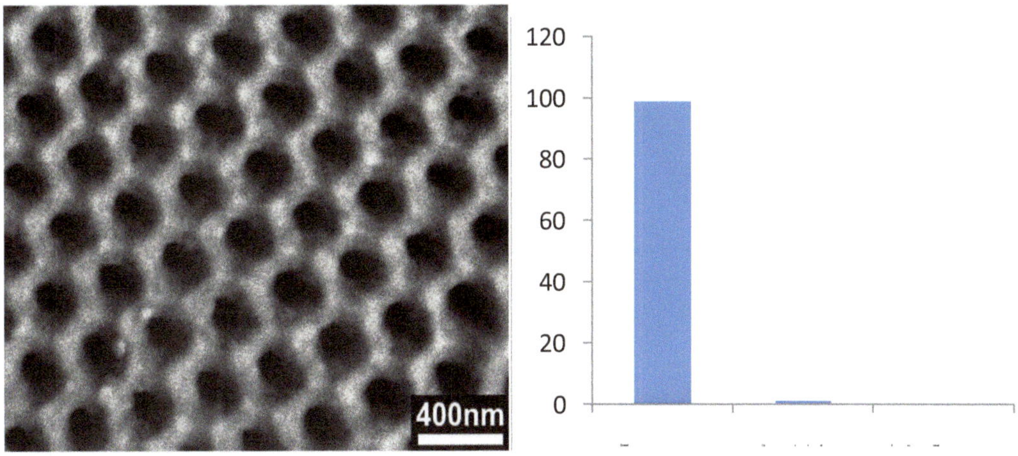

Figure 5. The surface morphology of modified porous silica with percentage frequency distribution chart.

576.38 nm, average closed pore width is 235.47 nm and enclosed pore width gave an average value of 105.74 nm. The average pore width is 305.86 nm. Open pores predominate the pore frequency distribution, it accounted for 62.85% of the total pore formed.

Ordered porous materials are more applicable compared to the disordered counterparts (Polarz et al., 2010). Few drops of trimethylbenzene (swelling agent) was added to the composition to change the packing parameters of the surfactant which had been reported to result in change of phase behavior (Polarz et al., 2010). Figure 5 showed the resultant pore formed as a result of presence of swelling agent. Rearrangement of the surfactant led to a uniform pore formation with 220.69 nm as width, predominantly a network of closed pores. Pore formation using templates system takes place via combination of templates to form a single molecule known as micelle. The micelle formed rearranges itself in the lyotropic phases to result into polymeric spheres. The spheres combine to form a network of spheres which is then incorporated into the material as pores (Polarz et al., 2010). In the formation of the disordered pores, the micelle rearrangement in the lyotropic phases resulted into irregular polymeric spheres. These spheres then combine to form a network of irregular spheres which resulted to disordered pores. The addition of swelling agent changed the agglomeration of the templates which then resulted in new micelles different from the previous micelles. These micelles resulted in a network of uniform polymeric spheres. The pathway for ordered and disordered pores for silica sourced from plant waste is shown in Figure 6.

Conclusion

Nanostructuring of the silica using templating concept resulted to macro porous silica. Disordered and ordered nanoporous silica was formed with pores width classified as macro pores. The ordered macroporous silica can be

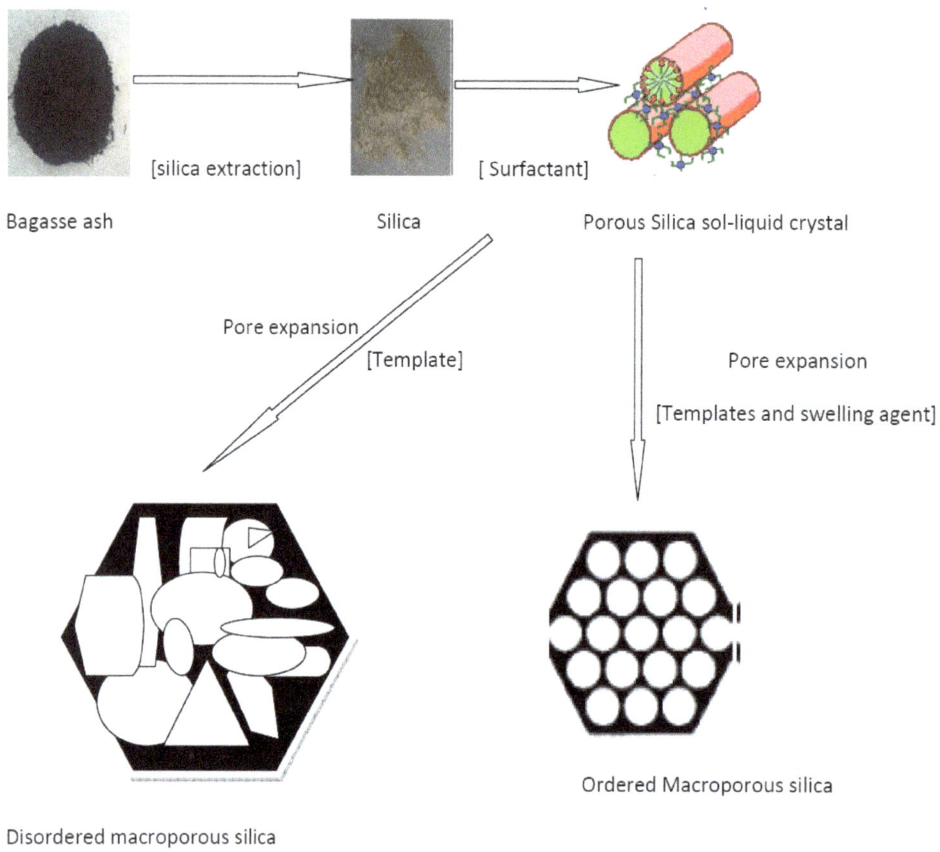

Figure 6. Schematic diagram of macroporous silica formation from bagasse ash.

used in selective filtration, industrial effluent treatment, catalyst carrier and drug delivery. The disordered porous silica obtained can be used for the treatment of industrial effluents and for filtration where the size of the expected residue is higher than the largest pore formed.

ACKNOWLEDGEMENTS

The authors are grateful to Centre for Energy Research and Development, Obafemi Awolowo University Ile-Ife and Chemistry Department, Federal University of Agriculture, Abeokuta, Nigeria for supporting this project. We appreciate the financial support from Third World Academy of Science, through the Grant No. 08-003 LDC/CHE/AF/AC-UNESCO FR: 3240184280.

REFERENCES

Affandi S, Setyawan H, Winardi S, Balgis R (2009). A Facile Method for the Production of High Purity Silica Xerogels from Bagasse Ash. J. Adv. Powder Tech., 20: 468-472.

Beck JS, Vartuli JC, Roth WJ, Leonowicz ME,Kresge CT, Schmitt KD, Chu CT, Olson DH, Sheppard EW, McCullen SB,Higgins JB, Schlenker JL (1992). A new family of mesoporous molecular sieves prepared with liquid crystal templates. J. Am. Chem. Soc., 114(10): 834.

Espindola-Gonzalez A, Martinez-Hernadez AL, Angeles-Chavez C,Castano VM, Velasco-Santos C (2010). Novel Crystalline SiO_2 Nanoparticles via Annelids Bioprocessing of Agro-Industrial Wastes. Nanoscale Res. Lett., 11671(010): 9654-9656

Iler RK (1968). The Chemistry of Silica. John Wiley New York.

Kalapathy U, Proctor A, Shultz J (2000). A simple method for production of pure silica from rice hull ash. J. Chem. Technol. Biotech., 75: 464-468.

Kalapathy U, Proctor A, Shultz J (2002). An improved method for production of silica from rice hull ash. Bioresoures technol., 73: 257-262.

Nittaya T, Apinon N (2008). Preparation of nanosilica powder from rice husk ash by precipitation method. Chiang Mai J. Sci., 35(1): 206-211.

Polarz S, Smarsly B (2002). Nanoporous Materials. J. Nanosci. Nanotech., 2(6): 581-612.

Randall QS, Joseph TH, SonBinh TN (2004) Prospects for nanoporous metal-organic materials in advanced separations processes. AIChE J., 50(6): 1090-1095.

Sayari A, Hamoudi S (2001). Periodic Mesoporous Silica Based Organic Inorganic Nanocomposite Materials. Chemistry Mat., 13: 3151-3168.

Sing KSW, Everet DH, Haul RAW, Siemeiniewka T (1994). Physical and Biophysical Chemistry Division Commission on Colloid and Surface Chemistry including Catalysis. Pure Appl. Chem., 66(8): 1739-1758.

Worathanakul P, Payubnop W, Muangpet A (2009). Characterization for post treatment effect of bagasse ash for silica extraction. World Academy of Science, Engineering and Technology, p. 56.

3

Heavy metal accumulation in *Oreochromis niloticus* and *Clarias gariepinus* from an integrated chicken-fish system

J. C. Nnaji[1]*, A. Uzairu[2], C. E.Gimba[2] and J. A. Kagbu[2]

[1]National Institute for Freshwater Fisheries Research, New Bussa, Nigeria.
[2]Department of Chemistry, Ahmadu Bello University, Zaria, Nigeria.

Accumulation of cadmium (Cd), chromium (Cr), copper (Cu), lead (Pb), manganese (Mn) and zinc (Zn) in fish from integrated laying chicken-fish ponds was investigated in order to determine the possible health risks to humans from heavy metal contamination. The 20 m^2 fish ponds were stocked with *Oreochromis niloticus* and *Clarias gariepinus* in a 3:1 ratio and a stocking density of 8 fingerlings/m^2. Treatment 1 (T1) consisted of chicken integrated at the rate of 2000 birds/ha; Treatment 2 (T2), 3000 birds/ha; Treatment 3 (T3), 4000 birds/ha, while Treatment 4 (T4) was the control without integration. Fish in T4 were fed compounded feed at 5% of body weight, while those in T1, T2 and T3 were fed on chicken manure and spilled feed that fell directly into the ponds. The ponds were sampled monthly and fish growth and water quality parameters were measured accordingly. Results of the experiment showed that tissue metal concentrations for both species were lowest in the muscles and highest in the kidney and liver. Final metal concentrations in the muscles of both fish species in T1 and T4 were below FAO (1983) and EU (2001) the limits and the consumption of fish muscle from such systems is considered safe. However, there is risk of Pb concentrations accumulating above the recommended levels in *C. gariepinus* at 3000 chickens/ha and above.

Key words: *Oreochromis niloticus*, *Clarias gariepinus*, integrated, laying chicken.

INTRODUCTION

Integrated chicken-fish farming (ICFF) is a low-cost system of fish production which involves the combination of chicken farming with fish culture where the wastes (manure and spilled feed) from the chicken sub-system become an input into the fish sub-system (Sinha, 1985; Nnaji et al., 2009). It leads to a more efficient land and water utilization, effective waste recycling, reduction in operational costs and more income for small farmers which invariably lead to higher standards of living (Rancis et al., 2004). However, there are health risks to humans and aquatic organisms from the use of livestock wastes in fish production through the possible introduction of pathogens (Garrett et al., 1997; FAO, 2003; FAO/OIE/WHO, 2005). Edwards (2008) stated that fish and plants passively accumulate microbial contaminants on their surfaces, but these rarely penetrate into edible fish flesh or muscle except for trematodes (parasitic tissue flukes). Parasites may also affect the health of fish since they can reduce fish growth rate, resistance to disease and cause fish mortalities.

Another important, but largely neglected health risk is the possible transfer of toxic heavy metals from fish to man in the livestock-fish system. While there is a lot of literature on disease risk in ICFF, little literature exist on heavy metal accumulation in the system. Jackson et al. (2003) stated that poultry litter and animal wastes in general contain high concentrations of some trace elements and due to the persistence and toxicity of heavy metals, it is important that their profile in the system is

*Corresponding author. E-mail: dozis03@yahoo.com.

studied and monitored regularly. Heavy metals are contained in both the spilled chicken feed and manure and these are consumed by fish with the possible accumulation of these metals above desirable limits if fish is cultured in such systems, especially for long periods of time. Heavy metals like arsenic (As), zinc (Zn) and copper (Cu) are also added as feed supplements to boost poultry production and they may accumulate in the animals which is also transferred to aquatic systems when poultry is integrated with fish. Cang et al. (2004), in a study of poultry and livestock feeds and manure in Jiangsu province, China, reported that Cu, Zn, lead (Pb), Cadmium (Cd) and Chromium (Cr) concentrations in animal manures were high with Cu concentration in a manure sample as high as 1726.3 mg/kg. Alexieva et al. (2007) studied 152 samples of feed ingredients and compounded feed for pigs and poultry in Bulgaria and found that Pb and Cd levels of feeds were higher than the current European official regulations. This study aims to determine heavy metal loads in fish cultured in an ICFF system for a period of 12 months.

MATERIALS AND METHODS

Study area and site

New Bussa, Niger State, is in the semi-arid zone of Nigeria and is located between latitude 9° 51'N to 10° 55' N and longitude 4° 23'E to 4° 45'E. It has a tropical continental climate characterized by a shorter wet season (May to September) and a longer dry season (October to April) with a temperature range of 15°C to nearly 40°C. Higher temperatures are usually experienced between the months of February and April (the hot season) which is within the dry season period. Annual rainfall is about 1000 mm and the town is bordered by the Kainji Lake created by the damming of River Niger at Kainji. The experimental site was located at the Fish Farm of the National Institute for Freshwater Fisheries Research (NIFFR), New Bussa. NIFFR has a national mandate to research on all aspects of freshwater fisheries.

Experimental design

Eight earthen ponds, each of area 20 m^2 and depth 1.5 m, were constructed by excavation and six chicken sheds (floor area: 2 × 1 m) were constructed over six ponds with wood. The sheds were 1 m above the ponds and the sides of each chicken shed were covered with wire mesh while the floors were made with iron rod mesh. Treatment 1 (T1) consisted of a pond integrated at the rate of 2,000 laying chickens/ha; treatment 2 (T2) consisted of 3,000 laying chickens/ha; treatment 3 (T3) consisted of 4,000 laying chickens/ha and treatment 4 (T4) is the control treatment without integration. Each pond was stocked with a polyculture of *Oreochromis niloticus* and *Clarias gariepinus* in a ratio of 3:1 and a stocking density of 8 fingerlings/m^2. Point-of-lay chickens (NERA Brown strain, 18 week old) were also stocked in the sheds. Fish in T4 were fed commercial floating fish feed (41.30 to 42.80% crude protein content) twice daily at 5% of their body weight while fish in T1, T2 and T3 feed on fresh chicken manure and spilled chicken feed that fell directly into the ponds.

Sampling and analyses

Faecal droppings and spilled chicken feed were collected separately for a period of 24 h weekly and were weighed. The monthly average of these weekly measurements gave the daily manure and spilled feed loading rates for each treatment.

Analytical grade reagents and chemicals were used for this study. Distilled-deionised water was also used in all relevant analyses. All analyses were done in triplicate. Procedural and reagent blanks were used and a clean laboratory environment was ensured during the analysis and preparation of solutions. Glassware and plastic containers were washed with liquid soap, rinsed with distilled water and then soaked in 10% HNO_3 solution for 24 h (Todorovic et al., 2001) and dried in a drying oven at 80°C for 5 h. Proximate analyses (moisture, crude protein, crude fat, crude fibre, nitrogen free extract and ash contents) of fish feed, chicken feed and air dried chicken manure were done monthly according to FAO (1994). Moisture and organic matter content of manure were determined monthly with the methods of Xing et al. (2008), pH and conductivity were determined with the methods of Igua and Huasi (2009); total nitrogen was analysed according to Bremner (1965) and Ademoroti (1996), while total phosphorus content of fresh manure was determined with the method of Agbenin (1995).

Water samples for physico-chemical analysis were collected at a depth of 10 cm with acid washed polyethylene bottles, while fish samples were collected with a drag net, the scales were removed and they were washed with clean water before analyses. Sample collection was between 6 and 8 am in the morning. A total of 40 *O. niloticus* and 20 *C. gariepinus* samples were cropped from each pond during each sampling period. Weights of fish were measured with analytical balance. Temperature and pH of water were measured with a pre-calibrated ATC portable pH/temperature meter (model PH-009 (III)). Conductivity was measured with a ELE conductivity meter (model DA-1) while dissolved oxygen was determined with Hach dissolved oxygen meter (model 50175). Temperature, pH and DO were measured daily, while ammonia, nitrate and nitrite were determined monthly according to standard methods (APHA, 1998).

Heavy metal analysis with flame atomic absorption spectrophotometry (FAAS)

The heavy metals of interest were Cd, Cr, Cu, Pb, manganese (Mn) and Zn. Aqueous stock solutions were prepared with the salts of these metals and six working standards were prepared in triplicate for each metal by serial dilution of the stock solutions. These and blank solutions were aspirated into a buck scientific atomic absorption spectrophotometer (model VGD 210). A calibration curve of absorbance versus concentration was prepared for each metal and used for the determination of metal concentrations in the samples. Fish samples were dissected immediately after collection and the organs of interest (muscle, gill, kidney, liver and gut) were separated. Fish organ samples were digested as described by Olaifa et al. (2004), while chicken feed and manure samples were digested according to Bamgbose et al. (2000). Standard conditions in line with the equipment manufacturer's recommendations were maintained throughout the FAAS analysis.

Statistical analysis

Statistical analysis was done using the SPSS (version 13.0) for windows software package. Mean concentrations and standard deviations were calculated for each parameter. The results were also subjected to analysis of variance (ANOVA) and means were compared using Duncan multiple range test. Pearson correlation analysis was also performed to analyse and confirm the relationship between the parameters and chicken manure and feed inputs.

Table 1. Chemical composition of fresh laying chicken manure.

Parameter	Mean ± SD
Moisture (%)	75.81 ± 1.28
Organic matter (%)	28.60 ± 2.51
Electrical conductivity (μScm^{-1})	30.00 ± 2.39
pH	7.68 ± 1.05
Nitrogen (%)	1.72 ± 0.11
Phosphorus (%)	0.65 ± 0.25
Potassium (%)	1.41 ± 0.69

Table 2. Mean daily manure and feed loading rates (g day^{-1}).

Month	Manure loading rate			Feed loading rate			
	Treatments			Integrated Treatments (spilled chicken feed)			Control (fish feed)
	T1	T2	T3	T1	T2	T3	T4 (5% of body weight)
October	186.042 ± 7.688	291.590 ± 1.755	392.215 ± 1.209	11.630 ± 1.824	17.850 ± 0.636	24.265 ± 1.506	1.058
November	187.630 ± 2.629	291.620 ± 5.338	387.630 ± 5.870	11.505 ± 1.084	16.870 ± 1.504	24.165 ± 1.791	6.592
December	184.225 ± 3.577	312.311 ± 4.972	390.732 ± 1.806	11.738 ± 0.961	15.984 ± 1.126	23.374 ± 1.387	10.453
January	188.951 ± 2.830	286.544 ± 2.760	385.518 ± 6.754	11.970 ± 0.692	17.045 ± 0.915	24.715 ± 1.087	15.652
February	187.940 ± 3.261	289.362 ± 1.007	387.700 ± 2.142	10.385 ± 1.457	16.530 ± 0.768	24.698 ± 1.207	20.610
March	186.383 ± 1.605	289.074 ± 2.356	385.755 ± 2.339	10.528 ± 1.306	16.805 ± 0.462	24.563 ± 0.731	28.129
April	186.482 ± 7.294	287.100 ± 3.970	390.306 ± 3.502	12.354 ± 0.918	16.852 ± 0.357	24.522 ± 1.256	33.882
May	185.659 ± 5.581	287.263 ± 5.884	390.890 ± 2.099	13.485 ± 0.902	16.343 ± 0.641	24.373 ± 1.504	40.455
June	183.266 ± 9.270	291.510 ± 4.771	394.675 ± 5.857	13.583 ± 0.359	17.123 ± 0.864	22.130 ± 1.139	46.629
July	188.975 ± 5.279	299.860 ± 10.150	403.462 ± 8.685	14.074 ± 0.399	18.628 ± 0.794	25.326 ± 0.956	56.618
August	190.511 ± 2.352	297.236 ± 5.723	404.838 ± 10.594	13.930 ± 1.099	19.266 ± 1.182	27.663 ± 1.070	66.925
September	192.064 ± 1.840	300.449 ± 5.428	404.428 ± 4.154	14.009 ± 0.495	18.650 ± 0.575	26.506 ± 1.269	78.121

RESULTS AND DISCUSSION

Results of proximate analysis showed that the crude protein (CP) content of chicken feed (CF) ranged from 11.77 to 19.20%; chicken manure (CM) from 18.00 to 21.90% and fish feed (FF) from 41.30 to 42.80%. CP of fish feed was significantly higher ($P < 0.05$) than those of CF and CM. Moisture content of CF had a range of 8.80 to 11.90%; chicken manure from 13.70 to 18.20%; fish feed from 8.10 to 8.85 %.

The chemical composition of fresh chicken manure is as shown in Table 1. Table 2 shows the mean daily manure loading rates (DMLR) into each treatment. For T1, DMLR was highest (192.06 ± 1.84 g/day) in the month of September and lowest (183.26 ± 7.69 g/day) in June. It was highest in September (300.44 ± 5.42 g/day) and lowest in January (286.51 ± 2.76 g/day) in T2. Highest values for T3 (404.838 ± 10.594 g/day) occurred in August and the lowest (385.518 ±

Table 3. Fish mortalities and percentage of survival.

Month	Treatments and mortalities								Total
	T1		T2		T3		T4		
	ON	CG	ON	CG	ON	CG	ON	CG	
January	3	–	1	–	6	–	1	–	11
February	–	–	2	–	15	–	1	–	18
March	–	–	3	–	–	–	1	–	4
June	3	–	5	–	5	–	–	–	13
July	–	–	–	–	1	–	–	–	1
August	2	–	3	–	8	–	1	–	14
TOTAL	8	–	14	–	35	–	4	–	61
Survival (%)	93.33	100.00	83.33	100.00	70.83	100.00	96.67	100.00	

ON = *O. niloticus* and CG = *C. gariepinus*.

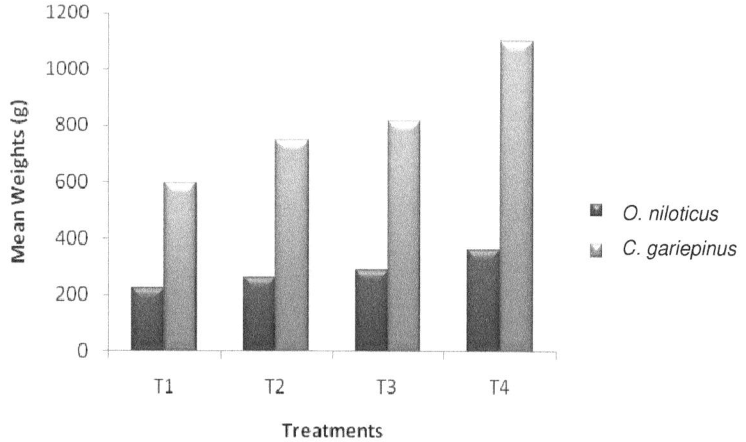

Figure 1. Mean weight gain of *O. niloticus* and *C. gariepinus*.

6.754 g/day) in January. DMLR varied significantly ($P < 0.05$) among the treatments, but showed non-significant variation ($P > 0.05$) across the months. T1 had four chickens integrated over a pond, while T3 had eight chickens. DMLR in T3 was slightly more than the double of T1 across the months. Daily feed loading rates (DFLR) for the four treatments are also shown in Table 2. The daily spilled feed loading rate (DSLR) for the integrated treatments depends on the activity and feeding rate of chickens, and the more active the chickens are, the more the amount of feed spilled. DSLR in T3 was significantly higher ($P < 0.05$) than that those of T1 and T2 across the months. DFLR for T4 was 5% of fish body weights calculated from the weights of fish measured during sampling.

Table 3 is for fish mortalities and percentage of survival. It indicates that sixty-one (61) *O. niloticus* mortalities were recorded in the course of the experiment. The highest mortality was recorded in February (18), while the lowest occurred in July (1). There were no fish mortalities in the October, November, December, April, May and September. However, no *C. gariepinus* mortality was recorded in any of the treatments. Values for percentage of survival of *O. niloticus* were 93.33, 88.33, 70.83 and 96.67% for T1, T2, T3 and T4, respectively. *O. niloticus* in T1 had the highest percentage survival among the integrated treatments, while T4 had the highest value overall. The percentage of survival value for *C. gariepinus* was 100% in all the treatments and this indicates the greater capacity of the later to withstand the sometimes unfavourable conditions in the integrated chicken-fish system.

Figure 1 shows the mean weight gain (MWG) for both fish species in the twelve month experimental period. The highest MWG of 1100.95 g was recorded for *C. gariepinus* in T4 while the lowest (221.14 g) was recorded for *O. niloticus* in T1. MWG of *C. gariepinus* was significantly higher ($P < 0.05$) than that of *O. niloticus* in all the treatments. Organic fertilizer use has a long tradition in tropical semi-intensive aquaculture and it increases fish yields through soluble and/or particulate pathways (Knud-Hansen et al., 1993). It is obvious from the mean weight gains (MWG) that the higher manure and spilled feed input into T2 resulted in the production of a higher fish biomass than in T1. The same scenario was obtained in T3 which had the highest mean weight gain among the integrated treatments. However, fish biomass production was highest in T4 which received compounded feed. This shows that *O. niloticus* and *C. gariepinus* production is higher with compounded feed than with chicken manure and spilled chicken feed. MWG for *C. gariepinus* was greater than those of *O. niloticus* in all the treatments and it indicates higher growth performance of *C. gariepinus* compared *O. niloticus* in an integrated chicken fish system.

Water temperature ranged from 21.4 to 33.4°C in T1 and from 20.2 to 33.4°C in T2. T3 and T4 had temperature ranges of 21.1 to 32.0°C and 19.4 to 32.0°C, respectively. Air temperature had a range of 24.9 to 36.6°C. The temperature range for optimum growth and performance of *O. niloticus* is 25 to 33°C (Guerrero, 1985). Water temperatures in the experiment were generally within this range, except in the months of December and January (the harmattan period) when temperatures were sometimes below the lower limit in all the treatments. In addition, water temperatures were sometimes above the upper limit in T1 and T2 in the month of March. *C. gariepinus* can grow and survive within a range of 23 to 35°C (UNEP, 1999) and temperature values were generally within this range.

Mean pH ranged from 6.41 to 7.99 in T1 and from 6.52 to 7.91 in T2. The lowest pH value in T3 was 6.56, while the highest value was 8.68. In T4, pH range was 6.15 to 7.80. The highest pH value in the experiment (8.68) was recorded in T3, while the lowest value of 5.26 was in T1 both occurring in the month of September. pH was not significantly affected ($P > 0.05$) by the treatments. Optimum pH for freshwater fish is 6.5 to 9 (Boyd, 1998). In this experiment, there were no great swings in pH and the ranges were generally within the 6.5 to 9 pH range recommended for warm water aquaculture. However, pH in all the treatments were sometimes below this optimum range in the month of September with T1 having the lowest experimental pH value 5.26 in September.

Mean dissolved oxygen (DO) values had ranges of 4.46 to 7.80, 3.30 to 8.17, 3.46 to 7.09 and 4.76 to 8.05 mg/L for T1, T2, T3 and T4, respectively. The lowest DO value of 3.30 mg/L was obtained in T2 in the month of September, while the highest value of 8.17 mg/L occurred in the same treatment in October. DO was affected significantly ($P < 0.05$) by the treatments (manure and spilled loading rates) and they were generally lower in T3 than in all other treatments. There is negative correlation between manure loading rate and DO level inT1 ($r = -0.333$), T2 ($r = -0.235$) and T3 ($r = -0.518$), but positive correlation existed between feed loading rate and DO (T1, $r = 0.192$; T2, $r = 0.151$; T4, $r = 0.204$). However, feed loading rate and DO were negatively correlated ($r = -0.369$) in T3. According to Swann (1990), dissolved oxygen (DO) is by far the most important chemical parameter in aquaculture. Low-dissolved oxygen levels are responsible for more fish kills, either directly or indirectly, than all other problems combined. Warm water fish require DO ≥ 5 mg/L for good growth and reproduction (Swingle, 1969). However, DO values in the experiment were in many cases, less than this benchmark in the integrated treatments. This scenario was more prominent in the months of December, January, March and September.

Total nitrogen (TN) had a range of 0.561 to 8.031 mg/L (mean 5.023 ± 2.368 mg/L) in T1 and a range of 0.470 to 11.120 mg/L (6.276 ± 3.396 mg/L) in T2. T3 had a TN range of 0.385 to 15.280 mg/L (mean 9.722 ± 5.275 mg/L), while T4 had a range of 0.435 to 7.010 mg/L (mean 4.003 ± 2.112 mg/L). TN was significantly affected ($P < 0.05$) by the treatments with higher values in T3 as compared to other treatments. Manure loading and TN were positively correlated in all the treatments with significant relationships in T3 ($r = 0.623$, $P < 0.01$). Nitrate range for T1 was 0.230 to 6.530 mg/L (mean 4.363 ± 2.138 mg/L) and 0.200 to 10.130 mg/L (mean 5.619 ± 3.199 mg/L) in T2. T3 had a range of 0.250 to 14.930 mg/L and a mean value of 8.618 ± 4.897 mg/L. Range and mean for T4 were 0.210 to 6.120 and 3.218 ±1.628 mg/L, respectively. The values for T4 were significantly lower ($P < 0.05$) than those of the integrated treatments. Nitrate concentration is rarely a problem in aquaculture. However, nitrate concentration is a pointer to abnormal levels of organic matter in fish ponds. In this study, the range for nitrate content in T3 exceeded the 2 to 10 mg/L range recommended by Boyd (1998) for aquaculture. This may point to too much manure and spilled feed load in the treatment.

Nitrite concentration ranged from 0.019 to 0.140 mg/L in T1 with a mean value of 0.070 ± 0.037 mg/L and from 0.022 to 0.223 mg/L (mean 0.107 ± 0.064 mg/L) in T2. The range for T3 was 0.018 to 0.319 mg/L (mean 0.152 ± 0.099 mg/L), while T4 had a nitrite range of 0.010 to 0.105 mg/L (mean 0.054 ± 0.029 mg/L). Nitrite concentration was affected significantly ($P < 0.05$) by treatment. The blood pigment, haemoglobin transports oxygen from the gills of fish to the body tissues where it is exchanged with carbon dioxide. Abnormal levels of nitrite in a fish ponds interferes with this process, because nitrite oxidizes haemoglobin to methemoglobin which cannot transport oxygen leading to the disease called methemoglonamia which in acute cases, results in asphyxiation of fish. In the experiment, Nitrite concentration for the month of September in T3 exceeded the recommended benchmark of <0.3 mg/L by Boyd (1998), but the concentration in all the treatments exceeded the <0.1 mg/L recommended by UNEP (1999) especially towards the end of the experiment.

Values for free ammonia concentration in water samples from each treatment indicates that it ranged from 0.012 to 0.074 mg/L in T1 and from 0.015 to 0.107 mg/L in T2. In T3 and T4, it had ranges of 0.020 to 0.156 and 0.009 to 0.062 mg/L. Mean values for ammonia

Table 4. Percentage recovery of heavy metals from spiking experiment.

Metal	Samples		
	Water	Sediment	Fish muscle
Cadmium	73.81 ± 1.94	90.16 ± 2.25	77.37 ± 3.07
Chromium	80.40 ± 3.92	94.30 ± 1.80	82.61 ± 2.55
Copper	85.66 ± 3.17	80.20 ± 2.18	75.33 ± 1.76
Lead	71.30 ± 1.26	92.75 ± 1.95	86.65 ± 2.18
Manganese	96.21 ± 2.14	95.00 ± 2.60	110.20 ± 3.74
Zinc	102.80 ± 4.05	105.30 ± 3.42	90.16 ± 1.92

Each value represents the mean and standard deviation for three replicates.

Table 5. Metal contents in chicken and fish feed samples (mg kg^{-1}).

Treatment	Metal	Oct.	Nov.	Dec.	Jan.	Feb.	Mar.	Apr.	May	Jun.	Jul.	Aug.	Sep.
Chicken feed	Cd	0.137	0.122	0.147	0.159	0.130	0.128	0.141	0.125	0.127	0.149	0.126	0.150
	Cr	0.330	0.321	0.470	0.446	0.385	0.502	0.540	0.411	0.552	0.510	0.362	0.438
	Cu	4.688	4.121	3.208	3.110	4.216	5.199	4.185	5.223	3.211	3.200	3.209	4.228
	Pb	0.744	0.693	0.890	0.792	0.623	0.770	0.881	0.809	0.736	0.917	0.900	0.845
	Mn	94.615	85.910	85.315	75.280	84.950	85.029	90.740	94.050	86.113	85.380	94.806	95.227
	Zn	74.110	70.140	74.320	72.832	77.270	76.384	68.10	67.447	68.009	65.285	69.406	72.033
Compounded fish feed	Cd	0.611	0.435	0.591	0.637	0.690	0.744	0.720	0.658	0.537	0.660	0.585	0.716
	Cr	2.941	2.504	3.082	3.114	2.536	2.687	2.255	1.906	2.071	2.735	1.688	2.502
	Cu	5.070	5.624	5.662	4.701	4.269	4.140	3.641	4.400	5.215	5.070	4.910	4.347
	Pb	3.300	2.974	1.745	1.951	2.374	2.883	3.109	2.956	2.313	3.164	3.217	2.920
	Mn	20.660	21.340	19.353	20.209	20.084	18.311	21.465	19.164	19.540	18.310	20.225	20.570
	Zn	82.452	80.251	88.713	76.905	80.557	85.075	80.182	79.110	78.633	77.009	79.571	80.293

concentration were 0.046 ± 0.021, 0.059 ± 0.031, 0.084 ± 0.046 and 0.032 ± 0.018 mg/L for T1, T2, T2 and T4, respectively. Ammonia concentration was significantly lower ($P < 0.05$) in the control (T4) compared to the integrated treatments. Unionized ammonia (NH_3) is produced in aquatic systems through direct excretion by fish and the decay of organic matter like manure and uneaten feed. At low levels (<0.1 mg/L NH_3) it is a strong irritant, especially to the gills of fish, but prolonged exposure to sub-lethal levels can lead to skin and gill hyperplasia and bacterial gill disease (Fishdoc, 2008). In this study, the concentration in T3 exceeded the <0.1 mg/L recommended value for fish culture from June to September. This may explain the relatively high mortality rate of *O. niloticus* in T3. Table 4 shows the percent recoveries calculated for each metal in water, sediment and fish samples from the spiking experiment. Zn had the highest recover Values (102.80 and 105.30%) from the water and

Table 6. Metal contents in fresh chicken manure samples (mg kg^{-1}).

Treatment	Metal	Oct.	Nov.	Dec.	Jan.	Feb.	Mar.	Apr.	May	Jun.	Jul.	Aug.	Sep.
T1	Cd	0.251	0.250	0.200	0.250	0.236	0.318	0.290	0.220	0.252	0.270	0.277	0.278
	Cr	0.865	0.660	0.741	0.681	0.620	0.720	0.735	0.541	0.610	0.630	0.549	0.750
	Cu	17.330	16.184	16.927	15.728	16.118	17.480	17.283	15.702	12.075	13.868	15.200	17.391
	Pb	2.230	1.930	2.311	2.050	1.830	1.910	2.261	1.640	1.629	1.614	1.610	2.155
	Mn	135.380	127.550	130.197	134.108	135.260	133.180	132.960	126.135	130.420	125.800	133.116	135.480
	Zn	100.274	90.575	91.920	95.600	100.201	96.587	98.172	88.160	93.115	90.780	98.235	100.104
T2	Cd	0.235	0.225	0.210	0.233	0.249	0.311	0.295	0.226	0.229	0.237	0.262	0.276
	Cr	0.850	0.685	0.736	0.607	0.680	0.843	0.800	0.642	0.680	0.630	0.600	0.754
	Cu	16.800	17.128	15.980	14.899	15.216	16.862	16.328	14.585	13.121	14.877	12.054	17.370
	Pb	2.272	1.968	1.930	1.765	1.820	2.103	1.980	1.720	1.790	1.631	1.709	1.982
	Mn	135.225	130.553	133.162	130.259	134.440	129.310	135.232	130.210	128.435	120.104	130.758	134.535
	Zn	98.390	92.730	100.430	98.900	101.230	93.628	98.230	95.173	97.110	96.935	100.340	100.146
T3	Cd	0.269	0.260	0.225	0.248	0.230	0.305	0.274	0.216	0.248	0.219	0.264	0.280
	Cr	0.871	0.683	0.727	0.625	0.710	0.786	0.808	0.656	0.607	0.638	0.536	0.750
	Cu	15.980	15.147	17.970	15.923	16.018	15.870	17.344	15.471	12.101	14.845	13.119	16.399
	Pb	2.246	1.944	1.814	1.926	1.801	2.266	2.326	2.018	1.770	1.640	1.863	2.127
	Mn	130.372	132.550	130.139	131.280	132.250	125.300	130.205	128.054	131.428	128.130	130.995	134.531
	Zn	100.418	93.722	92.380	99.746	99.200	97.629	97.867	95.195	94.072	91.580	96.350	99.370

sediment samples, while Mn had the highest recovery in the fish muscle sample respectively. The percentage recovery values are all above 70%, and so, the digestion method used and the FAAS analyses are reliable.

Metal concentrations in chicken feed and fish feed samples are shown in Table 5. Cd in chicken feed had a range of 0.137 to 0.159 mg/kg (mean 0.137 ± 0.012), Cr had a range of 0.321 to 0.552 mg/kg (mean 0.411 ± 0.148), Cu had a range of 3.110 to 5.223 mg/kg (mean 3.983 ± 0.789), Pb had a range of 0.623 to 0.917 mg/kg (mean 0.800 ± 0.091), Mn had a range of 75.280 to 95.227 mg/kg (mean 88.118 ± 5.940) and Zn had a range of 65.285 to 77.270 mg/kg (mean 71.278 ± 3.795). These values are lower than the values obtained by Islam et al. (2007), who in a study of three layer and broiler feeds in Bangladesh, found substantial levels of heavy metals in these feeds with ranges of 0.1852 to 0.0232 ppm for Cd; 20.6498 to 0.6019 ppm for Pb; 5.7875 to 0.0926 ppm for Cr; 302.2001 to 0.0695 ppm for Mn; 37.5725 to 0.0463 ppm for Cu and 422.3023 to 0.0232 ppm for Zn. Zn concentrations in this study were similar ($P > 0.05$) to Mn concentrations, but were significantly higher ($P < 0.05$) than the concentrations of other metals. Cd concentrations were the lowest in chicken feed samples. The profile for metal contents in chicken feed across the months was Mn>Zn>Cu>Pb>Cr>Cd.

Cd in fish feed ranged from 0.537 to 0.744 mg/kg (mean 0.632 ± 0.088), while Cr ranged from 1.688 to 3.114 mg/kg (mean 2.502 ±0.453). Cu had a range of 3.641 to 5.662 mg/kg (mean 4.754 ± 0.615), Pb had a range of 2.313 to 3.300 mg/kg (mean 2.742 ± 0.517), Mn had a range 18.310 to 21.465 mg/kg (mean 19.936 ± 1.035) and Zn had a range of 77.009 to 88.713 mg/kg (mean 80.729 ± 3.346). Zn concentrations were significantly higher ($P < 0.05$) than the concentrations of other metals in fish feed. The general profile for metal contents in fish feed

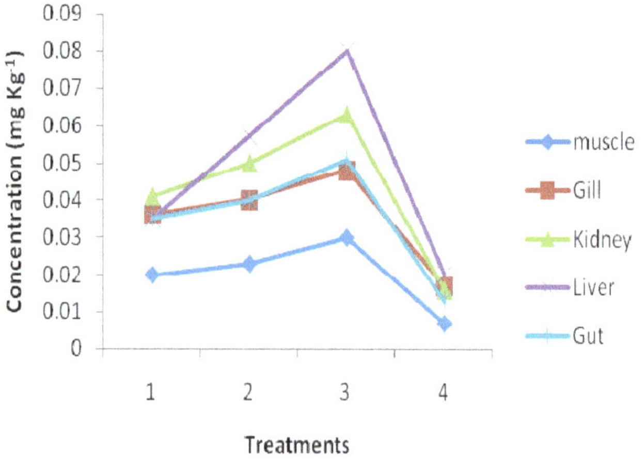

Figure 2. Final Cd concentrations in *O. niloticus* tissues.

across the months was Zn>Mn>Cu>Pb>Cr>Cd.

Metal contents in fresh chicken manure samples are shown in Table 6. Cd concentration ranged from 0.200 to 0.318 mg/kg in T1 with the highest value occurring in March and the lowest value in December. Cr ranged from 0.541 to 0.865 mg/kg, while Cu concentration had a range of 12.075 to 17.480 mg/kg which shows widely variable monthly concentrations of these metals in chicken manure. Pb content had a range of 1.610 to 2.311 mg/kg, Mn range was 125.800 to 135.480 mg/kg and Zn ranged from 88.160 to 100.274 mg/kg. Mean metal concentrations were Cd 0.258 ± 0.032, Cr 0.675 ± 0.093, Cu 15.941 ± 1.628, Pb 1.931 ± 0.269, Mn 131.632 ± 3.554 and Zn 95.310 ± 4.268 mg/kg. There was significant difference ($P < 0.05$) in the manure metal concentrations in T1. In T2, range of values for metal concentrations was: 0.210 to 0.311 mg/kg for Cd; 0.600 to 0.850 mg/kg for Cr; 12.054 to 17.370 mg/kg for Cu; 1.631 to 2.272 mg/kg for Pb; 120.104 to 135.232 mg/kg for Mn and 92.730 to 100.340 mg/kg for Zn. Mean metal contents were; 0.249 ± 0.031 mg/kg for Cd, 0.709 ± 0.087 mg/kg for Cr, 15.435 ± 1.645 mg/kg for Cu, 1.889 ± 0.185 mg/kg for Pb, 131.019 ± 4.203 mg/kg for Mn and 97.770 ± 2.757 mg/kg for Zn, respectively. Mn concentrations in manure samples of T2 were significantly higher ($P < 0.05$) than concentrations of other metals. Mean metal contents for the metals in T3 were; Cd 0.253 ± 0.027, Cr 0.670 ± 0.095, Cu 15.516 ± 1.621, Pb 1.978 ± 0.220, Mn 130.436 ± 2.413 and Zn 96.461 ± 3.031 mg/kg, while concentration ranges were; 0.219 to 0.305 mg/kg for Cd, 0.536 to 0.871 mg/kg for Cr, 12.101 to 17.970 mg/kg for Cu, 1.640 to 2.326 mg/kg for Pb, 125.300 to 134.531 mg/kg for Mn and 91.580 to 100.418 mg/kg for Zn. Across the months, Cd concentrations were significantly lower ($P < 0.05$) than those of other metals in T3. Comparison of means showed no significant difference ($P < 0.05$) in mean values of each particular metal in all the treatments. The general profile for metal contents in fresh chicken manure samples in all the treatments across the months was Mn>Zn>Cu>Pb>Cr>Cd.

In all the treatments, Mn and Zn concentrations were much higher than the concentrations of other metals and this may not be unconnected with high Mn and Zn concentrations in the chicken feed. Hanč et al. (2008) analysed chicken manure for metals and concluded that it contained 0.232 and 47.700 mg/kg of Cd and Cu, respectively. Khai et al. (2008), in an experiment to determine the effect of chicken and composted manure on soil chemical properties, determined the concentrations of Cu, Zn, Cd and Pb in chicken manure as 48.5, 263.0, 3.4 and 25.3 mg/kg, respectively. These values are higher than those obtained in the present study.

Metal concentrations in fish samples were analysed in December, three months after stocking, because the fish were stocked as fingerlings and were allowed to grow to a size that will enable the dissection and collection of adequate quantities of each tissue. Cd concentration in *O. niloticus* muscle ranged from 0.003 to 0.020 mg/kg in T1, while gill concentration ranged from 0.012 to 0.036 mg/kg. Concentration in the kidney had a range of 0.019 to 0.045 mg/kg, while concentrations in liver and gut were 0.011 to 0.035 mg/kg and 0.005 to 0.035 mg/kg, respectively. Cd concentrations were highest in the kidney and lowest in the muscle in T1. The range of Cd concentrations in T2 were; muscle, 0.003 to 0.023 mg/kg; gill, 0.016 to 0.040 mg/kg; kidney, 0.021 to 0.050 mg/kg; liver, 0.014 to 0.057 mg/kg and gut, 0.007 to 0.040 mg/kg. Final Cd concentration in T2 was highest in the liver and lowest in the muscle. Concentration ranges in T3 were; muscle, 0.005 to 0.030 mg/kg; gill, 0.019 to 0.048 mg/kg; kidney, 0.028 to 0.063 mg/kg; liver, 0.021 to 0.080 mg/kg and gut, 0.013 to 0.051 mg/kg. Cd liver concentrations were generally higher than others with lowest values occurring in the muscle. T4 received only compounded fish feed and Cd concentrations were; 0.002 to 0.007 mg/kg for muscle; 0.004 to 0.017 mg/kg for gill; 0.005 to 0.016 mg/kg for kidney; 0.004 to 0.020 mg/kg for liver and 0.002 to 0.013 mg/kg for gut. Final Cd concentration was highest in the liver and lowest in the muscle. Muscle, gill, kidney, liver and gut concentrations were significantly lower ($P < 0.05$) in T4 than in other treatments, while liver concentrations in T3 was significantly higher ($P < 0.05$) than in T1 and T2. Figure 2 shows the profile of final CD concentrations in the different treatments.

Ranges for Cr concentrations in the tissues of *O. niloticus* in T1 were; muscle, 0.012 to 0.039 mg/kg; gill, 0.017 to 0.052 mg/kg; kidney, 0.015 to 0.043 mg/kg; liver, 0.013 to 0.050 mg/kg and gut, 0.013 to 0.045 mg/kg. Final concentration was highest in the gill and lowest in the kidney. In T2, Cr range was 0.014 to 0.056 mg/kg in the muscle, 0.019 to 0.064 mg/kg in the gill, 0.017 to 0.048 mg/kg in the kidney, 0.025 to 0.057 mg/kg in the liver and 0.026 to 0.053 mg/kg in the gut. Ranges in T3 were; muscle, 0.018 to 0.080 mg/kg; gill, 0.025 to 0.090 mg/kg; kidney, 0.030 to 0.065 mg/kg; liver, 0.033 to 0.070 mg/kg and gut, 0.027 to 0.072 mg/kg. Muscle

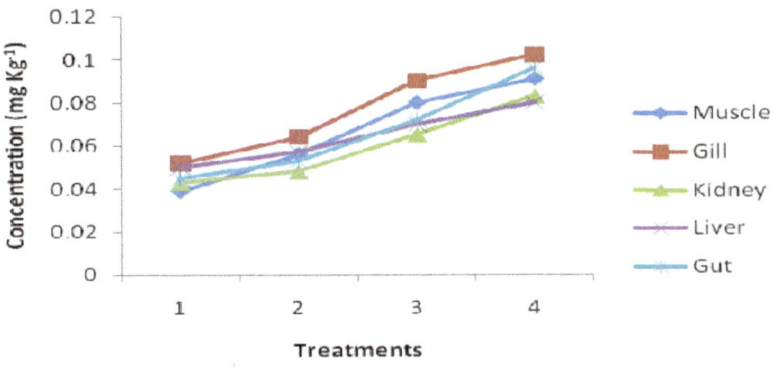

Figure 3. Final Cr concentrations in *O. niloticus* tissues.

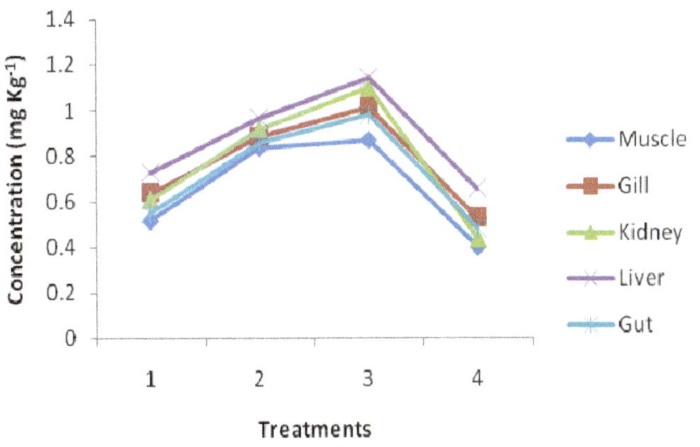

Figure 4. Final Cu concentrations in *O. niloticus* tissues.

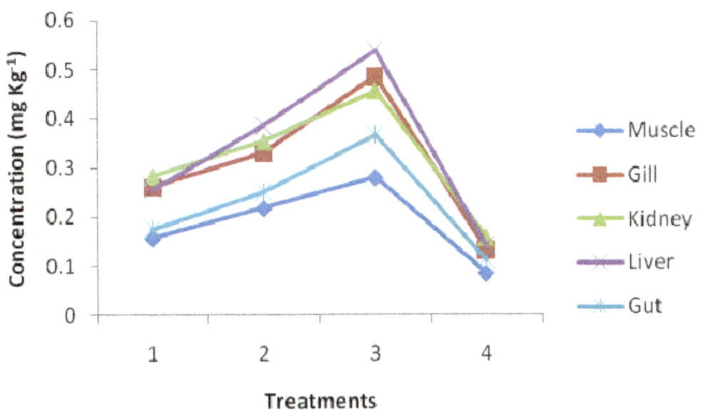

Figure 5. Final Pb concentrations in *O. niloticus* tissues.

concentration in T4 ranged from 0.021 to 0.091 mg/kg; gill, 0.030 to 0.102 mg/kg; kidney, 0.032 to 0.083 mg/kg; liver, 0.030 to 0.080 mg/kg and gut, 0.029 to 0.096 mg/kg. The highest concentrations were in the gill, while the lowest were in the muscle and liver. Cr concentrations were significantly affected ($P < 0.05$) by treatments with the lowest values recorded for the T1 which received the lowest manure and spilled feed loading and highest values in T4, the control. Final Cr concentrations in liver and gut were similar ($P > 0.05$) in the integrated treatments, but differed significantly ($P < 0.05$) from the final value in the control (T4). Profile of final Cr concentrations under different treatments is as shown in Figure 3.

Ranges for Cu concentrations in the tissues of *O. niloticus* for T1 were; muscle, 0.302 to 0.521 mg/kg; gill, 0.264 to 0.640 mg/kg; kidney, 0.225 to 0.611 mg/kg; liver, 0.241 to 0.730 mg/kg and gut, 0.278 to 0.550 mg/kg. Cu concentration was significantly higher ($P < 0.05$) in the liver than in other tissues. In T2, ranges for Cu concentration were muscle, 0.330 to 0.834 mg/kg; gill, 0.409 to 0.885 mg/kg; kidney, 0.437 to 0.919 mg/kg; liver, 0.610 to 0.966 mg/kg and gut, 0.495 to 0.860 mg/kg. There was no significant variation ($P > 0.05$) in metal concentrations for the tissues of T2 fish. Concentrations in T3 were 0.408 to 0.867 mg/kg for muscle, 0.562 to 1.014 mg/kg for gill, 0.618 to 1.100 mg/kg for kidney, 0.711 to 1.143 mg/kg for liver and 0.570 to 0.982 mg/kg. Muscle concentration was significantly lower ($P < 0.05$) than those of other tissues. T4 had the lowest metal concentrations with values for Cu in muscle having a range of 0.173 to 0.398 mg/kg, gill, 0.212 to 0.525 mg/kg; kidney, 0.250 to 0.437 mg/kg; liver, 0.294 to 0.655 mg/kg and gut, 0.257 to 0.480 mg/kg. Final Cu concentration in T4 was lowest in the muscle and highest in the liver. Cu concentration was significantly affected ($P < 0.05$) by the treatments with higher values in the tissues of fish from the integrated treatments than in the control treatment. The profile of final Cu concentrations under the different treatments is as shown in Figure 4.

Pb concentrations in *O. niloticus* tissues had ranges of muscle 0.010 to 0.157 mg/kg, gill, 0.031 to 0.262 mg/kg; kidney, 0.064 to 0.283 mg/kg; liver, 0.051 to 0.255 mg/kg and gut, 0.030 to 0.176 mg/kg in T1. In T2, Pb concentration ranges were 0.028 to 0.218 mg/kg in the muscle, 0.030 to 0.330 mg/kg in the gill, 0.125 to 0.354 mg/kg in the kidney, 0.076 to 0.385 mg/kg in the liver and 0.064 to 0.251 mg/kg in the gut. Muscle, 0.039 to 0.280; gill, 0.070 to 0.486; kidney, 0.156 to 0.457; liver, 0.102 to 0.539 and gut, 0.094 to 0.367 mg/kg were the ranges in T3, while the ranges in T4 were; muscle, 0.009 to 0.085 mg/kg; gill, 0.011 to 0.132 mg/kg; kidney, 0.025 to 0.158 mg/kg; liver, 0.022 to 0.140 mg/kg and gut, 0.018 to 0.112 mg/kg. Pb concentration was significantly affected ($P < 0.05$) by treatments with highest values in T3 which received the highest manure loadings and the lowest values in T4 which is the control treatment without manure input. Final Pb concentrations in the different tissues under the various treatments are shown in Figure 5.

Mn concentrations in *O. niloticus* muscle of T1 ranged from 0.230 to 1.620 mg/kg, gill, 0.516 to 2.811 mg/kg; kidney, 1.405 to 3.267 mg/kg; liver, 1.220 to 3.824 mg/kg and gut, 0.618 to 2.742 mg/kg. Mn concentration in the muscle was significantly lower ($P < 0.05$) than those of other tissues. Mn concentration in T2 had ranges of 0.427 to 2.591 mg/kg in the muscle, 0.854 to 4.166 mg/kg in the gill, 1.906 to 4.529 mg/kg in the kidney, 1.536 to5.440 mg/kg in the liver and 0.895 to 4.114 mg/kg in the gut. Ranges for tissue concentrations of Mn in T3 were;

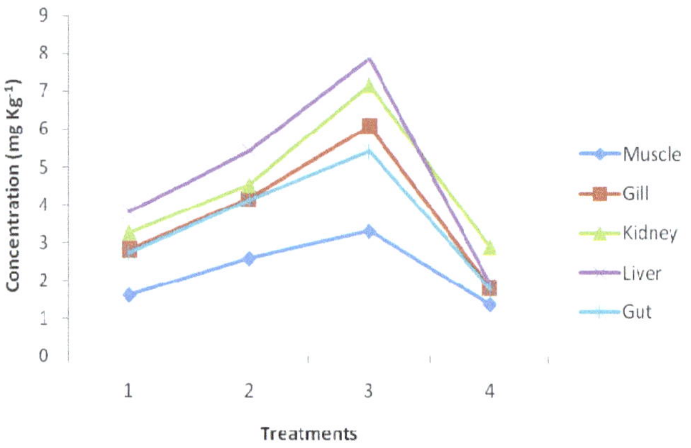

Figure 6. Final manganese concentrations in *O. niloticus* tissues.

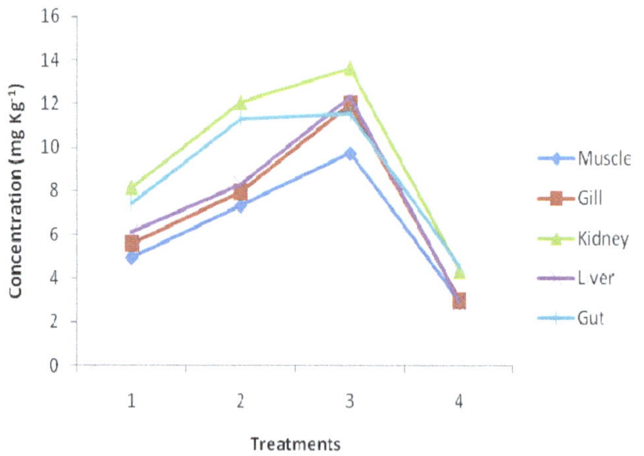

Figure 7. Final Zinc concentrations in *O. niloticus* tissues.

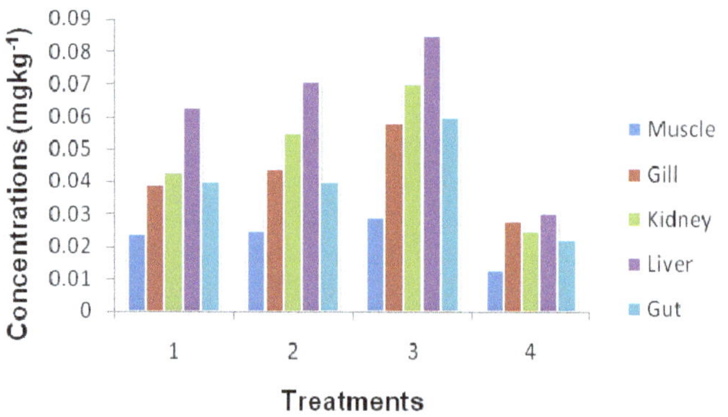

Figure 8. Concentration of cadmium in *C. gariepinus* tissues.

muscle, 0.523 to 3.305 mg/kg; gill, 1.015 to 6.101 mg/kg; kidney, 2.237 to 7.180 mg/kg; liver, 2.280 to 7.877 mg/kg and gut, 1.427 to 5.416 mg/kg. Tissue concentrations in T4 had ranges of 0.126 to 1.379 mg/kg for muscle, 0.288 to 1.812 mg/kg for gill, 0.791 to 2.897 mg/kg for kidney, 0.429 to 1.928 mg/kg for liver and 0.253 to 1.795 mg/kg for gut. Tissue concentrations in T3 were significantly higher ($P < 0.05$) than those of other treatments. Figure 6 profiles the final Mn concentrations in the various tissues. The highest Mn concentrations were in the livers while the lowest were in the muscles of fish in all the treatments.

Concentrations of Zn in the various tissues under the different treatments had the following ranges in T1: 1.250 to 4.970 mg/kg for muscle, 1.563 to 5.600 mg/kg for gill, 2.411 to 8.174 mg/kg for kidney, 1.826 to 6.129 mg/kg for liver and 2.521 to 7.433 mg/kg for gut. Zn concentration in *O. niloticus* muscle in T2 had a range of 1.630 to 7.341 while concentrations in gill, kidney, liver and gut had concentration ranges of 1.987 to 7.966, 2.836 to 12.080, 3.110 to 8.330 and 3.296 to 11.305 mg/kg, respectively. Values for kidney were significantly higher ($P < 0.05$) than those of other tissues in T1 and T2. Ranges for Zn concentration in muscle in T3 was 2.646 to 9.738, gill, 3.422 to 11.984 mg/kg; kidney, 4.470 to 13.640 mg/kg; liver, 3.728 to 12.330 mg/kg and gut, 4.007 to 11.538 mg/kg. T4 had the following ranges for Zn in fish tissues: muscle, 0.652 to 2.935 mg/kg; gill, 0.781 to 3.022 mg/kg; kidney, 1.052 to 4.370 mg/kg; liver, 0.910 to 2.926 mg/kg and gut, 1.260 to 4.542 mg/kg. Values for Zn concentrations in the tissues of T4 were significantly lower ($P < 0.05$) than those of other treatments. Final Zn concentrations in the five tissues of *O. niloticus* are profiled in Figure 7.

Concentration of Cd in the tissues of *C. gariepinus* was also determined. Muscle concentration in T1 ranged from 0.008 to 0.024 mg/kg while gill concentration ranged from 0.019 to 0.039 mg/kg. Kidney concentration had a range of 0.025 to 0.043 mg/kg while concentrations in liver and gut were 0.025 to 0.063 and 0.010 to 0.040 mg/kg, respectively. Cd concentrations were highest in the liver and lowest in the muscle in T1. The range of Cd concentrations in T2 were; muscle, 0.005 to 0.025 mg/kg; gill, 0.014 to 0.044 mg/kg; kidney, 0.018 to 0.055 mg/kg; liver, 0.032 to 0.071 mg/kg and gut, 0.011 to 0.040 mg/kg. Final Cd concentration inT2 was highest in the liver and lowest in the muscle. Cd concentration ranges in T3 were; muscle, 0.007 to 0.029 mg/kg; gill, 0.022 to 0.058 mg/kg; kidney, 0.033 to 0.070 mg/kg; liver, 0.044 to 0.085 mg/kg and gut, 0.016 to 0.060 mg/kg. Cd concentrations in T4 were; 0.002 to 0.013 mg/kg for muscle; 0.009 to 0.028 mg/kg for gill; 0.0015 to 0.025 mg/kg for kidney; 0.010 to 0.030 mg/kg for liver and 0.005 to 0.022 mg/kg for gut. Final Cd concentration was highest in the liver and lowest in the muscle. Muscle, gill, kidney, liver and gut concentrations were significantly lower ($P < 0.05$) in T4 than in other treatments, while kidney, liver and gut concentrations in T3 were significantly higher ($P < 0.05$) than those of T1 and T2. Figure 8 profiles of the final concentrations of Cd in the tissues of *C. gariepinus* under the different treatments.

Cr concentrations in the tissues of *C. gariepinus* had

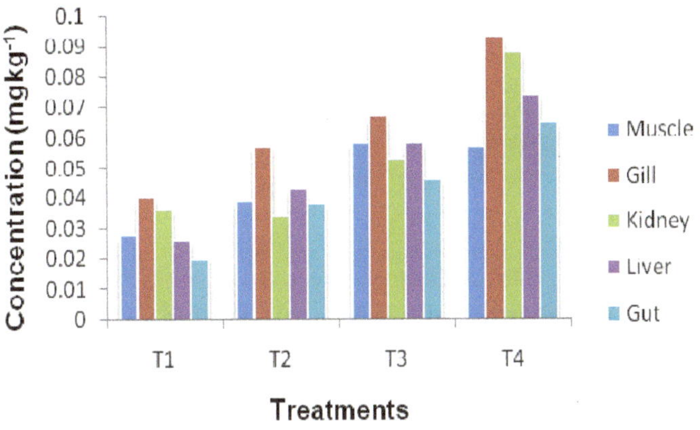

Figure 9. Concentration of chromium in *C. gariepinus* tissues.

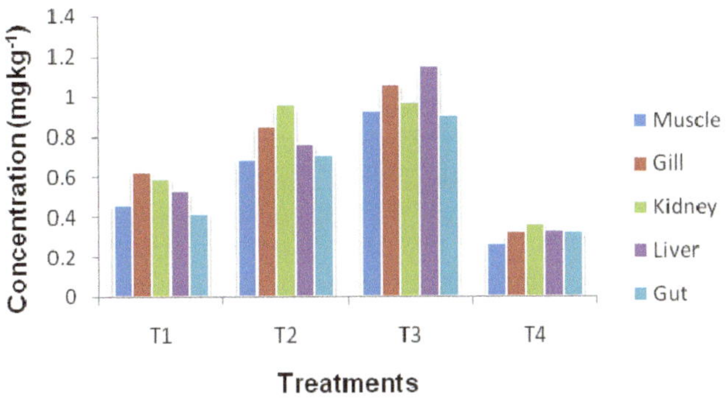

Figure 10. Concentration of Copper in *C. gariepinus* tissues.

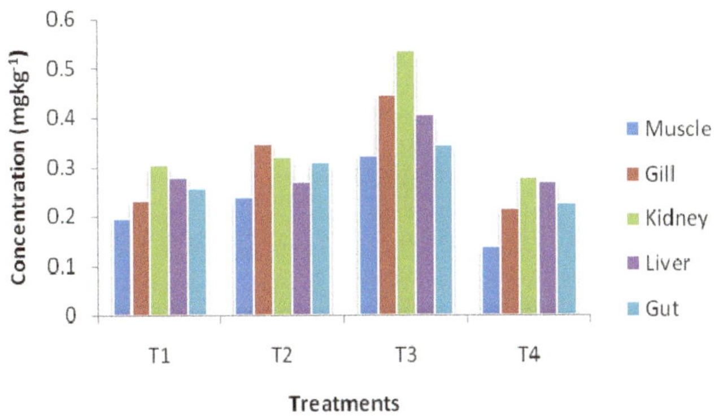

Figure 11. Concentration of lead in *C. gariepinus* tissues.

the following ranges in T1: muscle, 0.006 to 0.028 mg/kg; gill, 0.006 to 0.040 mg/kg; kidney, 0.003 to 0.036 mg/kg; liver, 0.004 to 0.026 mg/kg and gut, 0.005 to 0.020 mg/kg. Final concentration was highest in the gill and lowest in the gut. In T2, Cr range was 0.008 to 0.039 mg/kg in the muscle, 0.010 to 0.057 mg/kg in the gill, 0.008 to 0.034 mg/kg in the kidney, 0.005 to 0.043 mg/kg in the liver and 0.007 to 0.038 mg/kg in the gut. Ranges in T3 were; muscle, 0.011 to 0.058 mg/kg; gill, 0.010 to 0.067 mg/kg; kidney, 0.008 to 0.053 mg/kg; liver, 0.014 to 0.058 mg/kg and gut, 0.009 to 0.046 mg/kg. Muscle concentration in T4 ranged from 0.014 to 0.057 mg/kg; gill, 0.021 to 0.093 mg/kg; kidney, 0.015 to 0.088 mg/kg; liver, 0.013 to 0.074 mg/kg and gut, 0.010 to 0.065 mg/kg. The highest concentration in T4 occurred in the gill while the lowest was in the muscle. Cr concentrations were significantly affected ($P < 0.05$) by the treatments with higher concentrations occurring in the T4. Profile of Cr concentrations under the different treatments is as shown in Figure 9.

Range of Cu concentrations in the tissues of *C. gariepinus* in T1 were; muscle 0.194 to 0.460 mg/kg; gill, 0.188 to 0.593 mg/kg; kidney, 0.233 to 0.548 mg/kg; liver, 0.260 to 0.647 mg/kg and gut, 0.217 to 0.437 mg/kg. In T2, ranges for Cu concentration were muscle 0.266 to 0.685 mg/kg; gill, 0.279 to 0.850 mg/kg; kidney, 0.283 to 0.765 mg/kg; liver, 0.317 to 0.960 mg/kg and gut, 0.243 to 0.704 mg/kg. Cu concentration ranges in T3 were 0.295 to 0.929 mg/kg for muscle, 0.318 to 1.062 mg/kg for gill, 0.325 to 0.968 mg/kg for kidney, 0.301 to 1.154 mg/kg for liver and 0.290 to 0.907 mg/kg for gut. T4 had the lowest metal concentrations with values for Cu in muscle having a range of 0.093 to 0.264 mg/kg; gill, 0.138 to 0.327 mg/kg; kidney, 0.104 to 0.331 mg/kg; liver, 0.121 to 0.369 mg/kg and gut, 0.080 to 0.325 mg/kg. Cu concentration was significantly affected ($P < 0.05$) by the treatments with higher values in the tissues of fish from the integrated treatments. The profile of Cu concentrations under different treatments is as shown in Figure 10.

Ranges for Pb concentrations in T1 were; muscle, 0.008 to 0.169 mg/kg; gill, 0.024 to 0.234 mg/kg; kidney, 0.051 to 0.305 mg/kg; liver, 0.046 to 0.280 mg/kg and gut, 0.028 to 0.257 mg/kg. In T2, Pb tissue concentrations had ranges of 0.012 to 0.239 mg/kg in the muscle, 0.035 to 0.346 mg/kg in the gill, 0.068 to 0.320 mg/kg in the kidney, 0.051 to 0.265 mg/kg in the liver and 0.037 to 0.312 mg/kg in the gut. The ranges in T3 were muscle, 0.019 to 0.324 mg/kg; gill, 0.039 to 0.446 mg/kg; kidney, 0.077 to 0.535 mg/kg; liver, 0.060 to 0.408 mg/kg and gut, 0.041 to 0.345 mg/kg, while the ranges in T4 were; muscle, 0.010 to 0.137 mg/kg; gill, 0.021 to 0.214 mg/kg; kidney, 0.040 to 0.280 mg/kg; liver, 0.032 to 0.269 mg/kg and gut, 0.020 to 0.226 mg/kg. Pb concentration was significantly affected ($P < 0.05$) by treatments with highest values in T3 which received the highest manure loadings and the lowest values in T4. Final Pb concentrations in the different tissues under the various treatments are as shown in Figure 11.

Mn concentration in the muscle of *C. gariepinus* in T1 ranged from 0.781 to 3.430 mg/kg; gill, 0.953 to 4.315 mg/kg; kidney, 1.260 to 5.036 mg/kg; liver, 1.519 to 6.110 mg/kg and gut 0.835 to 4.128 mg/kg. Final Mn concentration was highest in the liver and lowest in the muscle. Mn concentration in T2 had ranges of 1.130 to 4.537 in the muscle, 1.295 to 7.019 in the gill, 1.009 to 9.239 in the kidney, 1.610 to 11.744 in the liver and 0.974

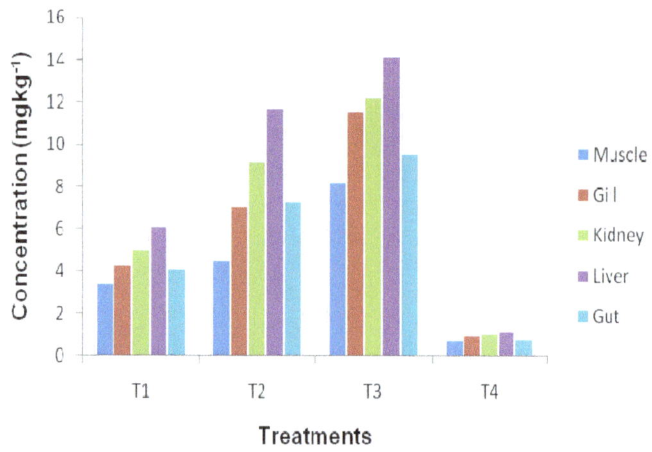

Figure 12. Concentration of manganese in *C. gariepinus* tissues.

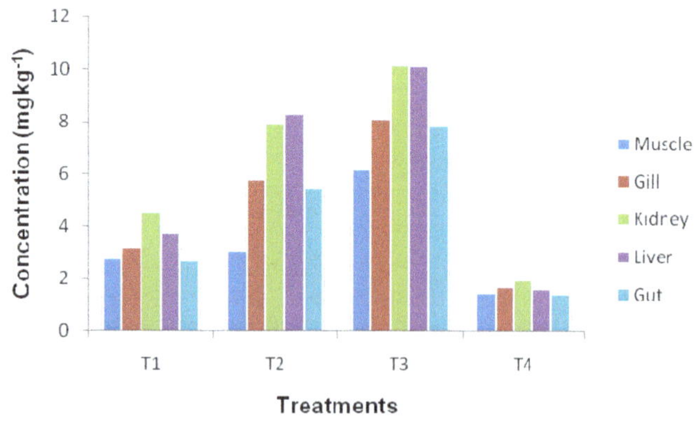

Figure 13. Concentration of Zinc in *C. gariepinus* tissues.

to 7.296 mg/kg in the gut. Range of values for tissue concentrations of Mn in T3 were; muscle, 1.485 to 8.243 mg/kg; gill, 2.611 to 11.560 mg/kg; kidney 1.014 to 12.284 mg/kg; liver, 2.151 to 14.155 mg/kg and gut, 1.600 to 9.570 mg/kg. Tissue concentrations in T4 had ranges of 0.164 to 0.720 for muscle, 0.209 to 0.975 for gill, 0.260 to 1.088 for kidney, 0.329 to 1.167 for liver and 0.157 to 0.782 mg/kg for gut. Mn concentrations in all tissues of T3 were significantly higher ($P < 0.05$) than the concentrations in corresponding tissues of other treatments. Figure 12 profiles the final Mn concentrations in the various tissues.

Zn concentration in *C. gariepinus* from T1 had ranges of 0.420 to 2.750 mg/kg for muscle, 0.561 to 3.174 mg/kg for gill, 0.840 to 4.512 mg/kg for kidney, 0.628 to 3.759 mg/kg for liver and 0.431 to 2.673 for gut mg/kg. Zn concentration in *C. gariepinus* muscle in T2 had a range of 0.517 to 3.066, while concentrations in gill, kidney, liver and gut had concentration ranges of 0.632 to 5.790, 1.039 to 7.930, 0.835 to 8.285 and 0.625 to 5.436 mg/kg, respectively. Ranges for Zn concentration in T3 were; muscle 0.794 to 6.180, gill 0.840 to 8.113, kidney 1.360 to 10.150, liver 1.045 to 10.133 and gut 0.703 to 7.849 mg/kg. T4 had the following ranges for Zn in fish tissues: muscle, 0.248 to 1.467 mg/kg; gill, 0.296 to 1.692 mg/kg; kidney, 0.511 to 1.974 mg/kg; liver, 0.322 to 1.607 mg/kg and gut, 0.230 to 1.420 mg/kg. Values for Zn concentrations in the tissues of T4 were significantly lower ($P < 0.05$) than concentrations of the same tissues in the integrated treatments. Final Zn concentrations in the five tissues of *C. gariepinus* are profiled in Figure 13.

Fish can serve as reliable indicators of metal pollution in the aquatic ecosystem they because absorb or ingest metals in waterbodies. The concentration of a metal in an organism is the product of equilibrium between the concentration of the metal in an organism's environment and its rate of ingestion and excretion (Idodo-Umeh, 2002). The gills and gut of fish are the main sites for heavy metal uptake (Annune and Iyaniwura, 1993), while the liver and kidney are known to accumulate high amounts of metals (Gbem et al., 2001). Abdel-Baki et al. (2011), in a study of metal concentrations in tilapia, stated that the average concentrations of Pb in different fish organs, like liver, gills, muscles and kidney were 60.7, 141.7, 39.7 and 276 ppb, respectively and that the kidney accumulated the highest concentrations, while the muscles accumulated the lowest. Cd concentration in *O. niloticus* from the four treatments in this study was highest in the kidney (except in T1 which had the highest concentrations in the liver) and lowest in the muscle. Cr concentrations in all the treatments were highest in the gills and lowest in the muscles for T1, gut for T2, liver for T3 and T4. Cu concentrations in all the treatments were highest in the liver and lowest in the muscles. Pb concentration in *O. niloticus* was highest in the kidney in T1 and T4 and lowest in the muscle. But in T2 and T3, liver concentrations were highest, while the lowest concentrations were in the muscle. Mn concentrations were highest in the liver in the integrated treatments, but had the highest value in the kidney tissue of T4, the control. The lowest Mn concentrations occurred in the muscle in all the treatments. Zn concentrations were higher in the kidney of *O. niloticus* as compared to other tissues, while the muscle had the lowest concentrations in the treatments except in T4, where the lowest concentration occurred in the liver. Yap et al. (2004) in a study of heavy metal concentrations in the muscles of *O. niloticus* from Kelana Jaya Fish Ponds, Malaysia, stated that Cd and Zn had ranges of 2.42 to 4.48 and 58.4 to 80.5 mg/kg, respectively.

Final Cd concentration in *C. gariepinus* of this study was highest in the liver and lowest in the muscle for all the treatments, but for Cr, the highest concentrations were in the gill and the lowest concentrations were in the gut with exception of T4 which had the lowest concentration in the muscle. Final Cu concentration was lowest in the muscle in T1, T2 and T4, while the gut had the lowest concentration of Cu in T3. The highest Cu concentrations in the treatments were T1, gill; T2 and T4, kidney; and T3, liver. Pb concentrations were higher in the kidney in T1, T3 and T4, while the highest Pb concentration in T2 occurred in the liver. Muscle concentrations of Pb had the lowest values in all the treatments. Mn concentrations had the highest values in the liver and the lowest values in the muscle, while Zn was more concentrated in the kidney in all the treatments except T2 which had the highest value in the liver.

It can be concluded that tissue metal concentrations for both species were generally lowest in the muscles and highest in the kidney and liver. Obasohan et al. (2008) analysed *C. gariepinus* Offal, gills, liver and muscle for Cu, Zn, Mn, Cd, Cr, Ni and Pb concentrations and concluded that tissue metal levels exhibited no particular trends, but seemed to be higher in offal and gills. Ekwu (2010) in a study of the distribution of Pb, Cd, Fe and Zn in the tissues of tilapia and *C. gariepinus* from a tidal fish farm, observed that the mean levels of heavy metals in gills, muscle, liver and intestines of catfish were Pb: 2.39 ± 1.24; Cd: 0.87 ± 0.83; Fe: 3.96 ± 1.55; Zn: 8.15 ± 3.18, while the mean levels in Tilapia were Pb: 0.035 ± 0.002; Cd: 0.61 ± 0.05; Fe: 1.09 ± 0.61; Zn: 2.17 ± 0.53 mg/kg. There were higher concentrations in liver, gills and muscles of *Clarias* than in Tilapia. In this study, Cd concentrations were higher in the muscles and liver of *O. niloticus* than in the same tissues of *C. gariepinus*. However, Cd values in gills, kidney and gut of *C. gariepinus* were higher than those of *O. niloticus*. Cr concentrations were generally higher in the tissues of *O. niloticus* than in the same tissues of *C. gariepinus*. There were no clear trends when Cu concentrations in the muscles and gills of the two species were compared, but Cu concentrations in the kidney, liver and gut of *O. niloticus* were generally higher than that of *C. gariepinus*. Pb concentrations in the tissues of *C. gariepinus* were higher than in the same tissues of *O. niloticus,* except in the muscle tissue which had higher values in the later. While Mn concentrations in the tissues of *C. gariepinus* were higher than in the same tissues of *O. niloticus*, the reverse was the case for Zn concentrations which were higher in the later.

The FAO (1983) limits for Cd, Cu, Pb and Zn in fish muscle are 0.5, 30, 0.5 and 30 mg/kg respectively. The European Union (2001) limits for Pb and Cd are 0.2 and 0.05 mg/kg wet weight, respectively. Concentrations of these metals in the tissues of *O. niloticus* were all below the FAO and EU limits except for Cd contents in the liver of fish in T2 and in the kidney, liver and gut of fish in T3 which received the highest chicken waste input. Metal contents in *C. gariepinus* were below FAO limits except Pb content in the kidney of *C. gariepinus* in T3 which was higher.

Final Pb contents in the tissues of the same fish in all the treatments were higher than the EU limits with the exception of the muscle tissue in T1 and T4 which had lower values than the standard. Cd contents in the tissues of *C. gariepinus* from all the treatments were generally higher than the EU limits, except the muscle tissues which had lower values.

Conclusion

Physico-chemical parameters were generally favourable in the four treatments of the experiment. Water temperatures were generally within the recommended ranges for optimum growth and survival of the two fish species. There were no great pH variations and pH values were generally within the desirable range for fish culture, but DO values were in many cases, less than this recommended lower limit, especially in the integrated treatments. The nitrate, nitrite and ammonia concentrations in T3 exceeded the recommended benchmarks and this led to the relatively high mortality rate of *O. niloticus* in T3. Indeed, water quality parameters were more adverse in T3 and *O. niloticus* production was higher in T2 than in T3 despite the fact that the later received higher manure and spilled feed loading.

Metal concentrations in fresh chicken manure showed a high monthly variation as compared to metal concentrations in chicken feed and compounded fish feed. Tissue metal concentrations for both species were lowest in the muscles and highest in the kidney and liver. Concentrations of Cd, Cu, Pb and Zn in the muscles of *O. niloticus* were all below the FAO (1983) and EU (2001) limits. However, Cd concentration in the liver of *O. niloticus* in T2 and in the kidney, liver and gut of fish in T3, were above the limits. Final Pb concentrations in the muscles of *C. gariepinus* in T2 and T3 were higher than the EU limits except a health risk is observed in humans if fish from such systems are consumed.

Final metal concentrations in the muscles of both fish species, T1 (2000 chickens/ha) and T4 (the control) were below international limits and the consumption of fish muscle from such systems operating within the combination ratio and culture period used for this study is considered safe. However, there is risk of Pb concentrations accumulating above the recommended levels in *C. gariepinus* at 3000 chickens/ha and above.

ACKNOWLEDGEMENTS

The authors are grateful to the management and staff of the National Institute for Freshwater Fisheries Research (NIFFR) where the experiment was conducted and to Messers Jafaru Isah, Gbenga Ogunseye, Nuhu Mohammed, Cletus Wonah, Usman B. Ibrahim, Usman A. Mohammed, Baba Yakubu Koro, Jimoh, Usman, Haruna Monai and the Industrial Training (IT) students from different tertiary institutions in Nigeria.

REFERENCES

Abdel-Baki AS, Dkhil MA, Al-Quraishy S (2011). Bioaccumulation of

some heavy metals in tilapia fish relevant to their concentration in water and sediment of Wadi Hanifah, Saudi Arabia. Afr. J. Biotechnol., 10(13): 2541-2547.
Ademoroti CMA (1996). Standard methods for water and effluent Analysis. Foludex Press Ltd. Ibadan, p. 37
Agbenin JO (1995). Laboratory Manual for Soil and Plant Analysis. Department of Soil Science, Ahmadu Bello University, Zaria, pp. 15-52.
Alexieva D, Chobanova S, Ilchev A (2007). Study on the level of heavy metal contamination in feed materials and compound feed for pigs and poultry in Bulgaria. Trakia J. Sci., 5(2): 61-66.
Annune PA, Iyaniwura TT (1993). Accumulation of two trace metals in tissues of freshwater fishes, *Oreochromis niloticus* and *Clarias gariepinus*. J. Aquat. Food Technol., 2(3): 5-18.
American Public Health Association (1998). American Water Works Association, Water Environment Federation. Standard Methods for Examination of Water and Wastewater (20th ed.), Washington D.C: APHA.
Bamgbose O, Odukoya O, Arowolo TOA (2000). Earthworms as bio-indicators of metal pollution in dump sites of Abeokuta city, Nigeria. Rev. Biol. Trop., 48(1): 5-9.
Boyd CE (1998). Water Quality for Pond Aquaculture. Research and Development Series No. 43. International Center for Aquaculture and Aquatic Environments, Alabama Agricultural Experiment Station, Auburn University, Alabama.
Bremner JM (1965). Total nitrogen. In Black CA (ed) Method of soil analysis, Part 2. American Society of Agronomy, Madison, pp. 119-178.
Cang L, Wang Y, Zhou D, Dong Y (2004). Heavy metals pollution in poultry and livestock feeds and manures under intensive farming in Jiangsu Province. China. J. Environ. Sci., 16(3): 371-374.
Edwards P (2008). Key issues in the safe use of wastewater and excreta in aquaculture. Guidelines for the Safe Use of Wastewater, Excreta and Greywater in Agriculture and Aquaculture, 3. WHO/FAO/IDRC/IWMI document, 3.
Ekwu AO (2010). Tissue distribution of heavy metals in fin fish from a tidal fish farm in South East Nigeria. In Ansa EJ, Fashina-Bombatta H, Ndimele E (eds) Proceedings of the 25th Annual Conference of the Fisheries Society of Nigeria, pp. 215-219.
European Union (2001). Commission Regulation as regards heavy metals, Directive, 2001/22/EC No 466.
Food and Agriculture Organization (FAO) (1994). Nutrition of fish and crustaceans: A laboratory manual. GCP/RLA/102/ITA project. Food and Agriculture Organization of the United Nations, Rome.
FAO (2003). Recycling of animal waste as a source of nutrients for freshwater fish culture within an integrated livestock system.
FAO/OIE/WHO (2005). Consultation on avian influenza and human health: Risk reduction measures in producing, marketing, and living with animals in Asia. FAO, Rome.
Fishdoc (2008). Fish are open systems.http://www.fishdoc.co.uk/water/open_systems.htm [Accessed 18 September 2008].
Food and Agriculture Organisation (1983). Compilation of Legal Limits for hazardous substances in fish and fishery products. FAO Fishery circular 464.
Garrett ES, Jahncke ML, Lima DSCAM (1997). Public and animal health implications of aquaculture. Paper presented at the International Life Sciences Institute Conference on Emerging Foodborne Pathogens: Implications and Control, Alexandria, Va., USA, March.
Gbem TT, Balogun JK, Lawal FA, Anune PA (2001). Trace metal accumulation in *Clarias gariepinus* (Tuegels) exposed to sublethal levels of tannery effluent. Sci. Total Environ., 271: 1-9.
Guerrero RD (1985). Tilapia farming in the Philippines: Problems and prospects. In Torres EB, Tan EO (eds) Philippines tilapia economics Smith. ICLARM Conf. Proc., 12: 3-14.
Hanč A, Tlustoš P, Száková J, Habart J, Gondek K (2008). Direct and subsequent effect of compost and poultry manure on the bioavailability of cadmium and copper and their uptake by oat biomass. Plant Soil Environ., 54(7): 271–278.
Idodo-Umeh G (2002). Pollution assessments of Olomoro water bodies using Physical, Chemical and Biological indices. PhD. Thesis, University of Benin, Nigeria, p. 485.
Igua P, Huasi L (2009). Effect of Chicken Manure, Tithonia diversifolia and Albizzia spp on Maize Plant Height and Dry Matter Production – Lessons Learnt in the Eastern Highlands of PNG. 17th International Farm Management Congress, Bloomington/Normal, Illinois, USA, pp. 240-251.
Islam SM, Kazi MAI, Hossain MM, Ahsan MA, Hossain AMM (2007). Propagation of heavy metals in poultry feed production in Bangladesh. Bangladesh J. Sci. Ind. Res., 42(4): 465-474.
Jackson BP, Bertsch PM, Cabrera ML, Camberato JJ, Seaman JC, Wood CW (2003). Trace Element Speciation in Poultry Litter. J. Environ. Qual., 32: 535-540.
Khai NM, Ha PQ, Vinh CN, Gustafsson JP, Oborn I (2008). Effects of biosolids application on soil chemical properties in peri-urban agricultural systems. VNU J. Earth Sci., 24: 202-212.
Knud-Hansen CF, Batterson TR, Mcnabb CD (1993). The role of chicken manure in the production of Nile tilapia, *Oreochromis niloticus* (L.). Aqua. Fish. Manag., 24: 483-493
Nnaji JC, Madu CT, Omeje VO, Ogunseye JO, Isah J (2009). An integrated chicken-fish system in concrete ponds. In: Proceedings of the 24th Conference of Fisheries Society of Nigerian (FISON), pp. 51-54.
Obasohan EE, Oronsaye JAO, Eguavoen OIA (2008). Comparative Assessment of the Heavy Metal Loads in the Tissues of a Common Catfish (*Clarias Gariepinus*) From Ikpoba and Ogba Rivers in Benin City, Nigeria. Afr. Scient., 9(1): 13-23.
Olaifa FE, Olaifa AK, Onwude TE (2004). Lethal and Sub-lethal Effects of Copper to the African Catfish (*Clarias Gariepinus*) Juveniles. Afr. J. Biomed. Res., 7(2): 65-70.
Rancis TF, Adma VPP, Amana TNR (2004). Integrated fish farming – A review. World Aqua., 35(3): 24-29.
Sinha VRP (1985). Integrated Carp Farming in Asian Country. Network of Aquacuture Centres in Asia, Bangkok, Thailand, p. 3.
Swann LA (1990). Basic Overview of Aquaculture: History, Water Quality, Types of Aquaculture, and Production Methods. Illinois-Indiana Sea Grant Program Extension Bulletin AS-457 and IL-IN-SG-E-90-2. Purdue University, West Lafayette, Indiana, p. 10
Swingle HS (1969). Methods of Analysis of waters: Organic matter and pond bottom soils used in fisheries research. Auburn University, Alabama, p. 119.
Todorovic Z, Poli P, Djordjevi D, Antonijevi S (2001). Lead distribution in water and its association with sediment constituents of the "Barje" lake (Leskovac, Yugoslavia). J. Serbian Chem. Soc., 66(10): 697–708.
UNEP (1999). Report on the Development and Harmonization of Environmental Standards in East Africa, 2: 1-4.
Xing L, Chen LJ, Han LJ (2008). Rapid Analysis of Layer Manure Using Near-Infrared Reflectance Spectroscopy. Poult. Sci., 87: 1281-1286.
Yap CK, Ismail A, Chiu PK (2004). Concentrations of Cd, Cu and Zn in the Fish Tilapia *Oreochromis mossambicus* Caught from a Kelana Jaya Pond. Asian J. Wat. Environ. Pollut., 2 (1): 65-70.

Adsorption of phenol on natural clay

M. Djebbar[1]*, F. Djafri[1,2], M. Bouchekara[1] and A. Djafri[2]

[1]Faculty of Chemistry, University of Technology, BP 1505, 31045 El-Mnouar, Oran, Algérie.
[2]University of Technology, Mascara, Algeria.

Natural clay being locally abundant and cheap material in Algéria can be easily activated to become a promising adsorbent for phenol removal from aqueous solution. The natural clay before and after activation was characterized using XRD and IR techniques. The effects of various experimental parameters, such as initial phenol concentration, temperature, pH, contact time and adsorbent done on the adsorption extent were investigated. Langmuir adsorption model was used for the mathematical description of the adsorption equilibrium and the equilibrium data fitted very well with this model. The activated natural clay had the monolayer adsorption capacity equal to 15 mg/g at pH value of 5.0 and 23°C, adsorption measurements show that the process is very fast and physical in natural clay. The extent of the phenol removal increased with the decrease in the initial concentration of the phenol and contact time of solution. The results showed that as the amount of the adsorbent was increased, the percentage (%) of phenol removal increased accordingly. Thermodynamic parameters showed that the adsorption of phenol on activated natural clay was exothermic.

Key words: Phenol, activated clay, adsorption isotherm, thermodynamic parameters, montmorillonites, adsorbent, x-ray diffraction.

INTRODUCTION

Organic contaminants from industrial waste that seriously threaten the human health and the environment, has been recognized as an issue of growing importance in recent years. Phenolic compounds are commonly produced in wastewater generated by petrochemical, oil refineries, coal conversion, steel plant, paint and phenol-producing industries (Aygiin et al., 2003), phenols are the most common in water, considered highly toxic (APHA, AWWA and WEF, 2006) even at low concentrations. Their high toxicity and difficulties in their biodegradation, generally closely related to functional group presence (e.g. chlorophenols), seriously threaten the human health and the environment, leading to the establishment of rigid limits on the acceptable environmental levels of phenol. They are found in effluents generated by various different industrial activities as well as associated with agricultural activities, especially due to the excessive use of chlorinated pesticides.

Adsorption process is a prominent method for removal of organic pollutants practically used by industries. Accordingly, abundantly reported findings were carried out on adsorption process using activated carbon. However, superior activated carbon is expensive and efforts to utilize low-cost adsorbent become a greal intention. Thus, the aim of this study was to explore the potential of modified montmorillonite adsorbent for adsorption of phénol and other pollution parameters, and some heavy metals. The performance of activated montmorillonite was compared with the natural montmorillonite. There are some literature references concerning the phenol adsorption on organobentonite (Juang et al., 2002; Vianna et al., 2004; Hall et al., 1996), but in our opinion they need further verification, especially in respect to the pH and its influence on the phenol uptake by adsorbent.

The prime objectives of the present investigation were to conduct the batch sorption tests, which included determination of sorption kinetics and isothermal

*Corresponding author. E-mail: djebbar_mustapha@yahoo.fr.

adsorption capacity of the sorbent and to assess the effect of pH on the removal of phenol from synthetic waste water.

MATERIALS AND METHODS

The clay used in this study was Algerian montmorillonite from Rousselin Mghnia (West Algeria). This Algerian montmorillonite clay was supplied by Chemical Ltd., Algerian Research Company of Non Ferreux (ENOF), Algeria. For this clay material, the cation exchange capacity (CEC) = 90 meq/100 g was assumed.

The natural clays were washed several times with distilled and deionised water and were completely dispersed in water. After 17 h at rest, the dispersion was centrifuged for one hour at 2400 rpm. The size of the clay particles obtained was < 2 µm.

These clay particles were dispersed in water and heated at 75°C in the presence of a solution composed of the sodium salts of bicarbonate (1 M), citrate (0.3 M), and chloride (2 M). The purpose of this operation was to eliminate inorganic and organic compounds, aluminium found in the inter-layer spaces and various free cations. Carbonates were removed by treatment with HCl (0.5 M) and chloride was eliminated after several washings. The organic matter was eliminated completely by treatment with H_2O_2 (30% v/v) at 70°C. The purified clay was dried at 110°C, than saturated with sodium (Na^+). To ensure complete transformation into the sodium form, all samples were washed several times with a NaOH solution (1 M). Purity of the natural clay samples before and after activation was tested by IR spectral analysis. An IR transmittance spectrum of he ground samples was obtained in the 4000 to 400 cm^{-1} range on Perkin-Elmer FT 1720 spectrometer.

Purity of the natural clay samples before and after activation was also determined by the analysis of the X-ray diffraction (XRD) spectra, which involved the identification and semiquantification of the characteristic peaks of the minerals present. A Philips diffractometer with D5000 Ni-filtered CuK, α radiation used was 1, 5406 Å. natural clay samples before and after activation with both randomly and preferred orientation were scanned over the interval 5 to 80°2θ at scanning speeds of 0.03°/5 s.

RESULTS AND DISCUSSION

Natural clay characterization

Results of the X-ray diffraction analysis for activated and activated clay are shown in Figure 1. Figure 1 clearly shows that the d spacing of clay increased from 10.27 to 12.11 Å, which could be attributed to the activated clay. Quartz (reflection at d = 2.89 Å, 2θ = 35.97°) and calcite (reflection at d = 2.89 Å, 2θ = 35.97°) are the major impurities. The purified sample with Na-exchange shows the position of the 001 (reflection at 12.11 Å, 2θ = 5.37°) characteristic of soduim and Kaolinite (reflection at d= 5.41 Å, 2θ = 19.03°).

Figures 2 and 3 present the spectra IR of natural and activated clay. The examination of these spectra revealed following characteristic absorption bands: The IR spectra of the clay samples were recorded over the spectral range 400 to 4000 cm^{-1}. In fact, IR techniques had been used by many researchers for identification of natural clay minerals (Hajjaji et al., 2001). The obtained characteristic bands of kaolinite appeared at 3618.67, 3641.59, 1100.39, 907.602, 830.38, 758.355, 521.24 and 456.84 cm^{-1} (Gadsden, 1975). The high intensity of the peak appearing at 1100.78 cm^{-1} was an indication of the large amount of this mineral in the sample. The band at 1100.78 cm^{-1} was attributed to Si-O stretching. The bands at 830.38 and 521.24, 456.84 cm^{-1} were assigned to Si-O-Al and Si-O-Mg, Si-O-Fe and bending vibrations respectively (Gadsden, 1975; Madejova, 2003). This indicated that most part of the layer charge resulted from trivalent (Al^{3+}, Fe^{3+}) to bivalent (Mg^{2+}) ion substitution in the octahedral sheet. These functional groups were present in silicate minerals such as kaolinite and montmorillonite. The band observed at 3412.41 cm^{-1} is assigned to stretching vibrations of adsorbed water molecules. Another characteristic band for bending vibrations of adsorbed water usually appeared at 1650 to 1600 cm^{-1} as a medium band (Gadsden, 1975). The stretching vibrations of the surface hydroxyl groups (Si-Si-OH, or Al-Al-OH) were found at 3618.67 and 3641.59 cm^{-1}.The characteristic bands of montmorillonite-Na were observed at 1036.39 and 456.20 cm^{-1} (Gadsden, 1975). The high intensity of the peak appearing at 1036.39 cm^{-1} was an indication of the large amount of this mineral in the sample. The band at 1036.39 cm^{-1} was attributed to Si-O stretching. The band at 456.20 cm^{-1} was assigned to Si-O-Si bending vibrations (Gadsden, 1975; Madejova, 2003).

Experimental procedure

Adsorption experiments were carried out by agitating 100 mg of natural clay with 100 ml of phenol solution of desired concentration and pH at 180 rpm, 23°C in a thermostated mechanical shaker (ORBITEK, Chennai). Concentration of phenol was estimated spectrophotometrically by monitoring the absorbance at 270 nm using UV-Vis spectrophotometer (Hitachi, model 8543).

The pH was measured using pH meter (3151 MWT, Germany). The samples were withdrawn from the shaker at predetermined time intervals and the phenol solution was separated from the adsorbent by centrifugation at 1,000 rpm for 15 min. The absorbance of supernatant solution was measured. Effect of pH was studied by adjusting the pH of phenol solutions using dilute HCl and NaOH solutions. Effect of adsorbent dosage was studied by agitating 100 ml of 5 mg/l phenol solutions with different adsorbent doses (1 to 5 mg) at equilibrium time.

The amounts of phenol sorbed by the adsorbent were calculated using the following equation (Denizli et al., 2005):

$$Q = (C_i - Ce) V/M \quad (1)$$

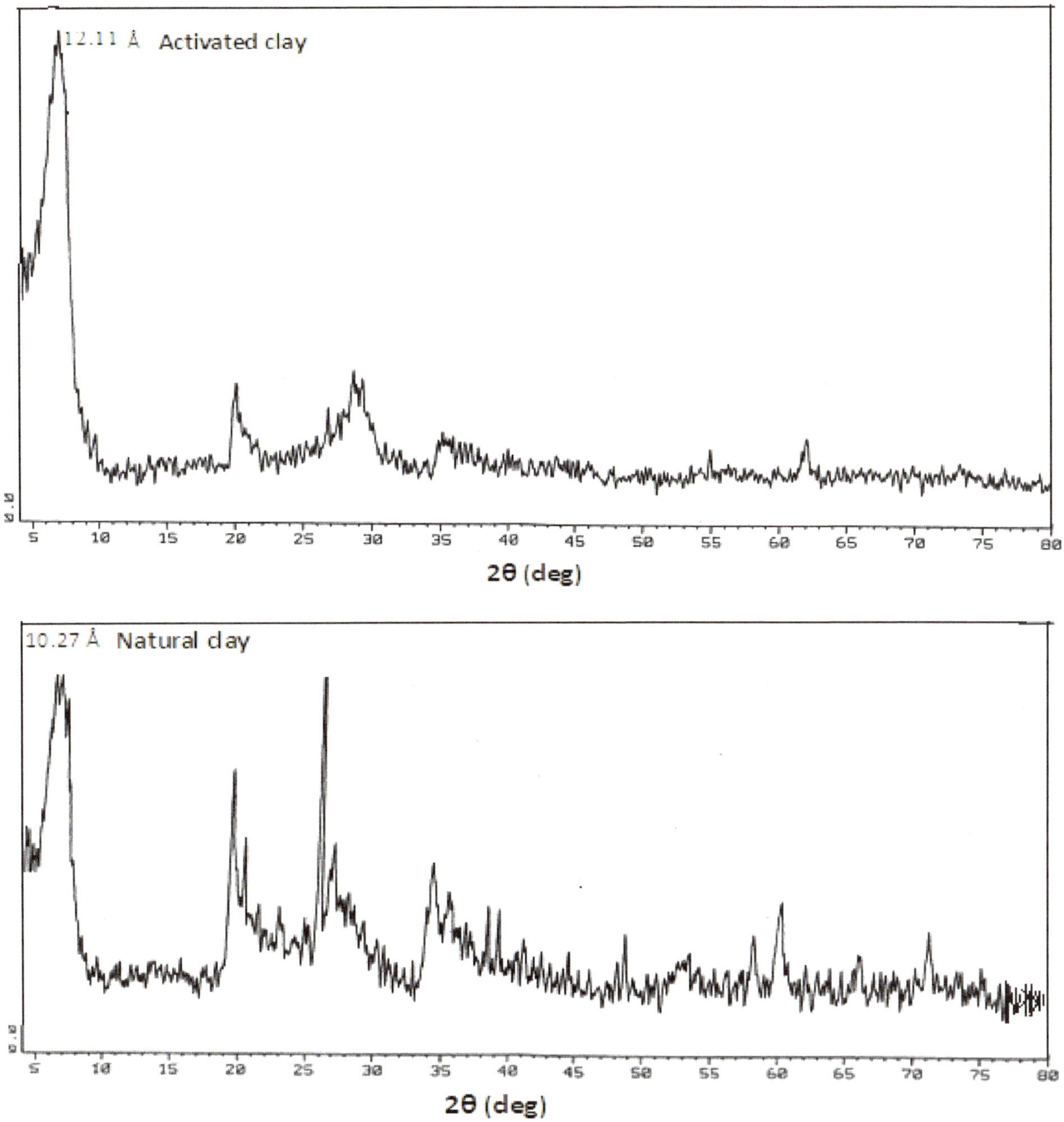

Figure 1. X-ray diffraction patterns of the natural clay and the activated clay.

Where Q (mg/g) is the amount of phenol sorbed by natural clay, C_0 and Ce (mg/L) are the initial and equilibrium liquid-phase concentration of phenol, respectively, V (L) the initial volume of phenol solution and M (g) the weight of the natural clay.

The percentage amount of adsorbate adsorbed on the adsorbent was calculated by Equation (I) where Ci and Ce are the percentage adsorption was calculated using the following equation:

$$\text{Percentage (\%) adsorption} = [(C_i - Ce) / C_0] \times 100\% \quad (2)$$

Adsorption isotherm

Langmuir isotherm model

Langmuir model assumes that the adsorption energy is constant and independent of surface coverage, adsorption occurs on localized sites with no interaction between adsorbate molecules and maximum adsorption occurs when the surface is covered by a monolayer of adsorbate (Langmuir, 1918). The following relation represents the linear form of the Langmuir isotherm

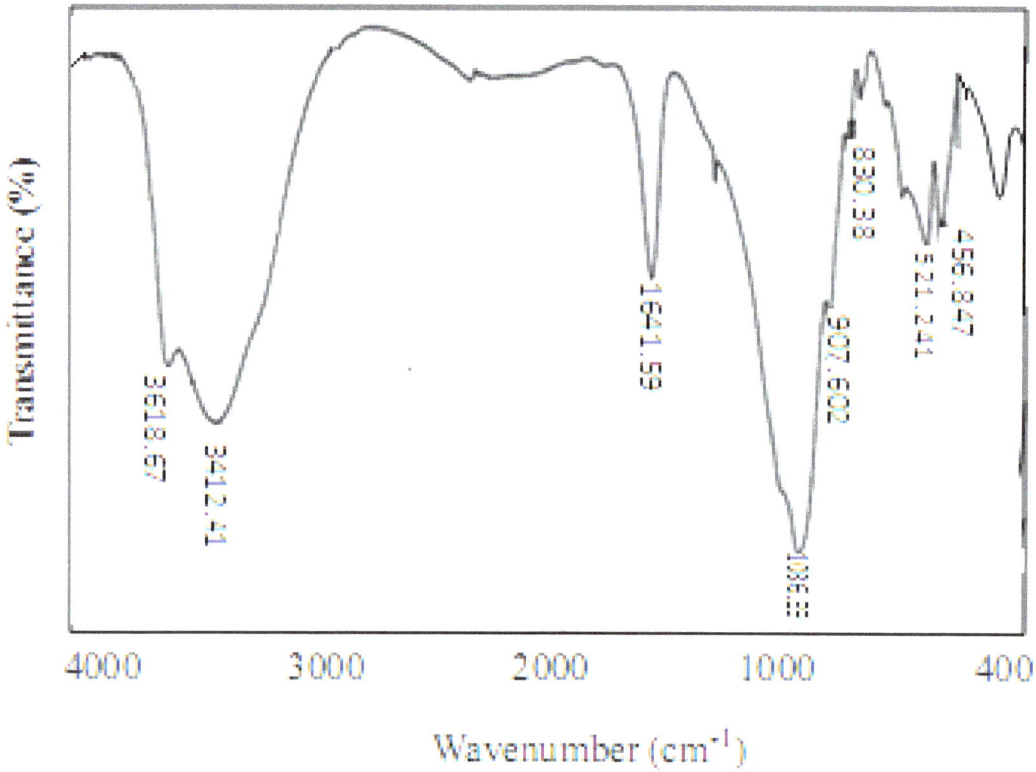

Figure 2. FTIR Spectra of activated clay.

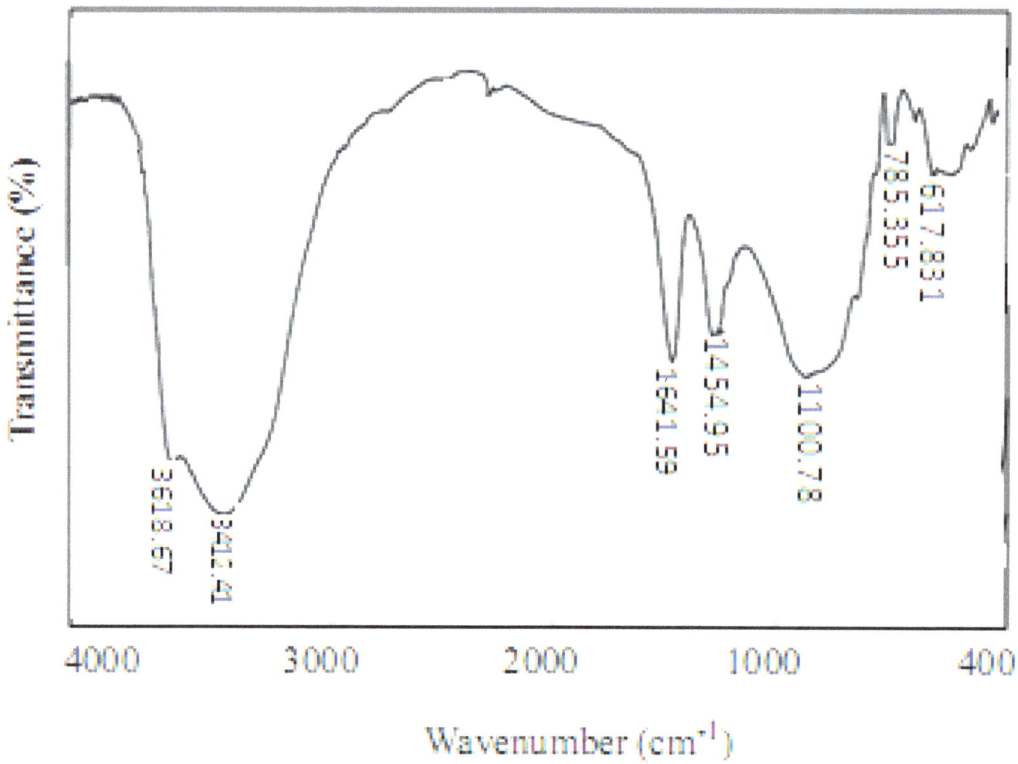

Figure 3. FTIR Spectra of natural clay.

model:

$$Q_e = Q_m \times K \times C_e/(1+K \times C_e) \quad (3)$$

where, Q_e is the isotherm amount adsorbed at equilibrium (mg/g), Ce is the equilibrium concentration of the adsorbate (mg/l), and qm (mg/g) and b (L/mg) are the Langmuir constants related to the maximum adsorption capacity and the energy of adsorption, respectively. These constants can be evaluated from the intercept and the slope of the linear plot of experimental data of l/qe versus l/Ce as shown in Figure 1.

Freundlich isotherm model

Freundlich model is predicated based on the assumption that maximum adsorption capacity consists of a monolayer adsorption. It occupied heterogeneous adsorption surface and active sites with different energy which is unlike the Langmuir model. The linear form of the Freundlich isotherm model is given by the following equation (Freundlich, 1926):

$$LnQ_e = LnK_F + 1/n\ LnC_e \quad (4)$$

Where K_F and l/n are Freundlich constants related to adsorption capacity and adsorption intensity, respectively of the sorbent. The values of K_F and l/n can be obtained from the intercept and slope, respectively, of the linear plot of experimental data of logqe versus logCe as shown in Figure 2.

Based on values of correlation coefficient, R^2 summarized in Table 3, the adsorption of phenol on activated and natural clay was described well by both Langmuir and Freundlich models. Both homogeneous and heterogeneous adsorption energy took place during the process. On the other hand, the Freundlich and Langmuir models were the best model to explain the adsorption behavior phenol on activated and natural clay.

The Freundlich model was the best to describe the adsorption equilibrium data at low concentrations and at different temperatures with R^2 = 0.999 (Equation 4), implying the presence of highly energetic sites where the molecules of phenol were adsorbed (Bras et al., 2005). After the point of inflection of the experimental data, the Freundlich isotherm predicted that the equilibrium adsorption capacity should keep increasing exponentially with increasing equilibrium concentration in the liquid phase. However, the experimental adsorption isotherm for phenol presented a long plateau at higher equilibrium concentrations, implying the presence of a monolayer. Thus, Freundlich model should not be used for extrapolation of this data to higher concentrations. Langmuir model (Equation 3) gave a better fit at higher concentrations (Figure 3) with a determination coefficient of R^2 = 0.998 (Table 2), implying the formation of a monolayer at higher concentrations, however, the deviation of this model at low equilibrium concentrations implied that the sites of adsorption on the surface of the adsorbent were heterogeneous.

For comparison, the parameter constants for adsorption of phenol on activated clay were also listed in Table 2. Based on values of correlation coefficient, R^2, the adsorption of phenol was best fitted by Langmuir model. In addition, it was clearly seen that the adsorption capacity, Q_{max} (mg/g) for adsorption of phenol on activated clay was significantly higher than adsorption on natural clay.

Besides, the values of Freundlich constant K_F taken as an indicator of adsorption capacity was also greater for adsorption of phenol on activated clay as compared to the adsorption on natural clay.

However, natural clay still showed its feasibility and it is a promising adsorbent since it is abundantly available and cheaper than commercial activated carbon. Basically, a good adsorption by activated natural clay could be explained additionally by its high surface area (Lin and Cheng, 2002).

Adsorption equilibrium

Effect of contact time

Figure 3 showed the effect of contact time on the removal of phenol by natural and activated clay. For different duration of time (ranging from 0.5 to 24 h), well-shaken phenol solution (100 ml, 5 ppm) was studied for equilibrium. The study revealed 5 and 5.5 h as the equilibrium time for natural and activated clay, respectively.

Natural and activated clay were considered fast because of a rapid increase of adsorbates adsorbed occurred at the first 120 min. Previous findings on the adsorption of phenolic compounds by various clay-based adsorbents had shown a wide range of adsorption time (Wu et al., 2001).

Therefore, the result in present study was in agreement with the other reported findings. Up to 70 and 60% of phenol, respectively was successfully adsorbed by activated and natural clay from the aqueous solution. This proved the feasibility of activated natural clay as an effective adsorbent (Halouli and Drawish, 1995; Song and Sandi, 2001).

Figures 4 and 5 displayed the adsorption isotherms at 23°C, of phenol from aqueous solutions onto natural and activated clay. As could be seen, there was a marked higher adsorption for activated clay than that for natural clay, reaching a maximum value of 18.86 mg/g for activated clay, 11.01 mg/g for natural clay, respectively.

Table 1. Characteristic of the selected adsorbate.

Name	Formula	Molecular weight	Henry's constant (bar m^3/moL)	Water solubility at 25°C (g/L)	Log K_{ow}	Observations
Phenol	C_6H_6O	94	4 X 10^{-7}	87	1.5	Aromatic semivolatilite weak acid, pK_a = 9.98

Table 2. Adsorption constants for adsorption of phenol on activated and natural clay at different temperatures.

Natural clay	Langmuir constants			Freundlich constants		
	Q_{max} (mg/g)	K_L(L/mg)	R^2	K_F (L/mg)	1/n	R^2
T = 296°K	11.095	1.170	0.997	0.589	0.4	0.997
T = 301°K	9.107	0.849	0.998	0.381	0.73	0.998
T = 306°K	7.630	0.508	0.997	0.271	0.98	0.998

Activated clay	Langmuir constants			Freundlich constants		
	Q_{max} (mg/g)	K_L(L/mg)	R^2	K_F (L/mg)	1/n	R^2
T = 296°K	18.86	1.710	0.997	0.703	0.266	0.998
T = 301°K	15.47	0.970	0.996	0.574	0.420	0.999
T = 306°K	12.51	0.586	0.940	0.456	0.594	0.998

Table 3. The thermodynamic parameters for the adsorption phenol on natural and activated.

Natural clay	Thermodynamic parameters		
	ΔH° (kJ/mole) mean value	ΔG° (kJ/mole)	ΔS° (J/k mole) mean value
T = 296°K		-3.293	
T = 301°K	-1.666	-3.311	5.510
T = 306°K		-3.340	

Activated clay	Thermodynamic parameters		
	ΔH° (kJ/mole) mean value	ΔG° (kJ/mole)	ΔS° (J/k mole) mean value
T = 296°K		**-2.611**	
T = 301°K	**-1.331**	-2.632	**4.339**
T = 306°K		-2.652	

The maximum adsorptions capacities obtained in this work were higher than those reported by Tahani et al. (1999) for the adsorption of phenol from water onto an Al-PILC (13 mg/g) and onto an inorganic-organic clay (IOC; quaternary amine-Al PILC) (29.2 mg/g).

Figure 6 showed an expanded view of the adsorption isotherm of phenol in the range of 0 to 5 ppm to show in detail the initial section of the isotherm. This Figure 7 showed a sigmoidal (S-shape) that reached a long plateau as the concentration increased. This isotherm could be classified as an S2 isotherm under the (Giles et al., 1960) classification system (cooperative adsorption) based on the initial shape of the isotherm. According to this classification in S2 isotherms, the more molecules of adsorbate were already adsorbed, the easier it was for additional amounts of adsorbate to become fixed to the surface of the adsorbent. This implied a side-by-side association between adsorbed molecules, helping to hold them to the surface and consequently making more stable the adsorbed layer.

Molecules of adsorbate that showed an S-shape isotherm (as in the case of phenol) were considered monofunctional where the attraction towards the surface of the adsorbent arose from its OH group, giving an arrangement of adsorbed molecules oriented perpendicularly to the surface. Under the experimental

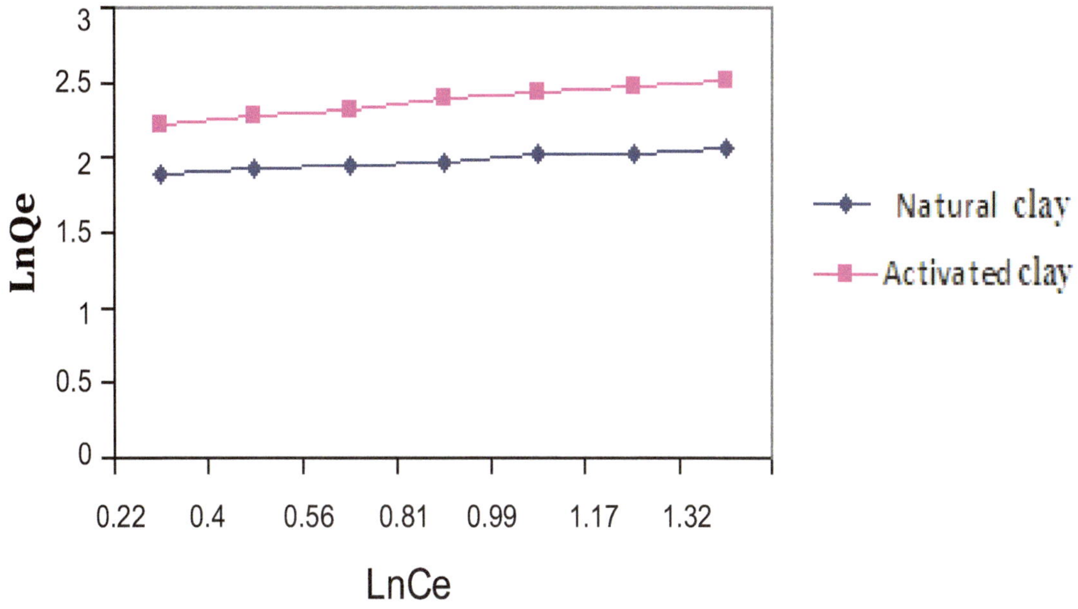

Figure 4. Freundlich isotherms for phenol adsorption on activated and natural clay modified montmorillonite at temperature of 23°C.

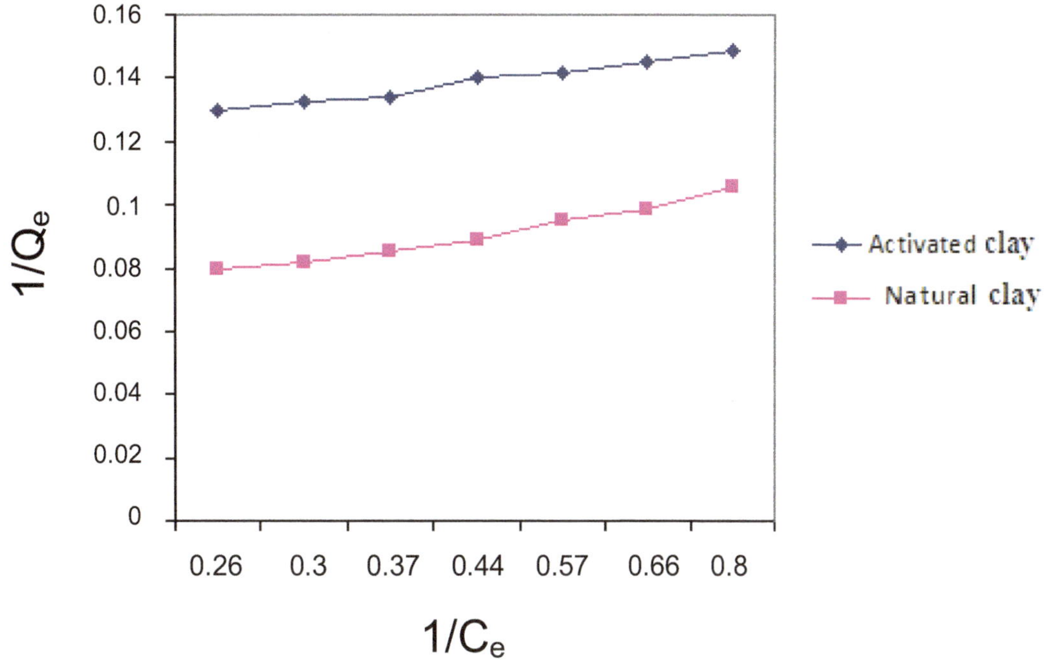

Figure 5. Langmuir isotherms for phenol adsorption on activated and natural clay at temperature of 23°C.

adsorption conditions, the pH was in the range of 5 to 6, below the pKa of phenol (pKa = 9.89) (Table 1) and below the pKa for silica (pKa = 9.42) (Goyne et al., 2002), therefore the phenol and the silanol groups of the surface of the adsorbent, were in a protonated form, making possible that the molecules of phenol were adsorbed onto the surface silanol groups through hydrogen bonds as illustrated in Figure 8. This mechanism was similar to that which was reported for the adsorption of phenol onto the surface of silica (Goyne et al., 2002).

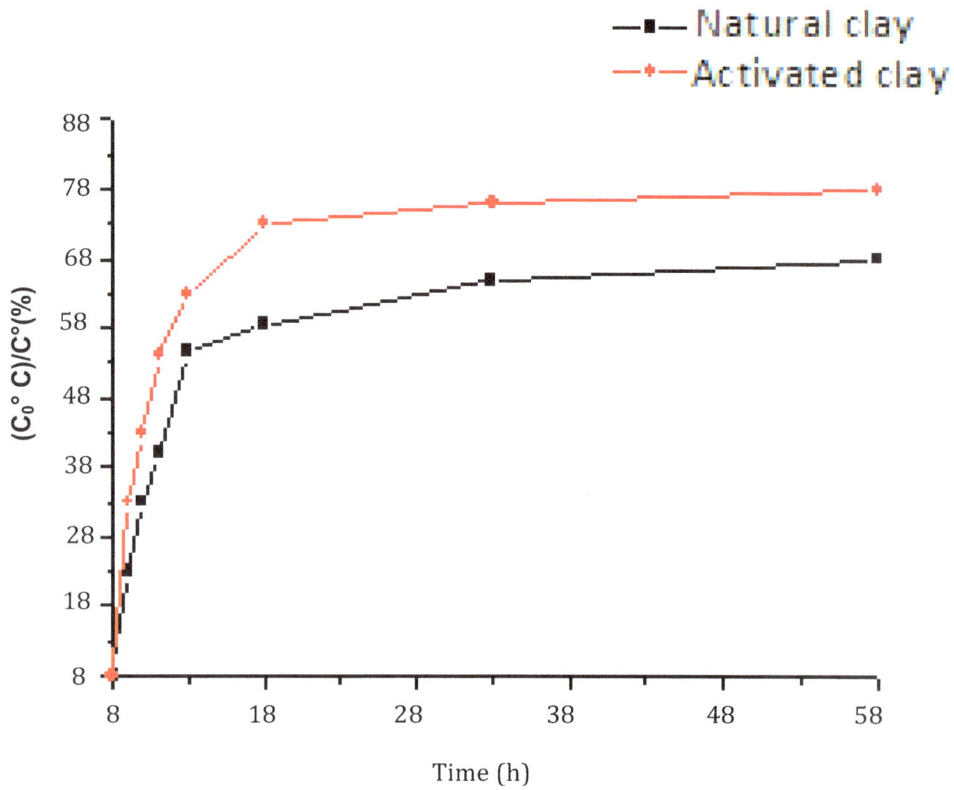

Figure 6. Effect of Time on phenol adsorption.

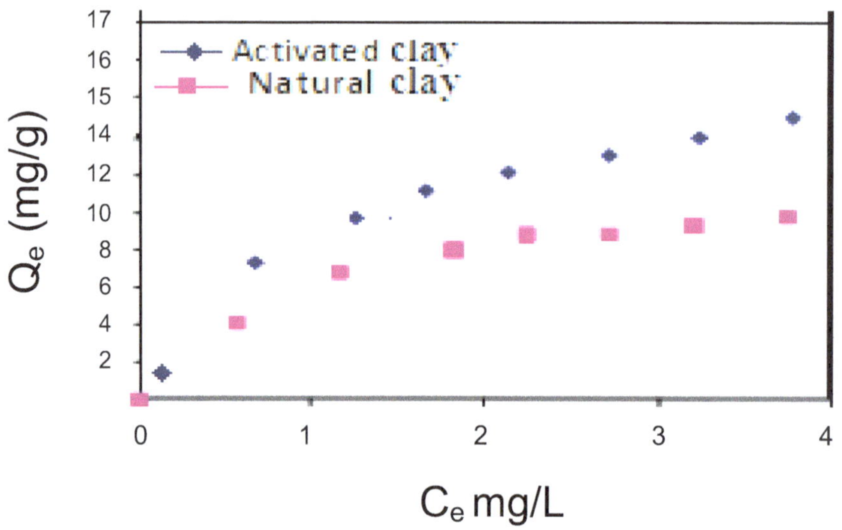

Figure 7. Sorption isotherm of phenol on natural and activated clay.

Effect of pH on phenol adsorption

Since phenol sorption had been reported to be affected by the pH of the adsorbate (Rengaraj et al., 2002), the adsorption of phenol by natural (NC) and activated clay (AC) were studied at various pH values of the phenol solution (100 ml, 5 ppm). The amount of phenol adsorbed showed a declining trend with higher as well as with lower pH, with maximum removal of phenol (up to almost 100% by both the adsorbents) at pH 5 (Figure 9). This

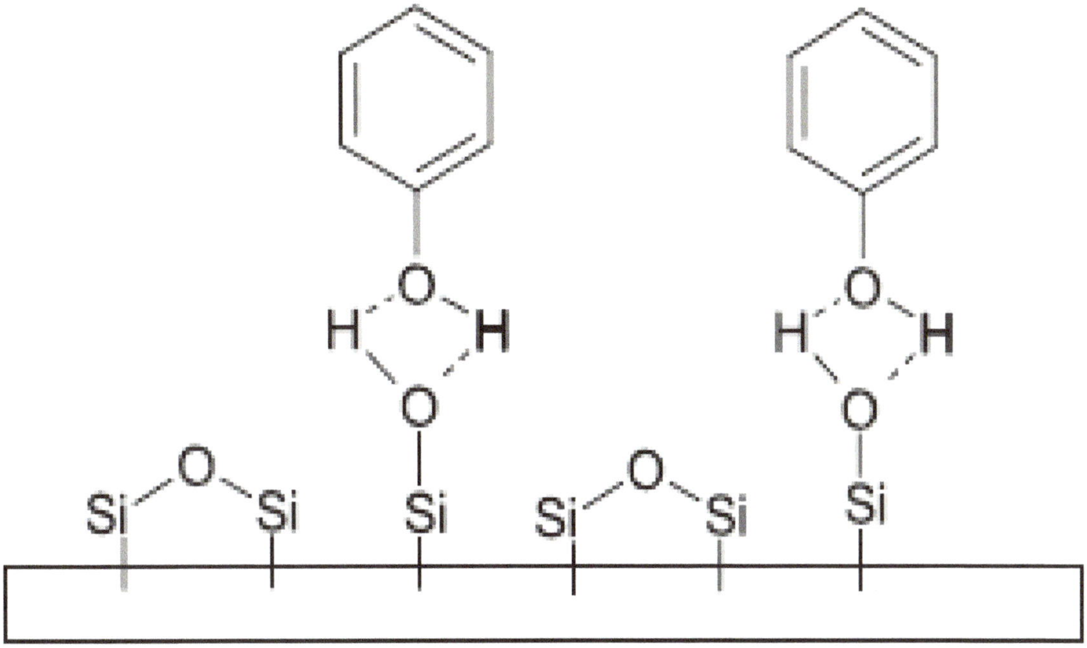

Figure 8. Proposed adsorption mechanism of phenol molecules onto the surface of natural clay through hydrogen bonds between OH group of the phenol and the silanol groups of the clay surface.

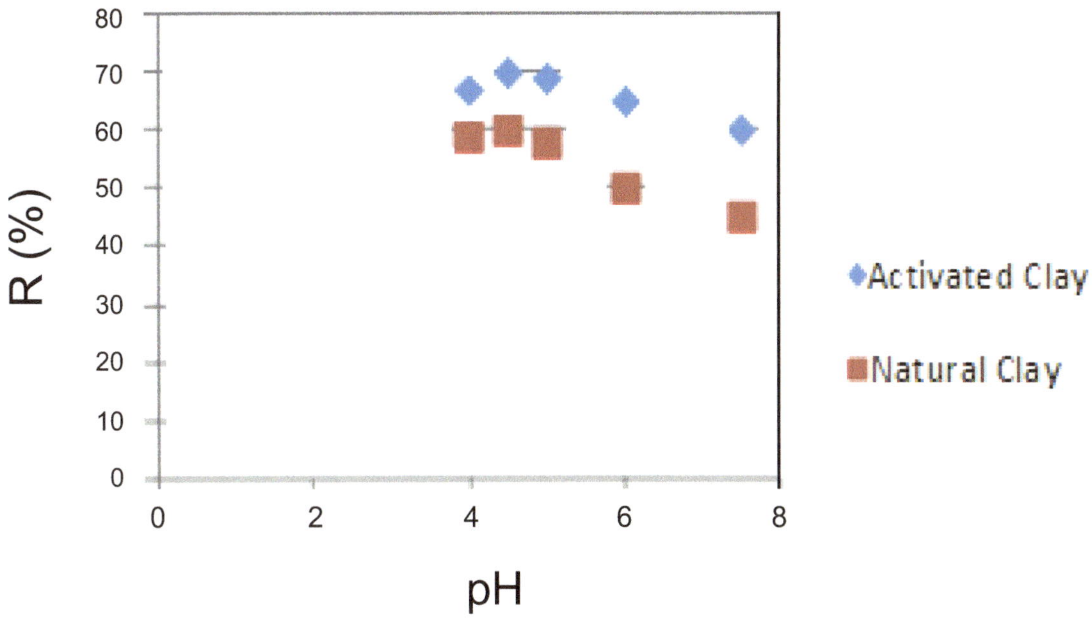

Figure 9. Effect of pH on phenol adsorption.

reduction of phenol-sorption might be because of the suppression by hydrogen ions (at lower pH), and hydroxyl ions (at higher pH) in addition to formation of various phenolic compounds at both acidic and alkaline conditions (Khalid et al., 2000; Singh et al., 1994).

ADSORPTION THERMODYNAMICS

Thermodynamic parameters

The thermodynamic parameters for the adsorption of

phenol by natural and activated clay such as the enthalpy change (ΔH^o), the Gibbs free energy change (ΔG^o) and the entropy change (ΔS^o) can be calculated from the variation of maximum adsorption with temperature (T) using the following basic thermodynamic relations (Seki et al., 2006; Al-kaim, 2005).

$$\Delta S° = (\Delta H°_{ads} / T - \Delta G° / T) \quad (5)$$

$$LnK_e = LnC_0 = - \Delta G° / RT \quad (6)$$

$$LnK_e = LnC_0 = \Delta S° / R - \Delta H°_{ads} / RT \quad (7)$$

According to Equation 7, the mean value of the enthalpy change due do the adsorption of phenol by natural and activated clay over the temperature range studied can be determined graphically by the linear plotting of In ke against 1/T using the least squares analysis shown in Figure 5. The mean enthalpy change can be determined from the slope of the straight line. The variation of Gibbs free energy and entropy change with temperature can be calculated using Equations 5 and 6, respectively, the results given in Table 3.

An important result obtained (Table 3) was that the Gibbs free energy (ΔG^o) was small and negative with its value decreased with increasing temperature. This indicated that the adsorption processes of Phenol by natural and activated clay could be enhanced by decreasing temperature. The values of entropy change ($\Delta S°$) were positive and remained almost constant with temperature. It was evident that structural changes in phenol and natural and activated clay occurred during the adsorption process. The negative values of enthalpy change ($\Delta H°$) showed the exothermic nature of the adsorption process of phenol by natural and activated clay.

The value of $\Delta H°$ for the adsorption was lower than 80 $kJ \cdot mol^{-1}$, suggesting the physical nature of the sorption (Seki and Yurdakoç, 2006), that is, physiosorption conducted with Van der Waals forces.

The enthalpy change ($\Delta H°$) of the adsorption (-1.666 and -1.333 KJ/mol) indicated physical adsorption of phenol on to natural and activated clay (ESSINGTON et al., 1994). The overall adsorption process seemed to be exothermic.

Conclusion

Natural clay was a potential and promising adsorbent for removal of phenolic compounds from aqueous solution. The results were compared with activated clay. The adsorption of phenol on actived clay was higher than adsorption on natural clay. However, Natural clay stood as low-cost adsorbent and it showed the feasibility to remove up to 60% of phenol, respectively for initial concentration.

The Freundlich equation agreed very well with the equilibrium isotherm also Langmuir equation gave accepted linearity. Finally, the adsorption of phenol on the natural clay surface was exothermic. The results revealed that the natural clay could be employed as adsorbent for treatment of aqueous waste streams.

ACKNOWLEDGEMENT

We would like to thank the University of Technology Faculty of Chemistry, Oran, Algérie, also, Faculty of Technology. University of Mascara for Lab. and Staff Facilities. We thank Prof. F. Djafri White for assistance in XRD and IR measurements and educational experience, and Prof. A. Djafri for helpful discussion. Finally, we offer a special thanks to Prof. M. Bouchekara for their guidance and support throughout this research.

REFERENCES

Al-kaim A (2005). Synthesis and characterization f zeolite- urea formaldehyde polymer complex and study of adsorption of some phenolic compounds from aqueous solution on its surface, M.Sc. Thesis, College of Science, University of Babylon.

APHA, AWWA, WEF (2006). Standard methods for the examination of water and wastewater, XX Edition. APHA, Washington.

Aygiin A, Karakag SY, Duman I (2003). Production of granular activated carbon from fruit stones and nutshells and evaluation of their physical, chemical and adsorption properties. Microporous and Mesoporous Mater., 66: I 89-I 95.

Bras I, Lemos LT, Alves A, Pereira MR (2005). Sorption of pentachlorophenol on pine bark, Chemosphere, 60: 1095-1102,

Denizli A, Cihangir N, Ttizmen N, Alsancak G (2005). Removal of chlorophenols from aquatic systems using the dried and dead fungus Pleurotus sajor caju. Bioresource Technol., 96: 59-62.

ESSINGTON M, Tiche B, Dutarrtre R, Figueras F (1994). Microporous Mater., 2: 525.

Freundlich H (1926). Colloid and Capillary Chemistry, Matheun and Co., Ltd. London.

Gadsden A (1975). In frared Spectra of Minerals and Related Inorganic Compounds. The Butter worth group, UK.

Giles CH, Mac Ewan TH, Nakwa SN, Smith D (1960). Studies in Adsorption. Part XI.* A System of Classification of Solution Adsorption Isotherms, and its Use in Diagnosis of Adsorption Mechanisms and in Measurement of Specific Surface Areas of Solids. J. Chem. Soc., 3: 3973-3993.

Goyne KW, Zimmerman AR, Newalkar BL, Komarneni S, Brantley SL, Chorover J (2002). Surface charge of variable porosity Al2O3 (s) and SiO2(s) adsorbents. J. Porous Mat., 9: 243-256.

Hajjaji M, Kacim S, Alami A, El-Bouadili A, El Mountassir M (2001). Chemical and mineralogical Characterization of a clay taken from the Moroccan Meseta and a study of the interaction between its fine fraction and methyleneblue. Appl. Clay Sci., 20: 1-12.

Hall KR, Eagleton LC, Acrivos A, Vermeulen T (1996). Pore-and solid-diffusion kinetics in fixed-bed adsorption under constant-pattern conditions, Ind. Eng. Chem. Fundam., 5: 212-223.

Halouli NM, Drawish (1995). Effects of pH and inorganic salts on the adsorption of phenol from aqueous systems on activated decolorizing charcoal, Sep. Sci. Technol., 30: 3313.

Juang RS, Lin SH, Tsao KH (2002). Mechanism of sorption of phenols from aqueous solutions onto surfactant-modified montmorillonite. J. Colloid Interface. Sci., pp. 254-234.

Khalid N, Ahmad S, Toheed A, Ahmad J (2000). A Potential of Rice

Husk for Antimony Removal. Appl. Radiat. Isotopes, 52: 30-38.
Langmuir I (1918). Adsorption of gasses on plane surfaces of glass, mica and platinum. J. Am. Chem. Soc. 40: 1361-1403.
Lin SH, Cheng MJ (2002). Adsorption of phenol and m-chlorophenol on organobentonites and repeated thermal regeneration. Waste Manage., 22: 595-603.1-7.
Madejova J (2003) FTIR techniques in clay mineral studies. Vibrational Spectroscopy, 31: 1-10.
Rengaraj S, Seuny-Hyeon M, Sivabalan R (2002). Agricultural Solid Waste for the Removal of Organics: Adsorption of phenol from Water and Wastewater by Palm Seed Coat Activated Carbon. Waste Manage., 22: 543-548.
Seki Y, Yurdakoç K (2006). Adsorption of Promethazine hydrochloride with KSF Montmorillonite. Adsorption, 12(1): 89-100.
Seki Y, Yurdakoç K (2006). Adsorption of Promethazine hydrochloride with KSF Montmorillonite. Adsorption, 12(1):89-100.
Singh BK, Mishra NM, Rawat NS (1994). Sorption Characteristics of Phenols on Fly Ash and Impregnated Fly Ash. Indian Environ. Health, p. 36.
Song K , Sandi G (2001). Characterization of montmorillonite surfaces after modification by organosilane. Clays Clay Miner., 49(2): 119-125.
Tahani A, Karroua M, El Farissi M, Levitz P, Van Damme H, Bergaya F, Margulies L (1999). Adsorption of phenol and chlorine derivatives PLICS and organo-PLICS. J. Chem. Phys., 96: 464-469.
Vianna MMGR, Franco JHR, Pinto CA, Diaz FRV, Büchler PM (2004). Sorption of Oil Pollution by Organoclays and a Coal/Mineral Complex. Brazilian. Br. J. Chem. Eng., 21(02): 239.
Wu PX, Liao W, Zhang HF, Guo JG (2001). Adsorption of phenol on inorganic-organic pillared montmorillonite in polluted water. Environ. Int., 26: 401-407.

5

A report on soap making in Nigeria using indigenous technology and raw materials

Warra, A. A.

Department of Biochemistry, Kebbi State University of Science and Technology, P.M.B. 1144, Aliero, Nigeria.

In everyday life we use soap to wash dishes, clean clothes, or keep our bodies presentable to nose and eye. Soap therefore has numerous applications in our daily life. One of its great values is keeping our household a far better place to live and work. However, contrary to what one may think, soap was invented not only for the purpose of personal hygiene; rather, it was invented to solve other purposes. Colourful yarns were valued very early in the history of textiles; wool as it comes from the sheep is coated with a layer of grease that interferes with the application of dyes, soap was used to solve this problem. This report attempts to explore the technology of soap production in Nigeria using indigenous raw materials.

Key words: Soap, raw materials, technology, chemistry, quality control.

INTRODUCTION

There could be hundred different ways of explaining what really a soap is and what can be regarded as a soap; we can regard it as any cleaning agent, manufactured in bars, granules, flakes, or liquid form, made from a mixture of mostly sodium or potassium salts of various fatty acids of natural oils and fats.

In addition to basic raw materials, other substances of medicinal importance are added as ingredient to produce medicated soaps; there are other types of soap which are soaps of metals other than sodium and potassium called metallic insoluble soaps that are not used as cleaning agents, but are used for other purposes. Examples are, calcium and magnesium soap used as lubricants and driers, aluminum and chromium soaps for sizing paper, lithium stearate (a lithium soap) for thicken oils into grease etc.

In this context, our discussion will be limited to soaps as cleaning agents.

A short history

Throughout history people were known to have taken bath in herb waters and other additions to the bathing medium thought to be beneficial. Cleopatra of Egypt for example used mare's milk, honey and essential oils in her bathing rites.

Historical studies revealed that soap were utilized in both ancient Egypt and Babylonia 5000 years ago. Mixtures of animal fats and alkaline plant ash were used to produce soap (Phanseil, 1998). Ancient peoples were believed to have employed wood ashes and water for washing and to have relieved the resulting irritation with grease or oil. In the first century A.D., Pliny describes a soap of tallow and wood ashes used by Germanic tribes to brighten their hair. It is recorded that Babylonians were making soap around 2800 B.C and it was known to the Phoenicians around 600 B.C. These early references to

Table 1. Fatty acid profiles Shea butter versus competitors.

Oil type	Lauric (12:0)	Myristic (14:0)	Palmitic (16:0)	Stearic (18:0)	Oleic (18:1)	Linoleic (18:2)	Linolenic (18:3)
Cocoa butter			25	35	30 - 40	2 - 4	
Olive			12	2	72	8	1
Palm oil (husk)	44	16	8	3	5	2	
Coconut							
Shea butter							
Literature	0 - 0.5	0 - 1.6	3 - 9	30 - 50	41 - 50	4 - 11	0 - 7.5
Uganda			6.5	26.4	59.3	6.2	0.2
Nigeria			3.2	38.9	47.5	6.5	0.2
Burkina Faso			12.1	42.5	39.3	4.5	0.2
Mali			19	31.1	42.6	5.7	0.2

Source: (Ferris et al., 2001).

soap and soap making were for the use of soap in the cleaning of textiles fibers such as wool and cotton in preparation for weaving into cloth. From the long history of ancient civilizations until today, the basics of soap making has not fundamentally changed, meaning that the basic process has not changed. Hence, blending the old tradition with modern day knowledge of surface active agents, accuracy, combination of good and carefully selected ingredients mixed and stirred at the right temperature and time can with experience produce the finest soaps (Ellis, 2008)

In this work, attempt was made to explore the ways of improving indigenous technology for soap production by the use of available local raw materials.

TECHNOLOGY OF SOAP PRODUCTION UTILIZING INDIGENOUS RAW MATERIALS

Indigenous raw materials

Animal fat

Tallow is one of the principal fatty materials used in soap making. Romans were doing soap made from animal fats and wood ashes 2500 years ago (McGraw-Hill Encyclopedia of Science and Technology, 2007). Soaps made from animal fat contain more of 16- and 18- carbon fatty acids and are generally harder and easy to form into shapes (Robertson, 2006). Tallow oil include, palmitic acid ($CH_3(CH_2)_{14}CO_2H$, stearic acid $CH_3(CH_2)_{16}COOH$) and oleic acid ($CH_3(CH_2)_7CHCH(CH_2)_7COOH$

Shea nut fat

Shea nut solid fat (butter or stearin) is ideal for use in making cosmetics, soap and detergents (Fintrac, 1990). Shea nut fat is composed principally of triglycerides (triacylglycerols) containing an oleic acid moiety at the 2-position and saturated fatty acids, usually stearic or palmitic acids, at the 1- and 3-positions (Acquaye et al., 2001). Chemical analysis of Shea butter extracted from nuts samples from four African countries (Uganda, Nigeria, Burkina Faso and Mali) were conducted by the Ben Gurion University, Isreal, as part of the ongoing EU funded INCO project on Shea. Fatty acid analysis shows there is a high level of variability in Shea oils across Africa (Table 1). The Ugandan sample had a 59% oleic acid content compared with 47% for Nigeria and only 39% for Burkina Faso (Ferris et al., 2001).

Jatropha seed oil

Oil for soap making is the most profitable use; the fruit of Jatropha contain viscous oil that can be used for soap making (Openshaw, 2000). It is rich in palmitic acid, with high levels of hydrophobicity, and makes a soft, durable soap under even the simplest of manufacturing processes (Pratt et al., 2002). High oil content of *Jatropha curcas* indicated that *J. curcas* are suitable as non-edible vegetable oil feedstock in oleochemical industries (biodiesel, fatty acids, soap, fatty nitrogenous derivatives, surfactants and detergents (Akbar et al., 2009).

Castor seed oil

Abitogun et al. (2009) confirmed the presence of ricinoleic acids, oleic acid, palmitic acid, stearic acid and dihydroxylstearic acid; this is an indication of good quality that can be modified so as to be useful in cosmetics. This oil differs from all other commercial oils in being rich in ricinoleic acid (~90%, 12-hydroxyoleic). Compared with

Table 2. Some physical and chemical characteristics of oils extracted from sesame seed grown in Jigawa State, Nigeria* (Mohammed and Hamza, 2008).

S/N	Analysis/physical and chemical characteristics	W	R
1	Colour	White	Red
2	Iodine value (gI_2/100 g)	103	116
3	Oil content (%)	48	50
4	Specific gravity (g/cm^3)	0.915	0.923
5	Acid value (mg KOH/g)	0.5	0.45
6	Peroxide value (Meq KOH/g)	8	7.45
7	Saponification value (mg KOH/g)	189	191
8	Cyanide test	Negative	Negative

*The values are mean of three replicates.

the common vegetable oils, castor oil is more viscous, less soluble in hexane, and more soluble in ethanol, all as a consequence of the presence of the hydroxy acid. This hydroxy acid has several interesting properties by which it can be converted to useful products. $CH_3(CH_2)_5CHOHCH_2CH{=}CH(CH_2)_7COOH$ [ricinoleic acid (12-hydroxyoleic acid)]. Sulfation converts the hydroxyl group to a sulfate ($-OSO_2OH$) with improved surfactant properties. Apart from soap, it is the earliest anionic surfactant (Gunstone, 2005).One of its uses is in the manufacture of transparent soaps (Kochhar, 1998). The inferior qualities of the oil are frequently employed in India for soap making (Grieve, 2009). In Nigeria, castor bean was cultivated on over 6,000 ha across most of the states. The crop grows in the wild even where it is not cultivated. Benue, Gombe, Yobe, Cross Rivers, Kogi, Ebonyi, Kwara, Zamfara and FCT are the promising producers where the bulk of the hectares are cultivated. The output is estimated at twelve thousand metric tonnes (12,000 mt) (Raw Materials Research and Development Council, 2009).

Neem seed oil

Oil extracted from its seeds composed primarily of triacylglycerols of oleic, strearic, linoleic, and palmitic acids. The seeds yield 40% of a deep yellow oil, the well-known 'Margosa oil' (Girish and Shankara, 2008). Of all other industrial uses in India, neem oil has been a major ingredient in soaps for at least 50 years (National Research Council, 1992). The indigenous neem seed oil is presently obtainable at Saberg International Ltd., producers of proudly Nigerian Naija Neems (Neem oil) at Technology Incubation Centre, Birnin Kebbi, Kebbi State, Nigeria.

Sesame seed oil

Sesame is an important export crop in Nigeria, and Nigeria has a substantial role in the global sesame trade. Annual exports of sesame from Nigeria are valued at about US$20 million and Nigeria is the primary supplier of sesame seed to the world's largest importer, Japan (Chemonics International Inc., 2002). The seed has been called "queen of the oil seed crop" because of the high yield of oil and its quality. Apart from its uses in cosmetics and perfume, it is also used for the manufacture of soap (Gandhi and Taimini, 2009). The hot-expressed oil is used chiefly for soap making (Jamieson and Baughman, 1924). The work of Mohammed and Hamza (2008) justified some quality characteristics of Nigerian variety of sesame seed oil (Table 2).

TECHNOLOGY

There are five major steps for soap making process:

(a) Saponification,
(b) Soap washing or glycerol extraction,
(c) Fitting,
(d) Moulding,
(e) Drying.

Although in some simple hard soap making, the steps can be classified further into seven main steps:

(1) Getting the right mixture of oil/fat and alkali, called "proving",
(2) "Boiling down" – removing the unwanted water, and checking for "doneness",
(3) Treating with salt to remove water, impurities, and glycerin a process called 'graining' this step makes a good solid soap for washing clothes,
(4) Adding colouring agents (colourants) and perfumes,
(5) Pouring into moulds, called "setting",
(6) Breaking the "green" soap out of the mould and splitting it into finished sizes and,
(7) Drying and airing the "green" soap.

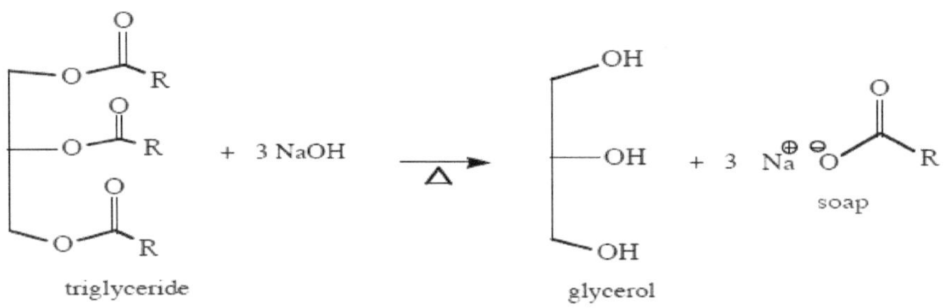

Figure 1. Saponification reaction. R is the long chain of carbon and hydrogen atoms.

Concise chemistry of soap making

Chemical reactions in soap making (that is, saponification)

In saponification reactions, esters are split into alcohols and salts of carboxylic acids (Figure 1). The word "Saponification" is derived from the latin word "saponins", meaning'soap' though the word has a wide application, for example, in papermaking, many components of wood pitch may become saponified under alkaline papermaking conditions (Mini-encyclopedia, 2008). Saponification is also widely used in more general term to refer to alkaline hydrolysis of any type of ester.

The overall reaction of triacylglycerol saponification can be thought of occurring in two steps. The first step is the hydrolysis of the ester linkages to produce glycerol and three fatty acid molecules: Fat or oil + $3H_2O_3$ fatty acids + glycerol. The second step involves a reaction between the fatty acid molecules and base (usually NaOH) in the alkali solution. This is an acid-base reaction that produces water plus salts: 3 fatty acids + 3NaOH → 3 fatty acid salts + $3H_2O$

Indigenous technology saponification

Mixing the fat or oil with base (alkali) can be done using a specially designed mixer. Small scale soap makers in Nigeria mostly use manual mixer.

Simple cold-process soap making procedure

For each soap formulation, pour directly 200 g/dm^3 NaOH solution into the stainless steel container (pot) containing the fat and oil in the ratio 1:1 (v/v). Warmed gently the fats/oil and pour into the stainless steel soap making pot followed by the alkali solution to form an intimate mix and then stir frequently for 10 to 15 min using wooden stirrer. Perfume and other ingredients can be added as additives at the last time before pouring the saponification mixture into moulds. After pouring, the soap is allowed to harden by air-drying for 24 h to obtain the soap bars, according to the method reported by Warra (2009) and the soap bars can then be observed for colour, texture, lathering and cleaning power.

Moulding and cutting

In this step, melted soap is poured into shaped-mould for it to "set" and harden in. Sometimes, the soap is "broken out" of the mould and "split" down to required sizes with a wire. The soaps are cut using cutting machine which can be made locally into different designs and size.

Stamping

This is the use of special stamp (either electric or manual) and trade marks on the products (soap products). Murjanatu Muhammad Abba illuminated the growing role of stamping in selling products in Nigeria when she wrote "This is very vital in selling products in Nigeria as most Nigerians buy name not the quality of the product" (Abba, 2008). Wooden or plastic, handmade stamp and stamp box or manual stamping machine are used.

Drying

Soap is normally obtained as having moisture content. Drying is required to reduce this water content of the soap to a carefully defined and controlled level, especially for toilet soap bars. Soap at control moisture content will generally produce, after, subsequent processing, a bar of the required appearance. This cannot be achieved until the soap has been aired for hours or up to about 1 month in the case of dry hard soap that takes longer to use up.

Packaging

This is the use of special materials to package products

to the market. The products are finally packed in cartons. Wrapping the soaps into nice paper or clean polythene will add greatly to its sales value. The cost associated with packaging material is an important factor in the cost of production. The cost packaging could be high for consumer products. Therefore, it is always important for a soap producer to estimate his packaging cost in order to represent his operating expenses which have great impact on the selling prize and profit or gain.

Safety precautions/hazards

1) Wear your gloves and goggles throughout the entire soap making.
2) If sodium hydroxide touches the skin, immediately rinse the affected area with running water
3) If a spill happens, take care of your skin first; rinse the area with cool water. Then apply vinegar to the area and rinse again.
4) Do not attempt to pour vinegar into your eye. Seek medical attention for any sodium hydroxide spill that had entered your eye.
5) Avoid contact with hot oils, as they can burn the skin. Consult a professional for assistance.
6) Avoid contact with essential and fragrance oils. Wear gloves and goggles while using them.
7) Flush the affected area with water.

Safety data and record keeping

Any soap making industry being it large or small scale should have documented information on safety categories and record keeping. The following parameters are some of the important references used.

Material safety data sheet (MSDS)

This is a reference that contains pertinent information related to the following safety categories for specific material: Manufacturer or vendor information, chemical composition, hazards and potential health effects, first aid measures, fire fighting measures, spill measures, handling and storage, exposure and protection, physical and chemical properties stability and reactivity, toxicological information, environmental information, disposal considerations, transportation information, and regulatory information. MSDS are usually available from chemical manufacturer, a commercial source, or a private library developed by the chemical plant.

According to Crowl and Louvar (2002), the industrial hygienist or safety professional must interpret the physical and toxicological properties to determine the hazards associated with a chemical. These properties are also used to develop a strategy for the proper control and handling of these chemicals.

Workplace safety

Workplace safety means having ingredients properly labeled and stored, maintaining a clean, orderly, well-lit, and well ventilated work area for making soap restricting the access of children and pets to the storage and work areas, having easy access to personal protective equipment: a telephone, a fire extinguisher, and running water.

Complaint file

This is a file that contains a record of a customer complaint, the date of the complaint, and the action taken by the soap maker.

Batch code

This is a numbered record of each batch of soap. The batch code sheet lists the company of manufacture, the name of the soap maker(s), the date made, the ingredients, weight numbers, any variation from the normal soap making procedure, the cure date or “do not use before” date, the expiration date, and any additional comments.

Ways of improving soaps quality

The use of glycerine

The presence of several polar-OH groups makes it strongly attracted to water, a feature that makes glycerine useful as a skin softener in products such as lotions, cosmetics, shaving creams, and liquid soaps.

The use of sequestrants

Many manufacturers use chelating molecules in their commercial soap products, often EDTA derivatives are used which helps to bind with any free calcium or magnesium ions to prevent soap scum (a process called sequestration). These also help to reduce fragrance loss, discolouration and rancidity.

Checking the soap pH

Checking the pH of soap is necessary not only for the purpose of improving soaps quality but to regulate the pH level which shall not contribute to the harshness of hands and skin. For the purpose of protecting public health, high pH levels in the 9 to 11 range or low in the 3 to 5 level are considered deleterious to the skin. This is in accordance with NAFDAC regulatory requirements on cosmetics, soaps and detergents (Umar, 2002).

The use of silicates

Sodium silicate [$Na_4(SiO_4)$] - an example of large family of silicates, (commonly known as 'water glass') is used commercially in detergents, where it maintains a constant pH and can degrade fats by hydrolysis.(Housecroft and Constable, 2006). Sodium metasilicate is a basic salt used in detergent, partly as a basic buffer and partly to keep dirt from settling back on to the fabric (Jones and Alkins, 2002). The SiO_3^{2-} ions are attached to dirt particles, giving the particles a negative charge and thereby preventing them from merging with others into larger, insoluble particles.

Perfuming and colouring

When the soap is melted for the last time before being poured out into moulds, it can be coloured and perfumed. It is best not to do this any earlier in the method, as the lye water could hurt the perfume or colouring. If you have perfumes or colouring, try them out on small amounts of soap.

Chemical analysis of fats and oils

To determine the industrial value of a particular oil or fat which is dependent on composition and purity, a number of chemical tests are carried out to have quality control over fats and oils used as raw materials for the production of soaps. Parameters: (1) saponification value (2) and iodine value (3) acid value

CONCLUSION

Soap production can be regarded as one area of business that is lucrative and needs only little capital to start with, and considering the vast available resources in Nigeria there is need to design and develop strategy on war footing in order to explore and utilize full benefits of these raw materials using the available indigenous technology.

REFERENCES

Abba MM (2008). Alleviating poverty though laundry and bar soap production in Daily Trust, Thursday, April 24, 2008. Media Trust Ltd. Abuja, Nigeria 8(94):30.

Abitogun AS, Alademeyin OJ, Oloye DA (2009). Extraction And characterization of Castor Seed Oil. Internet J. Nutr. Wellness 8(2).

Acquaye D, Smith M, Letchamo W, Angers P, Simon J (2001). Shea butter. Agrobusiness in Sustainable Natural African Plant Products pp.1-2. http://www.aspenwoodbathhandbody.com/Docs/sheabutter.pdf Accessed on 1/6/2009.

Akbar E, Yaakob Z, Kamarudin SK, Ismail M, Salimon J (2009). Charecteristic and Composition of *Jatropha curcas* Oil seed from Malaysia and its Potential as Biodiesel Feedstock. Eur. J. Sci. Res. 29(3):396-403.

Chemonics International Inc. (2002). Overview of the Nigerian Sesame Industry. Prepared for The United States Agency for International Development (USAID)/Nigeria. pp. 5-6.

Crowl DA, Louvar JF (2002). Chemical process safety fundamentals with applications. 2nd Edition. Prentice Hall PTR, New Jersey. pp. 74-75.

Ellis M (2008). Colonial soap making, it is history and techniques, Available at http:www.akasoft.com/soapfat/historycontent.html_ last accessed 7/6/2008.

Ferris RSB, Collinson C, Wanda K, Jagwe J, Wright P (2001). Evaluating the market Opportunities for shea nut and shea nut processed products in Uganda. Submitted to USAIDOctober, 2001. Natural Resources Institute. Accessed at http://www.foodnet.cgiar.org/projects/sheanut_Rep.pdf 27/6/2010.

Fintrac (1999). Market and technical survey: Shea nuts. Fintrac Inc Washington DC.

Gandhi AP, Taimini V (2009). Organoleptic and nutritional assessment of sesame (*Sesame indicum* L.) biscuits. Asian J. Food Agro-Industries 2(2):87-92.

Girish K, Shankara BS (2008). Neem- A Green Treasure. Electronic J. Biol. 4(3):102-111.

Grieve M (2009). A modern Herbal Castor oil plant. Available at http://www.botanical.com / botanical / mgmh / c /casoil 32.html. Accessed 4/4/2009. pp. 1-3.

Gunstone FD (2005). Vegetable Oils (ed) in Bailey's Industrial Oil and Fat Products, 6th Edition, Vol. 1. Edited by Fereidoon Shahidi. John Wiley & Sons, Inc. pp. 224-225.

Housecroft CE, Constable EC (2006). Chemistry: An introduction to organic, inorganic and physical chemistry. 3rd Edition. Pearson Education Limited, England. pp. 684, 704.

Jamieson GS, Baughman WF (1924). The chemical composition of sesame oil. J. Am. Chem. Soc. 46(3):775-778.

Jones L, Atkins P (2002). Biochemistry: molecules, matter, and change. 4th Edition. W.H. Freeman and Company New York. pp. 531, 874.

Kochhar SL (1998). Economic Botany in the Tropics.2nd ed. Macmillan India Ltd. McGraw- Hill Encyclopedia of science and technology (2007). vol. 16. 10th Edition. McGraw-Hill Companies Inc. New York pp. 586-587.

McGraw- Hill Encyclopedia of Science and Technology (2007) vol. 16 .10th Edition. McGraw-Hill Companies Inc. New York. pp. 586-587.

Mini-encyclopedia of paper making (2008). Retrieved on 12/2/2008 from http://www4. ncdu.edu/_hubble/equip unit/saponifi.htm.

Mohammed MI, Hamza ZU (2008). Physicochemical Properties of Oil Extracts from *Sesamum indicum* L. Seeds Grown in Jigawa State-Nigerian. J. Appl. Sci. Environ. Manag. 12(2):99-101.

National Research Council (1992). Neem: A Tree for solving Global Problems. National Academy Press, Washigton, D.C.

Openshaw K (2000). A review of *Jatropha curcas*: An oil plant of unfulfilled promise. Biomass and Bioenergy 19:1 Elsevier Science Ltd. 19:1-15.

Phanseil ON, Dueno E, Xianghong WQ (1998). Synthesis of exotic soaps in the chemistry laboratory in J. Chem. Educ. 75(5):612.

Pratt JH, Henry EMT, Mbeza HF, Mlaka E, Satali LB (2002). Malawi Agroforestry Extension Project Marketing and Enterprise Program. Main Report. Malawi Agroforestry Publication 47:44-46.

Raw Materials Research and Development Council (2009). Survey Report of Ten Selected Agro Raw Materials in Nigeria: Castor. Raw Materials Research and Development Council (Federal Ministry of Science and Technology), Abuja, Nigeria. P. 28.

Robertson DL (2006). Saponification and making soap. Experiment 8

Chemistry 102 Experiments Index. Retrieved from http://www.miracosta.cc.ca.us/home/dlr/102exp8.htm. on 22/3/2009.

Umar M (2002). Cosmetics, Soaps, Detergents and NAFDAC'S Regulatory Requirements. A paper presented at a Training workshop for small scale and medium scale Enter-Prizes organized by UNDP/JCSL and Ministry of commerce and Industry, Maiduguri, Borno State, Nigeria. pp. 1-3.

Warra AA (2009). Extraction and Chemical Analysis of Indigenous Shea nut oil. Proceedings of the 1st National conference of the Academic Staff Union of Polytechnics. Waziri Umaru Federal Polytechnic chapter, Birnin Kebbi, Nigeria. pp. 160-162.

6

Determination of some essential minerals in selected edible insects

Ajai A. I., Bankole M., Jacob J. O. and Audu U. A.

Department of Chemistry, Federal University of Technology, Minna Nigeria.

The essential minerals, Cu, Fe, Mg, Co, Zn, Na and K were determined in selected edible insects (Locust, Cricket, Grasshoppers and Termite) using standard analytical method. The result obtained from the study shows that Cu content ranged from 85.5 ± 3.40 mg/kg; Fe from 390.025 ± 31.61 mg/kg; Mg from 1534.84 ± 85.01 mg/kg; Co from 4.35 ± 0.28 mg/kg; Zn from 208.11 ± 63.90 mg/kg; Na from 223.25 ± 14.32 mg/kg and K from 325.69 ± 7.16 mg/kg respectively. The statistical analysis shows that Cu, Co, Na and K content in the studied insects were significantly different except for Fe, Mg and Zn which shows no significant difference at P ≥ 0.05. From the high essential minerals content obtained from this study, these insects could harnessed as alternative source of these minerals in order to help compliment the nutritional needs of an individual.

Key words: Minerals, edible, insects, locust, grasshoppers, termites.

INTRODUCTION

Insects are the most successful prolific group of organism in the animal kingdom, constituting of about 76% of known species of surviving animal (Yoloye, 2010). It is well-know that insects are an attractive and important natural source of food for many kinds of vertebrae animals, including birds, lizards, snakes, amphibians (toads, frogs, salamanders), insectivore and other mammals (Banjo et al., 2010). Insects have played an important part in the history of human nutrition in Africa, Colombia, Venezuela, Asia and Latin America (Ruddle, 2006; Chavunduka, 2010). They were equally important resources for Indians of North America, who like other indigenous groups, expended much effort in harvesting them (Finke et al., 2008). The aversion to insects as human food among Europeans is nothing more than custom and prejudices as rightly asserted by Kent (2002). Hundreds of species have been used as human food. Some of the more important groups include grasshopper, termites, caterpillars and beetle grubs. Insects are being market and consumed in Western and Delta region of Nigeria and road side-hawking are visible (Banjo et al., 2006; Ifie and Emeruwa, 2011).

The high consumption of insects as food can be attributed to their high nutrient content like protein (Yoloye, 2010), fats (Quin, 2005), vitamins and minerals (Akindele et al., 2009; Kodondi et al., 2010; Oliveira et al., 2007; Ledger, 2010). Also, consumption of insects as food reduces serum cholesterol and serves as haemostatic agent of tissue repairs and for accelerating the healing of wounds due to its chitin content (DeFoliart, 2007; Goodman, 2011). Other researchers in their review of different edible insects also reported the presence of various nutrient, micronutrients, proteins and vitamins in various edible insects and caution on the need to develop proper processing and decontamination method for the removal of the toxic, and pathogens and other

Table 1. Essential mineral content in selected edible insects (mg/kg).

Parameter	Locu	Cricket	Termites	Grasshoppers
Cu	99.04 ± 6.06[a]	69.05 ± 0.70[b]	77.77 ± 9.86[b]	73.02 ± 0.74[b]
Fe	574.75 ± 30.75[a]	519.00 ± 44.5[ab]	205.30 ± 32.45[c]	349.27 ± 123.21[bc]
Mg	1484.17 ± 98.74[a]	1538.77 ± 27.47[ab]	1400.17 ± 144.78[b]	1669.50 ± 25.24[a]
Co	8.55 ± 0.56[a]	2.07 ± 0.70[c]	0.15 ± 0.00[d]	5.45 ± 0.21[b]
Zn	160.37 ± 4.84[b]	256.55 ± 28.70[a]	159.3 ± 6.29[b]	256.92 ± 121.51[a]
Na	290.25 ± 7.42[a]	156.25 ± 5.30[c]	287.5 ± 21.21[a]	243.25 ± 9.54[b]
K	480.12 ± 7.24[a]	282.80 ± 17.88[c]	317.5 ± 10.60[b]	225.25 ± 7.07[d]

Values are mean duplicate determinations. Data with the same superscript along the same row are significantly different at $P \geq 0.05$.

antinutritional factors present in these insects if the consumer is to optimize the nutrient content of these insects (Birgit and Oliver, 2013).

Contamination of food products by heavy metals is becoming an unavoidable problem these days. Air, soil, and water pollution are contributing to the presence of harmful elements, such as Cadmium, Lead, Mercury, and Arsenic in foodstuffs. Most of these food stuffs are being consumed directly or indirectly by insects. The occurrences of heavy metals-enriched ecosystem components have impaired health of the population by the ingestion of foods contaminated by these harmful elements. The toxicity of heavy metals to man is mainly caused by their persistence in the environment and hinged strongly upon the chemical form in which they are ingested (Gorham, 2007).

Considering the high consumption rate of insects and their nutritional contribution to the growth and development of man and coupled with the fact that these insects migrate from one location to another and could be susceptible to heavy metal contamination through their feeding habits, there is therefore the need to constantly investigate the essential mineral content of some of these edible insects in order to ascertain if they are within the tolerable limit that will meet the nutritional needs of the consumer without causing any negative health effect.

MATERIALS AND METHODS

Sampling area and sample collection

Four types of edible insects, cricket (*Gymnogryllus lucens*), grasshopper (*Zonocerus variegatus*), termite (*Macrotermes bellicosus*) and locust (*Schisocerca gregaria*) were collected randomly from Gbaiko village, Kagara area and Gidan Kwano village of Niger State using an entomological net and some were handpicked. Insects of similar species were mixed together and sun- dried.

Sample pretreatment

All glass wares were carefully soaked with a solution of 10% nitric acid for 48 h and rinsed with deionized water. The samples were sun dried for one week to reduce the moisture content. The insects were dewinged before they were ground into powder with laboratory porcelain mortar and pestle. Each ground sample was stored in well- labeled air-tight glass containers and stored in an oven prior to digestion.

Digestion of sample

2.00 g of the sample was placed in a digestion tube and was predigested with 10 ml concentrated HNO_3 at 135°C until the liquor was clear. This was followed by the addition of 10 cm^3 HNO_3, 1 cm^3 $HClO_4$ and 2 cm^3 H_2O_2 and the heating continued at a temperature 0f 250°C for 1 h until the liquor became colourless. The digest was then slowly evaporated to near dryness (avoiding prolonged baking), cooled and dissolved in 1 m HNO_3. The digest was subsequently filtered through a Whatman No. 1 filter paper into a 100 ml volumetric flask and made up to volume with distilled water and stored for AAS analysis.

Elemental analysis

The elemental analysis of the samples were, performed using Atomic Absorption Spectrometer (Hitachi Z-8100, Japan) at the Central Laboratory for Research and Development, University Of Ibadan, Oyo State, Nigeria while Flame photometer (JENWAY PFP 7) was used to determine the K and Na content at the Department of Chemistry Laboratory, Federal University of Technology, Minna, Niger state.

Statistical analysis

The results obtained from the study were subjected to SPSS 15.00 statistical analysis using one way ANOVA and Duncan's multi range comparison test at 95% confident level.

RESULTS AND DISCUSSION

The result of essential minerals obtained from the selected edible insects studied using one way ANOVA and Duncan's multi range comparison test at 95% confident level are shown in Table 1. From the result in Table 1, the concentration of Cu in all the samples range between 69.5 ± 0.70 to 99.04 ± 6.06 mg/kg with locust having the highest concentration and cricket the least.

The finding from this study is comparable with that reported by Banjo et al. (2010), of between 73.72 to 99.04 mg/kg Cu in their analysed edible insects. But is at variance with that reported by Defoliart of between 100 to 120 mg/kg in other edible insects used as human food and as animal feeds (Defoliart, 2007). This variation can be attributed to ecological factors and variation in species.

The Iron content in the studied edible insects ranged between 205.30 ± 32.45 to 574.75 ± 30.75 mg/kg. These concentrations are lower than that reported by Bodenheimer (2005), of between 320 to 747 mg/kg. This variation could also be as a result of their feeding habits. Locust has the highest Fe value while termites the least Fe concentration. According to World Health Organization, Fe plays an important role as a heme molecule in red blood cells as it permits oxygen transport (WHO, 2006), but excessive Fe intake can also be a serious problem enhancing free radical activity in the body or damaging the liver (Ekop et al., 2010).

The magnesium concentration in the analysed edible insects ranged from 1400.17 ± 14.78 to 1669.50 ± 25.24 mg/kg, with termites having the least concentration of 1400.17 ± 14.78 mg/kg and locusts the highest 1669.50 ± 25.24 mg/kg. The need for magnesium in our diet is very important since Type 2 diabetes have been reported to be associated with low magnesium content in the body (WHO, 2005). These insects would be a better source and cheap means of obtaining this vital mineral from the food chain. It have also been reported that moderate magnesium supplements can help people with heart disease and osteoporosis, since excessive use of it can be toxic (Adeyemo et al., 2001). Generally, the Mg levels in all the samples showed no significant difference at $P \geq 0.05$ (Table 1).

The concentration of Co in the samples ranged between 0.15 ± 0.00 to 8.55 ± 0.56 mg/kg with locust, having the highest concentration of 8.55 ± 0.56 mg/kg and termites the least of 0.15 ± 0.00 mg/kg. The level of Co in all the samples was generally lower than that reported by Muyay (2009), of between 12.75 to 15 mg/kg in edible insects in Mexico. The differences are due to the geographical location and their feeding habits. Concentration of Zinc in the studied edible insects were; locusts 160.37 ± 4.84 mg/kg, cricket 256.55 ± 28.70 mg/kg, termites 159.30 ± 6.29 mg/kg and grasshoppers 256.92 ± 21.51 mg/kg respectively. This could be used to compliment the daily dietary Zinc need of 15 mg/day of an adult of 60 kg body weight (WHO, 2005). The concentration of Zinc does not show significant difference between the analysed edible insects at $P \geq 0.05$.

Considering the importance of sodium in human metabolism, Food and Agricultural Organization (FAO) has recommended a daily dietary allowance of 2400 mg of sodium per day (FAO, 2010). From the result of this study, sodium has a concentration of 290.25 ± 7.42 mg/kg in locust, 156.25 ± 5.30 mg/kg in cricket, 287.50 ± 12.21 g/kg in termites and 243.25 ± 9.54mg/kg in grasshoppers. These concentrations are lower than that reported by Banjo et al. (2010), of 1500 mg/kg in some edible insects. This variation can be attributed to differences in geographical locations as well as inter-elemental interactions (Quin, 2005).

Potassium levels for locust and termite species were 480.14 ± 7.24 mg/kg and 17.50 ± 10.60 mg/kg respectively. These values are significantly difference at $P \leq 0.05$. Those of the two, cricket and grasshopper species were 282.80 ± 17.88 mg/kg and 225.25 ± 7.07 mg/kg respectively. The result is very low compared with that reported by Chen et al. (2008), of 500 mg/kg. The variation could be attributed to the sampling location and the feeding habits of these insects. The recommended dietary allowance for K is 2,000 mg per day for the average adult (WHO, 2010). Generally, considering the high essential minerals content in the studied edible insects, they sourced and consumed in order to help compliment the deficiency of the nutritional needs of an individual.

Generally, the one way Anova and Duncan multi range comparative test, shows that edible insects, locust had the highest concentrations of most of the essential minerals analysed at $P \leq 0.05$ and at 95% confidence level and termites the least with variations between the other two insects. This goes to suggest that any consumer intending to source for these essential minerals in order to compliment his/her nutrient needs from these insects should concentrate more on locust.

Conclusion

Based on the findings it can be concluded that considering the high mineral content of these edible insects studied, their low cost and availability, that they can be harnessed and included in our diet in order to help compliment the deficiency of these minerals from other food sources.

REFERENCES

Adeyemo OE, Idowu OE, Ogabiela EE, Onianwa PC (2001). Copper and zinc contents of Nigerian foods and estimates of the adult dietary intakes. Food Chem. 72:87-95.

Akindele SO, Adibye FS, Ofuya TI (2009). Some edible insect species consumed by the people of Benue state, Nigeria. Pakistan J. Nutr. 8(7):946-950.

Banjo AD, Lawal OA, Fasunwon BT, Alimi GO (2010). Alkali and heavy metal contaminants of selected edible arthropods in southern Nigeria. Am . Ewras. J. Toxicol. Sci. 2(1):25-29.

Banjo AO, Lawal OA, Songonuga EA (2006). The nutritional value of fourteen species of edible insects in southwestern Nigeria. Afr. J. Biotechnol. 5(3):298-301.

Birgit AR, Oliver KS (2013). Nutritional composition and safety aspects of edible insects. Mol. Nutr. Food Res. 57(5):802-823.

Chavunduka DM (2010). Insects as a source of protein to the African. Rhodesian Sci. News 9:217-220.

Chen X, Feng Y, Zhang H, Chen Z (2008). Review of the nutritive value

of edible insects.
Defoliart GR (2007). The human use of insects as food and as animal feed. Bull. Entomol. Soc. Am. 35:22-35.
Ekop EA, Udoh AI, Akpan PE (2010). Proximate and Anti-nutrient composition of four edible insects in Akwa Ibom state, Nigeria. World J. Appl. Sci. Technol. 2(2):224-231.
FAO (2010). Review of food consumption data on edible insects: Food and Agricultural Organization/The International Network of Food Data Systems (INFOOD).
Finke MD, Defollart GR, Benevenga NJ (2008). Use of a four-parameter logistic model to evaluate the quality of the protein from three insect species when fed to rats. J. Nutr. 119:864-871.
Goodman WG (2011). Chitin: a magic bullet. Food Insects Newslett. 2(3):6-7.
Gorham JR ED (2007). Ecology and management of food-industry pests. FDA Tech. Bull. 4, Association of Official Analytical Chemists, Arlington, V.A P. 595.
Ifie I, Emeruwa CH (2011). Nutritional and anti-nutritional characteristics of the larva of *Oryctes monoceros.* Agric. Biol. J. N. Am. 2(1):42-46.
Kent G (2002). Africa's Food security under Globalization. AJENS. 2:22-29.
Kodondi KK, Leclercq M, Gaudin-Harding F (2010). Vitamin Estimations of three edible species of *Attacidae carterpillers* from Zarie, Int. J. Vitam. Nutr. Res. 57:333-334.
Ledger J (2010). The eight plague returneth: The locusts are coming! AFR. Wildl. 41:201-210.
Muyay T (2009). Les Insects Comme Aliments de L'Homme, Publication ser.II Vol. 69. CEEBA, Bandundu, Zaire, P. 177.
Oliveira JFS, DE Carvalho JP, De Sousa R, Simao MM (2007). The Nutritional value of four species of insects consumed in Angola. Ecol. Food Nutr. 5:91-97.
Quin PJ (2005). Foods and Feeding Habits of the Pedi, Witwatersrand University, Johannesburg, Republic of South Africa, P. 278.
Ruddle K (2006). The human use of insects examples from the Yukpa. Biotropica 5:94-101.
World Health Organization (2005). Health Effect of Lead- exposure- a review of the literature and a risk estimate .Scand J work Environ. Health 1998; 24(suppl. 1):1-51.
World Health Organization (2006). Biological half-time of Cadmium in the Blood of workers after cessation of exposure. Scand J work Environ. Health 1983; 9:327-31.
World Health Organization (2010). Cadmium Environmental Health Criteria, vol.134. Geneva.
Yoloye VL (2010). Basic Invertebrate Zoology. 1st edn. University of Ilorin Press, Ilorin. P. 192.

Production and characterization of activated carbon from selected local raw materials

Yusufu M. I.[1]*, Ariahu C. C.[2] and Igbabul B. D.[2]

[1]Department of Food Science and Technology, Federal Polytechnic, Idah, Nigeria.
[2]Department of Food Science and Technology, University of Agriculture, Makurdi, Nigeria.

Quality indices of activated carbons from cattle bone (CB), coconut-shell (CS) and wood carbon (WC) as influenced by mode of activation (heat and acid) were studied using the non-activated carbon samples as control. Preliminary investigation was carried out to determine optimum condition for carbonization and activations. The iodine values, ash, bulk densities, moisture, hardness and porosity analysis of the samples were evaluated using standard methods. Heat (H) activation was at 950°C for 3 h to produce heat activated bone carbon (HCB), and 850°C for $2^1/2$ h for heat activated coconut-shell (HCS) and wood carbon (HWC) while concentrated phosphoric acid (1:4) carbon: acid (w/w) was at 600°C for 2 h to produce acid activated wood (AWC), acid activated coconut-shell (ACS) and acid activated bone (ACB) samples. The activation yield for the acid activation ranged from 40.40 to 46.21% while heat activation ranged from 36.30 to 41.10%. Iodine values for heat activated carbons ranged from 828.39 (HCB) to 995.38 (HWC) mg/g. That of acid activated carbons ranged from 834.94 (ACB) to 914.17 (AWC) mg/g as compared with 66.2 to 222.7 mg/g for non-activated carbons (controls). Hardness of heat activated carbons ranged from 88.5 to 93.5% while that of acid activated carbons ranged from 62.0 to 89.5%. Bulk densities ranged from 0.40 to 0.79 g/cm^3. The moisture for heat activated carbons ranged from 4.7 (HCB) to 8.4% (HWC) while that of acid activated carbons ranged from 5.0 (ACB) to 9.4% (AWC). The volume activities for the activated carbon samples ranged from 228.7(HCB) to 324.9 (HWC). The ash for bone carbons ranged from 80.6 to 85.9% while those of coconut shell and wood ranged from 1.1 to 3.5%. The carbons presented Brunauer, Emmett and Teller (BET) surface areas between 718 and 1018 m^2/g, and micropores and mesopores with volumes between 0.004 and 0.776 cm^3/g. This study indicated that efficient activated carbons could be obtained from local raw materials by controlled activation with acid or heat for home and industrial purposes.

Key words: Quality indices, optimum condition, carbonization, activation.

INTRODUCTION

The importance and relevance of activated carbon to an ever growing society cannot be overemphasized considering its enormous uses. Its uses range from liquid phase to gaseous phase applications. Activated carbons are highly porous and adsorbent materials. They have wide applications in domestic, commercial and industrial settings (Mendez et al., 2006). In the food industry, activated carbon is used in de-colourization, deodorization and taste removal. It is used to remove heavy metals and organic contaminants from liquids. Activated carbon is used in water de-chlorination and processing of foods. It is also used in medicine for adsorption of harmful chemicals and drugs. In gas cleaning applications, activated carbon is extensively used in air filters at industrial level as well as in general air conditioning application (Oyo and Igbokwe, 2001; Diets, 1990), (Inamullah et al., 2008), (Elliot et al., 1989).

*Corresponding author. E-mail: yusufu_ikagu@yahoo.com.

ABBREVIATIONS: ACB, Acid activated bone carbon; **ACS,** acid activated coconut shell carbon; **AWC,** acid activated wood carbon; **HCB,** heat activated bone carbon; **HCS,** heat activated coconut shell carbon; **HWC,** heat activated wood carbon; **CAC,** commercial activated carbon.

Basically, there are two different processes for the preparation of activated carbon, the so-called physical or thermal and chemical activation. The first step is carbonization of the raw material and is performed in an oxygen free environment at temperature of about 700°C (Dileck and Oznuh, 2008). Carbonization means obtaining charcoal from the raw material and the produced charcoal having low surface area is not an active product. During the process of carbonization, volatile components are removed and the residual carbonaceous material is then activated in air at temperatures higher than 800°C, or in the presence of an activating agent at lower temperatures. According to Sang et al. (1997), activation can be physical or thermal with the use of air, carbon dioxide or water vapour at high temperature or a chemical process employing activating agents such as phosphoric acid (H_3PO_4) or zinc chloride ($ZnCl_2$). As reported by several authors, activation temperature significantly affects the production yield of activated carbon and also the surface area of activated carbon. The temperature used as low as 200°C (Haimour and Emeish, 2006). In chemical activation, the raw material or carbonized sample is impregnated with an activating agent and heated in an inert atmosphere (Srinvaskannan and Mohammad, 2004). During the activation process, the spaces between the elementary crystallites become cleared of less organized, loosely bound carbonaceous material. The resulting channels through the graphitic regions, the spaces between the elementary crystallites, together with fissures within and parallel to the graphitic planes, constitute the porous structure, with large internal surface area (Rodriguez and Solano, 1989). Sang et al. (1997) reported that at temperatures above 1000°C, the reaction in the pores become prohibited and limiting the reactions to the outer layer of the carbon particle. Guo and Lua (2001) mentioned that the characteristics of activated carbon depend on the physical and chemical properties of the raw materials as well as activation method used.

The physical properties of carbon are important in determining which form is best suited for specific application. For instance, abrasion resistance or hardness of activated carbon will be important if the carbon is to be used in application where frequent back-washing is required. Density characteristics can also be a major consideration for specific applications. Densities vary with the raw materials. A fewer quantity of carbon with a lower density will fit into a given container as compared to a carbon with high density. This is significant because while a container may require less carbon weight of low density to make a fill, its contaminant removal performance may be severely reduced as compared to a high-density carbon. Iodine value is an important characteristic of activated carbon. The iodine value gives a measure of the micro pore volume of carbon and it approximates the total internal surface of the carbon. The ash level reflects the purity of the carbon. The ash content can be important in water filtration applications. Activated carbon with high phosphate ash can brings about cloudy water during use because the ash could combine with metal ions to form magnesium or calcium precipitates (Ecologix systems, 2008). The performance of activated carbon is indicated by its adsorptive characteristics, which is derived from the specific surface area, porosity of the carbon (Akinyemi and Taiwo, 2004). Micro pores are formed in the interlayer spacing with widths in the range of 0.34 – 0.8 nm. It is the micro pores in activated carbon which have the greatest influence upon gas adsorption while macro pores and mesopores are important in transport of fluids to and from the micro pores (Hu and Srinivasan, 2001). The knowledge of the physical and activity characteristics of activated carbons will help in determining where each can be better applied. The goal of the study was to produce and characterize activated carbon from cattle bones; coconut-shell and wood.

MATERIALS AND METHODS

Source/Sample preparation

Coconut shells, waste pieces of wood Prosopis *africana* were respectively collected from coconut and wood sellers in Idah, Kogi state, Nigeria while cattle bones were collected from an abattoir opposite Federal Polytechnic, Idah, kogi state, Nigeria. Each material was crushed manually to small chips of approximately equal sizes (1.0 to 0.8 cm). A commercial activated carbon (Chemivivon, Sussex, UK) were purchased from Charlitech Laboratories, Makurdi, Nigeria.

Carbonization and activation

Carbonization was as described by Gimba et al. (2001). The carbonization time was varied between 0.5 and 3 h at 600°C to determine optimum carbonization for each raw material by fixed carbon measurement without ash formation. Acid activation was carried out as described by Tao and Xiaoqin (2008). The final carbons were weighed after drying to determine the final product yield:

% Final product yield = weight of carbon from furnace after drying × 100 (1)

Analytical methods

Iodine values were determined as described by Okuo et al. *(*2008). This involves the titration of 0.1 M sodium thiosulphate solution against 60.0 ml of activated carbon sample free aliquot solution (prepared by centrifuging 0.5 g of activated carbon sample in 75.0 ml of 0.086 M iodine solution using 5 ml of freshly prepared starch indicator. The iodine value (IV) was calculated using the formula:

IV (mg/g) = Y- X÷Y × V÷W × M (126.9) (2)

Where Y = volume of thiosulphate for blank, X = Titer value, V = volume iodine solution used, W= Weight of sample. M = molarity of iodine solution.

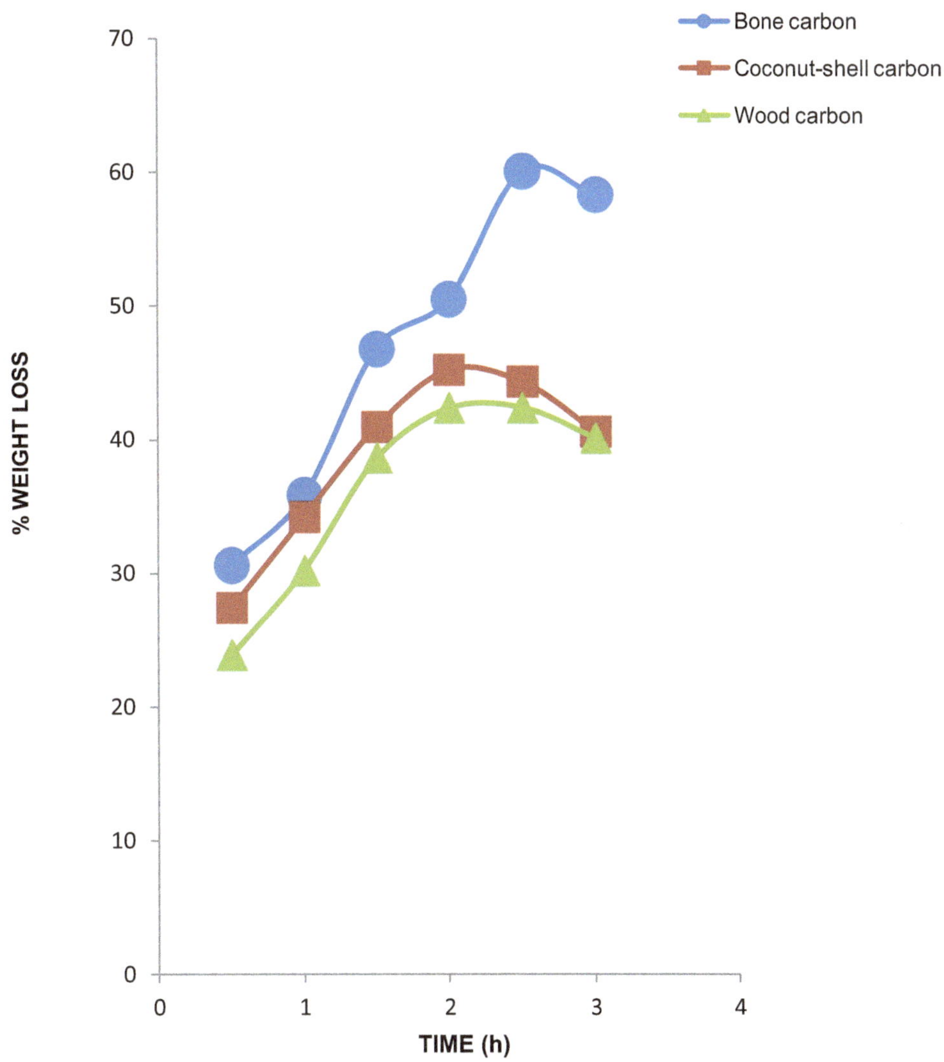

Figure 1. Effect of carbonization time on carbon weight loss.

Carbon hardness was determined according to ASTM (2002) method. This involves measuring 2 g of carbon sample in a standard sieve no 40. The sieve was shaken for 1h using mechanical shaker (Millano Italia-type) at 160 strokes per minutes with four steel balls (6.0 mm diameter each) inside the sieve. The amount of activated carbon retained in the sieve after 1h was expressed in percent. Bulk density was determined according to the method of Okaka and Potter (1976). Five-gram quantity of each activated carbon was weighed in a graduated glass cylinder and tap until a constant volume is noted. The density was calculated in g/cm^3. Moisture content was determined by the air-oven method as reported by AOAC (1994). Ash content was determined according to AOAC (1994).

Surface area and porosity characterization

This was performed by the adsorption of N_2 at 77K using micrometric surface area analyzer (ASAP2010) Micrometric Inc. USA. Before the adsorption of N_2, the samples were subjected to degasing for 2 h at final pressure of 133.32 × 10^{-4} pa. The total pore volume (V_T) was estimated from the volume of N_2 (as liquid) adsorbed at relative pressure of (p/p°) of 0.95. The microspore volumes were calculated from adsorption isotherms and mesopore was calculated as the difference between macro pore volume and micro pore volume. Barrett, Joyner, Halenda (BJH) method was employed for the evaluation of the pore size distribution (Results not shown).

Statistical analysis

Experimental results were analyzed statistically. Means were compared using test of significant difference (Steel and Torrie, 1980).

RESULTS

Carbonation/activation effects

Figure 1 shows the results of percentage weight loss for activated raw materials during carbonization at 600°C in

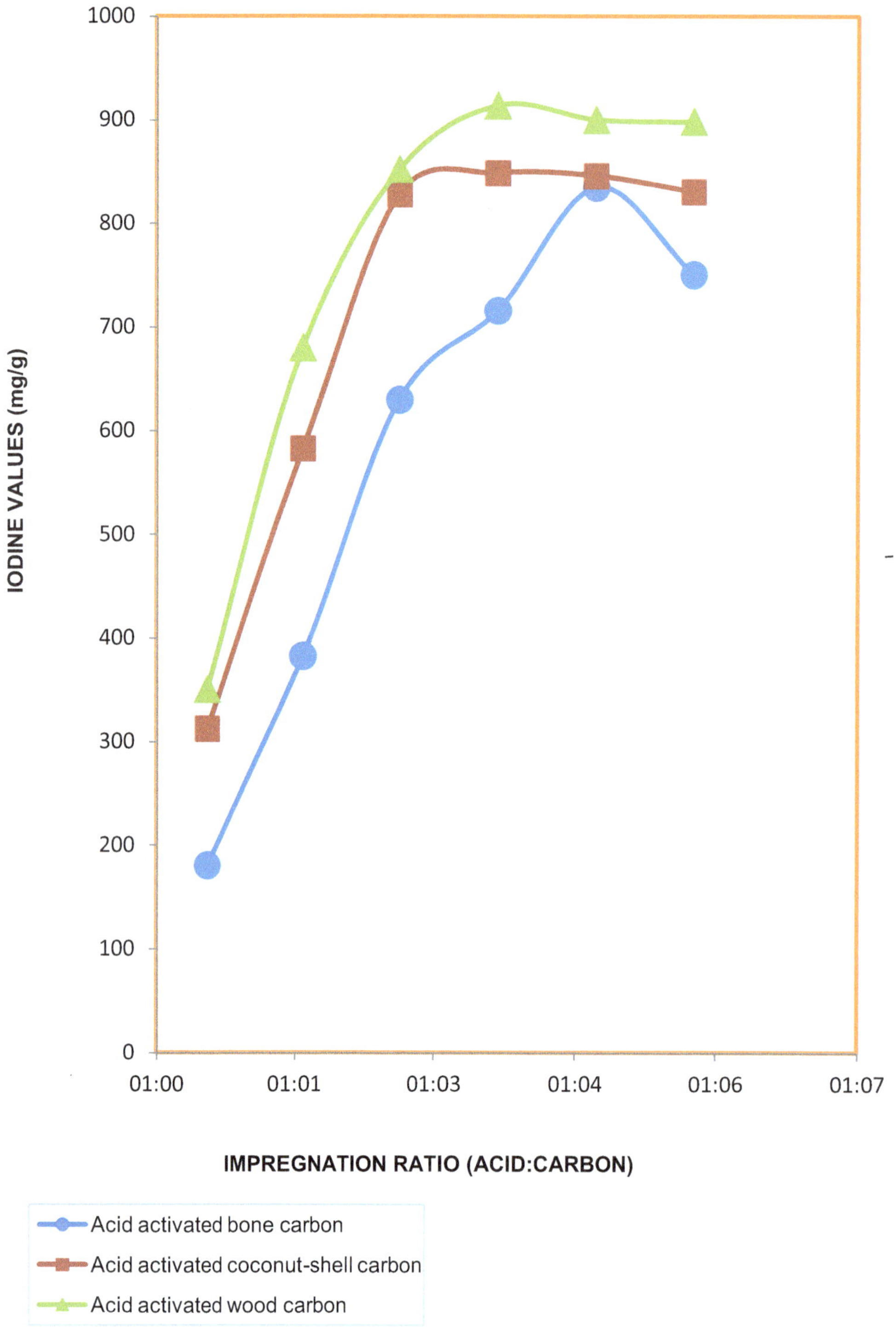

Figure 2. Effect impregnation ratio on iodine values of acid activated carbon.

a furnace. Increase in carbonization time resulted increased in weight loss for all the carbon samples. The increased in weight loss reached a maximum value at a carbonization time of 2 h for coconut-shell and 2.5 h for wood and for bone. Figure 2 shows the results of iodine values for activated carbons using phosphoric acid activation. Increase in impregnation ratio resulted increased in iodine values for all the carbon samples. The increased in adsorptive capacity reached a maximum value at an impregnation ratio of 1:3 for coconut-shell and wood and 1:4 for bone carbon. Figure 3 shows the variation of iodine values with time using heat (thermal)

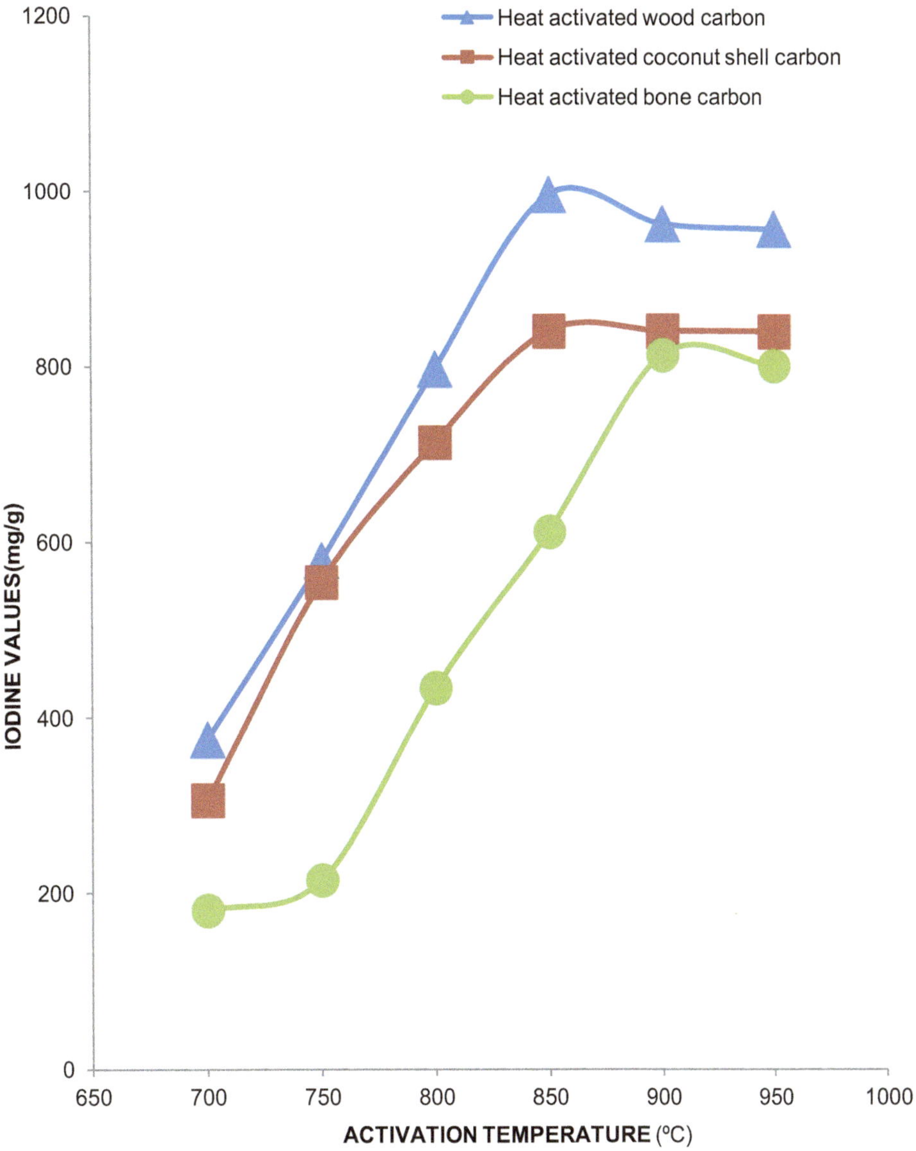

Figure 3. Effect of activation temperature on iodine values of heat activated carbon.

activation. Increase in carbonization temperature resulted in increased in iodine values for all the carbon samples. The increased in iodine values reached a maximum value at activation temperature of 850°C for coconut-shell and wood carbon and 900°C for bone. The results of final activation yield and volatile constituents of carbon raw materials after activation processes are shown in Table 1. The percent carbon yield for heat activated bone, coconut-shell and wood were 40.10, 39.30, and 36.30% while for acid activation percent fixed carbon were 46.21, 40.60 and 40.4, respectively. The volatile constituents were 60.2, 58.9, 63.1, 53.8, 59.2 and 58.6% respectively. The fixed carbon content of bone and wood were significantly affected by method of activation. The volatile constituents were not significant by carbon source and mode of activation.

Iodine values and ash

The iodine values and ash content of acid and heat activated carbon are shown in Table 2. The iodine values were significantly ($P \leq 0.05$) affected heat and acid activation process. The non-activated carbons had iodine values (66.16 to 222.69 mg/g) which is significantly lower than (828.39 to 995.38 mg/g) and (834.94 to 914.17 mg/g) for heat and acid activated carbons respectively. The heat activated wood carbon had significantly higher iodine value than the acid activated wood carbon. There is no significant difference between iodine values of bone

Table 1. Activated carbon yield and volatile losses.

Material	Treatment	Constituent (%)	
		Carbon	Volatile
Bone	Acid	46.21a	53.76a
	Heat	41.10b	60.20a
Coconut shell	Acid	40.62b	59.20a
	Heat	39.30b	58.91as
Wood	Acid	40.40b	58.60a
	Heat	36.30c	63.10a

Values are means of triplicate determinations. Means within a column with the same letters are not significantly ($P \leq 0.05$) different. Acid : Carbon ratio was varied between 1:1 and 1:6 to determine the optimum activation for each of the raw material by iodine value measurement. Heat : Heat activation temperature was varied between 700 to 950°C to determine optimum activation by iodine value measurement.

Table 2. Physico-chemical characteristics of carbon samples.

Material	Treatment	Physicochemical characteristics				
		IV	Ash (%)	Hardness (%)	BD	Moisture (%)
Bone	Acid	834.94a	82.4b	89.5b	0 .79a	5.0a
	Heat	828.39a	80.6c	93.5a	0.78a	4.7a
	Non	66.15b	85.9a	90.0b	0.78a	4.6b
Coconut shell	Acid	848.69a	2.8b	81.0c	0.68a	6.4a
	Heat	840.18a	2.4b	96.5a	0.66a	5.1
	Non	125.67b	3.5a	90.5b	0.60a	4.4b
Wood	Acid	914.17b	1.8a	62.8b	0.47a	9.4a
	Heat	995.38a	1.1b	88.5a	0.48a	8.4b
	Non	222.69c	2.1a	87.5a	0.40a	7.3c

Values are means of triplicate determinations. Means within a column with the same letters are not sig. ($P \leq 0.05$) different. IV = Iodine values, BD = bulk density, ash, hardness and moisture were determined using standard methods.

and coconut-shell carbons in both activation processes. The total ashes for bone, coconut-shell and wood carbon were 89.9, 3.5 and 2.0%, respectively. The total ash for acid activated carbons was higher than those of heat activated carbons. For both activation methods, the ash for the bone carbon was higher than those of coconut-shell and bone wood carbon.

Hardness, density and moisture

Hardness, carbon density and moisture values are presented in Table 2. The hardness of the non-activated carbons was 90.0, 90.5 and 87.5% for bone, coconut-shell and wood respectively. The hardness of the carbons increased during heat activation and decreased during acid activation. The increase was significant for bone and coconut-shell carbon .The hardness of heat activated carbons decrease in the order: coconut shell < bone < wood and the acid activated carbon in the order: bone < coconut-shell < wood. The density of the non-activated was 0.78 g/cm^3, 0.60 g/cm^3 and 0.40 g/cm^3, respectively for bone, coconut-shell and wood carbon. Generally, activations increased densities of the carbons and there was no significant ($p \geq 0.05$) difference in densities.

The moisture content was 4.6, 4.4 and 7.3% for bone, coconut-shell and wood, respectively. The moisture content increased during activation processes. There was no significant difference between bone carbon activated with acid and bone carbon activated with heat. Acid and heat activated wood carbon had significantly higher moisture than coconut-shell and bone carbon.

Activated carbon porosity

The results of BET surface area, macropore, mesopore and micropore volumes of the produced activated carbon

Table 3. Porosity analysys of produced activated carbons and a commercial activated carbon.

Samples	S_{BET} (m^2/g)	Porosity parameter		
		V_P (cm^3/g)	V_{micro} (cm^3/g)	V_{meso} (cm^3/g)
ACS	912c	0.711a	0.746a	0.025a
HCS	895d	0.768a	0.728a	0.040a
AWC	1018a	0.679a	0.663a	0.016a
HWC	1015b	0.779a	0.771a	0.008a
ACB	782e	0.662b	0.641b	0.021a
HCB	718f	0 .656b	0.652b	0.004b
CAC	1022a	0.782a	0.770a	0.012a
LSD (0.05)	5.175	0.115	0.113	0.083

S_{BET}= BET surface area, V_P = Macropore ,V_{micro}= Micro pore ,V_{meso}= Mesopore volume.

and the commercial activated carbon are presented in Table 3.The surface area for the carbons ranged from 718 to 1018 m^2/g. The activated carbons differed significantly in surface area. There were no significant difference in surface area of CAC and AWC. The macropore and mesopore volume were statistically the same for wood and coconut shell carbon. The activated carbon from bone presented significantly lower macropore and micropore volumes. The mesopore volume of heat activated carbon was significantly lower than mesopore volume for all the carbons.

DISCUSSION

Characterization of activated carbons

The results of percentage weight loss for activated raw materials during carbonization presented in Figure 1 shows that increase in carbonization time resulted in increased weight loss for all the carbon samples. The increased in weight loss is as a result of increase volatilization (loss of volatile matter) of heat labile substances which reached a maximum value at a carbonization time of 2 h for coconut-shell and 2½ h for wood and bone. The reduction in weight loss after the maximum values is an indication of the beginning of activation of the already formed carbon. Similar observations were reported (Oyo and Igbokwe, 2001), Sang et al. *(*1997) for other materials. According to these authors, during carbonization, a primary pore structure is developed followed by development of porosity and reduction in weight losses.

The results of iodine values for activated carbons using phosphoric acid activation at 600°C (Figure 2) indicated that increase in impregnation ratio resulted in increase in iodine values for all the carbon samples. The increase in iodine values is a result of development of porosity in the carbons which reached a maximum value at material to acid ratio of 1:3 for coconut-shell and wood and 1:4 for bone. The dropping of adsorptive capacity of the carbons after the maximum values is an indication of the beginning of structural deformation of the already formed carbon. In a related study, Thanittha (2005) reported an optimum condition of activation of bone: zinc chloride (1:3 w/w) equivalent to 120% of impregnation ratio at 800°C with 3 h of holding time which gave a yield of 44.8% with iodine number of 901 mg/g.

Heat activation (Figure 3) shows that increase in carbonization temperature resulted in increase in iodine values for all the carbon samples. The increase in iodine values is due to development of pores and increased porosity in the carbons, which reached a maximum value at activation temperature of 850°C for coconut-shell and wood carbon and 900°C for bone. The reduction in adsorptive capacity of the carbons after the maximum values is an indication of the beginning of structural deformation of the already formed activated carbons. This observation is in conformity with those of Ajayi and Olawale (2009) who observed that in the production of activated carbon from *Canariun schweinfurthi* Nutshell using chemical process, when the temperature increased from 600 to 900°C, the iodine values increased. It is evident that the activation temperature had a significant influence on the iodine number. The iodine number was observed to increase at higher activation temperatures. However, at activation temperature around 950°C and above the iodine number practically decreases. Thus it is deduced that the optimum activation temperature is around 850 - 900°C for H_3PO_4 activation which is also considerable in terms energy saving. The optimum condition were similar to those suitable for corn cob (Tsai et al*.,* 2001), various nutshells, chickpea husk and cork waste (Jun'ichi et al., 2002a, b; Carvalho et al., 2004).

The activated carbon yield (Table 1) for heat activated bone, coconut-shell and wood were 40.10, 39.30, and 36.30%, respectively while the yield for acid activation were 45.21, 40.60 and 40.40% respectively. These results are in agreement with the findings of other workers (Palnut, 2000; Savova et al., 2001; Aygum et al*.,* 2003) who reported carbon yield of 39.99 to 55.44% for most activated carbon raw materials. It can be seen that the local raw materials studied in this present work have high carbon yield. Product yield is an important measure of the feasibility for preparing activated carbon from a give precursor. According to Guo and Rock straw (2006), Guo and Lua (2011), significant product yield differences can be observed depending on the origin of carbon material and activation process.

The results shown in Table 2 indicated that iodine values were significantly affected by activation for all the carbons. The results indicates that acid is the best activating agent for both bone and coconut-shell carbon while heat is the best activating agent for wood carbon. In all cases, the iodine values of the carbons decrease in the order: wood < coconut-shell < bone. The iodine values reported in this study is in conformity with iodine values 600 to 1100 mg/g for water treatment carbons reported by Thanittha (2005). The observed analytical differences in iodine values could be attributed to differing adsorptive characteristics, which is derived from the specific surface area, pore size and pore volume of the carbon (Akinyemi and Taiwo, 2004). When appropriate activation conditions are provided, the oxidizing action of the oxidizing agent does not consists of indiscriminate removal of successive layers of atoms from the carbon matrix. Instead, the oxidation selectively erodes the surface in order to increase the surface area, develop greater porosity and leave the remaining atoms in the carbon matrix arranged in configurations that have specific affinity for particular adsorbents (Hassler, 1988).

The ash results show that activation reduces the ash levels of carbon. The reduction in ash could be attributed to volatilization of some inorganic constituents at higher temperature during activation that could lead to a reduction in ash and also acid could react with the mineral constituents of carbon and thereby reducing the ash during activation. Acid treatment was reported to have been used to reduce high ash levels of sub-bituminous and lignite carbon (Calgon carbon corporation, 2008). Acid activated carbons had higher total ash than the heat activated carbons. These could be attributed to acid residuals, which may remain with the carbon as part of the ash. This is in consonance with the results reported for sub-bituminous and lignite carbons treated with acid (Calgon Corporation, 2008). In both activation methods, the bone carbon had higher ash than the wood and coconut-shell. The higher ash content of bone carbon could be attributed to high inorganic component of bone charcoal (Ebonex Corporation, 2006).

The heat activated carbons shows higher hardness than the acid activated carbons (Table 2). This suggests that higher temperatures treatment had significant effect on hardness of carbon materials. The hardness ranged from 62.00 to 96.5%, indicating high abrasion resistanceand could be applied where frequent back-washing is required. The result of hardness for coconut-shell activated carbon is in consonance with the hardness of 95.0 to 99.0 reported for granular coconut - shell activated carbon (Ecologix systems, 2008). The densities of the carbons increase in the order: bone; coconut shell; wood. The difference in densities of the activated carbons can be attributed to differences in the carbon origin. The densities observed in this work are in close agreement with values of 0.35 to 1.2 g/m^3 reported for other agricultural raw materials by Oyo and Igbokwe (2001), Bansal et al. (1988).

The activated carbons exhibited different moisture uptake abilities following activations. Moisture uptake increased in the order: bone; coconut-shell; wood irrespective of activation method. This observation is in agreement with studies by Bansal et al. (1988) who reported moisture contents of 10% for acid activated wood and 5.0% for granulated acid activated coconut-shell carbon. The porosity analysis presented in Table 3 shows that the surface area for the activated carbons fall within the range of 600 to 1200 m^2/g reported for most widely used commercial activated carbons (Ng et al., 2002). The high surface area of the carbons is important since it limits the amount of adsorb ate which can be adsorbed.

The surface areas were higher for acid activated carbons and this is expected because chemical activation normally develops more porosity and gives high surface area when compared with thermal activation (Ajayi and Olawale, 2009). The increase BET surface areas and microspore (0.746 to 0.641 cm^{3}g^{-1}) structure of acid activated carbons was brought about by oxidation of the carbons by phosphoric acid. The surface areas (718 to 1018 m^2/g) reported for this study was higher than surface areas (305 to 539 m^2/g) reported for dew and enzyme retted flax shives activated with phosphoric acid (Marshall et al., 2007). The macrospore volume of the carbons ranged from 0.779 for HWC to 0.656 cm^{3}g^{-1} for HCB and the mesopore volume ranged from 0. 008 for HWC to 0.040 cm^{3}g^{-1} for HCS. The micropore and mesopore volumes of the carbons are comparable to micropore of 0.058 cm^{3}g^{-1} and mesopore 0.453 cm^{3}g^{-1} reported for activated carbon from brewer's spent grain (Mussatto, 2009).

Conclusion

This study indicates that efficient activated carbons can be obtained from bone, wood and coconut shell by controlled activation with phosphoric acid or heat for a number of industrial and residential applications.

REFERENCES

Ajayi OA, Olawale AS (2009). A comparative study of thermal and chemical activation of *Canarium Schweinfurthii* Nutshell. J. Appl. sci. Res., 5(12): 2148-2152.

Akinyemi OP, Taiwo EA (2004). Production of activated carbon from agricultural wastes. Proceedings of Nigerian Society of Chemical Engineers held in Warri- Nigeria.

AOAC (1994). Official Methods of Analysis. 14th edn. Association of Official Analytical Chemists. Washington DC.

ASTM (2002). Activated Carbon Standards. American Society for Testing and Materials. Online. Fritz publication. Http: // www.fritz.com.

Aygum A, Yenisoy KS, Duman I (2003). Production of granular activated carbon from fruit stones and nutshells and evaluation of their physical, chemical and adsorption properties. Microporous and Mesoporous Mater., 66: 189-195.

Bansal C, Donnet JB, Stoeckli F (1988). Active carbon, New York: Marcel Dekker. pp. 23-45.

Calgon Carbon Corporation (2008). Activated Carbon. What is it, how does it work. Online information Bulletin. www.calgon carbon .com.

Carvalho AP, Gomes M, Mestre AS, Pires J, Brotas de (2004). Activated carbons from cork waste by chemical activation with K_2CO_3: Application to adsorption of natural gas component. Carbon. 42: 667-691.

Diets VR (1990). Bibliography of solid adsorption and United States cane sugar refineries and bone char manufactures. National Bureau Standards. Washington DC.

Dileck C, Oznuh AY (2008).Production and characterization of activated carbon from bituminous coal through chemical activation. Afr. J. Biotechnol., 7(2): 3703 – 3710.

Ecologix Environmental Systems (2008). Coconut-shell and Wood activated Carbon. Online. Http:// www .ecologixsystem.Com.

Elliot C, Colby T, Hicks H (1989).Charcoal lung bronchiolitis Obliterants after aspiration of activated charcoal. Chect, 96(3): 672-674

Gimba CC, Olayemi JY, Siyaka ZM (2001). Comparative studies on the adsorption of dyes using activated carbon derived from *Arachis hypogee and Cocos nucifera* shells. Proceedings of Chemical Soc. of Nigeria. Book of abstracts, p. 99.

Guo J, Lua AC (2001). Characterization of adsorbents prepared from oil palm shell by thermal activation for removal of gaseous pollutants. Material Sci. 55: 334-339.

Guo J, Rockstraw DA (2006). Physicochemical properties of carbons prepared from pea nut by phosphoric acid activation. Bio resource Technol., 98: 1513-1528

Haimour NM, Emeish N (2006). Utilization of date stones for production of activated carbon using phosphoric acid. Waste management. 26(6): 651-660

Hassler J (1988). Purification with activated Carbon. 1st edn. Chemical Publishing. pp 20-26.

Hu Z, Srinivasan MP (2001).Mesoporous high surface area activated carbons. Microporous Mesoporous Mater., 34: 267–275.

Inamullah B, Khadija Q, Kazi RA, Abdul KA (2008).Preparation and characterisation of chemically activated almond shells by optimization of adsorption parameters for removal of chromium vi from aqueous solutions. World Academy of Science, Eng. Technol., 34: 199-204

Jun'ichi H, Toshihide H, Isao T, Katsuhiko M, Farid NA (2002a). Preparing activated carbon from various nutshells by chemical activation with K_2CO_3. Carbon; 40: 2381-2386.

Jun'ichi H, Toshihide H, Katsuhiko M, Vincent GG (2002b). Activated carbon from chickpea husk by chemical activation with K_2CO_3: preparation and characterization. Microporous Mesoporous Mater., 55: 63-68.

Marshall WE, Wartelle LH, Akin DE (2007). Flax shive activated carbon for metal remediation. Bioresource, 2(1):82 - 90

Mendez MOA, Lisboa ACL, Coutinhoand AR, and Otani C. (2006). Activated Petroleum Coke for natural gas storage. J. Braz. Chem. Soc., 17(6): 1144-1150.

Mussatto SI, Fernandes M, Rocha GJM, Orfao JJM, Teixeira JA, Robarto K (2009). Production and characterization of activated carbon from brewer's spent yeast lignin. Bioresour. Technol., 101(2010): 2450-2457.

Ng C, Marshal R, Rao M, Rishipal RB, Jack NL, Ralph JA (2002). Granulated activated carbons from agricultural by- products: Process description and estimated cost of production. LSU Ag center. Bulletin Number 881.

Okaka JC, Potter NN (1976). Functional properties and storage properties of cowpea- wheat flour blends in bread making. Food Sci. J., 42: 822-834.

Okuo JM, Akapaja EO, Eguono WO (2008). Kinetics of cd^{2+} and Se^{4+} ions removal from solution by the fruiting bodies of white rof fungi (plenrotus tuberregium) J. Chem. Soc. Niger., 33(1): 56-64.

Oyo KB, Igbokwe PK (2001). Production of activated carbon from coconut-shell. J. Chem. Soc. Nig., 26(1): 91-94.

Palnut S (2000). Production of activated carbon from coconut shells for adsorption of chromium(vi). MSc. Thesis in chemical engineering. Bangkok. King Mong university of Technology. Thombori. pp. 5-80

Rodriquez R, Salano P (1989). Preparation and characterization of activated carbon from agricultural wastes. Carbon 32(3):1422-1425.

Sang CK, In KH, Kyung AP (1997). Preparation and performance of the Briquette type activated carbon based bituminous coal. J. Ind. Eng. Chem., 3(30): 218-222.

Savova D, Apak E, Ekinci E, Yardim F, Petrov N, Budinova T, Razvigorova M Minkovo V (2001). Biomass conversion to carbon adsorbents and gas. Biomass Bioenerg., 21: 133-142

Srinvaskannan C, Mohammad ZA (2004). Production of activated carbon from rubber wood sawdust. Biomass Bioenerg. 27: 89-96.

Steel RGD, Torrie JW (1980).Principles and procedures of statistics. biometric approach. 2nd edn. Auckland New Zealand. McGraw Hill.

Tao XU, xiaoqin LIU (2008). Peanut shell activated carbon: Characterization, surface modification and Adsorption of Pb^{2+} from aqueous Solution. Chinese J. Chem. Eng., 19(3): 401- 406.

Thanittha K (2005). Upgrading cowpea ground waste as activated carbon using chemical activation with zinc chloride and potassium carbonate. MSc. Thesis submitted to Mahidol University, Thailand. pp. 1-78.

Tsai WT, Chang CY, Wang SY, Chang CF, Chien SF, Sun HF. (2001).Cleaner production of carbon adsorbents by utilizing agricultural waste corn cob. Resour. Conserv. Recycling, 32: 43-53.

8

Extraction of mercury (II) from sample water by Schiff base-chitosan grafted multi-walled carbon nanotubes

Ali Moghimi

Department of Chemistry, Varamin (Pishva) Branch Islamic Azad University, Varamin, Iran.
E-mail: alimoghimi@iauvaramin.ac.ir.

A functionalized Schiff base-chitosan-grafted multi-walled carbon nanotubes (S-CS-MWCNTs) material has been studied on pre-concentration and determination of trace mercury. In this process, the effects of solution pH solution, elution conditions on pre-concentration of trace mercury were studied and the effect of interfering ions was also investigated. A novel S-CS-MWCNTs solid-phase extraction adsorbent was synthesized by covalently grafting a Schiff base-chitosan (S-CS) onto the surfaces of oxidized MWCNTs. The stability of chemically modified S-CS-MWCNTs especially in concentrated hydrochloric acid was then used as a recycling and pre-concentration reagent for further uses of S-CS-MWCNTs. The application of S-CS-MWCNTs for sorption of a series of metal ions was performed by using different controlling factors such as the pH of metal ion solution and the equilibration shaking time by the static technique. Hg(II) was found to exhibit the highest affinity towards extraction by these S-CS-MWCNTs phases. The pronounced selectivity was also confirmed from the determined distribution coefficient or dissociation constant (K_d) of all the metal ions, showing the highest value reported for Hg(II) to occur by S-CS-MWCNTs. The potential applications of S-CS-MWCNTs for selective extraction of Hg(II) to occur from aqueous solution were successfully accomplished as well as pre-concentration of low concentration of Hg(II) (60 pg ml^{-1}) from natural tap water with a pre-concentration factor of 100 for Hg(II) off-line analysis by cold vapor atomic absorption analysis.

Key words: Pre-concentration, Hg(II), Schiff base-chitosan-grafted multiwalled carbon nanotubes (S-CS-MWCNTs).

INTRODUCTION

Mercury (Hg) has been considered as a human health hazard because it may cause kidney toxicity, neurological damage, paralysis, chromosome breakage and birth defects (Manahan, 1994). Mercury itself and compounds are included in all lists of priority pollutants and different regulations and guidelines have been developed limiting their levels in water and sediments (Hayes, 1997). But it plays an important role in science and technology and inevitably exists in the environment (IARC, 1993; Jin et al., 2003; Seiler et al., 1998). The detection of Hg(II) has long held the attention of the analytical community and, as such, a large number of protocols have arisen (Smith 1963; Kantipuly et al., 1990; Greenway and Townshend, 1993; Manzoori et al., 1998). On the other hand, the toxicity level of mercury is becoming lower and lower, the direct determination of mercury at sub-microgram per liter level is suffered from the matrix interferences. It is evident that the use of separation and pre-concentration procedures is still often necessary before the determination step, despite recent advances in analytical instrumentation (Pereira and Arruda, 2003). Recently, the solid-phase extraction (SPE) is being utilized for pre-concentration of heavy metals due to its flexibility, economical and environmental-friendly, absence of emulsion, speed and simplicity, sampling in the field, safety and ease of automation (Thurman and Mills, 1998; Moghimi, 2010; Tabarzadi et al., 2010; Moghimi et al., 2011). But the selectivity associated with SPE using solid support such as active carbon (Moghimi et al., 2011), ion-exchange resins (Swain et al., 2000), chelating fibers (Chang et al., 2001; Yang et al., 2004) and chelating resins (Tewari and Singh, 2000; Guo et al., 2004a, b; Zhou et al., 1983) and so forth is still to be improved for specific element.

Table 1. Operating conditions for ICP-MS.

Nebulizer	Babington
Spray chamber	Quartz double pass
RF power (kW)	1.5
Carrier gas flow rate (L min^{-1})	0.9
Makeup gas flow rate (L min^{-1})	0.15
Sample uptake rate (mL min^{-1})	0.3
Detection mode	Auto
Integration time (s)	0.1
Analytical masses	Hg

Now a days, metal ion imprinted polymers (MIIP) have been investigated as highly selective sorbents for SPE in order to concentrate and clean up samples prior to analysis.

One potential application that has recently attracted widespread interest is their use for clean up and enrichment of analytes present at low concentrations in complex matrices (Bae et al., 1998; Daniel et al., 2003; Rao et al., 2004; Liu et al., 2004). This element was selected for evaluation of the procedure because Hg(II) is one of the principal heavy metals of analytical interest due to its extreme toxicity even at relatively low concentrations (Moghimi, 2010; Bingye et al., 2012). This work investigates the applications of Schiff base-chitosan-grafted multi-walled carbon nanotubes (S-CS-MWCNTs) for selective extrac-tion and solid phase pre-concentration of Hg(II) from aqueous and natural water samples.

EXPERIMENTAL

Reagents and materials

MWCNTs PET (filaments: 130 and dtex 170) made in Textile Engineering Department, Amir Kabir University of technology, Tehran, Iran. Analytical grade nitrate salts of Acrylic acid, acryl amid, benzoyl peroxide, acetone, Hg, Mn, Fe and Hg(II), Li, Na, K, Mg, Ca, Sr, Ba, Zn, Cd, Pb(II), Ni(II), Co(II) and Cu(II)of reagent grade were of the highest purity. Ultra pure organic solvents were obtained from E. Merck, Darmstat-Germany, high Purity deionized water was used throughout the experiments and 3-chloro propyl trimethoxysilane was received from Aldrich Chemical, USA. Organic solvents were dried according to conventional methods. For all solutions double distilled water was used and the buffer solutions were prepared from 1.0 M sodium acetate to which different volumes of 1.0M HCl; HNO_3 were mixed and the pH-value of the resulting solution was adjusted with the use of a pH- meter. CS (Sangon Biotech Co., Ltd. Shanghai, China) and 4- methyl-benzaldehyde (Aladdin Reagent Co., Ltd., Shanghai, China) were used in this work. Oxidized MWCNTs (MWCNTs-COOH) were provided by the Chengdu Organic Chemicals Co., Ltd. (Chengdu, China). The lengths of the MWCNTs-COOH were in the range of 10 to 30 nm and their outer diameters were in the range of 20 to 30 nm (Zargaran et al., 2008). The optimum operating conditions and measurement parameters for ICP-MS are listed in Table 1.

Synthetic procedures

The synthesis of the S-CS-MWCNTs is illustrated in Figure 1.

Synthesis of S-CS

Purified CS was obtained according to the method reported in the literature (Wu et al., 2007). CS (1000 mg) was dissolved in 80.0 mL of Figure 1.

0.2% (w/w) aqueous acetic acid solution was allowed to stand overnight, and then 4.0 ml of 4-methylbenzaldehyde in methanol (2:1, v/v) were slowly added. After stirring for 24 h at room temperature, the resulting mixture was washed with ethanol several times, filtered, and dried in a vacuum oven at 40°C for 8 h. The obtained yellow powder was S-CS.

Synthesis of acyl-chloride-modified MWCNTs (MWCNTs-COCl)

MWCNTs-COOH (80 mg) were dispersed in 40.0 ml of thionyl chloride. The suspension was stirred for 24 h at 70°C. The residue was separated by filtration, washed with N,N-dimethylformamide (DMF), and dried under a vacuum at 80°C for 8 h to obtain MWCNTs-COCl.

Synthesis of S-CS -grafted MWCNTs-COOH

MWCNTs-COCl (50 mg) and S-CS (400 mg) were added in 30.0 mL of DMF. The mixture was stirred at 100°C for 48 h under a nitrogen atmosphere and filtered through a 0.22 µm microporous membrane. To remove the physically adsorbed compounds, the residue was washed with 0.2% acetic acid, ethanol, and DDW. The S-CS-MWCNTs adsorbent was obtained after drying overnight in an oven at 40°C.

Column preparation

S-CS-MWCNTs (40 mg) were packed into an SPE mini-column (6.0 cm × 9 mm i.d., polypropylene). A polypropylene frit was placed at each end of the column to prevent loss of the adsorbent. Before use, 0.5 mol L^{-1} HNO_3 and DDW were passed through the column to clean it.

Apparatus

The pH measurements were conducted by an ATC pH meter (EDT instruments, GP 353) calibrated against two standard buffer solutions of pH 4.0 and 9.2. Infrared spectra of S-CS-MWCNTs were carried out from KBr pellet by a Perkin-Elmer 1430 ratio recording spectrophotometer. Atomic absorption analysis of all the metal ions except Hg(II) were performed with a Perkin-Elmer 2380 cold vapor atomic absorption spectrometer. Hg(II) determinations were performed by a Varian Spect AA-10. Raman spectro-photometer analysis was performed with a Perkin-Elmer.

Adsorption and removed processes of metal ions

The adsorption ions onto S-CS-MWCNTs for Ni (II), Cu (II), and Pb(II) ions were investigated using the batch method. Experiments were carried out in an Erlenmeyer flask at the desired pH and 25°C temperature. The flasks were agitated on a shaker for 2 h. The amount of adsorbed metal was determined by the difference between the initial metal ion concentration and the final one after

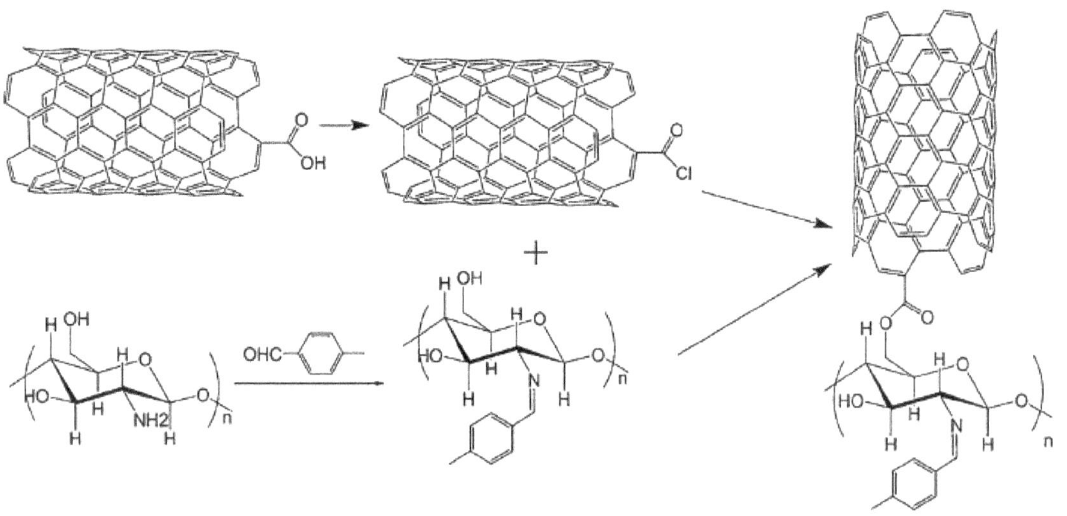

Figure 1. Schematic of synthesis and structure of S-CS-MWCNTs.

equilibrium (IARC, 1993). The concentration of ions was determined with a ICP-MS. The efficiency of metal ions recovery was estimated by the sorption yield (R%) and the q (mg/g) was calculated as:

$$R= (C_0- C_t)/C_0 \times 100 \quad (1)$$

$$q = (C_0 - C_t)/G \times V \quad (2)$$

where C_0 is the initial metal ion concentration (mg/l), C_t is the ion concentration after the adsorption period, V is the volume of solution L and G is the dry mass of the S-CS-MWCNTs sample (in gram). The metal ions adsorbed on the S-CS-MWCNTs were then removed by placing 0.1 g of metal loaded MWCNTs in 10 ml of 1M HNO_3 solution for 30 min (Wu et al., 2007).

Stability studies

The stability of S-CS-MWCNTs phases in different buffer solutions (pH 1 to 6) and concentrated hydrochloric and nitric acids were studied by batch equilibration. In this procedure, 500 mg of the phase was mixed with 50 ml of the selected solution in 100 ml measuring flask and automatically shaken for 5 h. The mixture was filtered, washed with 500 ml water and dried in an oven at 80°C. Around 100 mg of the treated phase was added to 1.0 ml of 0.1 M Hg(II) and 9.0 ml of 0.1 M sodium acetate and the mixture was shaken for 30 min by an automatic shaker. The percentage of hydrolysis of Poly (ethylene terephthalate) from the surface of S-CS-MWCNTs phases in different acidic solutions was calculated from the determined μmol g^{-1} value of each treated phase.

Sorption studies

Determination of metal capacity values (μmol g^{-1})

The determination of metal capacity of 13 metal ions, viz. Ba(II) Ca(II), Co(II), Zn(II), Cu(II), Fe(III), Hg(II), Mg(II), Mn(II), Ni(II), Pb(II) and Hg(II) as a function of pH was studied by the static technique. Then 100 mg of the dry phase was added to a mixture of 1.0 ml of 0.1 M metal ion and 9.0 ml of the buffer solution (pH 1 to 10 and 0.1 M sodium acetate) in 50 ml measuring flask. The mixture was then automatically shaken for 30 min, filtered, washed with 50 ml water and the unbound metal ion was subjected to complexometric titration using the proper buffer and indicator solutions and/or atomic absorption analysis. The effect of shaking time on the percentage extraction of metal ions was also studied for only Hg(II) by the static technique. In this, 100 mg of the S-CS-MWCNTs was added to 1.0 ml of 0.1M Hg(II) and 9.0 ml of 0.1M sodium acetate in 50 ml measuring flask and automatically shaken for the selected period of time (1, 5, 10, 20, 25, 30 and 35 min). The mixture was filtered, washed with 50 ml water and the free metal ion was determined as described earlier.

Determination of the distribution coefficient

About 100 mg of the S-CS-MWCNTs phase was mixed with 50 ml of the metal ion (1 $mgml^{-1}$) in a 100 ml measuring flask and shaken for 3 h by an automatic shaker. The mixture was filtered, washed with water and diluted with 2% nitric acid solution in order to fit in the linear dynamic range of each metal ion. A standard solution for each metal ion was also prepared in a similar way.

Percentage removal of Hg(II) from aqueous solutions

One liter of Hg(II) solution, containing 10, 50 and 100 ng ml^{-1} was passed over a column (Bingye et al., 2012) packed with 500 and 1000 mg each of S-CS-MWCNTs. The flow rate was adjusted to 2.0 ml min^{-1}. The eluents were collected and 5ml was diluted with 20 ml of 2% nitric acid solution and subjected to ICP-MS.

Pre-concentration of Hg(II) from aqueous and natural tap water

Two liters sample solution spiked with 20 pg ml^{-1} of Hg(II) in both double distilled water (DDW)and natural tap water were prepared and passed over a column packed with 1000 mg of S-CS-MWCNTs with a flow rate of 2 ml min^{-1}. Then 10 ml concentrated hydrochloric acid (10.0 M) was then passed over the phase and adsorbed metal ion to desorb the bound-Hg(II). The desorbed metal ion was directly determined by ICP-MS. A standard solution and blank aqueous and tap water samples were also prepared and determined for evaluation.

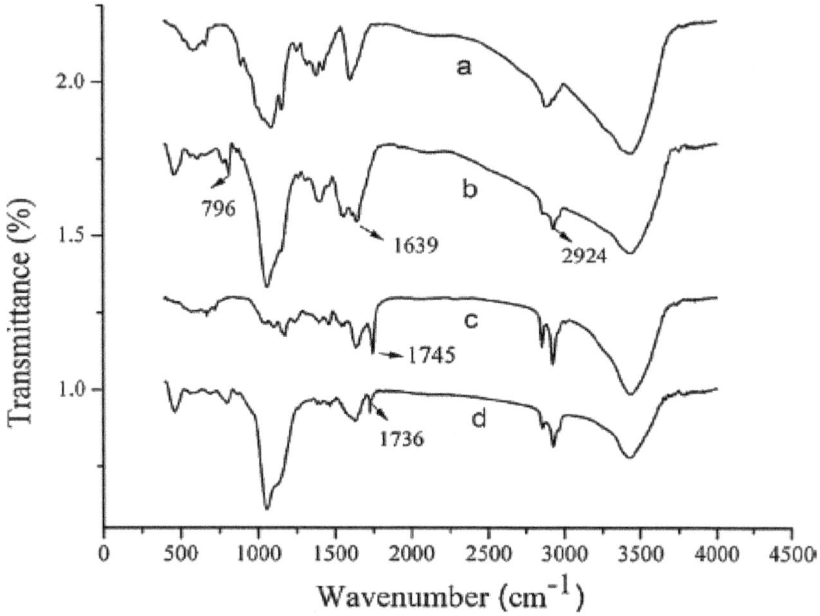

Figure 2. FT-IR spectra of (a) CS, (b) S-CS, (c) MWCNTs-COOH and (d) S-CS-MWCNTs.

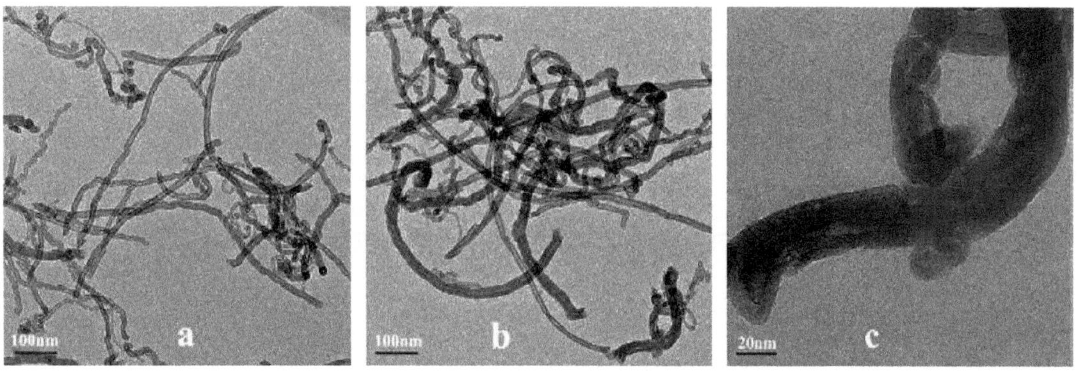

Figure 3. TEM images of (a) MWCNTs-COOH, (b) S-CS-MWCNTs and (c) S-CS-MWCNTs magnification.

RESULTS AND DISCUSSION

The FT-IR spectra of pure CS, S-CS, MWCNTs-COOH, and S-CSMWCNTs were recorded. Their IR adsorption frequencies are shown in Figure 2a–d. A comparison of the FT-IR spectra of CS and S-CS showed that a new band at 1639 cm^{-1} appeared in the S-CS spectrum. This band was attributed to the C = N stretching vibration and indicated that the Schiff base had been successfully prepared. As a result of CS reacting with 4-methyl-benzaldehyde, several new peaks also appeared in the S-CS, and they were assigned as follows. The peak observed at 2924 cm^{-1} was attributed to the stretching vibration of C H in CH_3. The peak at 796 cm^{-1} resulted from the C H bending vibration of the benzene ring. In the MWCNTs-COOH spectrum, there was a strong absorbance at 1744 cm^{-1}, which corresponds to the C = O of COOH. However, in the S-CS-MWCNTs, the absorbance at 1744 cm^{-1} was absent, and there was an absorbance at 1736 cm^{-1}, which indicated that the -OH of the CS had reacted with the MWCNTs to form ester bonds. The results confirmed the successful immobiliza-tion of S-CS on the surfaces of the MWCNTs. The surface microstructures of the grafted composites were observed using TEM. The TEM image of the MWCNTs-COOH is shown in Figure 3a. The average diameters of the MWCNTs-COOH were 20 to 30 nm. The TEM image of the S-CS-MWCNTs (Figure 3b) shows that their diameters were obviously larger than those of the MWCNTs, and we could clearly see that some of the S-CS was grafted onto the MWCNTs. In the higher-magnification TEM image (Figure 3c), a core–shell structure of S-CS-MWCNTs was clearly observed. The core of MWCNTs in the center was wrapped in grafted

Table 2. Metal capacity values determined in μmol g^{-1a}.

pH	Hg(II)	Pb(II)	Fe (II)	Mn(II)	Fe(III)	Co(II)	Ni(II)	Cu(II)	Cd(II)	Ca	Mg	Ba
NaOAc	248	59	52	20	-	14	43	63	19	23	20	20
6	217	45	36	58	-	35	64	91	39	21	23	28
5	136	33	23	69	10	50	69	74	20	35	2l	29
4	55	25	15	50	44	25	38	60	10	30	19	10
3	34	12	10	28	30	14	38	45	5	25	10	10
2	23	7	8	18	23	10	25	25	0	15	5	6
1	2	3	3	13	3	4	15	14	0	2	0	2

[a] Values are based on n = 3 with standard deviation of 4.

S-CS, proving that the MWCNTs were coated with the polymer. These coated polymer layers could reduce the strong attractive interactions among the nanotubes and prevent aggregation of the MWCNTs, and could also increase the number of adsorption sites for metal ions.

Stability studies

The stability of the newly synthesized S-CS-MWCNTs phases was performed in different buffer solutions (pH 1, 2, 3, 4, 5, 6 and 0.1 M sodium acetate) in order to assess the possible leaching or hydrolysis processes. The values of the metal capacity determined revealed that the highest one corresponds to Hg(II), this ion was used to evaluate the stability measurements for the S-CS-MWCNTs phase (Chang et al., 2001). The results of this study proved that the S-CS-MWCNTs is more resistant than the chemically adsorbed analog especially in 1.0, 5.0 and 10.0 M hydrochloric acid with hydrolysis percentage of 2.25, 6.10 and 10.50 for phase, respectively.

Thus, these stability studies indicated the suitability of phase for application in various acid solutions especially concentrated hydrochloric acid and extension of the experimental range to very strong acidic media which is not suitable for other normal and selective chelating ion exchangers based on a nano polymeric matrix (Manzoori et al., 1998). Finally, the S-CS-MWCNTs phases were also found to be stable over a range of 1 year during the course of this work.

Metal capacity in various controlling factors

The metal capacity values determined in μmol g^{-1} for the S-CS-MWCNTs in different buffer solutions were studied to evaluate the pH effect of metal ion on the extractability of the S-CS-MWCNTs phase. Table 2 compiles the μmol g^{-1} values for the 12 tested metal ions, viz) Hg(II), Mn(II), Fe(III), Co(II), Ni(II), Cu(II), Fe (II), Cd(II), Ca, Mg, Ba and Pb(II). Several trends can be observed and outlined from the data given. First, is the strong dependence of μmol g^{-1} extracted values from the metal ion solution for most tested metal ions on the pH-value (Bae et al., 1998; Tao et al., 1998). The maximum value was found to be mainly at higher pH-values (pH 4 to 6 and 0.10 M NaOAc). Second, is the strong affinity of the S-CS-MWCNTs phase for extraction and removal of Hg(II) from aqueous solution compared to other tested metal ions, as shown by the higher μmol g^{-1} values by S-CS-MWCNTs phases (Tao et al., 1998).

This behavior of S-CS-MWCNTs –loaded sulfur containing compounds for selective extraction and removal of Hg(II) from aqueous and natural water sample is well documented (Guo et al., 2004a, b) and reported based on different governing rules (Tewari and Singh, 2000). Third are the notably high μmol g^{-1} values determined for chemically S-CS-MWCNTs phase in comparison with values found as given in Table 2. The comparison between the metal sorption properties of chemically and physically- S-CS-MWCNTs phases has been extensively studied (Daniel et al., 2003) and the results presented in this work are consistent with the surface activity of the donor atoms responsible for metal ion interaction, sorption, extraction and selective removal. In the case of the physically adsorbed phase, some of these donor atoms are involved in physical adsorption processes with the active surface, leading to the minimization of the reactivity of such donor atoms for metal interaction and binding processes. The product, S-CS-MWCNTs, in this case is tuned with the active donor atoms (N) directed with the capability and accessibility for fast and direct interactions with the free metal ion present in solution. Fourth, are the general orders of metal capacity values for all tested metal ions by the two phases which are in many respects consistent and similar. Therefore, the conclusion drawn from this section can be outlined as the high superiority of phase for selective extraction of Hg(II) as well as the higher metal uptake behavior of S-CS-MWCNTs phase. The effect of shaking time on the percentage extraction of metal ions at various equilibration time intervals (1, 5, 10, 15, 20, 25 min) was also studied and evaluated as μmol g^{-1} and correlated to that determined at 30 min shaking time. Figure 4 represents the percentage extraction versus shaking time in min and clearly reflects the rapid

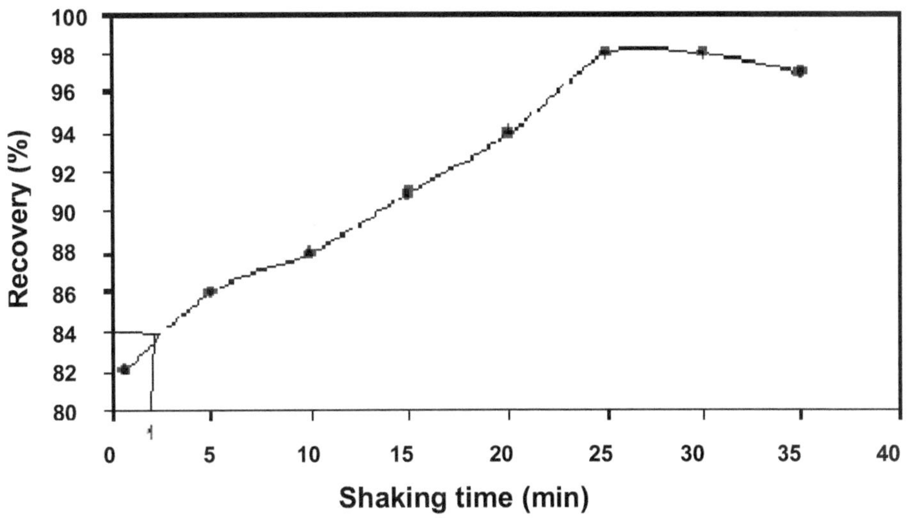

Figure 4. Effect of shaking time (min) on the percentage extraction of Hg(II) by the S-CS-MWCNTs phases.

Table 3. Distribution coefficient (K_d) values of various metal ions.

Metal ions	K_d
Mn(II)	90
Fe(III)	87
Co(II)	68
Ni(II)	57
Cu(II)	85
Cd(II)	530
Cu(II)	228
Hg(II)	2400
Pb(II)	120

exchange equilibrium between S-CS-MWCNTs phase and Hg(II). One minute shaking time was found to be sufficient to establish 84% of the determined µmol g^{-1} value at 30 min whereas 10 min shaking time led to 88% extraction. The data and results presented in this section reveal the superiority of S-CS-MWCNTs phase as previously declared in the stability studies.

The distribution coefficient (K_d) data of the tested metal ions with the two newly S-CS-MWCNTs phase are summarized in Table 3. It is evident that Hg(II) is the strongest sorbed metal ion by S-CS-MWCNTs phase. The distribution coefficient values of Hg(II) by the loaded S-CS-MWCNTs phase were found to be much higher than those reported for ion exchange resins containing S-CS-MWCNTs derivatives (Manzoori et al., 1998). In addition, the K_d values for Hg(II) by S-CS-MWCNTs phase were found to come on the second place after Hg(II) which behavior can be interpreted on the basis of the affinity of both nitrogen and hidroxyl donor groups present in S-CS-MWCNTs for binding with Hg(II) (Guo et al., 2004a, b). On the other hand, the various tested metal ions as shown in Table 3 were found to exhibit lower tendency to bind with S-CS-MWCNTs phase judging from the comparable low distribution coefficient values determined for these metal ions. The higher K_d value for Hg(II) and the lower ones for the other metal ions, except Hg(II), provide an additional evidence for the suitability of these two newly S-CS-MWCNTs phase for selective extraction of Hg(II) from aqueous solutions. It is also noteworthy that the conclusion drawn from the evaluation of the K_d values by S-CS-MWCNTs phase is consistent with the reported data.

Effect of pH

According to the recommended procedure solutions (the volume of each 100 mL) containing 0.5 lg mL^{-1} of Hg(II) was taken. Their pH values were adjusted in range 1.5 to 10 with HNO_3, 0.01M acetate and/or phosphate buffer solutions and the mixture was shaken for 5 h. The 0.1 g of S-CS-MWCNTs in different pH value, namely pH 1.5, 2, 3, 4, 5, 6, 7, 8, 9 and 10 Hg(II) was determined. It also can be seen in Figure 5, the sorption quantity of Hg(II) increases with the pH value increases in the studied pH ranges. Below pH 2.0, the sorption quantity of Hg(II) is very low since the functional groups were protonated. Owing to hydrolysis at higher pH value, pH 4.0 was chosen as the optimum pH for further studies.

Selective pre-concentration of Hg(II) from natural water for off-line ICP-MS

This study was undertaken in order to evaluate the potential application of S-CS-MWCNTs phase for

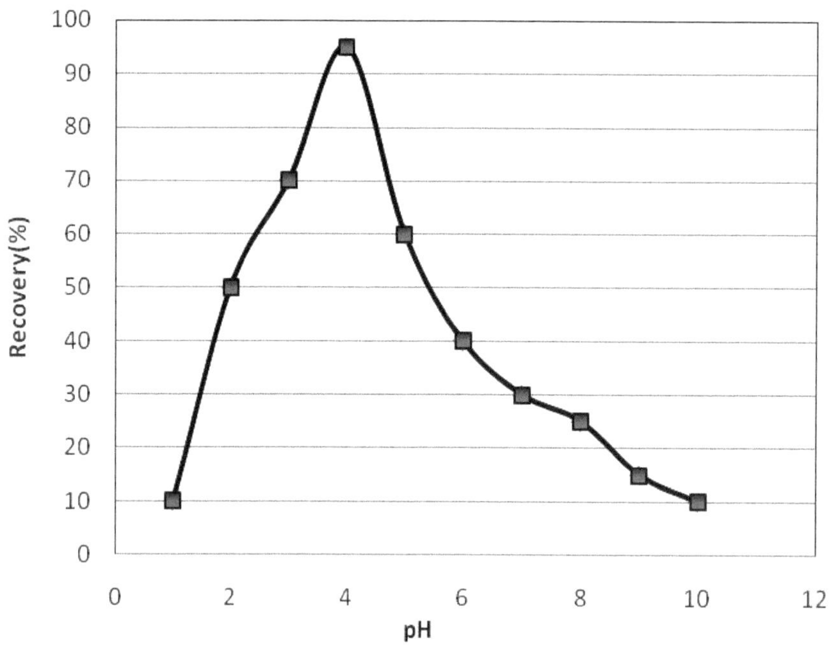

Figure 5. Effect of pH sorption of Hg(II) onto S-CS-MWCNTs. shaking time: 5 h, temperature: 25°C.

pre-concentration of trace levels of Hg(II) in natural water samples. Drinking tap water was used without prior treatments as an example and compared with double distilled water (DDW) to evaluate and investigate the matrix effect. Both drinking tap water and DDW (2 L) were spiked with 20 pg ml^{-1} of Hg(II). Several pre-concentration reagents are well known and extensively examined for desorption of the bound metal ions from the surface of the stationary phase and these include mainly, hydrochloric and nitric acid, thiourea HCl (Manzoori et al., 1998). However, some of these reagents are usually characterized by adsorption on the surface of S-CS-MWCNTs which lead to severe change in the nature of packing material as well as non reproducible results if the column used in future applications. The efficiency of metal ion pre-concentration especially Hg(II), by 10.0 M HCl acid is found to be high without causing any notable change to the chemical nature of the organic S-CS-MWCNTs.

Therefore, 10.0 ml of 10.0 M HCl was used for the elution of the adsorbed Hg(II) from the column bed. The pre-concentration factor targeted from this study is 100 as given. As the results indicate, the off-line detection results of the eluted and pre-concentrated Hg(II) are very good with a satisfactory pre-concentration factor which can be further increased to 100-fold by simply increasing the water sample volume to 5 L instead of 2 L. Moreover, natural tap water sample was found to give very close results to that reported for DDW sample and this comparison indicates that the matrix effects of the dissolved inorganic and organic matters played an insignificant role in the aimed selective extraction, removal and pre-concentration of Hg(II) by S-CS-MWCNTs phase.

Percentage removal of Hg(II) from aqueous solution

The use of a column technique is a common procedure for extraction, separation and selective extraction of metal ions from various aquatic systems (Pereira and Arruda, 2003). The column technique is characterized by major advantages over the batch or static equilibration method that is the possible application to large sample volumes (Chang et al., 2001; Moghimi, 2007; Yang et al., 2004; Ensing and Boer, 1999). This property enables the pre-concentration of metal ions at very low trace levels. The percentage removal of metal ions from aqueous solutions is essential for the evaluation of the method described and suggested here. This is mainly dependent on several well known factors such as the type and amount of packing sta-tionary and mobile phases and the flow rate of the mobile phase (Daniel et al., 2003). In this study, we attempted to evaluate the percentage recovery of Hg(II) with different spiked concentrations, namely 10, 50 and 100 ng ml^{-1} from 1 L of 0.1 M NaOAc solution by the application of two different amounts (500 and 1000 mg) of S-CS-MWCNTs phase packing. The results of the percentage removal of Hg(II) from aqueous solutions are presented in Table 4 which clearly demonstrate the suitability and validity of S-CS-MWCNTs phase for removal and extraction of Hg(II) with ICP-MS. In addition, the effect of packing amount of silica gel phase is also evident in Table 3 where the near completion of

Table 4. Percentage removal of Hg(II) from aqueous solutions by S-CS-MWCNTs[a].

Hg(II) spiked (ng ml^{-1})	Phase (mg)	Percentage removal
10	500	94±2
50	500	96±3
100	500	97±4
10	1000	99±3
50	1000	95±2
100	1000	96±3

[a]Values are based on triplicate analysis.

Table 5. Preconcentration of Hg(II) from DDW and natural tap water samples[a].

Phase (mg)	Sample volume (ml)	Hg(II) spiked (pg ml^{-1})	Pre-concentration reagent	Pre-concentration factor	Hg(II) detected[a] (ng ml^{-1}) with ICP-MS	Hg(II) detected[a] (ng ml^{-1}) with CV-AAS	Percentage removal
1000	2000 Tapwater (Tehran)	20	10.0 ml of 10.0 M HCl	200	3.93±2.4	3.99±2.9	98.4±2.3
	2000 DDW				3.95±2.6	3.90±3.1	98.6±2.1

[a] Values are corrected for blank concentration of water samples and based on triplicate analysis.

Hg(II) removal was accomplished by the use of 1000 mg phase. The results of the pre-concentration of Hg(II) from aqueous solutions are presented in Table 5 of Hg(II) from DDW and natural tap water samples.

Conclusions

In summary, a selective and sensitive method for the determination of trace levels of Hg(II) based on S-CS-MWCNTs as a sorbent for the determination of the analytes in biological samples as well as natural water samples. The sorption capacity for Hg(II) is 44.01 mg g^{-1} that is higher than most of the sorbents. The faster sorption and desorption are other distinct advantages. In comparison to the commonly immobilized in silica gel, the present sorbent has a superior pre-concentration factor (100-folds) for Hg(II). In addition, the preparation of S-CS-MWCNTs is relatively simple and rapid. However, the S-CS-MWCNTs sorbent has some drawbacks include lower reusability and stability, which can be repeated only 10 cycles. This may make it difficult to put into practical use and should be a subject of further investigations.

REFERENCES

Bae SY, Zeng X, Murray GM (1998). Micell-Mediated Extraction for Preconcentration and Determination of Copper and Zinc in Real Samples. J. Anal. At. Spectrom. 13:1177-1183.

Bingye D, Meirong C, Guozhen F, Bing L, Xv D, Mingfei P, Shuo W (2012). Review of Pb availability and toxicity to plants in relation with metal speciation; role of synthetic and natural organic ligands. J. Hazard. Mater. 219-220:103-110.

Chang X, Yang X, Wei X, Wu K (2001). Study on the mechanism of silver nanoparticles on the modified PP fiber. Anal. Chim. Acta 450: 231-236.

Daniel S, Gladis JM, Rao TP (2003). Selective solid-phase extraction of lead(II) from biological and natural water samples using surface-grafted lead(II)-imprinted polymers. Anal. Chim. Acta 488:173-179.

Ensing K, Boer T (1999). Tailor-made materials for tailor-made applications: application of molecular imprints in chemical analysis. Trac-Trend. Anal. Chem. 18:138-145.

Greenway GM, Townshend A (1993). Monitoring the earliest amyloid-b oligomers via quantized photobleaching of dye-labeled peptides. Anal. Proc. 30:438-445.

Guo Y, Din B, Liu Y, Chang X, Meng S, Tian M (2004a). Solid phase extraction of iron and lead in environmental matrices on amberlite xad-1180/pv. Anal. Chim. Acta 504:319-325.

Guo Y, Din B, Liu Y, Chang X, Meng S, Liu J (2004b). Removal of heavy metals from artificial metals contaminated water samples based on micelle-templated silica modified with pyoverdin. Talanta 62:207-213.

Hayes RB (1997). The carcinogenicity of metals in humans. Cancer Causes Control 8:371-376.

IARC (1993), International Agency for Research on Cancer Meeting of the IARC working group on beryllium, mercury, mercury and exposures in the glass manufacturing industry Scand. J. Work. Environ. Health 19:360-367.

Jin YH, Clark AB, Robbert JCS, Al-Refai H, Taylor JA, Kunkel TA, Resnick MA, Gordenin DA (2003). Diethyl sulfide modified silica and calix[4]pyrrole chelating resin: Synthesis and mercury(II) cation retention properties. Nat. Genet. 34:326-331.

Kantipuly C, Katragadda S, Chow A, Gesser HD (1990). Thiol-functionalized MCM-48: An Effective Absorbent of Mercury Ions Talanta 37:491-498.

Liu Y, Chang X, Wang S, Guo Y, Din B, Meng S (2004). Preconcentration of trace elements by using 1-(2-Pyridylazo)-2-naphthol functionalized Amberlite XAD-1180 resin and their determination by FAAS. Anal. Chim. Acta 519:173-179.

Manahan SE (1994). Environmental Chemistry, sixth ed., Lewis Publishers, Ann Arbor, pp. 185-192.

Manzoori JL, Sorouraddin MH, Shabani AMH (1998). Coordination of mercury(II) to gold nanoparticle associated nitrotriazole towards sensitive colorimetric detection of mercuric ion with a tunable dynamic range. J. Anal. At. Spectrom. 13:305-311.

Moghimi A (2007). Selective Pre-concentration and Solid Phase Extraction of Mercury(II) from Natural Water by Silica Gel-loaded (*E*)-*N*-(1-Thien-2′-ylethylidene)-1,2-phenylenediamine Phase. Chin. J. Chem. 25(10):1536-1542.

Moghimi A, Keyvani P, Abdouss M (2011) Modified Nano Polyacrylonitrile Fiber Using for Preconcentration of Hg(II) From Sample Water. Aust. J. Basic Appl. Sci. 5(9):1164-1171.

Moghimi A (2010). Preconcentration of mercury by adsorption of its diphenylthiocarbazone complex on an silica gel-immobilized Schiff base diphenylthiocarbazone (DPC) column and spectrophotometric determination. Orient. J. Chem. 26(3):753-761.

Pereira MG, Arruda MAZ (2003). On-line preconcentration system using a microcolumn packed with Alizarin Red S-modified alumina for zinc determination by flame atomic absorption spectrometry. Microchim. Acta 141:115-122.

Rao TP, Daniel S, Gladis JM (2004). İyon baskılamaya dayanan katı-faz ekstraksiyon uygulamaları [Solid-phase extraction applications based on ion-imprinting]. Trac-Trend. Anal. Chem. 23:28-34.

Seiler HG, Sigel A, Sigel H (1998). Handbook on Toxicity of Inorganic Compounds, Marcel Dekker, New York.

Smith H (1963). Procedures used at the National Bureau of Standards to determine selected. Anal. Chem. 35:635-341.

Swain K, Raje N, Kayasth S (2000). Çift baskılı polimerlerle Cd(II) iyonlarının sulu ortamdan uzaklaştırılması [Removal of Cd(II) ions from aqueous solution via double-imprinted polymers. *Anal. Chim. Acta* 405:297-301.

Tao G, Willie SN, Sturgeon RE (1998). Determination of total mercury in biological tissues by flow injection cold vapour generation atomic absorption spectrometry following tetramethylammonium hydroxide digestion. Analyst 123:1215-122.

Tabarzadi M, Abdouss M, Hasani SA, Shoushtary AM (2010). Ion adsorption studies of micro and nano acrylic fibers modified by ethanolamine. Ionen Adsorption von mit Ethanolamin modifizierten Mikro- und Nano-Acylfasern. Mat.-wiss. u.Werkstofftech. 41(4):221-225.

Tewari PK, Singh AK (2000). ANew Chelating Resin for Preconcentration and Determination of Mn(II), Ni(II), Cu(II), Zn(II), Cd(II), and Pb(II) by Flame Atomic Absorption Spectrometry. Analyst 125:2350-2355.

Thurman EM, Mills MS (1998). Solid-Phase Extraction: Principles and Practice, Wiley, New York.

Wu ZG, Feng W, Feng YY, Liu XH, Xu XH, Fujii A, Ozaki M (2007). Preparation and characterization of chitosan-grafted multiwalled carbon nanotubes and their electrochemical properties. Carbon 45:1212–1218.

Yang D, Chang X, Liu Y, Wang S (2004). Chemically Modified Submicron Silica Particulate Extractants for Preconcentration of Mercury(II). Mikrochim. Acta 147:219-225.

Zargaran M, Shoushtari AM, Abdouss M (2008). Diametric swelling and hydrophilic characteristics of fibrous acrylic ion exchanger. J. Appl. Polym. Sci. 110:3843-3847.

Zhou TZ, Qi DY, Zhang CP (1983). Flame Atomic Absorption Determination of Ni^{2+}, Cu^{2+} and Co^{2+} Ions in Some Water and Food Samples After Cloud Point Extraction Using a Thio Schiff-Base as a New Complexing Agent. Acta Chim. Sin. 41:237-242.

9

A new approach for acetylation of wood: Vinyl acetate

Nilgül Özmen* and Nihat Sami Çetin

Kahramanmaraş Sütçü İmam University, Faculty of Forestry, Kahramanmaraş, Turkey.

The chemical modification of wood has been the subject of research for many decades. The most favoured modification reaction involves the acetylation of wood with acetic anhydride. Reacted acetate provides dimensional stability by bulking the cell wall polymers and reduces further swelling when the woods come into contact with water. In addition that chemical modification also provides biological resistance. With acetylation, it is possible to extend at least twice or three times service life of wood as compared to control one. Acetylation of wood with acetic anhydride is not subjected to only academic studies, but also has become commercial process since 2003. The purpose of this study is to show the new methods for acetylation of wood. Turkish pine (*Pinus brutia*) sapwood was successfully acetylated by the new transesterification reaction with vinyl acetate. Under identical conditions, Turkish pine sapwood was also modified with acetic anhydride to compare with the new technique. The results of acetylated wood with acetic anhydride and vinyl acetate were compared, and characterized by Infrared (FTIR) and ^{13}C cross polarisation-magic angle spinning nuclear magnetic resonance (CP MAS NMR) analysis.

Key words: Acetylation, acetic anhydride, vinyl acetate, Turkish pine, chemical modification.

INTRODUCTION

Wood is renewable resources and has many utilisation areas. Unfortunately, some of the wood properties such as dimensional instability, biodegradability, flammability etc. limit the utilisation areas (Fengel and Weneger, 1989).

Chemical modification can be defined as a chemical reaction between reactive parts of wood cell wall polymers (hydroxyl groups generally) and a chemical reagent, with or without catalyst, to form a covalent bond between these two. All of the methods investigated to date have involved the chemical reaction of a reagent with the cell wall polymer hydroxyl groups. The chemical modification of wood with various reagents including anhydrides (such as acetic, succinic, maleic, propionic, butyric, hexanoic, crotonic and methacrylic anhydrides), isocyanates, formaldehyde, acetaldehyde, epoxides (such as ethylene or propylene oxide, glycidyl methacrylate, allyl glycidyl ether etc.) has been the subject of research for many decades (Stamm and Tarkow, 1947; Banks and Lawther, 1994; Kumar, 1994; Hon, 1996; Rowell, 1983, 1991, 2006; Cetin and Ozmen, 2001; Hill, 2006).

While there has been a lot of research conducted with various chemical reaction systems; many of them focused on the acetylation of wood with acetic anhydride (Çetin et al., 2011). In spite of the vast amount of research on the acetylation of both bulk wood and wood particles (for composites), commercialisation has been limited (Rowell, 2006).

Acetylation has become commercial product recently. In Netherlands, Titan Wood is the first company to produce acetylated wood in commercial scale (its name is Accoya wood). Their production capacity is about 24.000 m^3 per annum. It was expected that in near future, acetylated wood will replace the more toxic preservative treatments in many application (Kattenbroek, 2005).

Acetylation with acetic anhydride has got some disadvantage such as being expensive and requires significant capital investment. In addition, acetic acid is produced as by-product at the end of the reaction and must be removed from modified materials.

Scots pine sapwood has recently been successfully acetylated with a new transesterification reaction using

*Corresponding author. E-mail: nilgulozmen@hotmail.com.

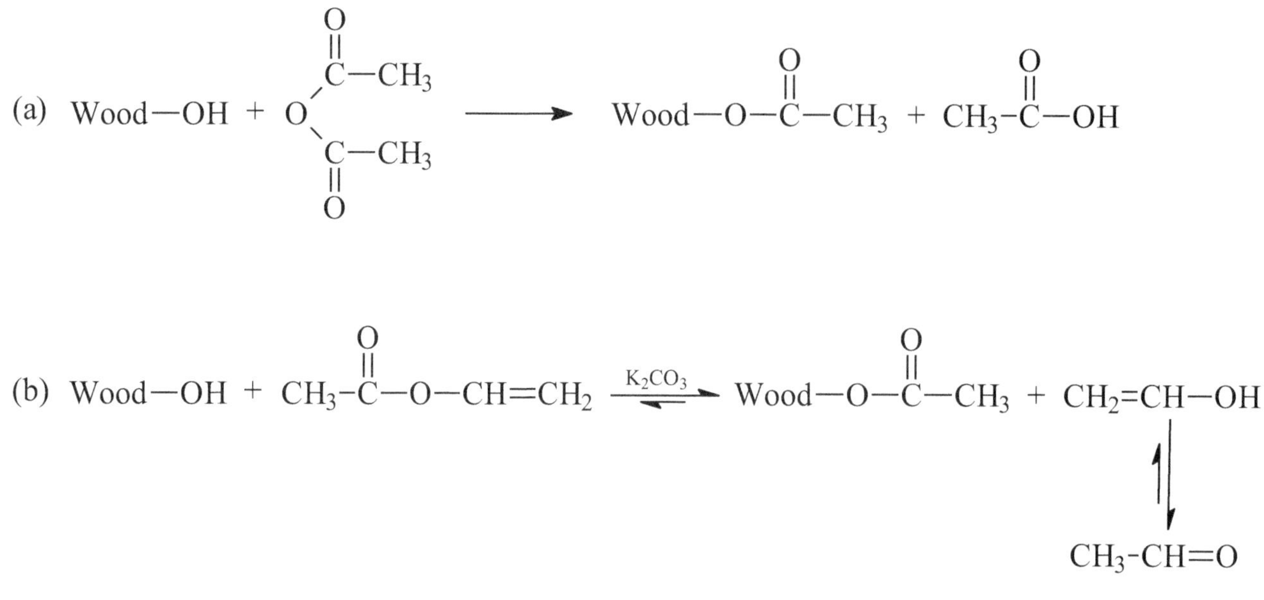

Figure 1. The reaction between acetic anhydride (a) or vinyl acetate (b) and wood.

vinyl acetate and a potassium carbonate catalyst (Çetin et al., 2011). The main advantages of this new technique are that vinyl acetate is cheaper than acetic anhydride and acetaldehyde is produced as a by product. Acetaldehyde is a non-acidic, low boiling point (bp= 21°C) compound that is easily removed from modified wood after reaction.

In this study, Pine sapwood (*Pinus brutia*) samples were acetylated using vinyl acetate and acetic anhydride. Modified samples were characterized by weight percent gain (WPG) calculations, FTIR and ^{13}C CP-MAS NMR spectroscopy analysis.

MATERIALS AND METHODS

Acetic anhydride, vinyl acetate, potassium carbonate and dimethyl formamide were purchased from Merck. All chemicals were used as received without further purification. Deionized water was used in all experiments.

Turkish pine (*P. brutia*) sapwood was ground using a hammer mill and sieved to a size of 40 mesh. Following that the wood flour was extracted with the standard solution (4/1/1, v/v, toluene:acetone:ethanol) using a Soxhlet extractor for 6 h. Extracted wood flour was oven-dried at 103°C overnight. Oven-dried wood flour was transferred to a desiccator and allowed to cool to room temperature.

Acetylation of wood flour

Extractive free Turkish pine sapwood flour was transferred to a round bottom flask containing dimethylformamide (DMF) solution with acetic anhydride (AA) or vinyl acetate. All reactions were performed at 100°C. Reaction conditions were derived from earlier publication (Çetin et al., 2011). For each reaction, 1 g wood flour was used and each set of conditions were repeated three times. At the end of the reaction, all modified samples were Soxhlet extracted with deionised water for 6 h and then with toluene:acetone:ethanol mixture (4/1/1, v/v) for 6 h. This was done to remove excess unreacted chemicals and by products. The Soxhlet thimble and contents were oven-dried overnight at 103°C transferred to a desiccator containing phosphorus pentoxide until cool, then weight gain levels were calculated. The weight percentage gain (WPG) was calculated according to following equations.

$$WPG(\%) = \frac{W_2 - W_1}{W_1} x\ 100$$

W_1 = before treatment sample weight
W_2 = after treatment sample weight

Infrared spectroscopy

Infrared absorption spectra of acetylated and unmodified wood flour were obtained with the potassium bromide (KBr) technique, using a Shimatzu 8400 s FT-IR spectrometer, at a resolution of 4 cm^{-1} (40 scans). In each case, 1% w/w of oven dry wood flour was dispersed in a matrix of KBr and pressed to form pellets.

^{13}C and NMR CP-MAS analysis

Solid state ^{13}C CP-MAS NMR spectra of Turkish pine wood flour were performed at room temperature on a Bruker DPX-400 NMR spectrometer (Bruker), using MAS rates of 4 and 8 kHz, a frequency of 100.61 MHz for ^{13}C NMR.

DISCUSSION

The reaction mechanism between wood hydroxyl groups and AA or VA is shown in Figure 1. As shown in Figure 2,

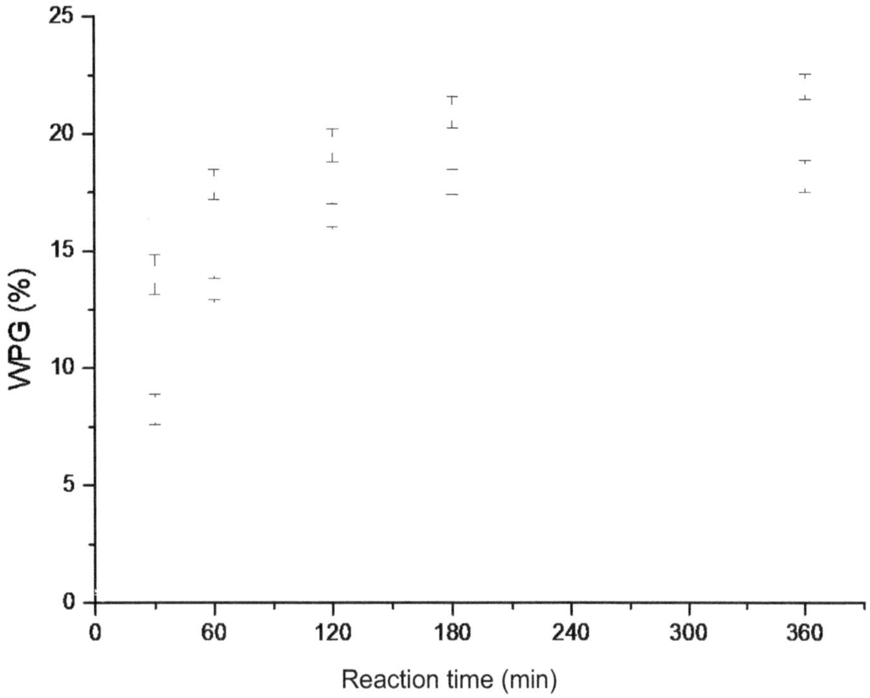

Figure 2. Reaction of Turkish pine with acetic anhydride (square) or vinyl acetate (triangle) at 100°C up to 6 h.

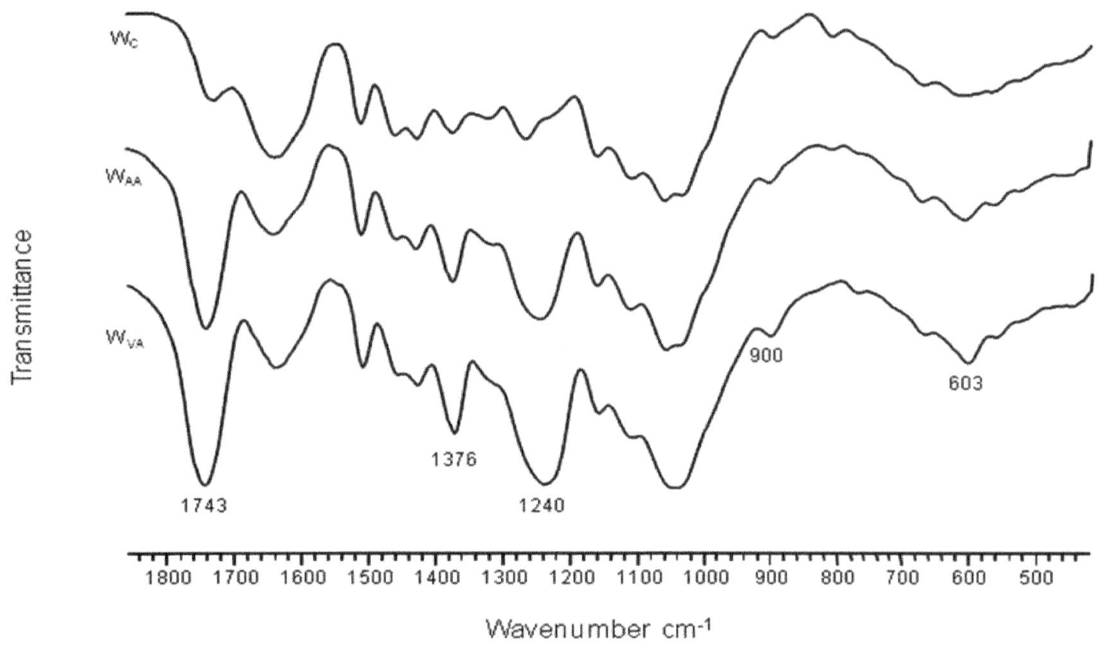

Figure 3. FTIR spectra of untreated (W_C), acetic anhydride modified (24% WPG) (V_{AA}), vinyl acetate modified (20% WPG) (W_{VA}) Turkish pine flour.

approximately 22 and 18% WPG levels were ultimately achieved for 6 h reaction with acetic anhydride and vinyl acetate, respectively.

The proof of the reaction between acetic anhydride or vinyl acetate and Turkish pine was obtained by analyzing the samples with FTIR spectroscopy (Figure 3). No

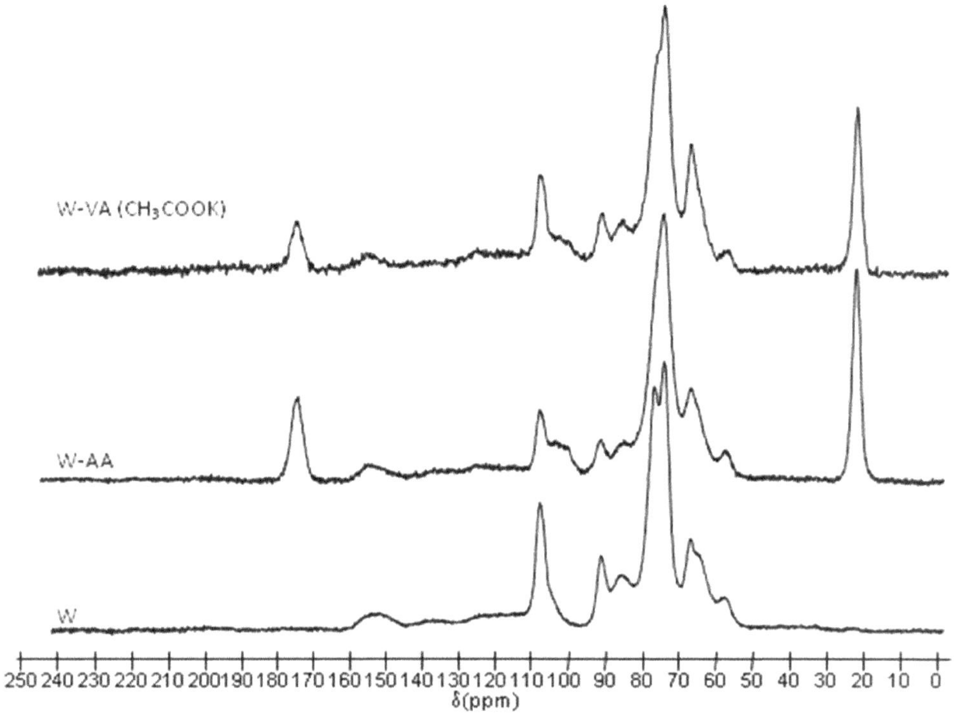

Figure 4. ^{13}C CP-MAS NMR spectra of unmodified (W), acetic anhydride modified (W-AA), vinyl acetate modified (W-VA) Turkish pine flour.

spectral difference was noted between two acetylation methods. Acetylated samples (W_{AA} and W_{VA}) were easily identified in the FTIR spectra, by emergence of a carbonyl stretching vibration at 1743 cm^{-1} (νC=O) in the spectra confirmed the formation of ester bonds after reactions with AA and VA (Silverstein et al., 1991). In addition to the C=O at 1743 cm^{-1}, the intensity of the band at 1240 cm^{-1} also increased and was associated to the C-O stretching vibration (ν C-O) of the acetyl moieties. The intensity of the bands located at 1376 cm^{-1} was also enhanced after acetylation. This band was attributed to the C-H bending (δs C-H) vibrations of the methyl groups introduced. The intensity of the band at 603 cm^{-1} also increased and was associated to some vibrations of the grafted methyl groups.

The FTIR observation was confirmed by the ^{13}C NMR spectrum of the samples. The ^{13}C NMR spectra of unmodified Turkish pine are shown in Figure 4 (W). As shown in Figure 4, the dominant pattern of the ^{13}C NMR spectra is that of the carbohydrates, namely C_1 (108 ppm), C_4 crystalline (90 ppm), C_4 amorphous (85 ppm), C_2 (76 ppm), C_3/C_5 (73 ppm), C_6 crystalline (65 ppm), C_6 amorphous (63 ppm) (Boonstra et al., 1996; Çetin et al., 2005; Çetin et al., 2011). The signal at 59 ppm indicates the lignin methoxy groups. The aromatic groups of the lignin appear at 150 ppm as wide band.

After acetylation of Turkish pine wood with acetic anhydride or vinyl acetate, the methyl band of the acetyl group at 21 ppm and the carboxylic group at 170 ppm show the acetyl groups on the wood components. There was no difference between the VA or AA modified wood spectra.

Conclusion

In this study, the catalysed acetylation of Turkish pine with AA or VA was investigated with using potassium carbonate catalyst. The acetylation of wood has been confirmed by WPG calculations, FTIR spectroscopy and ^{13}C CP-MAS NMR analysis. In the comparison between AA and VA modification results, the AA modification gave higher WPG values than VA modification at the same reaction conditions.

REFERENCES

Banks WB, Lawther JM (1994). Cellulosic polymers blends and composites. In Gilbert RD (ed). Derivitisation of Wood in Composites. Hanser/Garder Publications Inc., Cincinnati, pp. 131-155.

Boonstra MG, Pizzi A, Tekely P, Pendlebury J (1996). Chemical modification of Norway spruce and Scots pine. A ^{13}C NMR CP-MAS study of the reactivity and reactions of polymeric wood components with acetic anhydride. Holzforschung, 50(3): 215-220.

Cetin NS, Ozmen N, Birinci E (2011). Acetylation of wood with various catalysts. J. Wood Chem. Technol., 31: 142-153.

Cetin NS, Ozmen N, Tingaut P, Sebe G (2005). New transesterification reaction between acetylated wood and tetrametoxysilane: A feasibility study. Eur. Polym. J., 41: 2704-2710.

Cetin NS, Ozmen N (2001). Dimensional changes in Corsican pine sapwood due to reaction with crotonic anhydride. Wood Sci. Technol., 35(3): 257-267.

Fengel D, Wegener G (1989). Wood: Chemistry, Ultrastructure, Reactions. Water de Gruyter, Berlin.
Hill CAS (2006). Wood modification: Chemical, thermal and other processes. John Wiley and Sons, Ltd., England.
Hon DN-S (1996). Functional natural polymers: A new dimensional creativity in lignocellulosic chemistry. In Hon DN-S (ed.) Chemical Modification of Lignocellulosic Materials. Marcel Dekker Inc., New York.
Kattenbroek B (2005). How to introduce acetylated wood from the first commercial production into Europe. In Militz H, Hill CAS (eds.) Wood Modification: Process, Properties and Commercialisation. The Second European Conference on Wood Modification, Göttingen, Germany, pp. 398-403.
Kumar S (1994). Chemical modification of wood. Wood Fibre Sci., 26(2): 270-280.
Rowell RM (1983). Chemical modification of wood. For. Prod. Abst., 6(12): 363–382.
Rowell RM, Youngquist JA, McNatt D (1991). Composites from Recycled Materials. In Proceedings of 25th International Particleboard/Composite Materials Symposium W.S.U., Pulman, Washington, pp. 301-314.
Rowell RM (2006). Acetylation of wood: Journey from analytical technique to commercial reality. For. Prod. J., 56(9): 4–12.
Silverstein RM, Bassler GC, Morill TC (1991). Spectrometric identification of organic compounds (5th ed.). New York: John Wiley and Sons.
Stamm AJ, Tarkow H (1947). Dimensional stabilisation of wood. J. Phys. Colloid Chem., 51: 493-505.

10

Effect of additive for efficient combustion of cracked furnace oil

Arun K. Singh* and Satish Kumar

Indian Institute of Petroleum, Dehradun 248005, India.

Present communication deals with evaluation of various additives, formulations and chemicals that are either commercially available or synthesized in authors' laboratory for cracked residual fuel oil. The objective of this study is to assess the suitability of these additives and chemical components with regard to sediment and deposit inhibiting characteristics and study the effect of the additives on actual fuel combustion so that suitable type of additives can be identified for cracked residual fuel oil. The additives tested were found to cause considerable reduction in total sediments almost 10 times and in un-filterable residue by 2.5 times. Low air pressure (LAP) burner test in an actual furnace showed significant improvement in terms of combustion efficiency (15.4%), fuel economy (26.7%) and particulates in flue gases (40%). It was concluded that the quality of cracked furnace oil can be improved to an acceptable performance level by the help of proper additive.

Key words: Combustion-efficiency, cracked residue, fuel-economy, furnace-oil, fuel-quality.

INTRODUCTION

Furnace oil is a heaviest commercial fuel that can be obtained from petroleum crude oil, either as a distillate or a residue and it is burned in a furnace or boiler for the generation of heat or used in an engine for the generation of power. It is also known as fuel oil, residual fuel, marine fuel or bunker oil. Flash point of furnace oil should be approximately 40℃. Furnace oil is made of long cha ins hydrocarbon; particularly alkane, cycloalkane and aromatics (Wikipedia, 2010). The world's furnace oil or residual fuel oil production in 2004 to 2005 was estimated around 2500 Million Metric Ton (Fuel oil, 2010). Even then, residual furnace oil is the most neglected hydrocarbon fuel. While in developed countries there is a tendency to use distillate fuel oils and up grade the residue so as to reduce sediments, asphaltenes, metal content and sulfur in the residual fuel oils. In developing countries, industries are only using residual hydrocarbon fractions, which are not suitable as furnace oils (Danilov et al., 2003). Furnace oils as a result of thermal and catalytic cracking, delayed coking and visbraking in heavy oil unit contain large quantities of mono and di cyclic olefins, polynuclear aromatics hydrocarbons, heterocyclic components, asphaltenes, sediments, metal and other contaminants (around 0.7 to 1.5% as against the required value of 0.25%) (Murphree, 1951). The oils usually have high pour points and require preheating. These characteristics cause heavy deposits during storage and causing heavy clogging of fuel lines (Notes, 1984).

The deposits are formed at the burner nozzles and fire side of the tubes also. Heavy corrosion is caused due to high sulfur content of the fuel. The soot content is high in flue gases and fuel economy is generally poor due to reduced-flow and insufficient atomization at the burner nozzles (Jones et al., 1958). These tendencies can be curbed by using suitable additives, which not only reduce deposits, sludge formation, polymerization and corrosion during storage but also improve the combustion efficiency and fuel economy of the fuel. A comparison of furnace oils characteristics in Table 1 of the furnace oils of European (BS) and Indian (IS) (BIS, 2002) origins is showing lack of control on various parameters such as sulfur, sediments, asphaltenes and olefins in furnace oils of Indian origin, which is possible only if the refinery

*Corresponding author. E-mail: arun@iip.res.in or drak.singh13@gmail.com.

Abbreviations: LAP, Low air pressure; **MFA,** multifunctional additive; **BS,** European standard; **IS,** Indian standard; **HV,** high viscosity.

Table 1. Specification for furnace oil.

Characteristics	Standard	(IS- 1593 – 88) 1993 grade				(IS –Tentative) 1997 grade		
		LV	MV1	MV2	HV	LV	MV	HV
Kinematic viscosity at 50°c max	IS	80.0	125	180	370	80	125-180	370
Kinematic viscosity at 80°c max	BS	13.5	35	-	85	-	-	-
Sediments % by mass – max	IS	0.25	0.25	0.25	0.25	0.15	0.25	0.25
Sediment % by mass – max	BS	0.15	0.25	-	-	0.10	0.20	0.20
Sulfur % by Mass – max	IS	3.50	4.00	4.0	4.5	2.0	4.0	4.0
Sulfur % by mass – max	BS	3.50	4.00	-	4.0	1.0	2.5	2.5
Ash % by mass	IS	-	-	-	-	0.1	0.25	0.25
Ash % by mass	BS	0.10	0.15	-	0.25	0.1	0.15	0.15
Bosch	BS	-	-	-	-	0.85	0.90	1.0

Note: IS – Indian standard, BS – British standard first came in 1982, now ISO 8217 from 2005.

operations are suitably geared to bring about these changes. The unstable components of cracked furnace oil are mainly responsible for thermo-oxidative degradation, which leads to polymerization in the form of sediment formation (Nixon, 1962). The Environmental Protection Agency of United States of America put an act in January 1995, which requires deposit control additives be added to all furnace-oil. This type of additive is a detergent additive that acts as a cleansing agent in small passages in the fuel injectors. This in turn serves to ensure a consistent air and fuel mixture that will contribute to better fuel efficiency. Thus, furnace-oil additives are designed to meet the challenges created by changing emissions regulations and fuel standards, and variable fuel quality. Fuel quality is affected by desulphurization, fuel instability and increases deposits in exhaust tracts, boilers and heat exchangers. As per international standard practice, all furnace oil must have at least one antioxidant and one detergent-dispersant or multifunctional additive (MFA) (Ford, 1968).

The aim of present work is to improve operational efficiencies in residual fuel oil and cracked furnace oil burning by developing suitable additive. This fuel additive includes an antioxidant and MFA for residual fuel oil for improvement in fuel combustion quality. Thermo-oxidation phenomena are the main cause for degradation of furnace oil and formation of non-soluble fine particles. Further, its deposition as sediment is phenomena of colloids or coagulation. Additives for furnace oil should be selected after considering both activities. The oxidation reactions in hydrocarbon proceed via three separate but related mechanisms (Dunn, 2005):

$R^{\bullet} + O_2 \rightarrow ROO^{\bullet} \longrightarrow ROO^{\bullet} + RH$ (I) RH = Hydrocarbon, R = Hydrocarbon radical

$ROO^{\bullet} + RH \rightarrow ROOH + R^{\bullet}$ (II) $ROO^{\bullet}$ = peroxy radical, ROOH =Hydro peroxide

The process involves (I) build up of hydro- peroxide moieties and other free radicals (II) a free radical reaction with various hetero atom species (III) occurred such as condensations between organic sulfur, nitrogen, acidic and aldehyde species. The following reactions are involved in radical formation in oxidative deterioration of hydrocarbons. These general equations apply to both saturated and unsaturated radicals.

| |

$ROO^{\bullet}$ + >C=C< → ROO – C – $C^{\bullet}$ – (III)

| |

O

$ROO^{\bullet}$ + >C=C< → / \ $RO^{\bullet}$ + >C – C< (IV)

Peroxy radicals are thus mainly responsible for propagation of reaction chain. The inhibition of oxidation can be accomplished either by removal of peroxy radicals for breaking the oxidation chain reaction or by obviating free radical formation (Tenox, 2010).

The well known and widely used antioxidants of the chain breaking class are alkyl phenols and aromatic amines. Recently Mannich bases/ guanidine derivatives and hydroxy aromatic nitrogen heterocyclic like azoles, azines and imidazolines derivatives have taken precedence over these classes of compounds because of their better effectiveness even at higher temperatures at which phenols and amines become ineffective. It is believed that these materials function by donating a hydrogen atom from their active centers to peroxy radicals; the initial reaction is (Aluyor et al., 2008):

$ROO^{\bullet} + AH \rightarrow ROOH + A^{\bullet}$; $ROO^{\bullet} + A^{\bullet} \rightarrow ROOA$

The antioxidant properties of amine derivatives such as azoles, azines, guanidine and urea derivatives function by a combination of several modes of action such as oxygen and proton scavenging and hydro peroxide decomposition, azoles and azines are effective even in stabilizing the oxidation due to trace metallic components

(Barclay et al., 1999; Pedrielli et al., 2001). Recent work has been conducted on azine, azole and amide. These additives are highly effective in gum inhibition and prevention of oxidation at higher temperatures (Schwetlick et al., 2002; Antioxidant, 2010). Oxidative decomposition of hydrocarbon produces colloidal insoluble particles. The stability of colloidal particles is opposed by Vander Waals attractive force. Thus, in the absence of any forces opposing the Vander Waals force of attraction, the particles must be coagulating rapidly. When MFA present, coagulation does not takes place. Repulsion between colloid particles may be electrostatic with adsorption of ionized surfactants on the particles or steric phenomena of repulsion in the case of non-ionic or undissociated surfactant or polymers (Glavati et al., 1979). Detergent-dispersant additives or MFA can solubilize oil-insoluble polymerized products formed due to oil and fuel decomposition or at least keep it as dispersion.

This action prevents the precipitation and deposition of colloidal materials on metal surface in form of sludge, varnish and carbon deposit. MFA also form film on metal surface to avoid such deposition. Although a number of formulations have been described as furnace oil additives in the market but largest volume of additives sold in the market consists of the magnesium derivatives of sulfonates as furnace oil additives because of their ability to reduce smoke, act as corrosion inhibitors and detergents for maintaining over all cleanliness of burner parts (Sereda et al., 1981). However, residual fuel composition is being burnt in several stationary plants and furnaces having large quantities of asphaltenes do contain special formulations containing amine type of dispersants cum sedimentation inhibitors which are directly purchased by the consumers. Recent attention is being drawn towards clean fuel combustion and fuel economy. Although the use of combustion improvers and emission control additives are still limited but these are likely to grow at a rapid rate in future due to increasing stringency in control of emissions from thermal power plants oil fired boilers and trawlers. Present communication deals with evaluation of various additives, formulations and chemicals that are either commercially available or synthesized in authors' laboratory for cracked residual fuel oil. The objective is to assess the suitability of these additives and chemical components with regard to sediment and deposit inhibiting characteristics and study the effect of the additives on actual fuel combustion so that suitable type of additives can be identified for cracked residual fuel oil.

MATERIALS AND METHODS

The cracked furnace oil (without additive) was collected from refinery. The additives, which are reported in this paper, are of three types given Table 2. Generally, alkylation was done to provide oil solubility in the compounds. In this way, either pure compounds or their derivatives were used in the present study. The methods of derivative preparation were as follows:

N-alkylation

Amine compound (1mole), alcohol (1mole) and zerolite 225 catalyst (1%) refluxed in toluene with water trap for 8 h. The water of reaction was separated and the product cooled, filtered and toluene distilled off. Yield 70 to 80%.

C-alkylation

Compound (1 mole), olefin (4 moles) and anhydrous aluminium chloride (1mole) was taken in a reactor. Anhydrous aluminium chloride was added in several small portions to control the temperature between 60 to 80℃. The reaction was carri ed out for 10 h. The reaction product was poured on ice and hydrochloric acid mixture and kept overnight then extracted by petroleum ether 60 to 80℃. The solvent and olefin recovered by distillat ion. Yield 60 to 70%.

Mannich base

Amine (1 mole) and phenol (1 mole) was taken in benzene and cooled to 20℃ with constant stirring. To the stirred mixture, formaldehyde (1mole) was added drop wise in 30 min with temperature maintained at 25 to 30℃. The benzene wa s removed to get the required mannich base. Yield 95%.

Other branded additive obtained from market. It is a common practice to dope the furnace oil with additives in concentration range of 50 to 200 mg/liter. However, it was noticed during experiments that for this type of furnace oil, optimum dose would be around 1000 mg/liter (Stabilizer-500 mg/L + detergent /dispersant-mg/liter). So, the results on 1000 mg/L dose and only valuable information's are reported in this paper.

Characteristics of furnace oil

A study of Physico Chemical characteristics of the cracked furnace oil stream (Table 3) showed that asphaltenes content of the sample as determined by IP 53/90 method was 0.8% which was quite high. A qualitative study of the furnace oil stability and deposit forming tendencies was made by heating it to 100°C and spotting it on a whatman no. 1 filter paper placed in a blotter test frame and dried at 100°C in an oven for 1 h. The spot had a dark thick inner ring, which indicated the presence of large quantities of thermo-oxidative unstable constituents. This indicated that the cracked furnace oils besides containing large amounts of asphaltenes also contains unstable constituents and, therefore, require not only dispersant additives but also thermo-oxidation stabilizers (Nalco, 1963).

Ageing test

The aging was carried out by IP 390/90 and 375/86 in duplicate. 25 ± 1 g of aliquot of the homogenized sample was poured into the conical flask, attached with the air condenser by means of the cork, and the sample was placed in the well of the aging bath for 24± 0.25 h at 100±0.5°C (IP, 1990; ASTM, 1982). The flask from bath was removed; air condenser was replaced by the rubber stopper and shaken vigorously until all the sludge was uniform. To check this, flask was inverted and bottom/wall of the flask was examined for any sludge deposit, after draining off the oil. Deposits from the wall/ bottom of the conical flask were removed by scraping with the spatula. Content of the flask was re-shaken and the specified procedure (IP 375 / 86) was initiated within 1 min. For each test two

Table 2. Source of three types of additives.

Source of additives	Compounds	Code	Remarks
Synthesized in author's lab by using chemicals from Merck and Aldrich	• Mannich base of dodecyl phenol and poly isobutylene (PIB) succinimide of pentaethylenehexamine	D1	Detergent- Dispersant
	• Di dodecyl benzene sulfonate salt of pentaethylenehexamine	D2	-do-
	• Di dodecyl benzene sulfonate salt with tetraethylene pentamine	D3	-do-
	• Di dodecyl benzene sulfonate salt with triethenetetramine	D4	-do-
	• Phenylene diamine PIB succinimide mannich base	D5	-do-
	• PIB succinimide of pentaethylenehexamine	D6	-do-
	• Di dodecyl phenylene diamine	S1	Stabilizer
	• 5- amino tetrazole and dodecyl phenol adduct	S2	-do-
	• C- Alkylated dodecyl iH-benzotriazole	S3	-do-
	• 2,5, di dodecyl amino phenol	S4	-do-
	• N,N' di dodecyl gunadine derivative	S5	-do-
Pure compound	Octadecyl amines	S6	Merck
Branded Additives	• Commercial additive pack	C1	Nalco
	• Commercial additive pack	C2	Lubrizole
	• Commercial additive (2,6,di,tert. butyl-4methyl phenol)	C3	Merck

Table 3. Characteristics of residual fuel oil used for study

Characteristics	Method	Values
Kinematic viscosity CST at 100°C	ASTM D445/71	15.66
Kinematic viscosity CST at 40°C	ASTM D445/71	236.61
Water content % (Vol. /Vol.)	ASTM D95/83	nil
Ash (Wt. %)	ASTM D482/91	0.11
Sediments (existing)	IP 53 / ASTM D	4730.6
Rel. density at 20°C	ASTM D445/71	0.9582
Sulfur (Wt. %)	ASTM D129	1.42
Asphaltenes	IP 143	-
Sediment (potential) by (Wt. %)	IP 390 and 375	1.00
Carbon (wt %)		86.88
Hydrogen (wt %)		11.62
Pour point °C	ASTM D97	30
Flash point °C	ASTM D92	99.8
Calorific value (Kcal/kg)	ASTM D240	97.67

filter media were dried at 110℃ for 20 min. in the oven. Each paper were rapidly transferred separately to a numbered weighing bottle and allowed to cool in the cooling vessel up to room temperature (5 to 10 min). Pan balances were used for each weighing bottle plus filter medium by the tare method against an empty similar bottle, to nearest 0.001 g. The Whatman GF/A filter media are fragile and

Table 4. Study on additives for furnace oil stability.

Additives – 1000 mg/lit	Total sediment % In mg/100 ml	Unfilterable in mg / 100 ml
	Test IP 390/ 370	Long storage
Undoped fuel oil – Blank	920	3000
Alkyl phenol succinimide - D1	130	1598
Alkyl aryl hexa-amine sulfonate - D2	110	1382
Alkyl aryl tetra-amine sulfonate - D3	120	274
Alkyl aryl tri-amine sulfonate - D4	120	1485
Phenylene diamine succinimide - D5	80	1443
Alkyl penta amine succinimide - D6	40	2332
Hindered phenylene diamine - S1	180	1880
Tetrazole phenol adduct - S2	150	1738
Alkylated benzotriazole - S3	150	2052
Hindered amino phenol - S4	160	2002
Gunadine derivative - S5	150	-
Alkylated amines - S6	260	-
D6 + S3 (50:50)	190	-
D6 + S2 (50:50)	160	-
D6 + S1 (50:50	170	-
D5 + S3 (50:50)	110	-
D4 + S6 (50:50)	90	-
D4 + D5 (50:50)	110	1155
D1 + D6	80	-
D2 + S3	100	1358
D1 + D2	140	1556
D3 + S3 (50:50)	80	1125
D3 + S4 (50:50)	100	1215
Commercial additive pack -C1	340	1638
Commercial additive pack -C2	380	2624
Commercial additive pack -C3	190	1950

Note: name of additive is indicative and short.

were handled with care. The aged furnace oil sample prepared as described above and weighed to the nearest 0.01 g.

A portion of sample was poured (approximately 11 g) into a 30 ml beaker. Vacuum source was connected and vacuum was applied to an absolute pressure of 40 ± 2 kPa (minimum). The content of the beaker was heated up to 100 ± 2°C. Then the content at 100 ± 2°C was transferred to the center of the filter medium, taking care that no sample touched the walls during transfer. The beaker was reweighed to the nearest 0.01 g. The quantity transferred should be 10 ± 0.5 g. When filtration was not completed in 25 min., the test was discontinued and repeated again using 5 ± 0.3 g of sample. If filtration was not completed in 25 min. the result was reported as "filtration exceeds 25 min." When the filtration was completed and the upper filter medium appeared dry, the steam and vacuum continued for further 5 min. The steam supply was discontinued and passing tap water through the coils cooled the apparatus. The filtration unit was carefully washed with two portion of 25 ± 1 ml of the solvent from a syringe or graduated wash bottle with a fine nozzle, care was taken to remove any adhered sample from the wall of the upper part of the apparatus. Top portion of the filtration unit was carefully removed and the rim of the filter was washed with 10 ± 0.5 ml of the solvent in similar manner. Finally the whole of the filter medium area was washed with 10 ± 0.5 ml of n-heptane. Mass percentage of total sediment was calculated and reported in Table 4.

Thirteen weeks long storage test

Because fuel was stored at reduced temperature, it was allowed to come to ambient temperature (DuPont, 1964). It was made certain that entire fuel sample was at least 5°C above its cloud point before proceeding, to dissolve any separated wax. Furnace oil sample was filtered through glass-fiber filters. Storage oven was adjusted to a temperature of 50±1°C. 400 ml of filtered fuel was placed into each bottle. The period for this accelerated storage test was 13 weeks. After 13 weeks of storage bottles were removed from the storage oven and allowed to cool to 21 to 27°C in a dark place for 4 to 24 h. After cooling the fuels were analyzed by filtering it through Gooch crucible containing glass-fiber filters. Bottles were rinsed by n-heptane. Filter papers were weighed after drying and gum solvent was poured in to bottles for adherent gum determination. Results were reported in Table 4.

Combustion test

A horizontal cylindrical oil fired furnace fitted with a low air pressure (LAP) atomizing burner (Figure 1) has been used for the evaluation work, which are commonly used in India (Madan et al., 1999). The furnace consists of a circular cylindrical refractory lined combustion chamber (75 cm dia x 500 cm length) having a conical exit

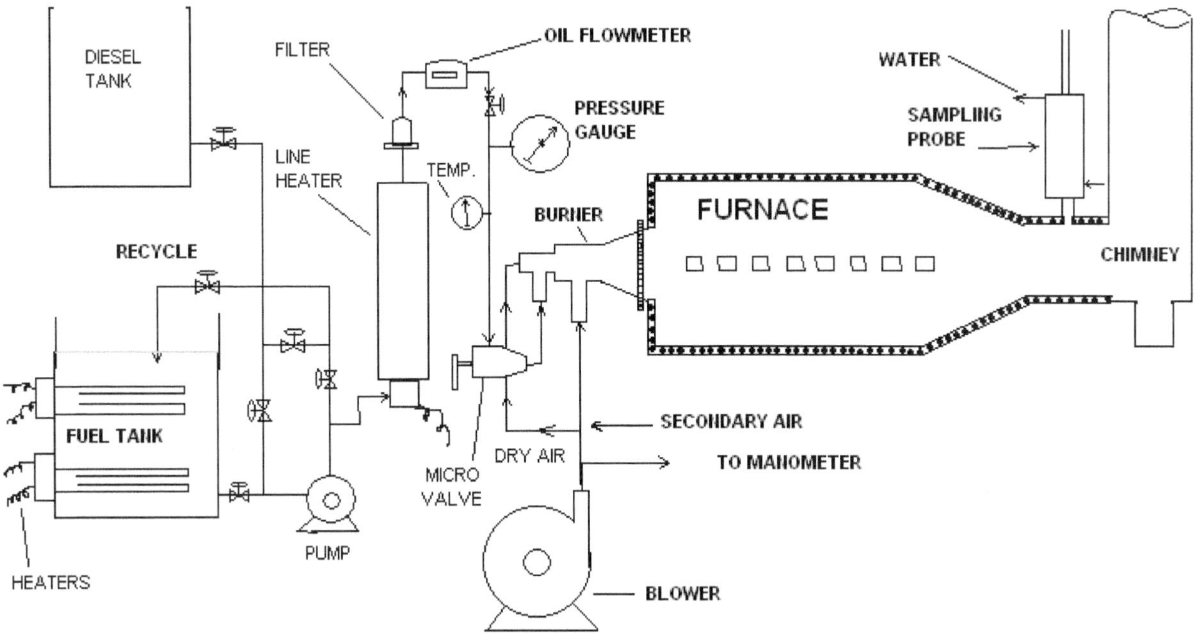

Figure 1. Experimental setup for combustion evaluation.

Table 5. Combustion study on additives for furnace oil stability.

Additive	Combustion efficiency Higher CO_2 %		Fuel economy less coke – less air		Particulate less smoke		Remarks
	CO_2 %	CE % increase	Air %	FE % increase	Smoke no	PM % decrease	
Blank	13	-	15	-	2.5	-	
C3	13.5	3.80	14	6.7	3.0	- 20	Good for FE only
D6+S2 (50:50)	15	15.4	11	26.7	1.5	40	Good for all
D3+S3 (50:50)	14	7.70	19	-26.7	3.5	-40	Good for all
D6	14	7.70	19	-26.7	3.5	-40	Improve CE only
S2	14	7.70	15	0	2.5	0	Improve CE only
D1+ D6 (50:50)	15	15.4	11	26.7	0.5	80	Improve CE only

CE- combustion efficiency, FE- fuel efficiency, PM – particulate matter.

connected to a chimney with natural draught. A suitable damper has been provided in the chimney to control furnace pressure. The burner assembly has been mounted on the front plate of the furnace. Flue gas was sucked through a water-cooled sampling probe provided in the chimney and was analyzed for CO_2% and 'Bacharach Smoke-Number' at regular intervals. Exhaust gas analyzed at a point in the chimney where no flame was visible by using 'Fyrite indicator' and 'Bacharach Smoke Meter'. Furnace wall temperature measured using Pt-Pt-Rh thermocouple. Specifications of burner were: capacity 10 to 65 lit/ h Turn down ratio 7:1; Furnace oil pressure-1 kg/cm^2; Air pressure 700 mm WG max. ; Excess air 5 to 10%, Furnace oil preheats temperature 80 to 90°C. The characteristics of furnace oil used for the evaluation are reported in Table 3. The additives were doped in furnace oil at 80 to 90°C with thorough mixing for 2 h. Initially the burner fired with HSDO for about 2 h to preheat the furnace and then switched over to untreated furnace oil. Fuel burning rate has been maintained by keeping pressure of the oil constant and with the help of micro valve provided near the burner. After the furnace wall temperature got stabilized (above 1000°C), measurement for the products of combustion were made by sucking flue gases through the sampling probe provided in the chimney. These measurements have been repeated at regular intervals while maintaining smoke level of flue gases at 2 to 3 'Bacharach Smoke Number'. Results were reported in Table 5.

RESULTS

This study was conducted into two parts. The first part related to the laboratory study for screening of the stabilizer additives and second part for combustion quality study at the pilot plant level. Table 1 indicates that BS is for fuel oil of better quality in comparison to IS. In

BS sediment old limit is 0.15 / 0.25 / - against IS limit of 0.25 and future BS limit is -/ 0.1/ 0.2 / 0.2 against IS 0.25 / 0.15/ 0.25/ 0.25. It means, the furnace oil available in India is of lower quality in comparison to European furnace oil even though both are satisfying the respective standards. It is a matter of further concern that the sample of furnace oil taken for study has poor quality than IS standard. It is indicated in Table 3 that analysis of furnace oil sample shows the kinematic viscosity 236.61cst at 40℃ and 15.66 cst at 100℃ indicates that it is high viscosity (HV) grade oil. It is meeting all specification except sediment, which is in higher side, 4.73 against prescribed limit of 0.2. It is possible only due to unstable component of furnace oil, which was oxidized during transportation and storage. Thus common furnace oil in India needs more effective additives than what was developed on furnace oil of European standard.

DISCUSSION

Aging test

Screening study was carried out with more than 100 compounds but only few are discussed here. Table 4 showing the results of evaluation of furnace oil performed under aging test in terms of total sediments with a number of additives and their combinations in cracked furnace oil samples. It was observed that dispersant additives D1, D2, D3, D4, D5 and D6 all reduce total sediment as determined by IP 390 and 375 after aging from 920 mg/100 ml in undoped furnace oil to the range 190 to 80 mg/100 ml which is much below the requisite value of 250 mg/100 ml. Stabilizer type of additives S1, S2, S3, S4 and S5 all reduce total sediment. Combinations of dispersant and stabilizers (D6+S3), (D6+S2), (D6+S1), (D5+S3), (D4+S6), (D4+D5), (D1+D6), (D2+S3), (D1+D2), (D3+S3) and (D3+S4) all reduce total sediment. Commercial additive pack C1, C2 and additive S6 are not so effective. It means alkylated amines will not be effective on aging phenomena. Phenol types of stabilizer (S3, S4 and C3,) are not so effective. Phenols are not effective in combination also. Succinimides are very effective as single additive or in combination. D1-Mannich base is better than phenol but less effective than succinimide. Sulfonates (D2, D3 and D4) are behaving similarly. So, first indication is, nitrogen carrying compounds (succinimides) are more effective than sulfur or oxygen carrying compounds (Sulfonates). Second, increase in number of nitrogen will be benefited. Order of effectiveness is $(NH^+_3)<(O\text{-}C\text{-}N)<(\text{-}NH_2)<(\text{-}NH\text{-})<(\text{-}N\text{-}C_2\text{-}N\text{-})<(\text{-}N\text{-}C_2\text{-}N\text{-}C_2\text{-}N)<(\text{-}N\text{-}C_2\text{-}N\text{-}C_2\text{-}N\text{-}C_2\text{-}N\text{-})<(\text{-}N\text{-}C_2\text{-}N\text{-}C_2\text{-}N\text{-}C_2\text{-}N\text{-}C_2\text{-}N\text{-})<(\text{-}N\text{-}C_2\text{-}N\text{-}C_2\text{-}N\text{-}C_2\text{-}N\text{-}C_2\text{-}N\text{-}C_2\text{-}N\text{-})$.

This trend indicates that chain length of multi-amine further increased, then a super additive will be produced, this can easily stabilize even furnace oil of poor quality with highly unstable component. D6- PIB succinimide of pentaethylene hexamine is most effective. Therefore; this additive can be used successfully for improving the stability of cracked furnace oil at higher temperature. This test simulates the accelerated oxidation of furnace oil in presence of catalyst (metals).

Long term storage stability

Study was carried out with more than 100 compounds but only few are discussed here. Long term storage stability simulates auto-oxidation behavior of furnace oil in contact with air. Characteristics for 13 weeks at 50°C as per procedure and its results in Table 4 shows that dispersant additive such as D2, D3, D4 and D5 are likely to give extremely good performance in the actual field application also. It can be also inferred that stabilizers S1, S2, S3, S4, S5 and S6 all reduce the value of sediment in fuel oil at 1000 mg/lit concentration from 3000 to 2332 - 274 mg/100 ml which is reasonably good from the point of view of fuel stability characteristics as reflected by storage test. However, 13 week long term storage stability test shows that only stabilizers S2, S3 and S5 are likely to perform satisfactorily alone. Generally rest of these additives should preferably be used in combination with dispersant additives.

Table 4 gives the results of evaluation of combination of these additives in equal ratio with total additive dosage kept at 1000 mg/lit level in the fuel oil. Combinations of additives like (D4+S6), (D1+D6), (D2+D3) and (D3+D3) show synergistic performance characteristics. These combinations may prove highly effective and some of them commercially viable too. A pattern of behavior is also found in this test. Here, behavior of succinimides is still best. But behavior of D3 is surprising. It is not due to sulfonate portion because same effect is not seen in other sulfonate D2 and D4. It is not due to number in multi-amine chain as evident from D6. Then possibility is with structure of multi-amine chain (Figure 2). Teraethylene pentamine should have a different structural effect than other polyamine. It is a matter of further study. Both (aging test and long term test) was giving one important clue that cracked furnace oil is a highly unstable product. It starts degrading just after production. Once it degraded, it is very difficult to revert it or made it good fuel by any means. So, it advisable to use additive in furnace oil, just after its production to avoid major part of oxidation/ degradation.

Combustion study

Referring to Table 5, which give results of performance trials with undoped and doped residual fuel oil (with some of the additives prepared in bulk) on an LAP burner in an actual furnace. It can be observed that additives such as D6+S2 and D1+D6 sows significant improvements in

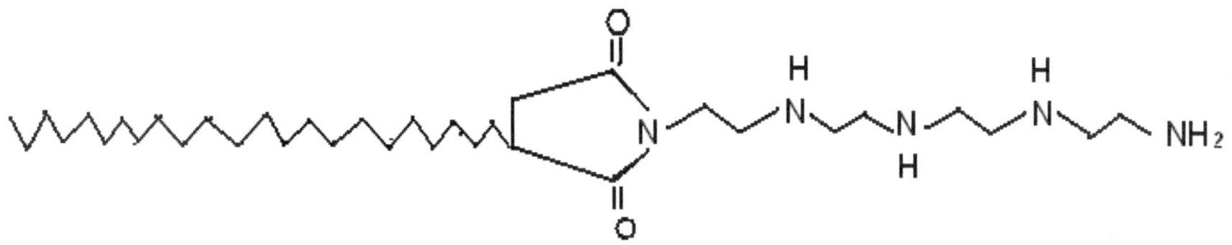

Figure 2. Molecular structure of PIB succinimide of tetraethylene pentamine.

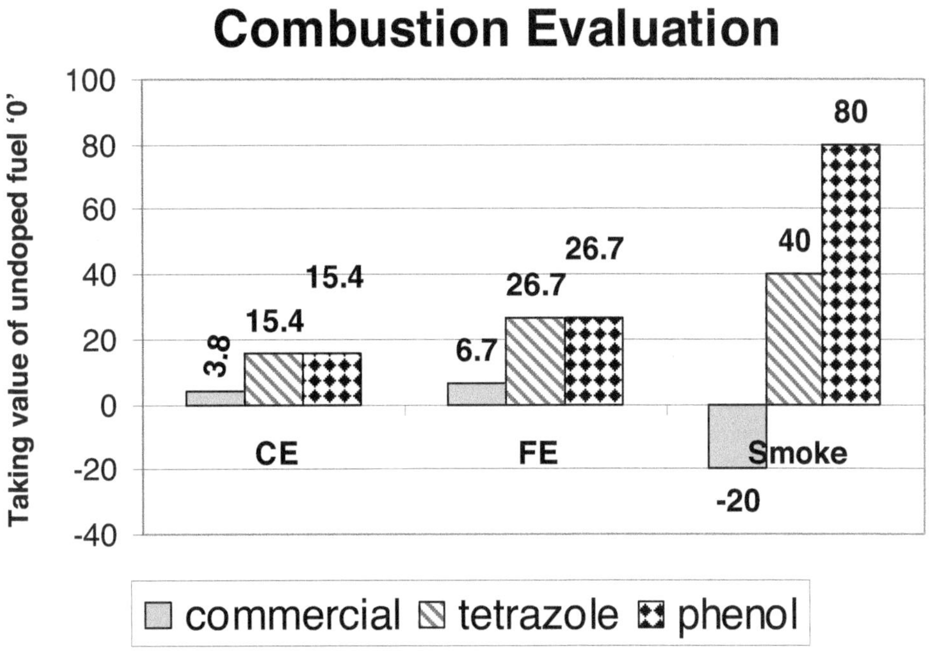

Figure 3. Combustion efficiency comparison. Note: Here in this graph Commercial means–C3, Tetrazole means– D6+S2 and phenol means– D1+D6.

terms of combustion efficiency (15.4% CO_2 in exhaust air), fuel economy (26.7% excess air), particulates in flue gases (Smoke No. reduction 40%) and deposit forming tendencies. Thus, succinimides with tetrazole or phenol in combination will be the best choice. In Figure 3, the performance of these two combinations was compared with undoped fuel. While additives such as C1 and S2 do not significantly differ from performance of undoped fuel oil except with respect to deposit forming tendencies. Additive D3+S3 and D6 perform poorly as compared to undoped fuel oil with respect to combustion efficiency and fuel economy but shows great deposit reducing tendencies. HV which was excellent performer in aging test and long storage test was not effective in combustion test. It is very clear that only succinimide or phenol will not be effective in actual practice. A combination of amine and phenol is needed for efficient combustion.

Succinimide can control oxidation, sediment, precipitation of polymerized fuel particles or deposit during transportation and storage. At low temperature and for longer period succinimide can stabilized the unstable component of furnace oil. Succinimide will resist the degradation. During combustion this resistance must be removed for quick and complete degradation or combustion. Here, phenol is acting as combustion controller. Succinimide is providing perfect atomized fuel and phenol is acting like a policeman for smooth combustion. This type of combination of additives may help in efficient combustion of furnace oil with more unstable component. It will be more beneficial to mix the additives in the refinery, immediately after its production. Cracked furnace oil is more unstable than distillates. So it needs high performing anti-oxidant. If the additives are doped in furnace oil at later stage, after transportation, then not only stabilizer but detergent-dispersant are also needed to disperse the already degraded product. It is better to add the additive pack during the production of furnace oil.

Conclusion

Cracked furnace oil not only has a high asphaltenes content but also thermo-oxidative unstable constituent which can cause large-scale deposits. Thus, general commercial additives, which are developed for good quality furnace oil, may not be able to enhance the combustion efficiency of the furnace oil having poor quality. A synergistic combination of dispersants and stabilizers (succinimide and phenol) are able to reduce the sediments to reasonably lower values of 80 to 100 mg/100 ml which are almost 10 times lower than the un-doped furnace oil. Many of these synergistic combinations of additives show almost 2.5 times the reduction in un-filterable residue after 13 weeks of storage at 50℃. It is possible to improve quality of cracked furnace oil to an acceptable performance level by doping it with a synergistic combination of additives at earlier stage/ during production.

ACKNOWLEDGEMENT

The grant of permission for publication and facilities for carrying out this work by Dr. M. O. Garg, Director, Indian Institute of Petroleum Dehradun and help from analytical persons are gratefully acknowledged.

REFERENCES

Aluyor EO, Ori-Jesu M (2008). The use of antioxidants in vegetable oils 12– A review. Afr. J. Biotechnol., 7(25): 4836-4842.

Antioxidant, Wikipedia-the free encyclopedia, accessed on Aug 30, 2010, http://en.wikipedia.org/wiki/Antioxidant.

ASTM (1982). Standard test method for tensile properties of plastics, Plastics (1), American Standard Test Methods, Consohocken USA,.

Barclay LRC, Edwards CE, Vinqvist MR (1999). Media effects on antioxidant activities of phenols and catechols. J. Am. Chem. Soc., 121: 6226-6231.

BIS (2002). Bureau of Indian Standard, New Delhi, India. IS 1593-1982. Furnace oil specification.

Danilov AM, Okina NG, Mitusova TN, Pugach IA (2003). Increase in the efficiency of use of furnace residual fuel oil. Chem. Technol. Fuels Oils, 39(1-2): 6-9.

Dunn RO (2005). Effect of Antioxidants on the Oxidative Stability of Methyl Soyate (Biodiesel). Fuel Proc. Technol., 86: 1071-1085.

DuPont (1964). 110°F Accelerated Fuel Oil Storage Test. Petroleum Laboratory Method F8-64, Petroleum Chemicals Division, E.I. duPont De Nemours & Co., Inc., Wilmington.

Ford JF (1968). Lubricating Oil Additives- A Chemist's Eye-View, J. Inst. Pet., 54: 198-209.

Fuel oil (2010). Multi Commodity Exchange of India Ltd. (MCX)– Furnace oil profile. accessed on Sept 3, 2010 http://www.mcxindia.com/ SitePages/ Contract Specification. aspx? ProductCode=FURNACE_OIL.

Glavati OL, Vipper AB, Glavati EV, Lukashevich KN, Marchenko AL (1979). Chem. Technol. Fuels and Oils: Operating mechanism of detergent dispersant additives, Springer, New York.

IP (1990). Standard Methods of content. vol I., Institute of Petroleum, UK,

Jones BA, Weil BH, Graham MH (1958). Progress in Petroleum Technology: Literature on the Combustion of Petroleum-derived Fuels, ACS publication.

Madan HK, Jaiswal UK, Kumar S (1999). Effect of Additives on Combustion performance of fuel oil, IIP Technical Report Comb-32A/99, Indian Institute of Petroleum: Dehradun, India.

Murphree EV (1951). Progrss in Petroleum Technology: Fluid Catalytic Cracking Process, ACS Publication.

Nalco (1963). Elevated Temperature Stability Test for Evaluating the Stability of Distillate Fuel oils. Nalco Technifax, Industrial Division TF9, Chicago, Ill., USA.

Nixon AC (1962). Autoxidation and Antioxidants: Autoxidation and Antioxidants of Petroleum. Lundberg WO (1962) Interscience, New York.

Notes (1984). On Heavy Fuel Oil, American Bureau of Shipping New York. Accessed on Sept 18, 2010., www.eagle.org/ .../Rules...HeavyFuelOil /Pub31_HeavyFuelOil.

Pedrielli P, Pedulli GF, Skibsted LH (2001). Antioxidant mechanism of flavonoids. Solvent effect on rate constant for chain-breaking reaction of quercetin and epicatechin in autoxidation of methyl linoleate. J. Agric. Food Chem., 49: 3034-3040.

Schwetlick K, Wolf D, Habicher (2002). Antioxidant action mechanisms of hindered amine stabilizers, Polymer Degradation Stability, 78(1): 35-40.

Sereda YaI, Pavlik TS, Gud LN (1981). Chemistry and Technology of Fuels and Oils, 17(10), Composition of Sulfonates in Detergent Dispersant Additives. Springer, New York, pp. 586-587.

Tenox (2010). Food-grade Antioxidants for Refined Vegetable Oils / High Performance Additives, Eastman Chemical Company, Kingsport, Tennessee, USA. Assessed on Aug 30, 2010., www.eastman.com.

Wikipedia (2010). Fuel oil- the free encyclopedia. accessed on Sept 17, 2010, http://en.wikipedia. org/wiki/ Fuel oil.

Comparative assessment of picolinic acid with common chemical preservatives in ginger juice during storage

Akinwande, B. A.[1], Adeoye, I. O.[2]* and Anjorin, I. B.[1]

[1]Department of Food Science and Engineering, Ladoke Akintola University of Technology, Ogbomoso, Oyo State, Nigeria.
[2]Department of Pure and Applied Chemistry, Ladoke Akintola University of Technology, Ogbomoso, Oyo State, Nigeria.

In consideration of the medicinal and health benefit of picolinic acid (PA), this research work compared its effectiveness as a preservative with potassium sorbate and sodium benzoate in ginger juice. Concentration of 0.07, 0.067 and 0.073% of PA, potassium sorbate and sodium benzoate, respectively, were added to ginger juice, packed in polyethylene terephthalate (PET) bottles and stored for 12 weeks at ambient condition. Chemical properties (total soluble solids, total phenolic content, total flavonoids, titratable acidity and ascorbic acid), pH as well as microbiological properties (total viable count and total mould count) of the juice were observed on a fourth night basis. Total phenolic content and total flavonoids were best retained in the sample with PA than others. Higher values for pH and total soluble solids were recorded in the sample preserved with potassium sorbate and sodium benzoate. Proliferation of microbial growth was best controlled by sodium benzoate. PA could be used in preference to the other two preservatives due to its ability to preserve bioactive substances as well as its health and medicinal advantages over others.

Key words: Ginger juice, preservative, picolinic acid, potassium sorbate, sodium benzoate.

INTRODUCTION

Preservatives are the substances that increase the shelf life of foods and beverages. Shelf life is determined by rates of growth of spoilage microorganisms and chemical degradation of food components. Preservatives inhibit one or both of these processes. Several chemical compounds are able to inhibit the growth of microorganisms. One of them, benzoic acid, is a well-known and commonly used food preservative that is present naturally in some materials (Marinova and Yanishlieva, 2003). Its derivatives such as sodium benzoate are also active compounds and are used commercially as food preservatives (Conrad et al., 1994). Potassium sorbate is another established food preservatives, its effectiveness is 5 to 10 times of sodium benzoate and is used in quantities in which there is no known adverse health effect (Westhoff and Franzier, 1998). However, only few of them grant claims for preservatives, which is why the search for ideal antimicrobial agent is continued.

Picolinic acid (PA) is a naturally occurring degradation product of L-tryptophan detected in the human body (Dazzi et al., 2001). It is synthesized in the liver, kidney, and other organs (Fernandez-Pol et al., 2001). PA was found to have a number of biological functions. It is a cheap and safe drug facilitating Zinc/Chromium ion absorption from the intestine because of its metal ion-chelating activity (Evans and Johnson, 1980). Its activity as a metal ion chelator and as a complex group with Chromium makes it useful as a dietary supplement for obese people, since it has a beneficial effect on reducing diabetes risk (Komorow et al., 2008).

Borawska et al. (2008) reported that PA and sodium picolinate show a high antimicrobial activity at pH 5.0 and 7.0 and that PAs and their salts may represent new potential food or cosmetic preservatives.

Ginger (*Zingiber officinale*), one of the oldest spices used worldwide for its peppery taste, preservatives power and typical aroma, has lots of medical benefits. Moreover, ginger contains a number of antioxidants such

*Corresponding author. E-mail: adeoyeid@gmail.com.

Table 1. pH of ginger juice during storage.

*Sample	Weeks						
	0	2	4	6	8	10	12
A	5.01^{a}	4.93^{b}	4.75^{c}	4.71^{d}	4.74^{cd}	4.74^{c}	4.76^{c}
B	5.10^{a}	5.03^{b}	4.91^{c}	4.87^{d}	4.89^{cd}	4.90^{c}	4.91^{c}
C	5.48^{a}	5.33^{b}	5.24^{c}	5.17^{d}	5.18^{d}	5.19^{d}	5.21^{cd}
D	5.23^{a}	5.14^{b}	5.07^{cd}	5.06^{d}	5.06^{d}	5.07^{cd}	5.08^{c}

Means followed by different letters across the rows are significantly different ($p < 0.05$) from one another. *A, Control sample; B, sample preserved with PA; C, sample preserved with potassium sorbate; D, sample preserved with sodium benzoate.

as beta-carotene, ascorbic acid, terpenoids, alkaloids and polyphenols such as flavonoids, flavones glycosides and rutin (Okezie et al., 1997). Ginger with its wide range of antioxidants can be a major source of natural or phytochemical antioxidants. The aim of this study was to carry out a comparative assessment of the preservative effect of PA with sodium benzoate and potassium sorbate in ginger juice during storage.

MATERIALS AND METHODS

Ginger (*Z. officinale*) root and sweetener were obtained from an open market in Ogbomoso. Polyethylene terephthalate (PET) bottles (50 ml size) were sourced from CASCADE Nigeria Limited (private company), Lagos State, Nigeria. PA, potassium sorbate and sodium benzoate were Sigma analytical reagents and were obtained from the Inorganic Chemistry Laboratory of Ladoke Akintola University of Techonology, Ogbomoso, Oyo State, Nigeria.

The ginger was sorted, cleaned, washed and crushed with mortar and pestle to extract the juice. The extracted juice was sieved with muslin cloth and diluted with portable water before the addition of sweetener. The ginger juice was divided into four portions. The first portion served as control without any preservative, 0.070% PA was added to the second portion, 0.067 potassium sorbate was added to the third portion, while 0.073% sodium benzoate was incorporated to the last portion. Each portion was pasteurized at 75°C for 30 min and hot fill into sterilized PET bottles. Sterilization of the bottles was carried out by immersing them in 7 ppm of chlorine solution for about 5 min. The filled bottles were kept on the shelf at room temperature and were sampled for analyses on fourth night basis in three replicates.

Chemical analysis

The pH and total soluble solids of the processed ginger juice were measured according to the method of AOAC (2000). Ascorbic acid content was also determined according to the procedures in the same method. The standard solution was prepared with 50 ml ascorbic and diluted to 100 ml with water. The sample and standard solution were titrated against dichloroindophenol solution to a pink endpoint lasting at least 10 s before the reading was taken. The result was expressed as mg ascorbic acid/ml sample. Titratable acidity was measured by addition of 10 ml of sample to 100 ml distilled water and titrating the solution with 0.1 N sodium hydroxide to an endpoint of light pink with 3 drops of phenolphthalein indicator (Height and Gump, 1995). The result was expressed as gram tartaric acid/100 ml.

The total phenolic content was determined by the Folin-Ciocalteu method. To 3.90 ml of water, 0.1 ml of the sample (10% v/v ginger juice) was added followed by addition of 0.5 ml of Folin-Ciocalteu reagent. After 3 to 6 min, 0.5 ml of saturated sodium carbonate (20 g of Na_2C0_3 in 100 ml of H_20) was added. After 30 min of vigorous mixing with a vortex mixer, a reading was taken at 725 nm (2120UV spectrophotometer). The result was expressed as gallic acid equivalents using a calibration curve with gallic acid as the standard (mg/L).

Total flavonoid content was determined according to the procedures outlined by Zhishen et al. (1999). 1 ml of diluted sample 1 ml sample/5 ml distilled H_20) was placed in a 10 ml flask. 4 ml distilled water was added, then 0.3 ml of $NaN0_2$ (5 g/100 ml distilled water) was added. After 6 min, 2 ml of 1 N NaOH was added and the solution was diluted to a total volume of 10 ml with distilled water. The absorbance of the solution was measured at 510 nm and flavonoid concentration was determined using a catechin calibration curve.

Microbiological analysis

The total mould count was determined according to the method of Adams and Moss (2008). 1 ml of the sample was pipetted into a sterile Petri dish and mixed with an appropriate volume of molten agar (potato dextrose agar) using pour plate method with incubation period of 48 h at 25°C. The plate was observed under electronic colony counter for microbial growth. The total bacteria count was also determined using the same method. 1 ml of the sample was pipetted into a sterile Petri dish and mixed with an appropriate volume of molten agar (plate count agar) using pour plate method with incubation period of 48 h at 35°C. The plate was observed under electronic colony counter for microbial growth.

Statistical analysis

All the analyses were done in three replicates. Statistical analysis of all the data was done with the Statistical analysis system (SAS) package (version 8.2 of SAS institute Inc., 1999). Statistically significant differences ($p < 0.05$) in all data were determined by analysis of variance (ANOVA) procedure, while Fisher's test was used in determining the least significant difference (LSD) for separation of the means.

RESULTS AND DISCUSSION

Chemical properties

Variation in pH values was observed in all the samples before storage as shown in Table 1. Highest value of

Table 2. Total soluble solids (%) of ginger juice during storage.

*Sample	Weeks						
	0	2	4	6	8	10	12
A	18.00^{f}	18.25^{e}	18.30^{de}	18.35^{d}	18.50^{c}	18.60^{b}	18.70^{a}
B	17.10^{e}	17.20^{d}	17.20^{d}	17.30^{c}	17.50^{b}	17.60^{a}	17.65^{a}
C	18.50^{e}	18.75^{d}	18.90^{c}	18.95^{c}	19.15^{b}	19.25^{a}	19.30^{a}
D	17.50^{f}	17.55^{f}	17.70^{e}	17.80^{d}	17.90^{c}	18.00^{b}	18.10^{a}

Means followed by different letters across the rows are significantly different ($p < 0.05$) from one another. *A, Control sample; B, sample preserved with PA; C, sample preserved with potassium sorbate; D, sample preserved with sodium benzoate.

Table 3. Ascorbic acid content (mg/100 ml) of ginger juice during storage.

*Sample	Weeks						
	0	2	4	6	8	10	12
A	9.75^{a}	8.75^{b}	7.50^{c}	6.50^{d}	5.50^{e}	4.75^{f}	4.50^{f}
B	12.65^{a}	10.1^{b}	9.00^{c}	8.25^{d}	7.50^{e}	7.00^{f}	6.25^{g}
C	12.65^{a}	10.5^{b}	8.75^{c}	8.00^{d}	7.50^{de}	7.00^{ef}	6.50^{f}
D	10.25^{a}	9.50^{b}	8.25^{c}	8.00^{d}	7.50^{e}	7.25^{f}	7.00^{g}

Means followed by different letters across the rows are significantly different ($p < 0.05$) from one another. *A, Control sample; B, sample preserved with PA; C, sample preserved with potassium sorbate; D, sample preserved with sodium benzoate.

5.23 was obtained in sample preserved with sodium benzoate, while least value of 5.01 was recorded in the control sample. The pH of all the samples decreased with increase in storage period till 6 weeks (with significant difference at $p < 0.05$) and increased thereafter till 12 weeks. Oroniran et al. (2011) also reported increase in pH value in soymilk that was preserved with spices during storage. Percentage reduction in pH at 6 weeks after storage was highest in control sample (5.99%), while values of 4.51, 5.66 and 3.25% were obtained in samples preserved with PA, potassium sorbate and sodium benzoate, respectively.

Total soluble solids content increased steadily, but with significant difference, in all the samples as storage progressed as shown in Table 2. This could be due to depolymerization, or cleavage of glycosidic linkages of the polysaccharides as storage continued. All the samples showed gradual decrease in the ascorbic acid content as storage time increased as shown in Table 3. Samples preserved with PA and potassium sorbate had highest value of 12.65 mg/100 ml at the beginning of the storage period, while control sample had the least value of 9.75 mg/100 ml. The percentage reduction from initial concentration ranged from 31.7% in the sample preserved with sodium benzoate to 53.8% in the control sample.

According to Zvaigzne et al. (2009), the content of Vitamin C in citrus fruit juices substantially decreased after 1 day of storage in a refrigerator. Osundahunsi (2008) reported similar degradation of ascorbic acid content in orange juice at both refrigerated and ambient storage conditions. Citrus juice concentrates that were stored at different temperatures for 8 weeks were also indicated to exhibit loss of ascorbic acid content (Burdulu et al., 2006). It was also indicated that change in the ascorbic acid content was observed at the end of Week 2 under both storage conditions and 40% reduction was observed after storage for 17 weeks at ambient condition.

The total titratable acidity of the samples were in the ranged of 1.60 g/100 ml in sample with sodium benzoate and 2.28 g/100 ml in sample preserved with PA (Table 4). Control sample and sample with potassium sorbate were in the range of 1.88 and 1.63 g/100 ml, respectively. It was observed that in all the samples, the total titratable acidity has a uniform increasing trend although sample that was preserved with PA recorded highest value at all sampling periods. Similar increasing trend was observed in soymilk sample that was preserved with local spices (Oroniran et al., 2011). The samples were significantly different from each other at $p < 0.05$.

Table 5 shows the results obtained for the total phenols in the samples. Except between Weeks 4 and 6, sample preserved with PA had highest values at the sampling times. Highest value of 12.50 mg/100 g was obtained in PA preserved sample at the end of the storage period, while least value of 9.50 mg/100 g was obtained in the control sample. There was a general decrease in the values obtained in all the samples with increase in storage time. Highest loss (24.0%) was recorded in control sample, while the sample preserved with sodium sorbate had least value (13.8%) for percentage loss followed by that preserved with PA (16.7%).

Total flavonoids in the samples reduced as the storage

Table 4. Total titratable acidity (g/100 ml) of ginger juice during storage.

*Sample	Weeks						
	0	2	4	6	8	10	12
A	1.33^{e}	1.60^{d}	1.78^{c}	1.85^{ab}	1.83^{b}	1.85^{ab}	1.88^{a}
B	1.94^{e}	2.08^{d}	2.25^{ab}	2.23^{bc}	2.20^{c}	2.23^{bc}	2.28^{a}
C	1.35^{b}	1.40^{b}	1.60^{a}	1.60^{a}	1.58^{a}	1.60^{a}	1.63^{a}
D	1.39^{c}	1.41^{c}	1.55^{b}	1.60^{a}	1.55^{b}	1.58^{ab}	1.60^{a}

Means followed by different letters across the rows are significantly different ($p < 0.05$) from one another. *A, Control sample; B, sample preserved with PA; C, sample preserved with potassium sorbate; D, sample preserved with sodium benzoate.

Table 5. Total phenolic content (mg/100 g) of ginger juice during storage.

*Sample	Weeks						
	0	2	4	6	8	10	12
A	12.50^{a}	12.00^{b}	11.50^{c}	11.00^{d}	10.50^{e}	10.00^{f}	9.50^{g}
B	15.00^{a}	15.00^{a}	14.00^{b}	14.00^{b}	13.25^{c}	13.00^{c}	12.50^{d}
C	14.50^{a}	14.75^{a}	14.50^{a}	14.00^{b}	13.00^{c}	12.75^{c}	12.25^{d}
D	13.00^{a}	13.00^{a}	12.25^{b}	12.00^{b}	11.25^{c}	11.00^{c}	10.50^{d}

Means followed by different letters across the rows are significantly different ($p < 0.05$) from one another. *A, Control sample; B, sample preserved with PA; C, sample preserved with potassium sorbate; D, sample preserved with sodium benzoate.

Table 6. Total flavonoids (mg/100 g) of ginger juice during storage.

*Sample	Weeks						
	0	2	4	6	8	10	12
A	22.00^{a}	18.50^{b}	16.25^{c}	16.50^{c}	15.00^{d}	14.50^{d}	13.50^{e}
B	28.50^{a}	25.50^{b}	24.50^{c}	24.25^{cd}	24.00^{d}	23.00^{e}	22.00^{f}
C	25.00^{a}	23.50^{b}	22.00^{c}	22.25^{c}	21.25^{d}	21.00^{d}	21.00^{d}
D	20.25^{a}	18.75^{b}	17.50^{c}	17.25^{c}	16.25^{d}	16.00^{d}	15.50^{e}

Means followed by different letters across the rows are significantly different ($p < 0.05$) from one another. *A, Control sample; B, sample preserved with PA; C, sample preserved with potassium sorbate; D, sample preserved with sodium benzoate.

Table 7. Total mould count (cfu/ ml) of ginger juice during storage.

*Sample	Weeks						
	0	2	4	6	8	10	12
A	3.0×10^{2}	4.0×10^{2}	1.1×10^{3}	1.9×10^{3}	2.6×10^{3}	2.9×10^{3}	3.2×10^{3}
B	1.0×10^{2}	2.4×10^{2}	3.5×10^{2}	4.7×10^{2}	7.3×10^{2}	1.0×10^{3}	1.8×10^{3}
C	1.0×10^{2}	2.0×10^{2}	4.2×10^{2}	5.5×10^{2}	7.9×10^{2}	1.2×10^{3}	1.5×10^{3}
D	1.0×10^{2}	1.3×10^{2}	2.4×10^{2}	4.8×10^{2}	5.6×10^{2}	7.0×10^{2}	1.0×10^{3}

Means followed by different letters across the rows are significantly different ($p < 0.05$) from one another. *A, Control sample; B, sample preserved with PA; C, sample preserved with potassium sorbate; D, sample preserved with sodium benzoate.

time progressed as shown in Table 6. At the end of storage time, the content ranged from 13.50 mg/100 g in control sample to 22.00 mg/100 g in sample preserved with PA. A similar decline has been reported (Zvaigzne et al., 2009) in another colour pigment (β-carotene) in citrus juices after 1 day of storage in a refrigerator. The decline was attributed to be due to the effect of light and oxygen on the juice during preparation process as well as during storage.

Antimicrobial properties

Lower count for total mould was obtained in all the samples with preservatives (range of 1.0×10^{2} to 1.8×10^{3} cfu/ml) as compared to the control sample (range of 3.0×10^{2} to 3.2×10^{3} cfu/ml) at each sampling period (Table 7). There was a general increase in the count in all the samples with increase in storage time. Highest values were obtained at the end of the 12 weeks storage period in all the samples. Sodium benzoate preserved sample had least mold count at all times except at Week 6 when least amount was obtained in PA preserved sample. PA was observed to have better preservative effect than potassium sorbate up to week 10.

The prominent organisms that were isolated from the samples include *Pseudomonas* spp., *Bacillus* spp., *Staphlococcus* spp., *Geotricum* spp. and *Saccharomyces* spp. at the end of storage period.

REFERENCES

Adams MR, Moss MO (2008). Food Microbiology. Ingram Publisher Services La Vergne USA.

Official methods of Analysis (AOAC) (2000). 17th ed. Association of Official Analytical Chemist,. Washington, D.C.

Borawska MH, Czechowska SK, Markiewicz, R, Pałka, J, Renata Świsłocka R, Lewandowski W (2008). Antimicrobial activity and cytotoxity of picolinic acids and selected picolinates as new potential food preservatives. Pol. J. Food Nutr. Sci. 58(4):415-418.

Burdulu HS, Koca N, Karadeniz F (2006). Degradation of vitamin C in citrus juice concentrates during storage. J. Food Eng. 74(2):211-216.

Conrad JS, Dawson SR, Hubbard ER, Meyers TE, Strothkamp KG (1994). Inhibitor binding to the binuclear active site of tyrosinase: temperature, pH, and solvent deuterium isotope effects. Biochemistry 33:5739-5744.

Dazzi C, Candiano G, Massazza S, Ponzetto A, Varesio L (2001). New high-performance liquid chromatographic method for the detection of picolinic acid in biological fluids. J. Chromatogr. Biomed. Sci. Appl. 10:61-68.

Evans GW, Johnson PE (1980). Characterization and quantitation of a zinc-binding ligand in human milk. Pediatr. Res. 14:876-880.

Fernandez-Pol JA, Hamilton PD, Klos DJ (2001). Essential viral and cellular zinc and iron containing metalloproteins as targets for novel antiviral and anticancer agents: implications for prevention and therapy of viral diseases and cancer. Anticancer Res. 21:931-58.

Height KG, Gump BH (1995). Red and white grape Juice concentrate component ranges. J. Food Comp. Anal. 8:71-77.

Komorow JR, Greenberg D, Juturu V (2008). Chromium picolinate does not produce chromosome damage. Toxicol. *In Vitro* 22:819-26.

Marinova EM, Yanishlieva NV (2003). Antioxidative action of quercetin and morin in triacylglycerols of sunflower oil at ambient and high temperatures. Food Chem. 81:189-197.

Oroniran OO, Olunlade BA, Otegbayo BO (2011). Spices as natural preservative in milk. In: Enujiugha, VN (ed.), Book of Abstracts, Nigerian Institute of Food Science and Technology Western chapter half yearly conference. Fed. Univ. Technol. Akure, Ondo State Nigeria. pp. 21-22.

Osundahunsi OF (2008). Kinetics of ascorbic acid degradation in aseptically packaged commercial orange juice produced locally in Nigeria. Bowen J. Agric. 2:1-7.

Westhoff DC, Franzier CW (1998). Food microbiology 4th edition. McGraw Hill publishing company, New Delhi. P. 215.

Zhishen J, Mengcheng T, Jianming W (1999). Determination of flavonoid contents in mulberry and their scavenging effects on superoxide radicals. Food Chem. 164:555-559.

Zvaigzne G, Karklina D, Seglina D, Krasnova I (2009). Antioxidants in various citrus fruit juices. ISSN 1392 –1231. Cheminė Technol. 52(3):56-61.

Preliminary screening of aflatoxin level in maize (*Zea mays* L.) in some selected markets in Benue State, Nigeria

S. T. Ubwa*, K. Asemave and G. O. Igbum

Department of Chemistry, Benue State University, Makurdi - Nigeria.

Preliminary screening of aflatoxin B_1 levels in maize (*Zea mays* L.) sold in some markets of Benue State, Nigeria, was carried out using thin layer chromatography (TLC). Concentrated chloroform extracts of 10, 15 and 40 µl were used in diethylether - methanol - water solvent system in the ratio of 96:3:1. Moisture contents of the stored and oven-dried fresh maize collected from the markets were in the range of 7.64 to 13.40% which are below the maximum recommended level (13.50%) for stored maize. The result of TLC revealed that aflatoxin B_1 was not detected, as no spot was observed in these samples. This suggests that aflatoxin B_1 contents of both the oven-dried fresh and stored maize are either negligible or below the detection limit of 3.125 µg/kg for the TLC method employed. Hence, aflatoxin B_1 levels in the samples, if present, fall below the permissible limit (4 to 5 µg/kg) by National Agency for Food and Drug Administration and Control (NAFDAC) - Nigeria. Further review of this work will be carried out by the authors using a more sensitive technique like high performance liquid chromatography (HPLC) to detect the exact status of aflatoxins in maize from these markets as they are representative of maize belt in Benue State so as to compare them with other world standards.

Key words: Maize, aflatoxin B_1, *Aspergillus flavus,* thin layer chromatography (TLC), moisture content.

INTRODUCTION

Maize is one of the staple foods in this part of the country. More often than not, the maize is allowed to sun-dry on the farm before harvesting amidst unpredictable weather conditions (Nigerian Stored Products Research Institute, 1982). Also, the manner of storage is worrisome, as the maize are stocked in sacks and heaped in stores either at home or in market places. This is capable of creating congenial moisture content in the maize which can highly favour the growth of aflatoxin. Considering the implication of this, it will be appropriate to determine the levels of aflatoxin in maize produced in these localities of the country. Aflatoxins are naturally occurring mycotoxins that are produced by many species of *Aspergillus*, a fungus, the most notable ones being *Aspergillus flavus* and *Aspergillus parasiticus*. Aflatoxins are toxic among the most carcinogenic substances known (Aflatoxin; From Wikipedia, the free encyclopedia (2012); Kerstin and Charity, 2011). After entering the body, aflatoxins may be metabolized by the liver to a reactive epoxide intermediate or hydroxylated to become the less harmful aflatoxin M_1 (AFM_1). Of all the known mytoxins - toxic substances produced by fungi, aflatoxins in maize and other food items such as milk, peanuts, or cottonmeal, etc, generate the most concern when they occur at high levels. Out of the 14 different types of aflatoxins produced in nature, aflatoxin B_1 is considered the most toxic and is produced by both *A. flavus* and *A. parasiticus* (Aflatoxin; From Wikipedia, the free encyclopedia (2012); Oliveira et al., 2009). Aflatoxin G_1 and G_2 are produced exclusively by *A. parasiticus.* Aflatoxins M_1 and M_2 were originally discovered in the milk of cows that fed on moldy grain. These compounds are products of a conversion process

*Corresponding author. E-mail: simon.ubwa@yahoo.com.

in the animal's liver (Aflatoxin; From Wikipedia, the free encyclopedia (2012); Oliveira et al., 2009; Chemical Hazards Evaluation, 2001; Ehrlich and Lee, 1984).

Growth of aflatoxins is dependent upon weather conditions before or after harvest. Preharvest aflatoxin contamination of maize is associated with drought and temperatures during grain fill. When soil moisture is below normal and temperatures are high, the number of *Aspergillus* spores in the air increases. These spores infect crops through areas of damage caused by insects. Once infected, plant stress occurs and the production of aflatoxin is favoured. Postharvest aflatoxin contamination can develop when grain is improperly managed through the drying and storage processes under favourable conditions of humidity and temperature (Oliveira et al., 2009).

Marco et al. (2008) reported the use of an accelerated solvent extraction (ASE) system followed by on-line solid-phase extraction liquid chromatography (SPE-LC) for the analysis of aflatoxins in corn and almonds. ASE uses high temperatures during extraction to speed-up the extraction process, while incorporating high pressure to maintain the solvents in their liquid state. The on-line SPE-LC approach automates sample clean-up and aflatoxin analysis, increasing throughput while decreasing labour.

Mustard oil, used for cooking in northern India, was analyzed spectrophotometrically and 33 of 100 samples were found to contain aflatoxins at levels of 55 to 87 ppb. An early unpublished study of peanut oil in China showed that 48% of 1172 samples were positive for aflatoxin B_1, and a more recent long-term survey in Fujian province found 66% of 323 samples aflatoxin-positive, with 71 samples exceeding the Chinese tolerance level of 20 ppb aflatoxin B_1. Aflatoxin contamination of olive oil has received the most attention but the results have been contradictory. Analysis of 50 Greek olive oils demonstrated the presence of aflatoxin B_1 in 72% of the samples but the highest level detected was 0.05 ppb, well below the European Union (EU) regulation of 2 ppb (Noreen et al., 2011).

Earlier work had reported aflatoxin levels in Spanish olive oils, and in a selection of both Greek and Spanish oils, of 13 to 155 ppb and 5 to 10 ppb, respectively. More recently, total aflatoxin levels of 0.006 to 0.04 ppb were found in 46% of 28 Sicilian olive oil samples examined, and of 20 experimental and 15 commercial samples analyzed by LC-mass spectrometry (LC-MS) only 3 of the latter were contaminated, but below the method quantification limits for individual aflatoxins.

High performance liquid chromatography (HPLC) method with fluorescence detection developed for simultaneous analysis of aflatoxin B_1 and ochratoxin A showed that only 3 of 30 olive oil samples from southern Italy and Morocco contained aflatoxin B_1 at 0.5 to 2.4 ppb, whereas 80% contained ochratoxin A (Noreen et al., 2011). Cano-Sancho et al. (2010) determined AFM_1 in 72 composites of milk, 72 composites of cheese and 72 composites of yoghurt from Catalonia. AFM_1 content was analyzed using an Enzyme-linked immunosorbent assay commercial kit.

The current acceptance level of total aflatoxin in maize, set by the United States' Food and Drug Administration (USFDA) is 20 µg/kg (Tara, 2005; Reddy and Salleh, 2011; Paul and McNeill, 2010). The European Commission has set maximum levels for aflatoxin B_1 between 2.0 and 8.0 µg/kg and for the sum total of all 4 of these toxins between 4.0 and 15.0 µg/kg in crops such as nuts, groundnuts, grains and dried fruits (European Commission Regulation, 2006; Marco et al., 2008). However, the present study is limited to preliminary investigation.

The researchers will carry out a review of the study to establish the exact concentrations of aflatoxin B_1 using a more sensitive technique.

MATERIALS AND METHODS

The following materials and facilities were utilized during this studies: Viewing cabinet, 270 × 270 mm base equipped with 1.5 W long wave ultraviolet (UV) lamp 366 mm, thin layer chromatography (TLC) plate (20 × 20 cm glass plates coated with 0.25 mm silica gel), methanol, n-hexane, chloroform, anhydrous diethyl ether (100%), acetone, silica gel, sodium chloride, aflatoxin B_1 standard solution (0.5 µg/ml). All the reagents were of British Drug Houses (BDH) analytical grade.

Sample collection

The stored maize was sun-dried before packaging. However, moisture content was not determined prior to storage. Adequate pest control measures were taken in the stores. The stores were well constructed with controlled ventilation and protected from rain and humidity. The samples were collected 4 months after storage. Sampling of commodities was conducted according to the methods employed by Bainton et al. (1980), Makun et al. (2010) and Hurburgh (2005). A total of 256 samples comprising 128 stored maize and 128 fresh maize were collected from major markets; Gboko, Naka, Makurdi, Gbajimba, Ihugh, Zaki-biam, Katsina-Ala and Adikpo in Benue State.

The samples were collected between 1st week of October and 2nd week of November, 2011. Four (4) sampling stations were mapped out in each market and 4 samples were randomly collected from each sampling station. The level of aflatoxins B_1 was determined in total of 8 composites/category by pooling all the 16 samples from each market. Fresh maize samples were collected immediately they were brought from the farms.

The stored maize samples were collected from major warehouses (stores) from the sampling stations. About 0.2 kg of each sample was collected into labelled polythene bags and transported to the laboratory where they were stored at -4°C until analysis. Fresh samples were immediately conveyed to laboratory and oven-dried at 105°C intermittently until constant weight was observed (after 6 h) to avoid mould growth.

Sample pre-treatment

The samples were reduced to fine particles, using milling machine, and a sieved (<2 mm). Then, the moisture content of each of the sample was determined. The samples were kept in Erlenmeyer flasks prior to extraction.

Determination of moisture content

A crucible was weighed empty and its weight noted as W_1. Then, some quantity of the milled maize was put into the dish and the combined weight was measured as W_2. The weighed sample was immediately transferred to the oven and dried at 105°C intermittently until constant weight was obtained after which it was allowed to cool for 1 h and reweighed.

The final weight was noted as W_3 (Nigerian Stored Products Research Institute, 1982). The process was repeated for all the samples. The moisture contents were then calculated for all the samples, using the formula:

$$\text{Percentage moisture} = \frac{\text{Loss in weight}}{\text{Weight of sample before drying}} \times 100$$

$$\text{Percentage moisture} = \frac{W_2 - W_3}{W_2 - W_1} \times 100$$

Preparation of TLC plate

30.0 g of Silica gel was weighed into 250 ml glass-stoppered Erlenmeyer flask, and 100 ml of distilled water was added. This was shaken vigorously for 1 min and poured into the applicator set at 0.25 mm thick. This was used to coat five 20 × 20 cm glass plate. The plates were left undisturbed for about 10 min, and it gelled. The coated plates were transferred into the oven and kept for 2 h at 80°C to dry. The plates were stored in the desiccator's cabinet until just before use.

Sample extraction

25 g each of the milled maize samples was weighed separately into 250 ml glass-stoppered Erlenmeyer flasks. 100 ml methanol-water (5.7: 1) was added to each of the flasks, stoppered and protected with a foil. The flasks were shaken vigorously for 30 min. The mixture was filtered through a Whatman No. 1 filter paper and the residue was discarded.

Clean up

20.0 ml of the filtrate was transferred into a 125 ml - separating funnel, and 20.0 ml of 10% sodium chloride solution was added and mixed. 12.5 ml of n-hexane was added and the mixture was shaken for 1 min. Phases were allowed to separate. The lower phase (aqueous) was drained into a second 125 ml separating funnel, while the upper phase (organic) was discarded (Alexander et al., 1997).

Chloroform extraction

To the aqueous phase in the separating funnel, 12.5 ml of chloroform was added and shaken for 1 min. Phases were allowed to separate and the lower phase (organic) was allowed to pass through a bed rock of sodium sulphate into a 250 ml Erlenmeyer flask. The above steps were repeated for all samples.

Concentration of sample extract

After the chloroform extraction, the extracts were concentrated by evaporating to dryness on a steam bath. The concentrated extract was sealed with a foil and kept in a coolant.

Spotting of the TLC plate

The evaporated extracts were each dissolved in 250 μl of chloroform. Various volumes of the samples extracts 10, 15 and 40 μl and aflatoxin standards were spotted on the start line of the TLC plate for all the samples. The spotting was done using 1000 μl syringe.

Plates development and viewing

The plates were developed in a development tank presaturated with recommended development solvent (diethylether-methanol-water in the ratio 96:3:1). The development was carried out in the dark (cupboard) to avoid photo degradation of aflatoxin. The spotted plate was placed vertically in the development tank and covered properly. It took 35 to 45 min for the solvent to reach the stop line (10 cm) from the base line. The plate was then removed and allowed to dry. This process was repeated for all the samples and the developed plates were viewed under a long wavelength UV lamp (366 nm).

RESULTS AND DISCUSSION

The results for moisture contents and levels of aflatoxin B_1 of maize samples obtained from selected markets in Benue State are shown in the Table 1.

The results of the moisture contents determination of the maize (Table 1) revealed that the moisture levels of the fresh oven-dried maize varied between 7.64 to 10.16%, while those of the stored maize varied between 9.96 to 13.41%. Stored maize samples collected from Ihugh and Adikpo markets recorded the lowest and highest moisture levels, respectively. However, these values were below the recommended moisture limit of 13.5% for stored maize (Eaton and Cropman, 1994; Nigerian Stored Products Research Institute, 1982). The low moisture contents of the maize observed in this study may have significant retardation effect on the growth of the fungus (*A. flavus*) responsible for the production of aflatoxin (Aflatoxin; From Wikipedia, the free encyclopedia (2012); Oliveira et al., 2009).

The TLC determination of the levels of aflatoxins in both the fresh and stored maize in Tables 1 revealed that aflatoxin was not detected for the volumes of 10, 15 and 40 μl of extracts spotted. This suggests that maize from areas sampled may be free from aflatoxin B_1 contamination or if present, the contamination may be below the detection limit of 3.125 μg/kg of the TLC method employed (Nigerian Stored Products Research Institute, 1982; Alexander et al., 1997). Furthermore, the low moisture levels of the maize may not be favourable for the growth of mould that enhances aflatoxin B_1 contamination.

Conclusion

The results of the preliminary investigation of aflatoxin levels in some fresh oven-dried and stored maize obtained from most prominent maize producing areas of

Table 1. Moisture contents and levels of aflatoxin B_1 of maize samples obtained from selected markets in Benue State.

Maize sample	Moisture content (%)	Aflatoxin B_1 (µg/kg)
Stored samples of Gboko	12.83	ND
Fresh samples of Gboko	10.16	ND
Stored samples of Naka	12.40	ND
Fresh samples of Naka	7.64	ND
Stored samples of Makurdi	11.70	ND
Fresh samples of Makurdi	9.70	ND
Stored samples of Gbajimba	11.90	ND
Fresh samples of Gbajimba	8.56	ND
Stored samples of Ihugh	9.96	ND
Fresh samples of Ihugh	8.53	ND
Stored samples of Zaki-biam	13.40	ND
Fresh samples of Zaki-biam	9.68	ND
Stored samples of Katsina-Ala	13.21	ND
Fresh samples of Katsina-Ala	9.63	ND
Stored samples of Adikpo	13.41	ND
Fresh samples of Adikpo	9.94	ND

ND, Not detectable.

Benue State, Nigeria revealed that the stored maize had lower residual moisture than the recommended 13.5% maximum moisture for maize storage. The low level of moisture contents of maize samples may significantly discourage mould infection and aflatoxin development in the maize. No aflatoxin B_1 was detected using the TLC method employed and this showed that even if aflatoxin B_1 were present, it may have been below the detection limit (3.125 µg/kg) of this technique.

RECOMMENDATION

Further study on this subject will be required using a more sensitive equipment such as HPLC with a detection limit of 0.10 µg/kg (Oliveira et al., 2009).

ACKNOWLEDGEMENTS

We wish to acknowledge the technical assistance we received from the staff of Chemistry Laboratory, Department of Chemistry, Benue State University, Makurdi, and Nigerian Stored Products Research Institute, Yaba, Lagos where this research was carried out.

REFERENCES

Aflatoxin; From Wikipedia, the free encyclopedia; http://en.wikipedia.org/wiki/Aflatoxin. Accessed on 08/03/2012.

Alexander RF, Baur MC (1997). Note on a two TLC procedure for determining Aflatoxin in maize. Cereal Chem. 54:699-794.

Bainton SJ, Coker RD, Jones BD, Morley EM, Nagler MJ, Turner RL (1980). Mycotoxin training manual; Tropical Product Institute, London. pp. 1-176.

Cano-Sancho, German M, Sonia R, Antonio J, Peris-Vicente J, Sanchis V (2010). Occurrence of aflatoxin M1 and exposure assessment in Catalonia (Spain). G. Cano-Sancho et al./Rev. Iberoam Micol. 27(3):130-135.

Chemical Hazards Evaluation (2001). Aflatoxin in Foods. Risk Assessment Studies, Report No. 5.

Eaton DL, Cropman JD (1994). The toxicology of aflatoxin, academic press New York. pp. 259-279.

Ehrlich KC, Lee LS (1984). Mycotoxins in grain dust; method for analysis of aflatoxin, occuatoxin A, Zearalenone, Vomitoxin, and secalonic acid. J. Assoc. Official Ann. Chem. 67:963-967.

European Commission (EC) commission regulation (EC) No 1881/2006 of 19 December, 2006: Setting maximum levels for certain contaminants in food stuffs.

Hurburgh CR (2005). Sampling and Testing For Aflatoxin in Corn (contribution of Dr. Charles R. Hurburgh, Jr., Professor of Agricultural Engineering, Iowa State University) http://www.extension.iastate.edu/Pages/eccrops/transfer/AflatoxinSamplingTesting.pdf.

Kerstin H, Charity M (2011). Aflatoxin control and prevention strategies in key crops of Sub-Saharan Africa. Afr. J. Microbiol. Res. 5(5):459-466.

Makun HA, Anjorin ST, Moronfoye B, Adejo FO, Afolabi OA, Fagbayibo G, Balogun BO, Surajudeen AA (2010). Fungal and aflatoxin contamination of some human food commodities in Nigeria. Afr. J. Food Sci. 4(4):127-135.

Marco K, Remco S, Fraser M, Brett M, Sheldon H, Bruce R (2008). Fast and Effective Determination of Aflatoxins in Grains or Food Using Accelerated Solvent Extraction followed by HPLC. http://www.dionex.com/en-us/webdocs/67572-LPN-2069-01-Aflatoxins-note.pdf.

Nigerian Stored Products Research Institute (1982). Maize; Storing your produce. Advisory Booklet 1:2-29.

Noreen M, Russell JM (2011). A Rapid Analytical Method for Determination of Aflatoxins in Plant-Derived Dietary Supplement and Cosmetic Oils. J. Agric. Food Chem. 58(7):4065-4070.

Oliveira CAF, Goncalves NB, Rosim, Roice E, Fernandes, Andrezza M (2009). Determination of Aflatoxins in Peanut products in the

Northeast Region of Sao Paulo, Brazil. Int. J. Mol. Sci. 10(1):174-183.
Paul V, McNeill S (2010). Aflatoxins in Maize.http://graincrops.blogspot.com/2010/10/aflatoxins-in-maize.html. Accessed on 02/04/2012.
Reddy KRN, Salleh B (2011). Co-Occurrence of Moulds and Mycotoxins in Maize Grains Used for Animal Feeds in Malaysia. J. Anim. Vet. Advan. 10(5):668-673.
Tara S (2005). "A Focus on Aflatoxin Contamination".United States National Agricultural Library, Food Safety Research Information Office. Retrieved May 8, 2012.

Studies on the adsorption efficiency of activated carbon for pesticide vapour

A. J. Hotton[1]*, J. T. Barminas[2] and S. A. Osemeahon[2]

[1]National Agency for Food and Drug Administration and Control P. M. B 1018 Jalingo, Taraba State, Nigeria.
[2]Department of Chemistry, Federal University of Technology, P. M. B. 2076, Yola, Adamawa State, Nigeria.

Chlorpyrifos pesticides are endocrine disruptors, inhibitors of acetyl cholinesterase, used to control pests and prime contributors to indoor pollution in retail stores. Seven selected locally sourced adsorbents; Bone charcoal (BCA), corncobs, cotton seed, kokoli, sawdust, coconut shell (CSA) and Muruchi were evaluated. Only BCA and CSA shows peaks and were further examined at adsorptive temperatures of 283, 293, 303, and 313K. Adsorbents physical properties that is, micro pore surface area, total pore volume; micro pore volume, particle density and true density were measured and compared. Results from gas chromatography and surface characterization revealed that the adsorptive capacities of carbon (CSA) at various temperatures were higher than that of carbon (BCA). The plots of Langmuir and Freundlich Isotherms were well fitted by the measured adsorption data. The parameters of adsorption equation for both Langmuir and Freundlich were determined for the two adsorbents, which indicates increased values of adsorptive capacity at low temperature and decreases at high temperature. This study shows that activated carbon made from coconut shell has high surface area, micro porous structure, high degree of surface reactivity and higher heterogeneous properties that suggest it to be a better adsorbent sufficient for the adsorption of chlorpyrifos vapour in an indoor environment.

Key words: Chlorpyrifos pesticide, activated carbon, adsorption isotherms.

INTRODUCTION

People usually think of air pollution as being outdoors, but air in your house or office could also be polluted. Source of indoor pollution include biological contaminants like mold and pollen, tobacco smoke, household products and pesticide gases. Indoor air pollution is caused by high levels of contaminants that become trapped indoors because of poor ventilation.

Pesticides are classed as semi-volatile organic compounds and include a variety of chemicals in various forms; exposure to it has been a concern to public health because of their persistence, toxicity and carcinogenicity (Febre et al., 2005). Research has documented the presence of biologic markers of pesticide exposure in adults and children (Arcury et al., 2006, 2007; Perara et al., 2006; Rauth et al., 2006; USEPA, 2006; Lu et al., 2008). Detectable level of pesticide have been reported in home dust, floors, children toys primarily in families residing in agricultural areas (Quadt et al., 2005).

Scientific field investigations have indicated that the air in homes and other building are more seriously polluted than the outdoor pollution. Study suggested that 80% of most people exposure to pesticide occurs indoors and the measurable level of up to a dozen pesticides have been found in the air inside home. The health effect of exposure includes irritation to eye, nose and throat, damage to central nervous system and kidney, headache, dizziness, muscular weakness and nausea. Chronic exposure to some pesticide can result to damage in the liver, endocrine, nervous systems and death. Both the active and inert ingredients in pesticide can be organic compound, therefore, both could add to the level of airborne organic in an indoor environment. A study of the effects of chlorpyrifos on human exposed over time showed that people on high levels have autoimmune antibodies that are common in people with autoimmune disorders. There is a strong correlation to chronic association with autoimmune disorder after exposure to

*Corresponding author. E-mail: anthonyhottons@gmail.com.

chlorpyrifos (Thrasher et al., 2002). Chloropyrifos pesticide belongs to the class of organophosphorus pesticide that is widely used and sold by agrochemical retailers in north -eastern Nigeria. Chlorpyrifos is a neurotoxin and suspected endocrine disruptor, and it has been associated with asthma, reproductive and developmental toxicity.

However, it is impossible to eliminate these chemical substances completely within an indoor retail outlet as long as the volatility of the pesticide, high temperature and humidity are in place. Nevertheless, it is necessary to limit the pesticide vapour to a considerable level. According to the US Environmental protection agency toxic release inventory (TRI) under title III of the CAAA, the organic vapours are to be regulated. The regulation requiring their emission sources to install maximum achievable control technology (MACTS). Current MACTS for organic vapors are condensation, incineration, carbon adsorption and liquid absorption. Of these technologies, carbon adsorption is a very common one because it offers some advantage over the others (Ruddy and Carroll, 1989; Spivey, 1988).

According to a recent survey in Northeastern Nigeria, 90% of agrochemical retailers were occupationally exposure to and carries a body burden of multiple pesticides in their system due to the sales, storage and handling of pesticides in retail stores (Hotton et al., 2010). To this end, this research work was geared toward developing an adsorbent that has the potential to adsorb pesticide vapour within an indoor retail environment. We examined and evaluated the adsorptive efficiency of seven selected locally sourced adsorbents; Bone charcoal (BCA), corncobs, cotton seed, kokoli, sawdust, coconut shell (CSA) and muruchi.

MATERIALS AND METHODS

Selected grades of cattle bones and coconut shell were locally sourced in Jalingo abattoir and market. They were washed in hot water and boiled for 10 min to remove fat and oil. The carbonized processes of the shell and bones include drying and heating in order to remove undesirable by-products such as ash and other hydrocarbon. The carbonization process is completed by heating the material at 400 to 600℃. The carbonized particle s were made "active" by exposing them to steam at high temperature for an hour. The steam burns off the decomposition products from the carbonization phase to develop a porous activate pore size. The carbonized charcoal were separately grated and sieved to fine particle size on a U.S standard mesh size No. 40 sieve (0.42 mm). Corncob, cotton seed, kokoali seed, murichi fibre, and sawdust were dried, grated and sieved.

Procedure for adsorption

Commercial pesticides (Chlorpyrifos) were obtained from Zee agrochemical Ltd Jalingo with an emulsion concentration of 25% EC. Different percentages of concentrations were prepared with distilled water, that is, 0.4, 0.6, 0.8 and 4%. 20 ml of the already prepared concentration of each pesticide were measured separately and placed at the centre of an enclosed heating glass bowl. 2 g each of the selected adsorbents were measured and placed within 20 cm^2 radius of the measured 20 ml pesticide in the heating bowl. The vaporization of the pesticide occurred at a given temperature. The adsorption temperature was performed at 283, 293, 303 and 313K which were regulated and maintained in a hot water bath, the process was allowed to stay for 2 h. The entire experiment was repeated for the pesticide and adsorbents at various concentrations and temperature. The respective adsorbents were collected after the exposure duration of 2 h and placed in a sealed glass vial label properly and stored at 4℃ u ntil analysis. Pesticide reference standards of organophosphorus pesticides were obtained from Niger- Sigma chemicals Ltd.

Surface characterization measurements

The surface characterizations of carbon were carried out only on the adsorbent of CSA and BCA for they were the only adsorbents that yielded peaks result by the gas chromatography (Gc) in this study. The particle size distribution was determined by Shimadzu Laser Diffraction particle size analyzer. The particle density was measured by a mercury displacement method using a mercury porosimeter (Autopore II 9200; Micromeritics, Inc GA, USA). The true density was determined by a helium displacement method using a pycnometer (Accupyc 1330; micromeritics, Inc, GA USA). The specific surface areas, total pore volume were determined from BET method and the adsorption volume of 0.95 relative pressure (adsorption /desorption apparatus ASAP 200 (Micromeritics Inc GA USA) using a continuous flow method.

Sample extraction, clean up and analysis

The adsorbent samples were analysed for organophosphorus compound using USEPA method 8141A by Gas chromatography-capillary column technique. Extraction was based on the method followed by Agarwal et al. (1976) with some modification using hexane: acetone (1.1) 20 ml (thrice) as solvent for extraction. Clean up was done by USEPA method 3620B by column chromatogram-phy. The extract were transferred to the column and eluted with hexane (10 ml), 6% diethyl ether in hexane (10 ml), 50% diethyl ether in hexane (10 ml) and finally with diethyl ether (10 ml). Eluent were collected and evaporated to dryness. Final samples were prepared in 2 ml hexane (HPLC grade). Quantification of organophosphorus pesticide adsorbed by the adsorbent was done by Shimadzu – 17A series GAS chromatography equipped with 63 NI selective electron capture detector. Quantitative analysis of OP residues in each sample was done by comparing the peak area with those obtained from a chromatogram of a mixed OP standard of known concentration. The average recoveries of fortified samples were exceeding 95%. An aliquot of reagent water was treated exactly as a sample including exposure to all glass ware, equipments, solvents and reagent used with the sample matrix. No analyte peak was detected in laboratory regents blank.

RESULTS

Table 1 shows the general trends of the adsorption capacity of the seven adsorbents that is, CSA, BCA, corncob, cotton seed, kokoali seed, murichi fibre, and sawdust conducted at different temperatures and concentrations.

At a constant temperature of 283 K and at varying

Table 1. Adsorptive capacity of the adsorbents used.

Temp (K)	Conc. (%)	CSA (mg/g)	BCA (mg/g)	Corn cob	Kokoli	Saw dust	Murichi fiber	Cotton seed
283	0.4	0.420	0.350	N.D	N. D	N.D	N. D	N.D
	0.6	0.63	0.450	N.D	N. D	N.D	N. D	N.D
	0.8	0.830	0.600	N.D	N. D	N.D	N. D	N.D
	4.0	2.500	1.800	N.D	N. D	N.D	N. D	N.D
293	0.4	0.290	0.172	N.D	N. D	N.D	N. D	N.D
	0.6	0.420	0.220	N.D	N. D	N.D	N .D	N.D
	0.8	0.560	0.250	N.D	N. D	N.D	N. D	N.D
	4.0	1.670	1.100	N.D	N. D	N.D	N. D	N.D
303	0.4	0.250	0.180	N.D	N .D	N.D	N. D	N.D
	0.6	0.360	0.250	N.D	N .D	N.D	N. D	N.D
	0.8	0.450	0.350	N.D	N .D	N.D	N .D	N.D
	4.0	1.250	1.200	N.D	N. D	N.D	N. D	N.D
313	0.4	0.190	0.150	N.D	N. D	N.D	N. D	N.D
	0.6	0.280	0.190	N.D	N. D	N.D	N .D	N.D
	0.8	0.360	0.250	N.D	N .D	N.D	N. D	N.D
	4.0	1.000	0.750	N.D	N. D	N.D	N .D	N.D

N.D = Not detected.

Table 2. Surface characterization of activated carbon CSA and BCA used.

Physical properties	Activated carbons CSA	Activated carbons BCA
Total surface area (N_2 BET Method) (M^2g^{-1})	1001	985
Micropore surface area (M^2g^{-1})	762	525
Total pore volume (Cm^3g^{-1})	0.4271	0.5525
Micropore volume (Cm^3g^{-1})	0.305	0.201
Particle density (gcm^3)	0.7672	0.6982
True density (gcm^3)	1.9250	1.6947

CSA = Coconut shell 'activated carbon, BCA = borne charcoal activated carbon.

concentrations, the adsorptive amount for CSA were 0.420, 0.630, 0.830 and 2.500 mg/g, respectively and 0.350, 0.450, 0.600 and 1.800 mg/g for BCA. While the remaining adsorbents that is, corn cob, cotton seed, kokoali seed, murichi fibre, and sawdust whose values were below the detection limit recorded as N.D in Table 1.

Table 2 shows the general trends of the adsorption capacity of the two adsorbent CSA and BCA at various concentrations and temperatures.

DISCUSSION

The results obtained from this study in Table 1 revealed the adsorptive values of CSA and BCA adsorbed at all levels of the varying temperatures and concentrations while the remaining five adsorbents displayed no results, as such they were not discussed further. The physical properties of the two adsorbents based on their surface characterization, shows that CSA has a large surface area property of 1001 M^2g^{-1} over the BCA with 985 M^2g^{-1}. Under an electron microscope in Figure 1, the high surface area structures of CSA are revealed. Individual particles are intensely convoluted and display various kinds of porosity; there may be many areas where flat surfaces of graphite-like materials run parallel to each other. The lager surface area is a measure of how much exposed area the adsorbent CSA has, expressed in square units. The finer the particle sizes of the activated carbon, the better the access to the surface area and the faster the rate of adsorption kinetics. The most fundamental property of the surface area is its additives. In vapour phase systems this needs to be considered against pressure drop, which will affect energy cost. Careful consideration of particle size distribution can

Figure 1. Activated carbon of CSA as viewed by an electron microscope.

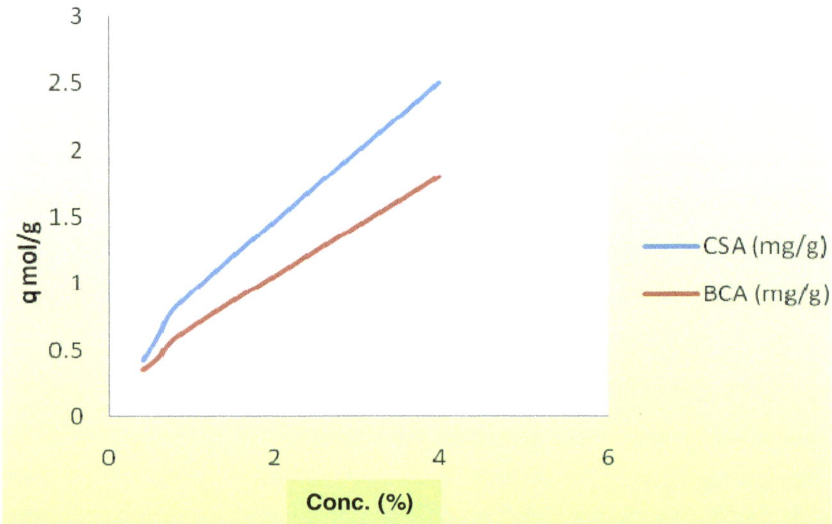

Figure 2. Plots of adsorbed amount against concentration at 283K for CSA and BCA.

provide significant operating benefits. The reaction rate or speed of reaction for a reactant or product in a particular reaction is intuitively a function of how fast or slow a reaction takes place. The rate of reaction increases as the surface area does. That is because more particles of the solid are exposed and can be hit by reactant molecules. This means that the more finely divided a solid or liquid reactant, the greater its surface area per unit volume, and the more contact it makes with the other reactant, thus the faster the reaction. The two adsorbents in this study displayed large surfaces but CSA appears to adsorb better than BCA, this may be attributed to the large proportion of surface area.

The micro pore surface area of CSA shell was found to be 762 m^2g^{-1} larger than carbon from BCA (525 m^2g^{-1}). The micro pore volume of carbon (CSA) appeared also to be higher with 0.305 cm^3g^{-1}, while the micro pore volume of BCA yielded a lower value of 0.201 cm^3g. These micro pores provide superb conditions for adsorption to occur, since adsorbing material can interact with many surfaces simultaneously. However, the results indicate that the values of the micropore surface area of CSA were higher than BCA which is a pointer that increasing the surface area of a substance generally increases the rate of a reaction. The particle density yielded 0.7672 gcm^3 CSA higher than the result obtained for BCA (0.6982 gcm^3). The result also shows that the true density of the two adsorbents in the study were 1.9250 for CSA and 1.6947 for BCA. Higher density provides greater volume activity and normally indicates better quality of the activated carbon.

However, the plot of the amount adsorbed for both adsorbents against concentrations as showed in Figures 2 to 4 shows the adsorbent CSA with a high adsorption

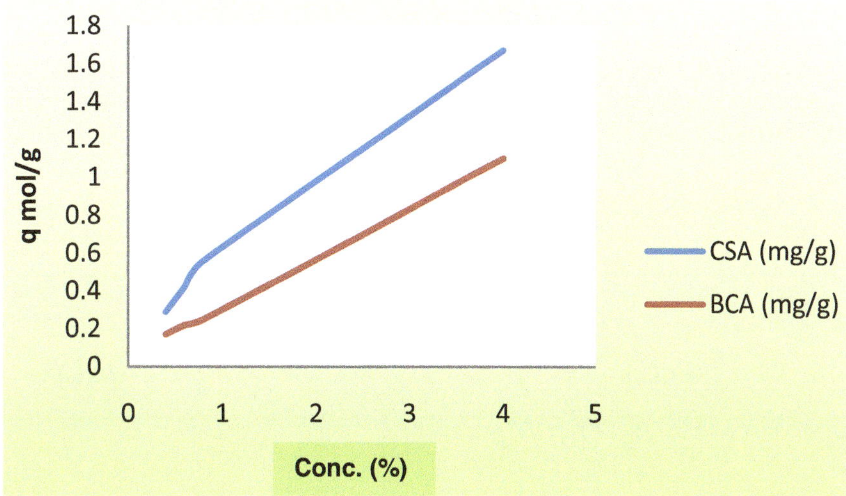

Figure 3. Plots of adsorbed amount against concentration at 293K for CSA and BCA.

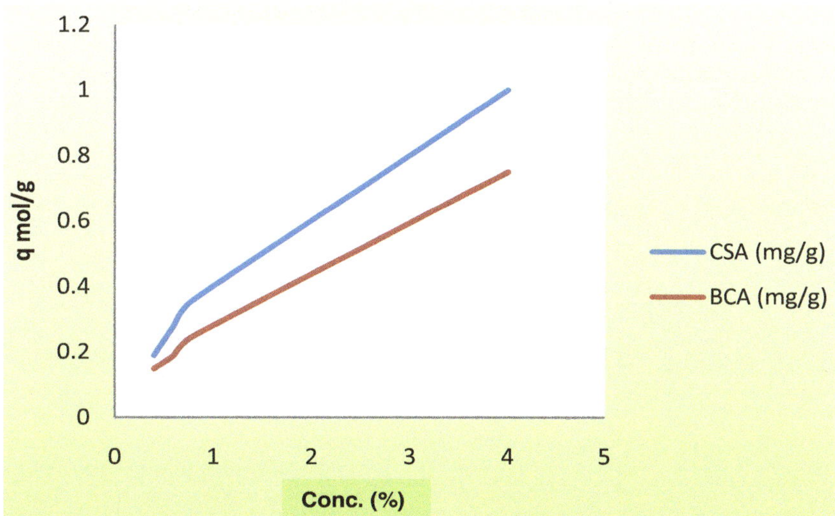

Figure 4. Plots of adsorbed amount against concentration at 313K for CSA and BCA.

rate than BCA. From the adsorption amount of BCA and CSA in Table 1, it can be deduced that the adsorption potential of CSA is of higher affinity than the BCA values. It was revealed also that the value of the adsorption capacity increases as the concentration increases at constant temperature for both adsorbent.

However, the overall trend at various temperatures and concentration shows that the Adsorbed amount decreases as the temperature increases. Generally, results indicate that the values of the surface areas, micro pore surface area, micro pore volume, particle density and true density support the fact that CSA in this studies exhibited a better heterogeneous distribution of surface energy for the adsorption of the adsorptive and that the micro pores are responsible for a large proportion of the surface area.

Adsorption isotherms

Since physical adsorption is non-specific every gas will adsorbed on the surface of active solid. How much gas adsorbed depends on the nature of the adsorptive, the surface area and temperature.

Adsorption of chlorpyrifos vapour on CSA and BCA at various concentrations and temperatures plots showed linear relationship that fits both the theories of Freudlich adsorption isotherm and the langmuir equation (Ching-Yuan, 1994; Ruthven, 1984; Susaki, 1991). The Langmuir adsorption isotherm equation can be

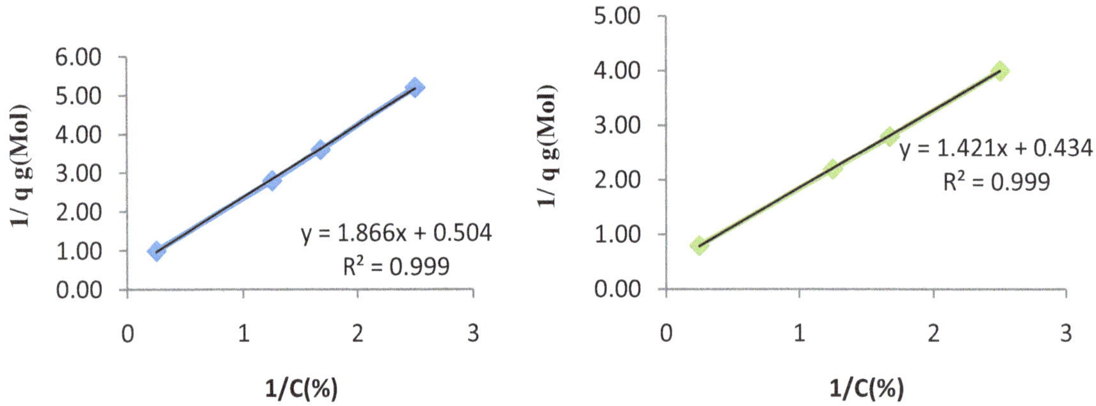

Figure 5. Langmuir plots on carbon CSA at 283 and 293K.

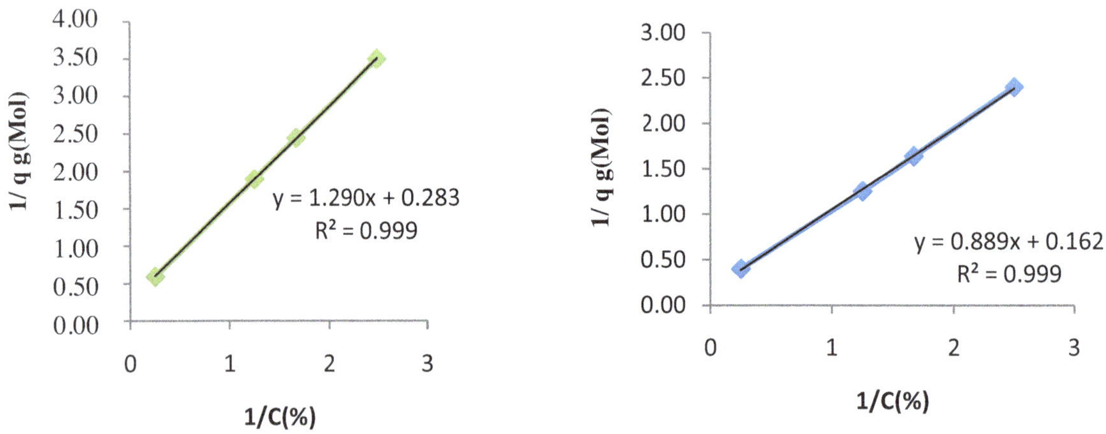

Figure 6. Langmuir plots on carbon CSA at 303 and 313K.

represented by:

$1/q = (1/q_m) + (1/q_m KC)$

Where q = the adsorbed amount (Unit mass of adsorbent), q_m = adsorption capacity at monolayer saturation, C = equilibrium concentration of adsorbate (pesticide concentration in solution), K = adsorption equilibrium constant

A plot of (1/q) versus (1/c) for Langmuir isotherm for both CAC and BCA yielded:

Slope = (1/q) k with $1/q_m$ = intercept

The plot of the Langmuir Isotherm shows a positive linear graph with the regression co-efficient justifying the fitness of the plotted experimental data which also agrees with the theory of adsorption as presented in Figures 5 and 6 for CSA. In order to find and calculate the values of the Langmuir isotherm parameters the least square method was adopted using the slope and the intercept to compute the values of q_m and K for both CSA and BCA.

Higher density and surface area provides greater volume activity and normally indicates better quality activated carbon. Physically, activated carbon binds materials by van der Waals force or London dispersion force. Micro pores of CSA provide superb conditions for adsorption to occur, since adsorbing material can interact with many surfaces simultaneously.

The evaluation of the values of Langmuir adsorption isotherm parameter computed from the plot, results shows that the adsorptive amount of CSA based on the isotherm parameter buttressed the fact that the carbon of CSA adsorbed higher than the carbon of BCA in Table 1. This also is in agreement with values obtained in Table 2 that shows adsorption potential of CSA over BCA. The results from this study shows that the values of K decreases with increase temperature and K increases with decrease in temperature which confirms that the adsorbents displayed a higher adsorption affinity at lower temperatures. Table 3 shows the evaluation of the values

Table 3. Langmuir parameters for adsorption of chlorpyrifos on activated carbon at various temperatures.

Adsorbent	Adsorption temperature (T)	Langmuir parameter		
		q_m (molg^{-1})	K	R^2
BCA	313	11.85	3.10	0.999
	303	13.62	3.40	0.998
	293	18.45	3.60	0.992
	283	20.6	3.80	0.996
CSA	313	19.8	3.27	0.999
	303	23	3.70	0.999
	293	35	4.55	0.999
	283	61.7	5.48	0.999

Table 4. Freundlich parameters for adsorption of chlorpyrifos on activated carbons at various temperatures.

Adsorbent	Adsorption temperature (T)	Freundlich parameter		
		K	n	R^2
BCA	313	3.5	3.35	0.999
	303	4.31	3.89	0.998
	293	2.7	4.57	0.996
	283	6.8	4.75	0.998
CSA	313	4.2	3.5	0.999
	303	5.9	2.9	0.999
	293	6.0	3.9	0.999
	283	8.5	3.6	0.999

of Langmuir adsorption isotherm parameter computed from the plot in Figures 5 and 6.

The Freundlich adsorption isotherm was also used whose equation was transformed into linear by taking the logarithm on both sides.

$$\ln q = \ln k + (1/n) \ln c$$

Hence, using the values of q and C in Table 1 a plot of In q versus In C yielded a straight line graph as shown in Figures 7 and 8. This implies that the adsorption process conforms to Freundlich adsorption isotherm.

The values of the parameter K and n obtained as shown in Table 4 reveals that the values of the parameter *n* decreases with increasing temperature. A higher value of K increases the adsorptive amount of the adsorbent. The value of *K* also agrees that carbon CSA values were higher than carbon BCA at the same temperature. The results confirmed and agreed also with our findings using the Langmuir adsorption isotherms, the Freundlich isotherm and the adsorptive amount.

The overall results reveals that both adsorbent adsorbed better at low temperature of 283 K yielding high values of 61.7 mol/g at a monolayer saturation of CSA and 20.6 mol/g monolayer saturation for BCA as shown in Table 3, relating the values of the monolayer saturation to the values obtained for the BET total surface area of CSA (1001) as shown in Table 2, it can be deduced that the surface area of CSA has the potential to adsorb the pesticide vapour at an amount of 61,700 mol/g.

In conclusion, our findings shows that it is achievable to control and limits chloropyrifos vapour to some considerable level in an indoor environment using coconut shell activated carbon adsorption technology. This study reveals that the adsorption capacity of carbon CSA is higher than that of carbon BCA because of its sufficient porosities and surface area which values agrees with the measured physical properties. The adsorption was more favorable at low temperature and shows that the adsorption is of physical reaction which can be reversible. The plots of Langmuir and Freundlich Isotherms were well fitted by the measured adsorption data.

ACKNOWLEDGEMENT

We acknowledged the management of National Agency

for Food and Drug Administration and control (NAFDAC) and the Department of Chemistry Federal University of Technology, Yola, Nigeria.

REFERENCES

Agarwal HC, Pilla MK, Yadav DV, Menon KB, Gupta RK (1976). Residues of DDT and its Metabolites in Human blood samples in Delhi India. Bull. World Health Org., 54: 349-351.

Arcury TA, Quandt SA, Rao P, Doran A, Snively BM, Davis S (2007). Organophosphate Pesticide Exposure in Farm Worker Family Members in Western North Carolina and Virgina: Case Comparisons: Hum. Organ, 64: 40-51.

Arcury TA, Grzywacz JG, Davis SW, Barr DD, Quandt S (2006). Organophosphorus Pesticide Urinary Metabolite levels of Children in Farm Workers Households in Eastern North Carolina: Am. J. Ind. Med., 49: 51-76.

Ching-Yuan C, Tsai WT (1994). Adsorption of Methylene Chloride Vapour on Activated Carbon. J. Chem. Tech. Biotechnol., 61: 145-151.

Fabra B, Both E, Olga K (2005). Analysis 0f the Insecticide Hexachlorocyclohexane Isomers in Biological Media: Environ. Chem. Lett., 3: 122-126.

Hotton AJ, Barminas JT, Osemeahon SA, Aboki T (2010). Monitoring of Cholinesterase Inhibition among Retailers of Agrochemical in Northeastern Nigeria; Eur. J. Sci. Res., 46(1): 028-035.

Lu C, Toepel K, Irish RA, Fenske DB, Barr R, Brav R (2008). Organic diets significantly lower children's dietary exposure to organophosphorus Pesticides. Environ. Health Perspect., 114(2): 260-263.

Perera F, Viswanathan S (2006). "Children's environmental health research- highlihts from the Columbia Center for Children's Envital Health Ann. NY. Acad. Sci., 1076: 15-28.

Quadt SA, Arcury TA, Rao TB, Snively DE, Cammann AM (2005). Agricultural and Residential Pesticides in Wipe Samples from Farmworker Family Residence in North Carolina and Virgina Carolina. Neurotoxicology, 26(4): 589-598.

Rauh VA, Garfinkel R, Perera FP, Andrews HF, Hoepner LD, Barr DB (2006). Impact of Prenatal Chlorpyrifos Exposure on Neurodevelopment in the First 3 Years of life Among Inner-City Children. Pediatrics, 118(6): 1845-1859.

Ruddy EN, Carroll LA (1983). Select the best VOC control strategy. Chem. Eng. Progress, 89(7): 28-35.

Ruthven DM (1984). Principles of Adsorption and Adsorption process. John Wiley, New York, NY, USA.

Spivey JJ (1988). Recovery of volatile organic from small industrial sources. Environ. Progress, 7: 31-40.

Susaki M (1991). Adsorption Engineering. Elservier, Amsterdam, The Netherlands.

Torres MC, Folgoso F, Reche AV, Garcia IM, Arcas F, Olea-Serrano (2006). Organochlorine pesticides in serum and adipose tissue of pregnant women in Southern Spain giving birth by cesarean section. Sci. Total Environ., 372: 32-38.

USEPA (2006). Measuring the Impact of the Food Quality Protection Act: Challenges and Opportunities. Report No. 20-6 – P-00028. Washington, DC: U.S. Environmental Protection Agency, Office of Inspector General Available:http://www.epa.gov./oig/reports/2006/20060801-2006-P-00028,pdf (accessed 1 August 2007).

Comparison of aromatics released via hydropyrolysis with their solvent extractable counterpart for biodegraded Nigeria crude oil

Ivwurie, W[1]*, Snape, C. E[2] and Sun, C[2]

[1]Department of Chemistry, Federal University of Petroleum Resources, Effurun, Delta State, Nigeria.
[2]Nottingham Fuel and Energy Centre, the University of Nottingham, University Park, Nottingham, United Kingdom.

The molecular profiles of aromatics generated through hydropyrolysis with the free (maltene) counterpart were compared, with a view to assessing the effectiveness of using molecular information for the aromatics bound in asphaltene matrix for polycyclic aromatic hydrocarbon source apportionment studies. The results obtained from gas chromatograph-mass spectrometry (GC-MS) total and selected ion molecular profiles of free and bound aromatics for Nigeria crude oil shows that the solvent extractible aromatics has either been heavily biodegraded or covered by humic materials arising from microbial activity and will therefore need further purification to resolve the aromatic hydrocarbons present. However, the aromatics released via hydropyrolysis (trapped in asphaltene matrix) are still preserved. The molecular profile of asphaltene derived aromatics is highly dominated by both alkylated parent and parent polycyclic aromatic hydrocarbons such as methyl fluoranthene and pyrene, Benzo (b/k) fluoranthene.

Key words: Crude oil, aromatics, hydropyrolysis.

INTRODUCTION

The Niger Delta oil is a medium crude with American petroleum institute (API) gravity greater than 20.6° and is naphthenic, non-waxy with pour point less than -25°C (-13°F) (Evamy et al., 1978). Studies have revealed that oils from throughout the Delta are from one source type such that one super family of oils occurs in the Niger Delta (Haack et al., 1997; Rooney et al., 1998). The processing of crude oil into different components and the combustion of these components as fuel liberates hazardous chemical substances into the atmosphere such as polycyclic aromatic hydrocarbons.

When polycyclic aromatic hydrocarbons come in contact with biological entities in particular and the environment in general, it carries negative consequences, given that they are genotoxic (have harmful effect on genes) and they could find their way into the atmosphere as aerosols or other forms in soil, water, or come directly in contact with man, other animals and plants. Jones et al. (1989) reported a steady increase in the polycyclic aromatic hydrocarbon content of an agricultural soil over the last century.

Hydrocarbons are the principal constituents of petroleum, usually exceeding 75%. The proportions of minor constituents such as sulphur, oxygen, nitrogen compounds, the organometallic complexes; and the organic salts vary considerably according to the type of petroleum. The non-hydrocarbon heteroatom compounds form the greater part of certain heavy crude oils (Tissot and Welte, 1984).

Saeed and Al-Mutairi (2000) in a comparative study on the composition of polycyclic aromatic hydrocarbons (PAHs) in the sea water-soluble fractions of different Kuwaiti crude oils reported that the total polycyclic aromatic hydrocarbons in the water-soluble fractions ranged from 171 to 2176 µg/L. Naphthalene and its homologs formed the bulk of the polycyclic aromatic hydrocarbons (ranging from 88.4 to 97.5% of the total polycyclic aromatic hydrocarbons). Significant levels of fluorene, phenanthrene and their homologs were also present in all of the water-soluble fractions. Polycyclic aromatic hydrocarbons with molecular weight higher than

*Corresponding author. E-mail: wivwurie@yahoo.co.uk.

than phenanthrene were also detected at very low concentrations.

However, where biodegradation of polycyclic aromatic hydrocarbons and biomarkers have occurred in analytical samples, a number of pyrolytic techniques can be used to generate molecular profiles for both correlation purpose and environmental studies, usually involving hydrous or anhydrous pyrolysis carried out in a closed system (Rubinstein et al., 1979; Larter, 1984; Cassini and Eglinton, 1986) as well as open system flash pyrolysis coupled with on-line identification of products.

A major drawback associated with these techniques is that they are notoriously poor at routinely providing quantitatively significant and /or generating biomarker/ hydrocarbon data, which accurately reflects actual, compositions of biomarkers and hydrocarbons bound in asphaltene matrix (Murray, 2001). Asphaltene hydropyrolysis has been found formidable in providing reliable data for correlation studies (Murray, 2001).

Pyrolysis, at high hydrogen pressures (>10 mPa) has been developed as an analytical pyrolysis technique that possesses the exclusive ability to release high yields of bound biomarkers from petroleum source rocks (Love et al., 1995, 1997).

This paper therefore compares the molecular profiles of solvent extractible crude oil aromatics with that generated via hydropyrolysis to ascertain whether the later can be effectively used for oil, oil correlation and environmental studies, where it is impossible to obtain credible data from the former.

MATERIALS AND METHODS

The asphaltenes prepared from Nigeria crude oil by precipitation with 40-fold excess of n-heptane (Jones et al., 1988) were mixed with sand or silica in the proportion of 50 mg of asphaltene to 1 g of silica. The mixture is then impregnated with an aqueous methanol solution (20% v/v) of ammonium dioxydithiomolybdate catalyst to give a nominal solution of 3-wt %. The catalyst was prepared by slow addition of ammonium heptamolybdate to ammonium sulphide solution. The heptamolybdate salt was then dissolved using a magnetic stirrer and solution stirred until a precipitate of ammonium dioxydithiomolybdate is formed. The precipitate was collected with the aid of a Buchner funnel and dried in a vacuum oven. The purpose of the ammonium dioxydithiomolybdate catalyst is to aid the release of aromatics trapped in asphaltene matrix, precipitated from the Nigeria crude oil. The sample is then dried in a vacuum oven at 60 to 70°C after which it is placed in a reactor tube. When attached to the hydropyrolysis set-up, it is resistively heated from ambient temperature (50 to 250°C) at 300°C min^{-1}, then to 500°C at 8°C min^{-1} maintaining a hydrogen pressure of 15 mPa and a flow rate of 10 dm^3min^{-1}. The liquid hydropyrolysate is then collected in a trap cooled with dry ice and recovered with dichloromethane (2 to 5 ml) for subsequent separation into aliphatic, aromatic and polar hydrocarbon components (Love et al., 1995; Murray, 2001; Meredith et al., 2004; Ivwurie et al., 2011).

The dichloromethane-soluble Nigeria crude oil and the hydropyrolysis product were dried, weighed and separated by silica gel adsorption chromatography into aliphatic, aromatic and polar fractions by successive elution with hexane to eliminate the aliphatics, dichloromethane- Hexane mixture (20% v/v) for the removal of the aromatic components and dichloromethane/ methanol mixture (50% v/v) to obtain the polar hydrocarbon components.

Characterization of polycyclic aromatic hydrocarbons was carried out by comparing the sample gas chromatograph-mass spectrometry trace with standard containing 16 polycyclic aromatic hydrocarbons listed by United States environmental protection agency.

Gas chromatography-mass spectrometry analysis was performed on a Fisons instruments 8000 gas chromatograph interfaced to a MD 800 mass spectrometer with a quadrupole mass analyzer (ionizing energy 70 eV, source temperature 280°C). The gas chromatograph capillary column is 50 m long; internal diameter is 0.32 mm and a film thickness of 0.25 µm, stationary phase is polyphenylmethylsiloxane. Laboratory base software written by VG was used for data acquisition and mass lynx for interpretation.

RESULTS

The total ion chromatogram for both solvent extractible and asphaltene derived polycyclic aromatic hydrocarbons from the Nigeria crude oil has been presented in Figure 1. Aromatics detected represent polycyclic aromatic hydrocarbons present in sample or co-elluting at similar retention time.

Selective ion resolution profile, m/z 202, 228 and 252 corresponding to fluoranthene/pyrene, Benzo(a) anthracene/ chrysene, and Benzo(b/k)fluoranthene/ Benzo(e)pyrene, respectively are also presented in Figure 2. Whereas, a neat resolution was obtained for the asphaltene derived aromatics, the solvent extractible aromatics were not resolved, being dominated by unresolved complex mixtures (UCM).

Diagnostic ratios for alkylated parent and parent polycyclic aromatic hydrocarbons are computed in Table 1. These ratios correspond to the ratios of peak areas obtained from the gas chromatogram of the aromatics shown in Figure 1. Diagnostic ratios are used to detect the sources of polycyclic aromatic hydrocarbons (Sun, 2001).

DISCUSSION

The efficacy of using hydropyrolysis for hydrocarbon source apportionment was tested using biodegraded Nigerian crude oil. The gas chromatogram for the free aromatic fraction of the Nigerian crude oil was dominated by unresolved complex mixture arising from severe biodegradation, which had taken place in the solvent extractible aromatics. Both aromatics were obtained from the Nigeria crude oil. This unresolved complex mixture usually arises from the accumulation of humic materials formed from the biodegradation process (Tissot and Welte, 1984) (Figure 1). However, the gas chromatograph profile obtained from asphaltene hydropyrolysis aromatics shows complex distribution of aromatics dominated by both alkylated and parent polycyclic aromatic hydrocarbons, consistent with gas

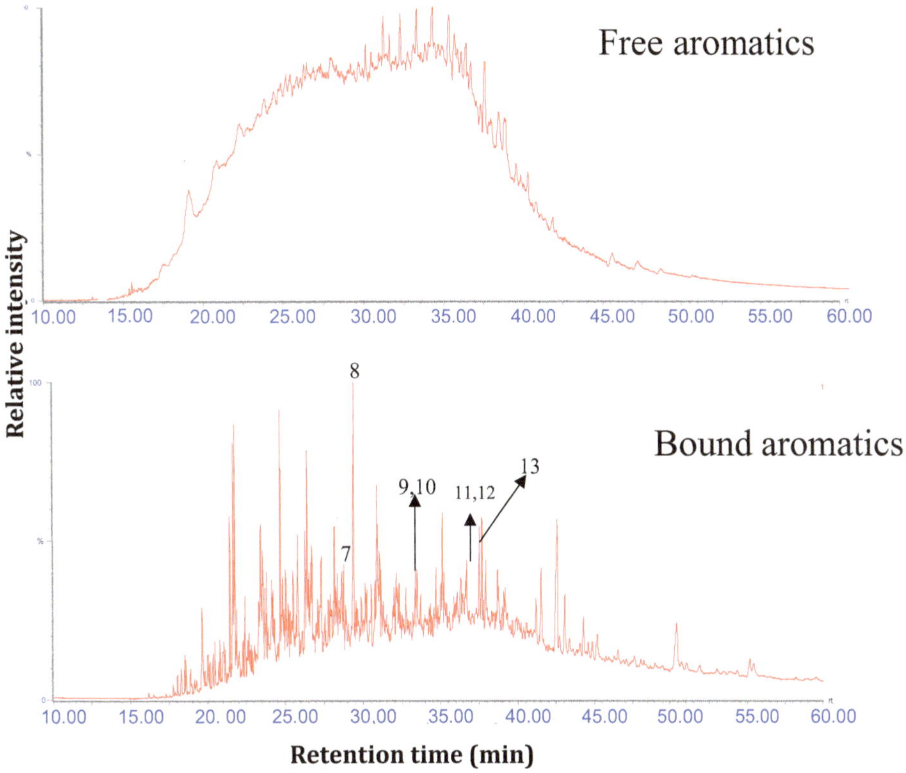

Figure 1. Total ion chromatogram for free and bound neutral aromatics from Nigeria crude oil. Key: 7 - Fluoranthene; 8 - Pyrene; 9 - Benzo(a)anthracene; 10 - Chrysene; 11 - Benzo(b)fluoranthene; 12 - Benzo(k) fluoranthene.

chromatographic profiles for crude oil generally (Tissot and Welte, 1984; Ivwurie et al., 2011). Alkylated parent polycyclic aromatic hydrocarbons such as methyl phenanthrene and fluoranthene, fluoranthene, pyrene and higher molecular weight polycyclic aromatic hydrocarbons were detected in the gas chromatographic profile of the asphaltene derived aromatics. This could not be found in the solvent extractible aromatic gas chromatographic profile of the Nigeria crude oil (Love et al., 1995).

Table 1 gives the diagnostic ratios for a few alkylated and parent polycyclic aromatic hydrocarbons from the bound aromatics for the Nigerian oil. No ratios could be calculated for the solvent extractible aromatics since polycyclic aromatic hydrocarbons could not be detected by the gas chromatograph due to severe biodegradation. Diagnostic ratios of 0.15, 0.09, 0.77 and 0.48 were obtained for fluoranthene (fluoranthene + pyrene), methylphenanthrene/phenanthrene, methylfluoranthene/ fluoranthene, and Benzo(e)pyrene/ (Benzo(e)pyrene + Benzo(a)pyrene, respectively. The diagnostic ratios for the free sample could not be computed due to unresolved complex mixture (UCM). The ratios obtained in Table 1 are within the range reported by Tissot and Welte (1984) and Masclet et al. (1987) for crude oil.

Selective ion m/z 202, 228 and 252 resolution corresponding to fluoranthene/ pyrene, Benzo(a)anthracene/ chrysene and Benzo(b/k) fluoranthene/Benzo(e)pyrene profiles of the free and bound aromatics obtained from gas chromatograph-mass spectrometer has been further compared in Figure 2. Selective ion resolution in gas chromatography-mass spectrometry is usually employed for both qualitative and quantitative treatment of chemical species of analytical interest (Love et al., 1997; Meredith, 2004). In this report, it was used to detect the presence of parent polycyclic aromatic hydrocarbons with m/z of 202, 228 and 252. Whereas, in the free aromatics, large hump corresponding to unresolved complex mixtures are seen, parent polycyclic aromatic hydrocarbons are well resolved in the asphaltene derived aromatics. This is consistent with the report of Love et al. (1998) and Murray (2001), that hydropyrolysis of asphaltene can provide credible data for correlation studies where signatures of the free (solvent extractible product) profile has been altered.

Conclusion

This paper compared the molecular distribution of free and bound aromatics from the Nigeria crude, as a primary source. Both the free and asphaltene derived aromatics were compared for the purpose of building

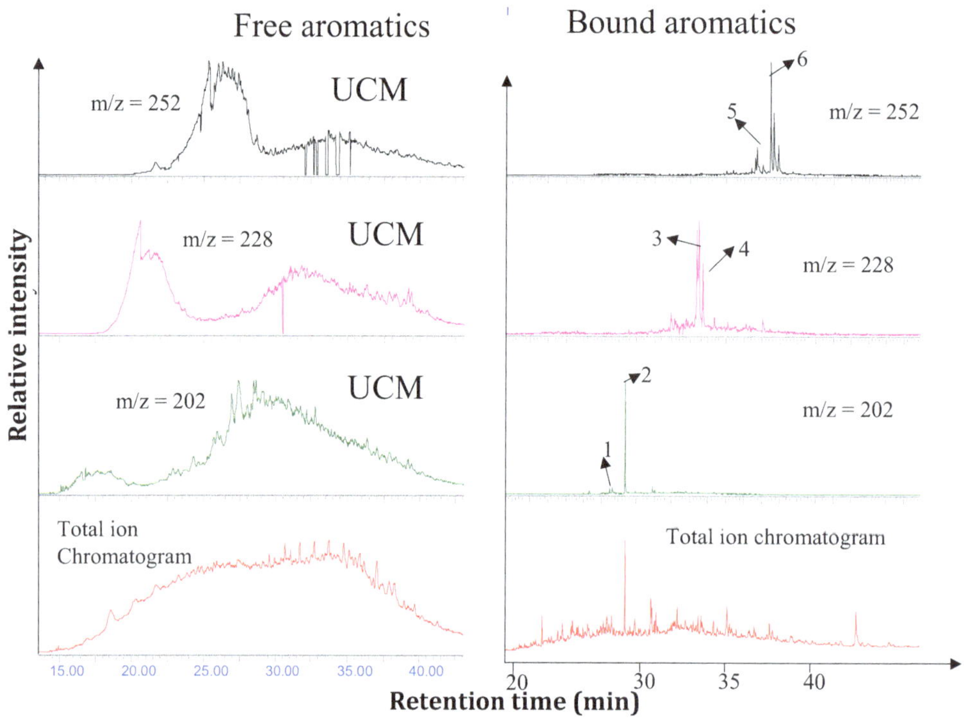

Figure 2. Comparison of GC-MS fragmentograms for m/z, 202, 228, 252 Nigeria crude oil PAHs (free and bound). Key: 1 - Fluoranthene; 2 - Pyrene; 3 - Benzo(a)anthracene; 4 - Chrysene; 5 - Benzo(b/k)fluoranthene; 6. Benzo(e)pyrene.

Table 1. Polycyclic aromatic hydrocarbon diagnostic ratios for Bound Nigeria crude oil aromatics.

Diagnostic ratio	Bound PAHs
Fl/(fl+py)	0.15
MePh/Ph	0.09
MeFl/fl	0.77
BeP/(BeP+BaP)	0.48

Values in the table represent ratios of peak areas of PAHs/alkyl derivatives.

credible molecular database for source apportionment.

The Nigeria crude oil aromatics were dominated by unresolved complex mixtures arising from bio-degradation, but those obtained from asphaltene hydropyrolysis were well resolved, showing the presence of both alkylated and parent polycyclic aromatic hydrocarbons from their gas chromatographic profile and the diagnostic ratios computed. The free Nigeria oil has been heavily biodegraded, but the asphaltene derived, well preserved.

REFERENCES

Cassani F, Eglinton G (1986). Organic Geochemistry of Venezuelan extra heavy oils. 1 Pyrolysis of asphaltenes: A technique for the correlation and maturity evaluation of crude oils. Chem. Geol., 56: 167-183.

Evamy BD, Haremboure J, Kamerling P, Knaap WA, Molloy FA, Rowlands PH (1978). Hydrocarbon habitat of tertiary Niger Delta. AAPG Bulletin, 62: 1-39.

Haack R, Sundararaman P, Dahl J (1997). Niger Delta petroleum system: AAPG/Asociação Brasiliera de Geòlogos Petròleo Joint Hedberg Research Symposium Meeting Abstracts, Rio de Janeiro, Brazil. Extended abstracts, unpaginated.

Ivwurie W, Snape CE, Sun C (2011). "Comparison of Nigeria crude oil solvent extractible and bound biomarker distribution for correlation studies. Int. J. Chem. appli., 3(1): 55-60.

Jones DM, Douglas AG, Connan J (1988). Hydrous pyrolysis of asphaltenes and polar fractions of biodegraded oils. In L. Matavelli and L. Novelli (eds). Advances in Organic Geochem. Pergamon Press, Oxford, pp. 15-40.

Jones KC, Stratford JA, Waterhouse KS, Furlong ET, Giger W, Hites RA, Schaffner C, Johnston AE (1989). Increases in the polynuclear aromatic hydrocarbon content of an agricultural soil over the last century. Env. Sci. Technol., 23: 95-100.

Larter SR (1984). Application of analytical pyrolysis techniques to Kerogen characterization and fossil fuels exploration/ exploitation. In: Voorhees K. (*Ed.*) Analytical Pyrolysis, Methods and Appl. Butterworth, London, pp. 212-275.

Love GD, Snape CE, Carr AD, Houghton RC (1995). Release of Covalently- bound alkane biomarkers in high yields from Kerogen via catalytic hydropyrolysis. Org. Geochem., 23: 981-986.

Love GD, McAulay A, Snape CE, Bishop AN (1997). Effects of Process variables in catalytic hydropyrolysis on the release of covalently bound aliphatics hydrocarbons from sedimentary organic matter. Energy and Fuels, 11: 522-531.

Love GD, Snape CE, Fallick AE (1998). Differences in the mode of incorporation and biogenicity of the principle aliphatic constituents of a Type 1 shale. Org. Geochem., 28: 797-911.

Masclet P, Bresson MA, Mouvier G (1987). Polycyclic aromatic hydrocarbons emitted by power stations and influence of combustion conditions. Fuel, 66: 556-562.

Meredith W, Russell CA, Cooper M, Snape CE, Love GD, Fabbri D, Vane CH (2004). Trapping hydropyrolysates on silica and their subsequent thermal desorption to facilitate rapid fingerprinting by GC-MS. Org. Geochem., 35: 73-89.

Murray IP (2001). Release of covalently bound biological marker molecules from sedimentary organic matter via catalytic hydropyrolysis. Unpublished PhD thesis, University of Strathclyde.

Rooney MA, Matava T, Nwankwo BC, Chung HM, Lambert-Aikhionbare DO (1998). Geochemical characterization of the hydrocarbons in the petroleum system of the southeastern Niger Delta (abs.). AAPG International Meeting Abstracts, Rio de Janeiro, Brazil, p. 343.

Rubinstein I, Spckerelle C, Strausz OP (1979). Pyrolysis of asphaltenes: a source of geochemical information. Geochimica et Cosmochimica Acta., 43: 1-6.

Saeed T, Al-Mutairi P (2000). Comparative composition of polycyclic aromatic hydrocarbons (PAHs) in the sea water-soluble fractions of different Kuwaiti crude oils. Advances in Environ. Research. 4: 141-145.

Sun C (2001). Sourcing PAH in the environment. Unpublished PhD thesis.University of Strathclyde.

Tissot BP, Welte DH (1984). Petroleum Formation and Occurrence. 2nd Edition. Springer- Verlag, New York, U.S.A, p. 699.

15

Free radicals and antioxidants: Myths, facts and mysteries

Abid A. Lone, Shaiq A. Ganai, Rayees A. Ahanger, Hilal A. Bhat, Tauseef A. Bhat, Imtiyaz A. Wani

Division Of Post Harvest Technology, Sher-e-Kashmir University of Agricultural Sciences and Technology- Shalimar, Kashmir, India.

Reactive oxygen species (ROS) and reactive nitrogen species (RNS) are well recognized for playing a dual role as both deleterious and beneficial species. Overproduction of ROS arising either from mitochondrial electron transport chain or excessive stimulation of NAD(P)H oxidase results in oxidative stress, a deleterious process that can be an important mediator of damage to cell structures, including lipids and membranes, proteins and DNA. The powerful action of antioxidants in preventing premature lipid oxidation in food suggests that the same compounds, when consumed in the daily diet, could unfold anti-oxidative/anti-ageing effects in the human body. Failure of supplemental beta carotene to prevent cancer and cardiovascular diseases (CHD) in intervention trials, suggest that the associations of that nutrient, reflect confounding, rather than cause and effect results. The role of oxidative stress in diseases, especially cancer and CHD has been overstated. In the light of recent physiological studies, it appears more advisable to maintain the delicate internal redox balance of the cell than to interfere with the antioxidant homeostasis by a non-physiological, excessive exogenous supply of antioxidants in healthy humans.

Key words: Antioxidants, free radicals, Redox reactions, reactive oxygen species (ROS), reactive nitrogen species (RNS) and phytochemicals.

INTRODUCTION

The field of antioxidants and free radicals is often perceived as focused around the use of antioxidant supplements to prevent human diseases. In fact, antioxidants/free radicals permeate the whole of life, creating the field of redox biology. Free radicals are not all bad, neither are antioxidants all good. Life is a balance between the two: Antioxidants serve to keep down the levels of free radicals, permitting them to perform useful biological functions without too much damage (Halliwell and Gutteridge, 2006). Yet some damage is inevitable, requiring repair systems to maintain cell viability and integrity. The causes of the poisonous properties of oxygen were obscure prior to the publication of Gerschman's free radical theory of oxygen toxicity in 1954, which states that the toxicity of oxygen is due to partially reduced forms of oxygen (Gerschman et al., 1954).

In the same year, observations of a weak electron paramagnetic resonance (EPR) signal attributable to the presence of free radicals in a variety of lyophilised biological materials were reported by Commoner et al. (1954). The world of free radicals in biological systems was soon thereafter in 1956, explored by Denham Harman who proposed the concept of free radicals playing a role in the ageing process (Harman, 1956). A second era of the research on free radicals in biological systems began in 1969, when McCord and Fridovich discovered the enzyme superoxide dismutase (SOD) and thus provided convincing evidence about the importance of free radicals in living systems (McCord and Fridovich, 1969).

A third era of free radicals in biological systems dates from 1977, when Mittal and Murad provided evidence that the hydroxyl radical, $OH^{\bullet}$, stimulates activation of guanylate cyclase and formation of the "second messenger" cyclic guanosine monophosphate (cGMP) (Mittal and Murad, 1977).

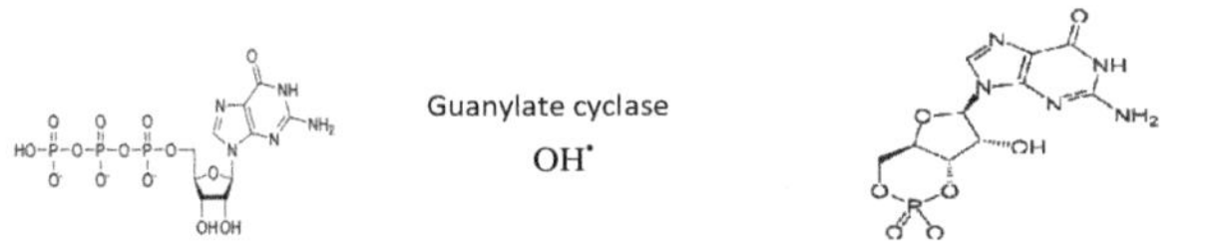

guanosine triphosphate (GTP) 3',5'-cyclic guanosine monophosphate (cGMP)

Since then, a large body of evidence has been accumulated that living systems have not only adapted to a coexistence with free radicals but have developed various mechanisms for the advantageous use of free radicals in various physiological functions. Oxygen free radicals or, more generally, reactive oxygen species (ROS), as well as reactive nitrogen species (RNS), are products of normal cellular metabolism. ROS and RNS are well recognised for playing a dual role as both deleterious and beneficial species, since they can be either harmful or beneficial to living systems (Valko et al., 2006, 2007). Beneficial effects of ROS occur at low/moderate concentrations and involve physiological roles in cellular responses to noxia, for example, in defense against infectious agents and in the function of a number of cellular signaling systems. One further beneficial example of ROS at low/moderate concentrations is the induction of a mitogenic response. The harmful effect of free radicals causing potential biological damage is termed oxidative stress and nitrosative stress (Kovacic and Jacintho, 2001; Ridnour et al., 2005; Valko et al., 2001). This occurs in biological systems when there is an overproduction of ROS/RNS on one side and a deficiency of enzymatic and non-enzymatic antioxidants on the other. In other words, oxidative stress results from the metabolic reactions that use oxygen and represents a disturbance in the equilibrium status of prooxidant/antioxidant reactions in living organisms. The excess ROS can damage cellular lipids, proteins, or DNA inhibiting their normal function. Because of this, oxidative stress has been implicated in a number of human diseases as well as in the ageing process. The delicate balance between beneficial and harmful effects of free radicals is a very important aspect of living organisms and is achieved by mechanisms called "redox regulation". The process of "redox regulation" protects living organisms from various oxidative stresses and maintains "redox homeostasis" by controlling the redox status *in vivo* (Droge, 2002).

The objective of this review is to provide an insight into the production and function of ROS and to examine the effectiveness of antioxidant supplementation in human subjects.

Reactive oxygen species (ROS)

Free radicals can be defined as molecules or molecular fragments containing one or more unpaired electrons in atomic or molecular orbital (Halliwell and Gutteridge, 2006; Gutteridge and Halliwell, 2010). These unpaired electron(s) usually gives a considerable degree of reactivity to free radicals. Radicals derived from oxygen represent the most important class of radical species generated in living systems (Miller et al., 1990).

Molecular oxygen (dioxygen) has a unique electronic configuration and is itself a radical. The addition of one electron to dioxygen, forms the superoxide anion radical ($O_2^{\bullet-}$) (Miller et al., 2005) as seen in Figure 2. Superoxide anion, arising either through metabolic processes or following oxygen "activation" by physical irradiation, is considered the "primary" ROS, and can further interact with other molecules to generate "secondary" ROS, either directly or prevalently through enzyme or metal-catalysed processes (Valko et al., 2005). Various pathways of ROS formation are outlined in Figure 2. The production of superoxide occurs mostly within the mitochondria of a cell (Cadenas and Sies, 1998). The mitochondrial electron transport chain, is the main source of ATP in the mammalian cell and thus essential for life. During energy transduction, a small number of electrons leak to oxygen prematurely, forming the oxygen free radical, superoxide, which has been implicated in the pathophysiology of a variety of diseases (Kovacic et al., 2005; Valko et al., 2004). Measurements on sub-mitochondrial particles suggest an upper limit of 1 to 3% of all electrons in the transport chain leaking to generate superoxide instead of contributing to the reduction of oxygen to water. Superoxide is produced from both Complexes I and III of the electron transport chain, and once in its anionic form, it is too strongly charged to readily cross the inner mitochondrial membrane. In Complex I (NADH dehydrogenase, also called NADH: ubiquinone oxidoreductase; EC 1.6.5.3) two electrons are removed from NADH and transferred to a lipid-soluble carrier, ubiquinone (Q). The reduced product, ubiquinol (QH_2) freely diffuses within the membrane, and Complex I translocates four protons (H^+) across the membrane, thus producing a proton gradient. Complex I, is one of the main sites at which premature electron leakage to oxygen occurs, thus being one of the main sites of production of harmful superoxide.

The pathway of electrons occurs as follows: NADH is oxidized to NAD^+, by reducing Flavin mononucleotide to $FMNH_2$ in a two-electron step. $FMNH_2$ is then oxidized in two one-electron steps, through a semiquinone intermediate. Each electron thus transfers from the $FMNH_2$ to a Fe-S cluster, from the Fe-S cluster to ubiquinone (Q). Transfer of the first electron results in the free-radical (semiquinone) form of Q, and transfer of the second electron reduces the semiquinone form to the ubiquinol form, QH_2. During this process, four protons are translocated from the mitochondrial matrix to the intermembrane space.

In Complex III (cytochrome *bc₁* complex; EC 1.10.2.2), the Q-cycle contributes to the proton gradient by an asymmetric absorption/release of protons. Two electrons are removed from QH_2 at the Q_O site and sequentially transferred to two molecules of cytochrome *c*, a water-soluble electron carrier located within the intermembrane space. The two other electrons sequentially pass across the protein to the Q_i site where the quinone part of ubiquinone is reduced to quinol. A proton gradient is formed by two quinol (4H+4e-) oxidations at the Q_o site to form one quinol (2H+2e-) at the Q_i site. In total six protons are translocated: two protons reduce quinone to quinol and four protons are released from two ubiquinol molecules. When electron transfer is reduced,by a high membrane potential or respiratory inhibitors such as antimycin A, Complex III may leak electrons to molecular oxygen, resulting in superoxide formation.

Recently, it has been demonstrated that Complex I-dependent superoxide is exclusively released into the matrix and that no detectable levels escape from intact mitochondria (Muller et al., 2004). This finding fits well with the proposed site of electron leak at Complex I, namely the iron–sulphur clusters of the (matrix-protruding) hydrophilic arm. In addition, experiments on Complex III show direct extra mitochondrial release of superoxide, but measurements of hydrogen peroxide production revealed that this could only account for <50% of the total electron leak even in mitochondria lacking Cu, Zn-SOD. It has been proposed that the remaining 50% of the electron leak must be due to superoxide released to the matrix. The hydroxyl radical, $OH^{\bullet}$, is the neutral form of the hydroxide ion. The hydroxyl radical has a high reactivity, making it a very dangerous radical with a very short *in vivo* half-life of approximately 10^{-9} s (Pastor et al., 2000). Thus when produced *in vivo* $OH^{\bullet}$ reacts close to its site of formation.

The redox state of the cell is largely linked to an iron (and copper) redox couple and is maintained within strict physiological limits. It has been suggested that iron regulation ensures that there is no free intracellular iron; however, *in vivo*, under stress conditions, an excess of superoxide releases "free iron" from iron-containing molecules. The release of iron by superoxide has been demonstrated for [4Fe–4S] cluster containing enzymes of the dehydratase-lyase family (Liochev and Fridovich, 1994). The released Fe^{2+} can participate in the Fenton reaction, generating highly reactive hydroxyl radical [$Fe^{2+}(Mn^{+})+H_2O_2 \rightarrow Fe^{3+}(Mn^{++}) + OH^{\bullet}+OH^{-}$, where Mn^{+} is any univalent metal cation].

Thus under stress conditions, $O_2^{\bullet-}$ acts as an oxidant of [4Fe–4S] cluster-containing enzymes and facilitates $OH^{\bullet}$ production from H_2O_2 by making Fe^{2+} available for the Fenton reaction (Valko et al., 2006; Leonard et al., 2004). The superoxide radical participates in the Haber–Weiss reaction ($O_2^{\bullet-} + H_2O_2 \rightarrow O_2 + OH^{\bullet}+OH^{-}$) which combines a Fenton reaction and the reduction of Fe^{3+} by superoxide, yielding Fe^{2+} and oxygen ($Fe^{3+} + O^{2\ \bullet-} \rightarrow Fe^{2+} + O_2$) (Liochev and Fridovich, 2002).The Fe–S cluster contains also iron responsive elements (IREs)-binding protein (IRE-BP). The iron-responsive element-binding proteins, also known as IRE-BP and IRBP and IRP, bind to IREs in the regulation of human iron metabolism. This Fe–S cluster has been implicated as the region of the protein that senses intracellular iron levels and accordingly modifies the ability of the IRE-BP to interact with IREs. IRE-BP produced in iron-replete cells has aconitase activity (Han et al., 2005).

In mammalian cells, oxidants are able to convert cytosolic aconitase into active IRE-BP, which increases the "free iron" concentration intracellularly both by decreasing the biosynthesis of ferritin and increasing biosynthesis of transferring receptors. The most realistic *in vivo* production of hydroxyl radical according to the Fenton reaction occurs when Mn^{+} is iron, copper, chromium, or cobalt. However, Rae and co-workers recently reported that the upper limit of so-called "free pools" of copper was far less than a single atom per cell (Rae et al., 1999). This finding casts serious doubt on the *in vivo* role of copper in Fenton-like generation of hydroxyl radical. Although Fenton chemistry is known to occur in vitro, its significance under physiological conditions is not clear, noting particularly the negligible availability of "free catalytic iron" due to its effective sequestration by the various metal-binding proteins (Kakhlon and Cabantchik, 2002).

However, organisms overloaded by iron (as in the conditions of hemochromatosis, b-thalassemia, hemodialysis), contain higher amounts of "free available iron" and this can have deleterious effects. "Free-iron" is transported into an intermediate, labile iron pool (LIP), which represents a steady state exchangeable, and readily chelatable iron compartment (Kakhlon and Cabantchik, 2002). Additional reactive radicals derived from oxygen that can be formed in living systems are peroxyl radicals ($ROO^{\bullet}$) as seen in Figure 2. The simplest peroxyl radical is $HOO^{\bullet}$, which is the protonated form (conjugate acid; *pKa* ~ 4.8) of superoxide ($O_2^{\bullet-}$) and is usually termed either hydroperoxyl radical or perhydroxyl radical.

Given this *pKa* value, only ~ 0.3% of any superoxide present in the cytosol of a typical cell is in the protonated form (De Grey, 2002). It has been demonstrated that hydroperoxyl radical initiates fatty acid peroxidation by two parallel pathways: fatty acid hydroperoxide (LOOH)-independent and LOOH-dependent (Aikens and Dix, 1991). The LOOH-dependent pathway of $HO_2^{\bullet-}$ initiated fatty acid peroxidation may be relevant to mechanisms of lipid peroxidation initiation *in vivo*. Xanthine oxidase (XO,

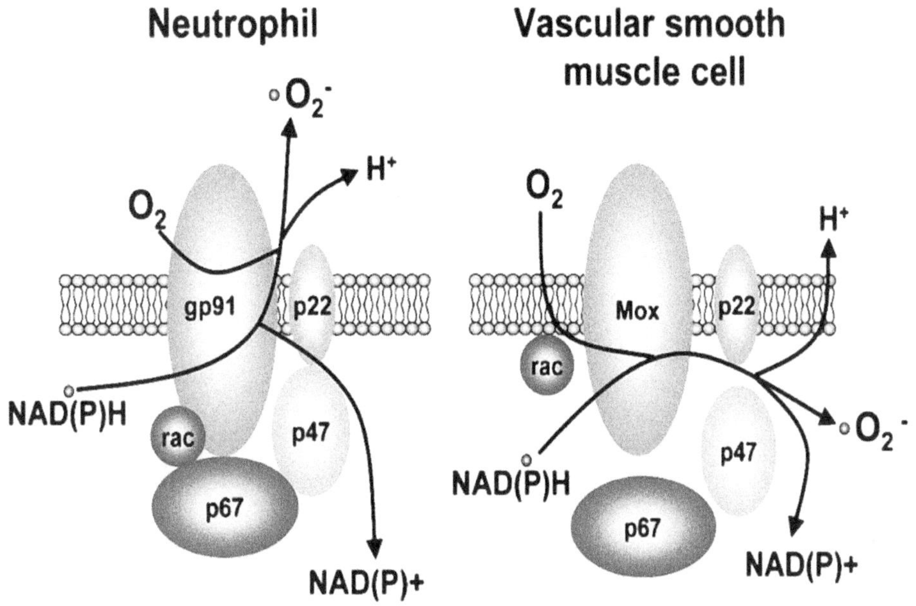

Figure 1. Structure of the NAD(P)H oxidase in neutrophils and vascular smooth muscles.

EC 1.1.3.22) and xanthine dehydrogenase (XD, EC 1.1.1.204) are interconvertible forms of the same enzyme, known as xanthine oxidoreductase (XOR). In purine catabolism, XOR catalyzes the oxidative hydroxylation of hypoxanthine to xanthine and subsequently of xanthine to uric acid. The following chemical reactions are catalyzed by xanthine oxidase:

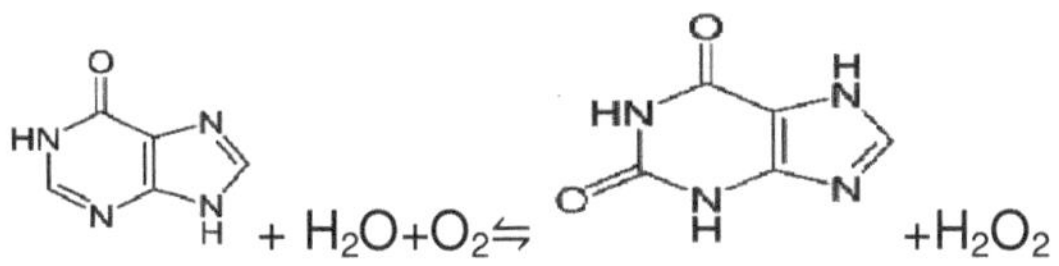

Hypo-xanthine Xanthine

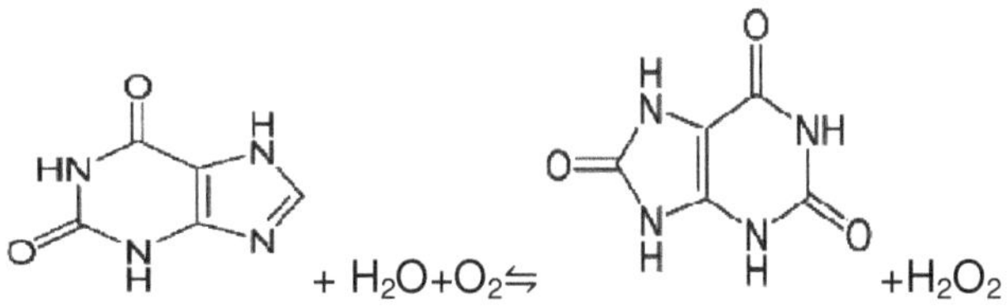

Uric acid acts as a potent antioxidant and free radical scavenger. XOR has, therefore, important functions as a cellular defense enzyme against oxidative stress. With both XO and XD forms, but particularly with the XO form, numerous ROS and RNS are synthesized (Vorbach et al., 2003). Thus, the synthesis of both an antioxidant (uric acid) and numerous free radicals (ROS and RNS) makes XOR an important protective regulator of the cellular redox potential. Xanthine oxidase can also act on certain other purines, pterins, and aldehydes. For example, it efficiently converts 1-methylxanthine (a metabolite of caffeine) to 1-methyluric acid, but has little activity on 3-methylxanthine. Peroxisomes are known to produce H_2O_2, but not $O_2^{\bullet-}$, under physiologic conditions (Valko et al., 2004).

Peroxisomes are major sites of oxygen consumption in the cell and participate in several metabolic functions that use oxygen. Oxygen consumption in the peroxisome leads to H_2O_2 production, which is then used to oxidize a variety of molecules. The organelle also contains catalase, which decomposes hydrogen peroxide and presumably prevents accumulation of this toxic compound. Thus, the peroxisome maintains a delicate balance with respect to the relative concentrations or activities of these enzymes to ensure the absence of net production of ROS. How the organelle maintains this equilibrium is unclear. When peroxisomes are damaged and their H_2O_2 consuming enzymes down regulated, H_2O_2 releases into the cytosol, which significantly contributes to oxidative stress. If a phagocytic cell such as the neutrophil is exposed to a stimulus, it has the ability of recognising the foreign particle and undergoing a series of reactions called the respiratory burst (DeCoursey and Ligeti, 2005).

Nicotine adenine dinucleotide phosphate (NAD(P)H) oxidase is best characterised in neutrophils, where its production of $O_2^{\bullet-}$ generates the respiratory burst necessary for bacterial destruction. The enzyme complex consists of two membrane-bound components, gp91phox and p22phox, which comprise cytochrome *b*558, the enzymatic centre of the complex. After activation, cytosolic components, involving p47phox, p67phox, p40phox and the small G coupled proteins, Rac and Rap1A, translocates to the membrane, to form the active enzyme complex shown in Figure 1. The nonphagocytic

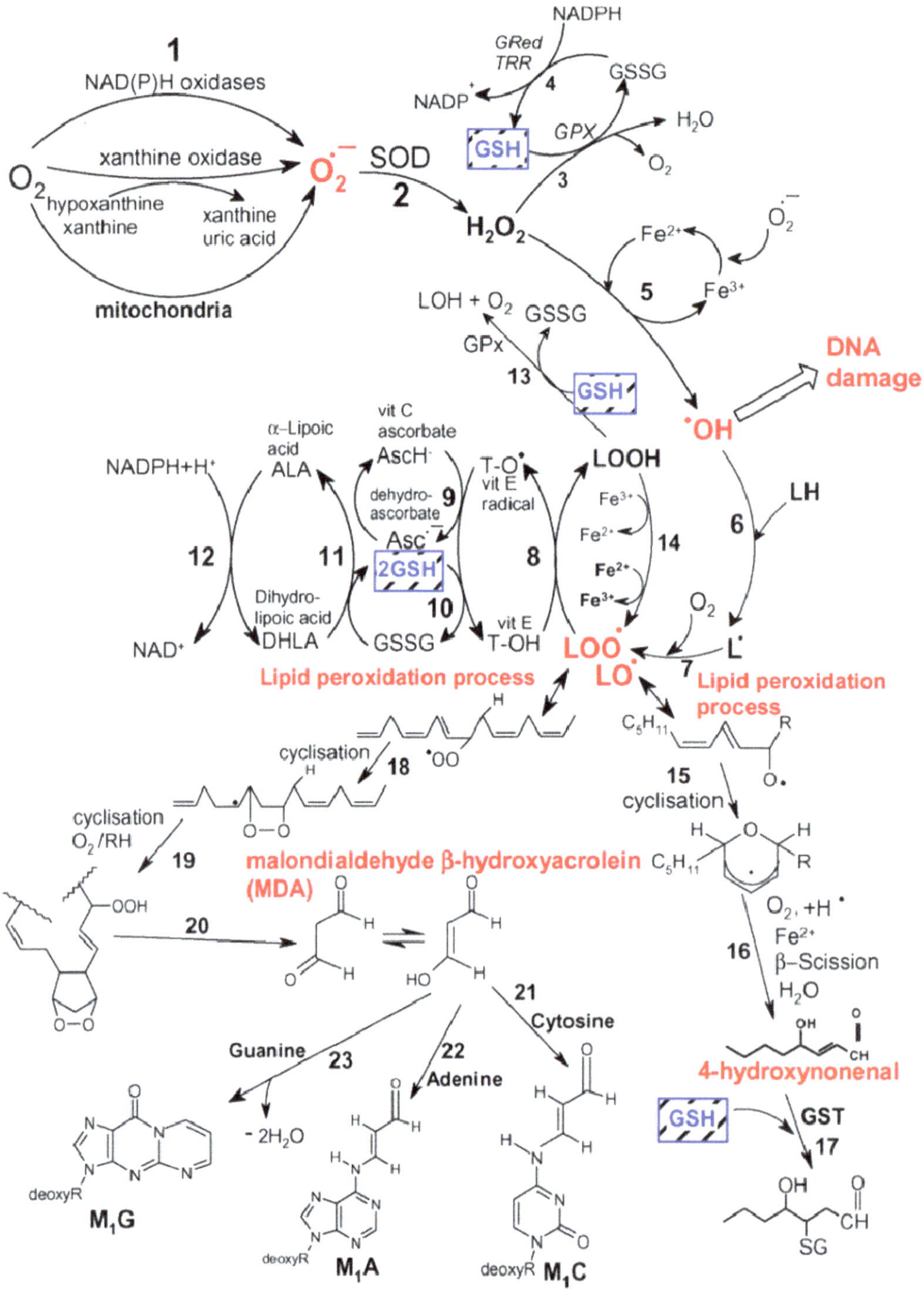

Figure 2. Pathways of ROS formation.

NAD(P)H oxidase, produce superoxide at a fraction (1–10%) of the levels produced in neutrophils and are thought to function in intracellular signaling pathways, as shown in Figure 1.

As can be seen in Figure 2, Pathways of ROS formation, the lipid peroxidation process and the role of glutathione (GSH) and other antioxidants (Vitamin E, Vitamin C, lipoic acid) in the management of oxidative stress, shows an unbalanced equation.

Reaction 1: The superoxide anion radical is formed by the process of reduction of molecular oxygen mediated by NAD(P)H oxidases and xanthine oxidase or non-

enzymatically by redox-reactive compounds such as the semi-ubiquinone compound of the mitochondrial electron transport chain.
Reaction 2: Superoxide radical is dismutated by the superoxide dismutase (SOD) to hydrogen peroxide.
Reaction 3: Hydrogen peroxide is most efficiently scavenged by the enzyme glutathione peroxidase (GPx) which requires GSH as the electron donor.
Reaction 4: The oxidised glutathione (GSSG) is reduced back to GSH by the enzyme glutathione reductase (Gred) which uses NADPH as the electron donor.
Reaction 5: Some transition metals (e.g. Fe^{2+}, Cu+ and others) can breakdown hydrogen peroxide to the reactive hydroxyl radical (Fenton reaction).
Reaction 6: The hydroxyl radical can abstract an electron from polyunsaturated fatty acid (LH) to give rise to a carbon-centered lipid radical (L•).
Reaction 7: The lipid radical (L•) can further interact with molecular oxygen to give a lipid peroxyl radical (LOO•). If the resulting lipid peroxyl radical LOO• is not reduced by antioxidants, the lipid peroxidation process occurs (reactions 18–23 and 15–17).
Reaction 8: The lipid peroxyl radical (LOO•) is reduced within the membrane by the reduced form of Vitamin E (T-OH) resulting in the formation of a lipid hydroperoxide and a radical of Vitamin E (T-O•).
Reaction 9: The regeneration of Vitamin E by Vitamin C: the Vitamin E radical (T-O•) is reduced back to Vitamin E (T-OH) by ascorbic acid (the physiological form of ascorbate is ascorbate monoanion, AscH–) leaving behind the ascorbyl radical (Asc•–).
Reaction 10: The regeneration of Vitamin E by GSH: the oxidised Vitamin E radical (T-O•) is reduced by GSH.
Reaction 11: The oxidised glutathione (GSSG) and the ascorbyl radical (Asc•–) are reduced back to GSH and ascorbate monoanion, AscH–, respectively, by the dihydrolipoic acid (DHLA) which is itself converted to _-lipoic acid (ALA).
Reaction 12: The regeneration of DHLA from ALA using NADPH.
Reaction 13: Lipid hydroperoxides are reduced to alcohols and dioxygen by GPx using GSH as the electron donor.
Lipid peroxidation process: Reaction 14: Lipid hydroperoxides can react fast with Fe^{2+} to form lipid alkoxyl radicals (LO•), or much slower with Fe^{3+} to form lipid peroxyl radicals (LOO•).
Reaction 15: Lipid alkoxyl radical (LO•) derived for example from arachidonic acid undergoes cyclisation reaction to form a six-membered ring hydroperoxide.
Reaction 16: Six-membered ring hydroperoxide undergoes further reactions (involving_-scission) to from 4-hydroxy-nonenal.
Reaction 17: 4-hydroxynonenal is rendered into an innocuous glutathiyl adduct (GST, glutathione S-transferase).
Reaction 18: A peroxyl radical located in the internal position of the fatty acid can react by cyclisation to produce a cyclic peroxide adjacent to a carbon-centered radical.
Reaction 19: This radical can then either be reduced to form a hydroperoxide (reaction not shown) or it can undergo a second cyclisation to form bicyclic peroxide which after coupling to dioxygen and reduction, yields a molecule structurally analogous to the endoperoxide.
Reaction 20: Formed compound is an intermediate product for the production of malondialdehyde. Reactions 21, 22, 23: Malondialdehyde can react with DNA bases Cytosine, Adenine, and Guanine to form adducts M1C, M1A and M1G, respectively.

Reactive nitrogen species (RNS)

$NO^{\bullet}$ is a small molecule that contains one unpaired electron on the antibonding 2π* y orbital and is therefore, a radical. $NO^{\bullet}$ is generated in biological tissues by specific nitric oxide syntheses (NOS's), which metabolises arginine to citrulline with the formation of $NO^{\bullet}$ via a five electron oxidative reaction (Ghafourifar and Cadenas, 2005).

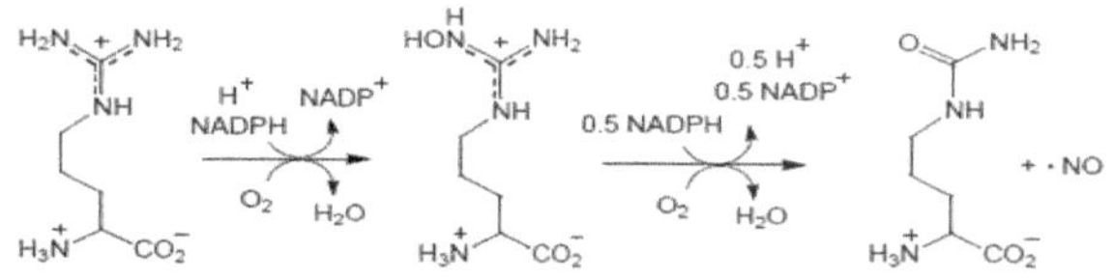

L-Arginine N-Hydroxyarginine L-Citrulline

Nitric oxide ($NO^{\bullet}$) is an abundant reactive radical that acts as an important oxidative biological signaling molecule in a large variety of diverse physiological processes, including neurotransmission, blood pressure regulation, defense mechanisms, smooth muscle relaxation and immune regulation (Bergendi et al., 1999). Due to its extraordinary properties, $NO^{\bullet}$ was acclaimed as the "molecule of the year" in Science Magazine (Koshland, 1992). $NO^{\bullet}$ has a half-life of only a few seconds in an aqueous environment. $NO^{\bullet}$ has greater stability in an environment with a lower oxygen concentration (half life >15 s). However, since it is soluble in both aqueous and lipid media, it readily diffuses through the cytoplasm and plasma membranes (Chiueh, 1999).

$NO^{\bullet}$ has effects on neuronal transmission as well as on synaptic plasticity in the central nervous system. In the extracellular milieu, NO• reacts with oxygen and water to form nitrate and nitrite anions. Overproduction of reactive nitrogen species is called nitrosative stress (Klatt and Lamas, 2000; Ridnour et al., 2004). This may occur when the generation of reactive nitrogen species in a system exceeds the system's ability to neutralize and eliminate them. Nitrosative stress may lead to nitrosylation reactions that can alter the structure of proteins and so inhibit their normal function. Cells of the immune system produce both the superoxide anion and nitric oxide during

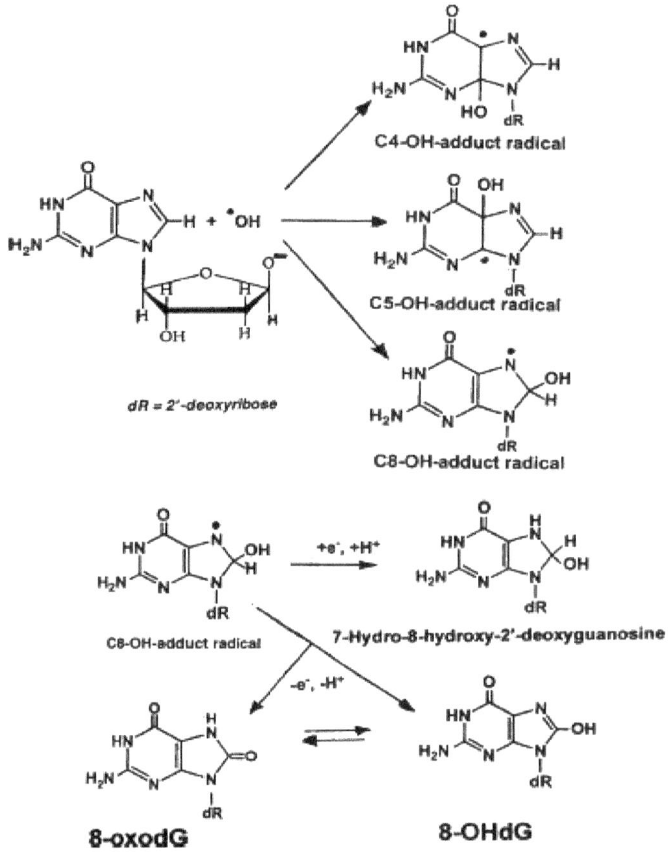

Figure 3. Reaction of 2_deoxyguanosine with hydroxyl radicals

the oxidative burst triggered during inflammatory processes. Under these conditions, nitric oxide and the superoxide anion may react together to produce significant amounts of a much more oxidatively active molecule, peroxynitrite anion ($ONOO^-$), which is a potent oxidising agent that can cause DNA fragmentation and lipid oxidation (Carr et al., 2000):

$$NO\bullet + O_2^{\bullet -} \rightarrow ONOO^- \quad (1)$$

Reaction (1) has one of the highest rate constants known for reactions of NO•, 7.0 × 109 $M^{-1}S^{-1}$. Thus NO• toxicity is predominantly linked to its ability to combine with superoxide anions. Nitric oxide readily binds certain transition metal ions; in fact many physiological effects of NO• are exerted as a result of its initial binding to Fe^{2+}-Haem groups in the enzyme soluble guanylate cyclase (sGC) (Archer, 1993).

$$Fe^{2+}\{sGC\} + NO^{\bullet} \rightarrow Fe^{2+}\{sGC\}\text{–}NO \quad (2)$$

The product is represented here as $\{Fe^{2+}\text{–}NO^{\bullet}\}$, however, $\{Fe^{3+}\text{–}NO^{-}\}$ is also commonly seen. The convention $\{FeNO\}^7$, where the superscript is the sum of the metal *d* electron count (here 6 or 5) and the occupancy of the relevant NO π orbital (here 1 or 2), is often employed to avoid specific assignment of oxidation states.

Oxidative damage to DNA, lipids and proteins

At high concentrations, ROS can be important mediators of damage to cell structures, nucleic acids, lipids and proteins (Valko et al., 2006). The hydroxyl radical is known to react with all components of the DNA molecule, damaging both the purine and pyrimidine bases and also the deoxyribose backbone (Halliwell and Gutteridge, 1999). The most extensively studied DNA lesion is the formation of 8-OH-G (8-hydroxy-2-deoxyguanosine) as shown in Figure 3. Figure 3 shows reaction of 2_deoxyguanosine with hydroxyl radicals, radical adducts followed by reduction to 7-hydro-8-hydroxy-2_deoxyguanosine, and by oxidation to 8-hydroxy-2-deoxyguanosine (8-OHdG) or its tautomer 8-oxo-7-hydro-2_deoxyguanosine (8-oxodG).

Permanent modification of genetic material resulting from "oxidative damage" incidents represents the first step involved in mutagenesis, carcinogenesis, and

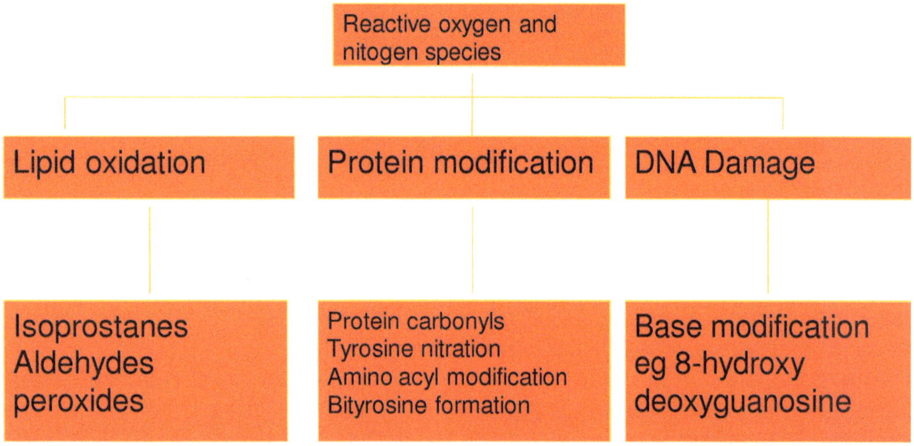

Figure 4. Various biomarkers of oxidative damage (adapted from Rice Evans, 1999).

ageing. It is known that metal-induced generation of ROS results in an attack not only on DNA, but also on other cellular components involving polyunsaturated fatty acid residues of phospholipids, which are extremely sensitive to oxidation (Siems et al., 1995).

Once formed, peroxyl radicals ($ROO^{\bullet}$) can be rearranged via a cyclisation reaction to endoperoxides (precursors of malondialdehyde) with the final product of the peroxidation process being malondialdehyde (MDA) (Wang et al., 1996) as shown in Figure 2. The major aldehyde product of lipid peroxidation other than malondialdehyde is 4-hydroxy- 2-nonenal (HNE). MDA is mutagenic in bacterial and mammalian cells and carcinogenic in rats. Hydroxynonenal is weakly mutagenic but appears to be the major toxic product of lipid peroxidation. Mechanisms involved in the oxidation of proteins by ROS were elucidated by studies in which amino acids, simple peptides and proteins were exposed to ionizing radiations under conditions where hydroxyl radicals or a mixture of hydroxyl/superoxide radicals are formed (Stadtman, 2004). The side chains of all amino acid residues of proteins, in particular cysteine and methionine residues of proteins are susceptible to oxidation by the action of ROS/RNS (Stadtman, 2004; Dalle-Donne et al., 2005).

Oxidation of cysteine residues may lead to the reversible formation of mixed disulphide between protein thiol groups (–SH) and low molecular weight thiols, in particular GSH (S-glutathiolation). The concentration of carbonyl groups, generated by many different mechanisms is a good measure of ROS-mediated protein oxidation. A number of highly sensitive methods have been developed for the assay of protein carbonyl groups (Dalle-Donne et al., 2005).

Advanced glycation end products (AGEs) is a class of complex products. They are the results of a reaction between carbohydrates and free amino group of proteins. The intermediate products are known, variously, as Amadori, Schiff Base and Maillard products, named after the researchers who first described those (Dalle-Donne et al., 2005). Most of the AGEs are very unstable, reactive compounds and the end products are difficult to be completely analyzed. The brown colour of the AGEs is probably related to the name of melanoidins initially proposed by Maillard (1912), and well known from food chemistry. The best chemically characterized AGEs compounds found in human are pentosidine and carboxyl methyl lysine (CML). Various biomarkers of oxidative damage are shown in Figure 4.

Antioxidants

Exposure to free radicals from a variety of sources has led organisms to develop a series of defense mechanisms (Cadenas, 1997). Defense mechanisms against free radical-induced oxidative stress involve: (i) preventative mechanisms, (ii) repair mechanisms, (iii) physical defences, and (iv) antioxidant defences. Enzymatic antioxidant defences include superoxide dismutase (SOD), glutathione peroxidase (GPx), catalase (CAT). Non-enzymatic antioxidants are represented by ascorbic acid (Vitamin C), tocopherol (Vitamin E), glutathione (GSH), carotenoids, flavonoids, and other antioxidants.

Under normal conditions, there is a balance between both the activities and the intracellular levels of these antioxidants (Liu, 2004). This balance is essential for the survival of organisms and their health. Various pathways for the management of oxidative stress by GSH and other antioxidants are shown in Figure 2. Here, we briefly describe the role of major thiol antioxidant and redox buffer of the cell, the tripeptide, glutathione (GSH) (Masella et al., 2005).

The oxidised form of glutathione is GSSG, glutathione disulphide. Glutathione is highly abundant in the cytosol

(1–11 mM), nuclei (3–15 mM), and mitochondria (5–11 mM) and is the major soluble antioxidant in these cell compartments. Because GSH is synthesized in the cytosol by the sequential action of glutamate-cysteine ligase and glutathione synthetase, its mitochondrial presence requires inner membrane transport. Two mitochondrial electroneutral antiport carrier proteins have been shown to have the capacity to transport GSH, the dicarboxylate carrier protein and the 2-oxoglutarate carrier protein.

Recently, it has been shown that externally added GSH is readily taken up by Mitochondria, despite the ~8mM GSH present in the mitochondrial matrix (Shen et al., 2005). It therefore appears that GSH is taken up against a concentration gradient. GSH in the nucleus maintains the redox state of critical protein sulphydryls that are necessary for DNA repair and expression. Oxidised glutathione is accumulated inside the cells and the ratio of GSH/GSSG is a good measure of oxidative stress of an organism (Jones et al., 2000). Too high a concentration of GSSG may damage many enzymes oxidatively. The main protective roles of glutathione against oxidative stress are according to (Masella et al., 2005): (i) glutathione is a cofactor of several detoxifying enzymes against oxidative stress, e.g. glutathione peroxidase (GPx), glutathionetransferase and others; (ii) GSH participates in amino acid transport through the plasma membrane; (iii) GSH scavenges hydroxyl radical and singlet oxygen directly, detoxifying hydrogen peroxide and lipid peroxides by the catalytic action of glutathionperoxidase; (iv) glutathione is able to regenerate the most important antioxidants, Vitamins C and E, back to their active forms; glutathione can reduce the tocopherol radical of Vitamin E directly, or indirectly, via reduction of semidehydroascorbate to ascorbate (Figure 2).

The capacity of glutathione to regenerate the most important antioxidants is linked with the redox state of the glutathione disulphide-glutathione couple (GSSG/2GSH) (Pastore et al., 2003). The basic mechanism of action of GSH is illustrated as follows:

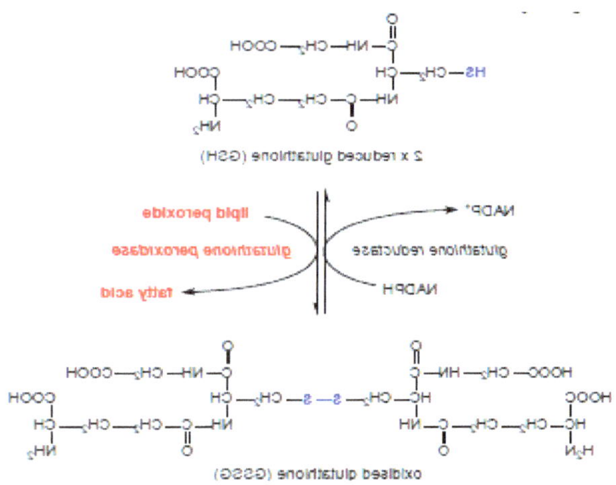

The various roles of enzymatic antioxidants (SOD, Catalase, and glutathione peroxidase) and non-enzymatic antioxidants (Vitamin C, Vitamin E, carotenoids, lipoic acid and others) in the protection against oxidative stress can be found in a numerous reviews and original papers (White et al., 1997) as shown in Figure 2.

ROS and mechanisms of maintenance of "redox homeostasis"

Free radicals and reactive diamagnetic species derived from radicals operate at low, but measurable concentrations in the cells. Their "steady state" concentrations are determined by the balance between their rates of production and their rates of removal by various antioxidants. Thus each cell is characterised by a particular concentration of electrons (redox state) stored in many cellular constituents and the redox state of a cell and its oscillation determines cellular functioning (Schafer and Buettner, 2001).

In recent years the term "redox state" has not only been used to describe the state of a redox pair, e.g. GSSG/2GSH, Asc•–/AcsH– and others, but also to describe more generally the redox environment of a cell (Butler, 2000; Schafer and Buettner, 2001). The redox state of a cell is kept within a narrow range under normal conditions, similar to the manner in which a biological system regulates its pH. Under pathological conditions, the redox state can be altered to lower or higher values. A 30mV change in the redox state means a 10-fold change in the ratio between reductant and oxidant species (Schafer and Buettner, 2001). The intracellular "redox homeostasis" or "redox buffering" capacity is substantiated primarily by GSH and thioredoxin (TRX). The glutathione (2GSH/GSSG couple) represents the major cellular redox buffer and therefore is a representative indicator for the redox environment of the cell (Droge, 2002; Schafer and Buettner, 2001).

Under enhanced oxidative stress conditions, GSSG content increases, this in turn increases the content of protein mixed disulphide. A significant number of proteins involved in signaling that have critical thiols, such as receptors, protein kinases and some transcription factors can be altered in their function by formation of mixed disulphides (Thannickal and Fanburg, 2000). In this regard, GSSG appears to act as a non-specific signalling molecule. The high ratios of reduced to oxidised GSH and TRX are maintained by the activity of GSH reductase and TRX reductase, respectively. Both of these "redox buffering" thiol systems counteract intracellular oxidative stress; in addition to antioxidant functioning in the cell, GSH and TRX are involved in cell signaling process (Droge, 2002; Thannickal and Fanburg, 2002).

In addition to GSH and TRX, there are other relatively low molecular weight antioxidants, that when present at high concentration, can significantly contribute to overall ROS scavenging activity (McEligot et al., 2005). These include various free amino acids, peptides, and proteins. Oxidised proteins are substrates for proteolytic digestion and contribute to maintenance of redox homeostasis in the cell (Dr¨oge, 2002). Oxidative modifications of proteins increase their susceptibility to proteolytic attack; proteolytic degradation is executed mainly by proteasomes. Proteolysis was estimated to increase more than 10-times after exposure to superoxide radical or hydrogen peroxide. It should be noted that proteins significantly vary in their susceptibility to oxidative damage. For example, intact proteins are less sensitive to oxidation than misfolded proteins.

The term redox signalling is used to describe a regulatory process in which the signal is delivered through redox reactions. Redox signalling requires that the steady state of "redox balance" is disturbed either by an increase in ROS formation or a decrease in the activity of antioxidant system(s). The regulated increase in free radicals (ROS/RNS) leads to a temporary imbalance that represents the physiological basis for redox regulation. Thus, physiological demonstration of redox regulation involves a temporary shift of the intracellular redox state toward more oxidising conditions.

Signalling mechanisms that respond to changes in the thiol/disulphide redox state involve: (i) transcription factors AP-1 and NF-κB; the transcription factor NF-κB was originally studied in cells of the immune system wherein it regulates cell survival (Baeuerle and Baltimore, 1996), it is widely expressed in the nervous system and exists in neurons in both an inducible and a constitutively active form (Kaltschmidt et al., 1993). NF-κB resides in the cytoplasm in an inactive form consisting of three subunits: p65 and p50 and an inhibitory subunit called IκB (Siebenlist et al., 1994). When IκB is bound to p50/p65, it is inactive; signals that activate NF-κB causes dissociation of IκB releasing p50/p65, which then translocates to the nucleus and binds to specific κB DNA consensus sequences in the enhancer region of a variety of κB-responsive genes (Mattson et al., 2000). In neurons, NF-κB is activated by various intercellular signals, including cytokines, neurotrophic factors, and neurotransmitters (Carter et al., 1996). (ii) bacterial OxyR; (iii) protein tyrosine phosphatases; (iv) Src family kinases; (v) JNK and p38 MAPK signaling pathways; (vi) insulin receptor kinase activity, and others (Droge, 2002; Aslund et al., 1999).

Under pathological conditions, however, abnormally large concentrations of ROS/RNS may lead to permanent changes in signal transduction and gene expression, typical for disease states. The process of redox signaling is adopted by various organisms including bacteria to induce protective responses against oxidative stress and to restore the original state of "redox homeostasis" after temporary exposure to ROS/RNS. For example, the production of NO• is the subject of direct feedback inhibition of NOS by NO•. Prokaryotes have several different signalling pathways for responding to ROS or to alterations in the intracellular redox state. Studies on *Escherichia coli* explored that low levels of ROS activate expression of several gene products involved in antioxidant defense including Mn- SOD, catalase, glutathione reductase, and others.

Several proteins that are synthesised in *Escherichia coli* after exposure to hydrogen peroxide are under the control of the OxyR locus. The OxyR protein controls protective responses against lethal doses of hydrogen peroxide or against killing by heat (Aslund et al., 1999). Hydrogen peroxide or an oxidative shift in the intracellular thiol/disulphide redox state converts the reduced form of OxyR (containing –SH groups) into its oxidised and regulatory active form containing –S–S– groups. The formation of disulphide bonds can be reversed by glutaredoxin and by thioredoxin. One of the best studied models of redox regulation in mammalian cells is the redox control of heme oxygenase-1 (HO-1) (Keyse and Tyrrell, 1989).

HO-1 induction in skin fibroblasts may serve as an inducible defence pathway to remove heme liberated by oxidants. The HO-1 protein and mRNA are strongly induced by ROS, UVA irradiation and various stressors; thus the inducibility of HO-1 mRNA in many tissues and various mammalian species has rendered HO-1 mRNA a useful marker for cellular oxidative stress at the mRNA level. As mentioned above, the cell cycle is characterized by fluctuations in the redox environment of a cell, mediated, in particular by intracellular changes in concentration of glutathione (Kern and Kehrer, 2005; Schafer and Buettner, 2001).

GSH has been shown to play a role in the rescue of cells from apoptosis; depletion of GSH, which renders the cellular environment more oxidising, was concomitant with the onset of apoptosis. Generally, a more reducing environment (maintained by elevated levels of glutathione and thioredoxin) of the cell stimulates proliferation and a slight shift towards a mildly oxidizing environment initiates cell differentiation. A further shift towards a more oxidising environment in the cell leads to apoptosis and necrosis. While apoptosis is induced by moderate oxidising stimuli, necrosis is induced by an intense oxidising effect (Evens, 2004; Voehringer et al., 2000). From the above discussion it is clear that the redox environment is the critical determinant for the trigger of apoptosis. Recent studies indicate that a knowledge of the mechanisms by which TRX, GSH, and Ref-1 maintain the intracellular "redox buffering" capacity can conveniently be used in the development of targeted cancer-preventive and therapeutic drugs (Evens, 2004).

OXIDATIVE STRESS AND DAMAGE

What happens if the balance between ROS and antioxidants is upset? Having too many ROS in relation to the available antioxidants is said to be a state of oxidative stress. Sies (1991), defined this term as a disturbance in the prooxidant-antioxidant balance in favor of the former, leading to potential damage. Such damage is often called oxidative damage, which has been defined as the biomolecular damage caused by attack of reactive species upon the constituents of living organisms (Halliwell and Whiteman, 2004).

Increased oxidative damage can result not only from more oxidative stress, but also from failure to repair or replace damaged biomolecules. Oxidative stress can result from decreases in antioxidant levels, e.g. mutations decreasing the levels of MnSOD; depletions of dietary antioxidants and other essential dietary constituents (e.g. copper, iron, zinc, and magnesium) can also cause it. For example, children with the protein deficiency disease,Kwashiorkor, suffer oxidative stress, involving low GSH levels (lack of sulfur-containing amino acids in the diet) and iron overload (inability to make enough transferrin.

Oxidative stress can also be due to increased ROS production for instance, by exposure to elevated O2; the presence of toxins that produce ROS (e.g. paraquat), or excessive activation of natural systems producing ROS, e.g. inappropriate activation of phagocytes (Halliwell and Gutteridge, 2006; Halliwell, 2000). What do cells do when under oxidative stress? It depends on the cell and the level of stress applied (Figure 4). Usually intracellular free Ca^{2+} levels rise; so do levels of iron catalytic for free radical reactions shown in Figure 4.

Several cell types respond to mild oxidative stress by proliferating, which can be good in wound healing but bad if it leads to tissue fibrosis (Cave et al., 2005). Cells may adapt to the stress by up-regulation of defense and/or repair systems. This may completely protect against damage, to some extent but not completely, or sometimes overprotect; the cells are then resistant to higher levels of oxidative stress imposed subsequently. Adaptation need not always involve increases in antioxidants; there can be decreases in ROS-producing systems, increases in other protective mechanisms (such as chaperones), or changes in oxidative damage targets (e.g. *E. coli* under oxidative stress, can replace a fumarase enzyme sensitive to inactivation by O_2 with one that resists O_2. Moderate oxidative stress usually halts the cell cycle, or can drive cells into senescence; the cell survives but can no longer divide. Severe oxidative damage, especially to DNA, may trigger death by apoptosis, necrosis, or mechanisms with features of both. Indeed, ROS act as triggers of apoptosis, and as participants in apoptosis induced by other mechanisms, in both plants and animals.

Figure 5, shows how cells respond to oxidative stress (Halliwell and Gutteridge, 2006). Stimulation of proliferation by low levels of RS (Reactive species) is associated with increased net phosphorylation of multiple proteins, often because of RS inactivate protein phosphatase enzymes (Kang et al., 2005) and sometimes because of increased protein kinase activity (Storz at al., 2004). The cell is generally a reducing environment, especially the mitochondria (GSH/GSSG>100) and cytosol (GSH/GSSG>100), but less so in the endoplasmic reticulum lumen (GSH/GSSG≈3), since a more oxidizing environment is required for optimal protein folding and disulfide bridge formation, ER (endoplasmic reticulum); HO-1, (haem oxygenase 1).

The following notes relate to the numbers in Figure 5: [1]'Reactive species' is a collective term that includes reactive oxygen, nitrogen, and halogen and sulfur species. For example, the term 'reactive oxygen species' (ROS) includes both oxygen-centered radicals (e.g. hydroxyl, superoxide, peroxyl) and certain non-radicals that are oxidizing agents and/or readily converted into radicals (e.g. peroxynitrite, hypochlorous acid). Similarly, reactive nitrogen species (RNS) includes the radicals' nitric oxide, and nitrogen dioxide as well as non-radicals such as nitryl chloride and nitroxyl anion. Many (but not all) RS are powerful oxidizing agents, capable of damaging DNA and other biomolecules. For a full description of RS and discussion of terminology, please see Halliwell and Gutteridge (2006).[2]Oxidative stress is a serious imbalance between RS levels and antioxidants, leading to potential damage. [3]Oxidative damage is the biomolecular damage caused by direct attack of RS. Its cellular levels are controlled by the balance between rate of damage and rate of repair or replacement of damaged biomolecules. [4]Caspase activity can also be modulated by changes in intracellular pH caused by RS (Parveiz and Clement, 2004; Akram et al., 2006).

ANTIOXIDANTS AND THE PREVENTION OF CANCER

Mechanisms of action

Cancer is the end point of a multistep process involving a sequence of events that occur over a period of years or even decades. DNA damage is considered to be one of the most important contributors to cancer. Much of this damage is oxidative in nature. A marker of mutagenic DNA damage would be useful in the estimation of cancer risk of various populations and in monitoring the effects of chemoprevention. Much of this damage is oxidative in nature. It has been estimated that a typical human cell experiences about 10000 oxidative "hits" to its DNA each day. DNA repair enzymes remove most of this damage, but not all of it. Oxidative lesions to DNA accumulate with age, and so does the risk of cancer. If a cell containing damaged DNA divides before its DNA can be repaired, the result is a permanent genetic alteration, the first step in carcinogenesis.

Cells that divide rapidly are more susceptible to carcin-

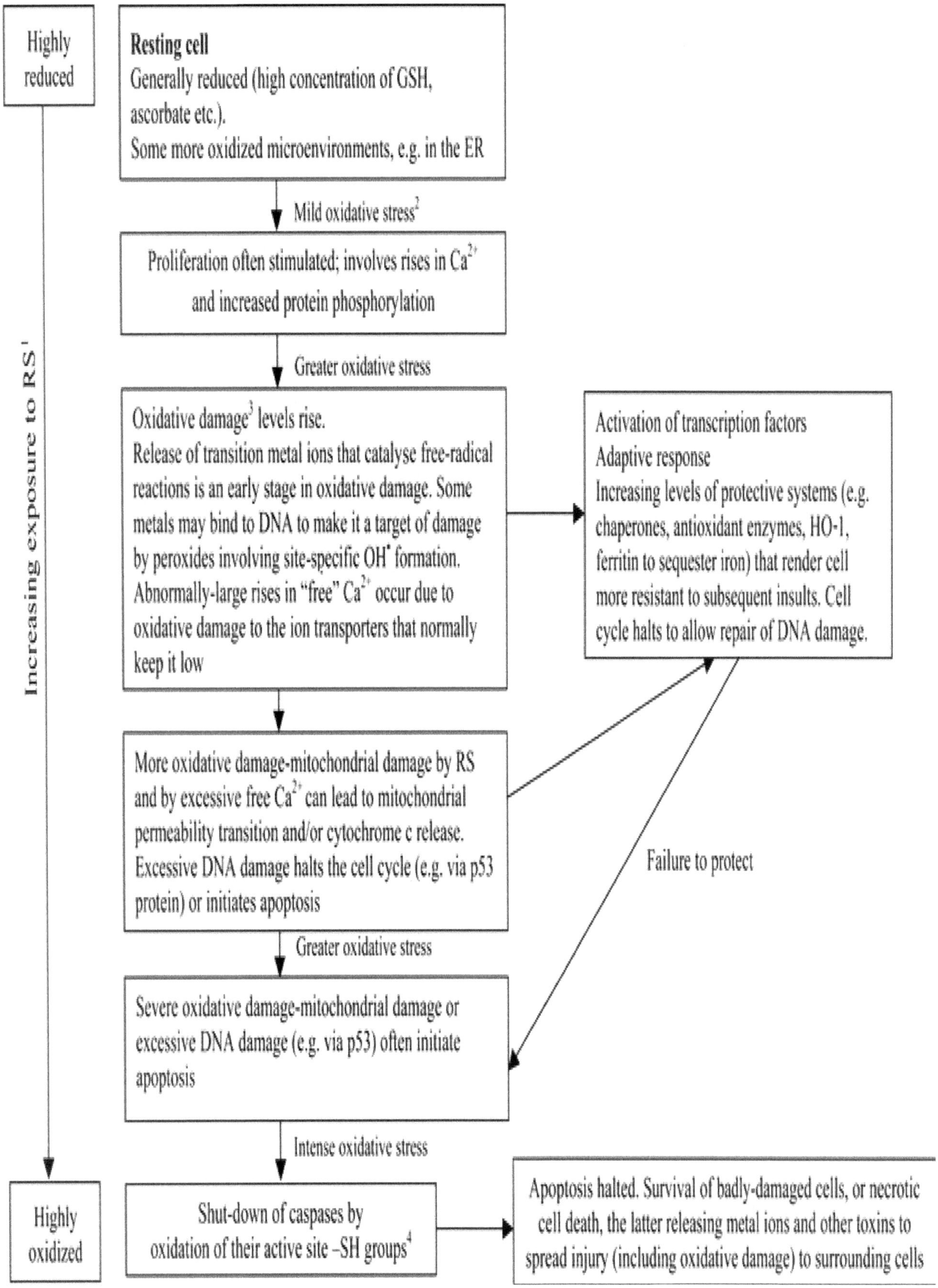

Figure 5. Cells respond to oxidative stress.

ogenesis than slowly dividing cells because there is less opportunity for DNA repair before cell division. Oxidants and antioxidants may also play a role in the later stages of cancer development. There is increasing evidence that oxidative processes contribute to the promotion stage of carcinogenesis, although the mechanisms for this are not well understood. Antioxidants may be able to cause the regression of premalignant lesions or inhibit their development into cancer.

Preliminary studies have indicated that some antioxidants, particularly ß-carotene, may be of benefit in the treatment of precancerous conditions such as oral leukoplakia (which may be a precursor of oral cancer). Some antioxidant nutrients may protect against cancer through mechanisms other than their antioxidant properties. For example, carotenoids may both enhance immune function and increase gap junctional communication (a type of interaction between cells that inhibits cell proliferation); both of these actions may be relevant to cancer prevention.

Tissue levels of antioxidants

Some epidemiologic studies have used biomarkers — physiological indicators of antioxidant status — rather than dietary data as measures of exposure. The most commonly used short-term biomarker is the concentration of an antioxidant in blood plasma. The findings of a major Chinese study support the concept that antioxidants may be protective against cancer. This study was conducted in 65 Chinese counties with an extraordinary diversity of disease patterns. Blood samples were collected from 100 randomly selected residents of each county; samples from members of the same community were then pooled for analysis. There was a consistent inverse correlation between cancer death rates in the 65 counties and blood levels of ß-carotene, vitamin C and selenium. For a wide range of cancers, associations between low death rates and high vitamin C levels were stronger than those for any other nutrient. High ß-carotene levels were associated with a reduced risk of stomach cancer. The EURAMIC study (European Community Multicentre Study on Antioxidants, Myocardial Infarction and Breast Cancer), now in progress, is using adipose tissue concentrations of vitamin E and ß-carotene and concentrations of selenium as biomarkers of antioxidant status. These biomarkers were chosen because they are believed to be an integrated measure of exposure over months rather than days. Epidemiologic studies in six European locations (Berlin, Germany; Coleraine, United Kingdom; Granada, Spain; Malaga, Spain; Zeist, The Netherlands; and Zürich, Switzerland) will compare levels of these biomarkers in breast cancer patients with those in otherwise similar women without the disease who live in the same areas.

Effects of individual carotenoids

In a few recent studies, epidemiologists have attempted to distinguish effects of specific carotenoids, rather than focusing on total carotenoids or ß-carotene. The results of these studies have suggested that several different carotenoids may be associated with reduced cancer risks. For example, in a study conducted in Hawaii, high dietary intakes of ß-carotene, α-carotene and lutein were each associated with reduced risks of lung cancer. Another study in the United States found that blood levels of total carotenoids, α- carotene, ß-carotene, cryptoxanthin and lycopene, but not lutein, were lower among women who later developed cervical cancer than among those who remained healthy. In the same U.S. population, blood levels of five major carotenoids were all found to be lower among individuals who later developed oral cancer than among controls.

Vitamin E

The evidence linking vitamin E and cancer risk is less extensive than that for vitamin C and carotenoids. Until recently, a lack of reliable information on the vitamin E content in foods has impeded epidemiologic studies of dietary vitamin E intake. Instead, most researchers have used blood vitamin E levels as a biomarker of vitamin E nutriture. The findings of these blood studies have been inconsistent; some have shown an inverse association between vitamin E levels and cancer risk, whereas others have found no association.

A large multicentre case-control study conducted by the U.S. National Cancer Institute, associated the use of vitamin E supplements with a 50% reduction in oral cancer risk. Dietary vitamin E and multivitamins had no significant effect. Dietary vitamin E intake in the United States is generally less than 15 international units (IU)/day (11 mg α-tocopherol equivalents [ATE]/day); multivitamins provide 30 IU/day (22 mg ATE/day); and most single-entity supplements sold in the United States contain at least 100 IU (74 mg ATE). This suggests the possibility that vitamin E may show a "plateau effect", meaning that below a certain critical dose it may not have a detectable inverse association with cancer risk. The plateau dose is likely to be greater than that which can be obtained through diet alone.

It is also possible, however, that the relationship between vitamin E supplementation and oral cancer seen in this study may not have been causal; people who chose to use supplements may have differed from nonusers in other ways that might affect oral cancer risk. At present, the use of supplements is an individual decision, and consumers receive little guidance about appropriate types and doses.

In general, individuals who choose to take supplements tend to make other efforts to improve their health as well, reflecting a generally health conscious lifestyle.

Intervention trials

Several clinical intervention trials designed to evaluate the effect of antioxidant supplementation on cancer risk are now in progress as shown in Table 1. Studies of this type are the best way to establish definitively the effects of a specific substance. However, cancer prevention intervention trials must involve large numbers of subjects and lengthy periods of treatment. Therefore, intervention trials are usually undertaken only when substantial evidence from other types of investigations suggests that an effect is likely to be demonstrated.

The Linxian trials were designed to determine the effect of nutrient supplementation on cancer risk in the general population. Participants in the trial either received various combinations of nutrient supplements or received placebos daily for 5 years. One of the supplements (a ß-carotene/vitamin E/selenium combination) was found to reduce stomach cancer deaths by 21%, total cancer deaths by 13% and total mortality by 9%. Thus, one or more of these antioxidants appear to be protective against cancer in the population studied. However, the results of this trial may not be directly applicable to Western populations, where both dietary habits and disease patterns are drastically different from those of this nutritionally deprived Chinese community. The other Linxian trial evaluated the effects of multivitamin and ß-carotene supplementation on oesophageal cancer risk in individuals who already had oesophageal dysplasia, a precancerous condition. More than 3.000 subjects received either supplements or placebos daily for 6 years. The results of the study were negative; supplementation did not reduce oesophageal cancer risk.

The Finnish trial, conducted by the Alpha-Tocopherol, Beta Carotene Cancer Prevention Study Group, evaluated the effect of vitamin E and ß-carotene on the incidence of lung cancer and other cancers in 29,000 men who had smoked for an average of 35.9 years. The men were randomly assigned to one of four regimens: ß-carotene (20 mg/day), vitamin E (50 mg/day as dl-α-tocopheryl acetate), both supplements together at the same levels, or an inactive placebo. Follow-up continued for 5 to 8 years, with lung cancer as the primary endpoint. Subjects taking ß-carotene either alone or with vitamin E were compared with subjects not taking ß-carotene. Contrary to expectations, the men who were taking ß-carotene showed a statistically significant 18% increase in lung cancer incidence. There was no evidence of an interaction between vitamin E and ß carotene with respect to lung cancer. ß-Carotene had some effect on the incidence of cancers other than lung cancer. The unexpected results of the Finnish trial should receive serious consideration, but the findings of a single study must always be placed in appropriate perspective. More studies are required.

Another cancer prevention trial, completed in 1994 in the United States, failed to show that a 4-year period of antioxidant treatment could prevent the occurrence of new colorectal adenomas in patients with a history of past adenomas. The 751 patients participating in this trial were randomly assigned to receive daily either ß-carotene (25 mg), vitamin E (400 mg) plus vitamin C (1 g), all three nutrients together, or an inactive placebo.

Cancer: Summary

In summary, both biochemical and epidemiologic studies have indicated that antioxidant nutrients and the foods that contain them may have important protective effects in the prevention of human cancer (Table 1). The evidence for a beneficial effect of fruits and vegetables is overwhelming. The evidence for protective effects from individual antioxidant nutrients is less definitive, and therefore other constituents of a diet may also play an important role (Greenwald and McDonald, 1999; Stephens et al., 1996; Van der Vliet, 2000). One major intervention trial, conducted in a non-Western population, has shown a beneficial effect of combined supplementation with ß-carotene, vitamin E and selenium.

In two recently completed trials in Western populations, however, one showed no benefit of supplementation with ß-carotene, vitamin E or vitamin C on prevention of colorectal adenomas, whereas in the other, supplementation with ß-carotene increased the risk of lung cancer in heavy smokers. More solid evidence on the relationship between antioxidant nutrients and cancer may be forthcoming within the next few years from other intervention trials in Western countries (Temple, 2000).

ANTIOXIDANTS AND CARDIOVASCULAR DISEASE

Mechanisms of action

Current theory suggests that oxidation may play a role in cardiovascular disease in two ways, one involving the long-term development of atherosclerosis and the other involving the immediate damage that occurs during a heart attack or stroke. Antioxidants may help counteract both of these processes.

LDL oxidation and atherogenesis

An increasing body of scientific evidence supports the concept that oxidation, possibly mediated by free radicals, may contribute to atherogenesis by transforming low-density lipoprotein (LDL) into an oxidised form. Oxidised LDL has been found in damaged arterial walls

Table 1. Some major intervention trials using antioxidants for cancer prevention.

Country	Cancer site	Study group	Sample size	Treatment	Result	References
China	Oesophagus, stomach and overall mortality	General population of a high risk area	30000	Various combinations of nutrients	(i) Stomach cancer reduction by 21% (ii) Total cancer deaths reduced by 13% (iii) Mortality reduced by 9%	NA
Finland (ATBC cancer study group)	Lung	Male long term smokers	29000	β-carotene and vitamin E (20 and 50 mg/ day respectively)	18% increase in lung cancer incidence	Heinonen et al., 1998
USA	All	Male physicians	22000	β-carotene (50 mg on alternate days)	No effect	Hennekens et al., 1996 a, b
USA (CARET)	Lung	Smokers and asbestos workers	17000	β-carotene (30 mg)and Vitamin A (Retinyl Palmiatae, 25000 IU)	Increase in lung cancer	Goodman et al., 1998
Australa	Skin	General population	1800	β-carotene (550 mg)	No effect	Green et al., 1998
USA	Head and neck	Subjects with previus cancers removed surgically	600	β-carotene	No effect	Mayne et al., 2000
Boston / USA	Breast cancer cells	-	-	Vitamin E	Prevention of tumour development	Schafer and Kerger, 2009
Finland	Prostate cancer	-	-	Vitamin E And catechol	Positive	Heionean et al., 1998
SELECT trial	Prostate cancer	Men	18000	(i) Vitamin E plus selenium (ii) Vitamin E plus selenium placebo (iii) Selenium plus Vitamin E placebo (iv) Double placebo (each tracked for 7-12 years)	(i) Vitamin E offered no prostrate cancer prevention (ii) 17 percent increased risk of cancer among who were assigned to take vitamin E (iii) Non-significant reduction in risk of cancer among men who were assigned to take vitamin E and selenium	Lipmaan et al., 2009; Klein et al., 2011

and has been shown to have several actions that could contribute to the initiation and progression of arterial damage. There is evidence from human studies that associates the extent of LDL oxidation with the extent of atherosclerosis. For example, in a study of Swedish heart attack victims, it was found that increased susceptibility of LDL to oxidation was associated with increased severity of atherosclerosis. Similarly, in a Finnish study, men with accelerated progression of atherosclerosis were found to have higher levels of antibodies to oxidised LDL than men who showed no progression of atherosclerosis. Additional evidence comes from short-term intervention trials in human volunteers, as described below in the Intervention Trials section.

Iscaemia/reperfusion injury

Another way in which oxidants and antioxidants may influence cardiovascular disease is through the process of ischemia/reperfusion injury, which is the immediate cause of tissue damage during a heart attack or stroke. All living tissues of the human body need oxygen to survive. If their supply of blood and oxygen is cut off (a situation called ischemia) they begin to die. Irreversible damage can be prevented only by reperfusion-the restoration of blood flow and reintroduction of oxygen. Ironically, although reperfusion is necessary for recovery, it can damage tissues further, because harmful oxygen free radicals are formed during the reperfusion process. The role of free radicals in heart attacks has been assessed in experimental model systems. Free radicals have been found to accumulate in isolated heart muscle that has been subjected to a temporary interruption of blood flow.

In some experimental models, treatments that inhibit the accumulation of free radicals have been shown to reduce the severity of damage to the heart muscle. In one human study, pretreatment with vitamin C was apparently of benefit to patients who underwent long periods of cardiac arrest during cardiopulmonary bypass surgery. The release of enzymes associated with ischemia was strikingly decreased in those patients who received vitamin C, indicating a reduction in cell damage.

Epidemiologic evidence

An increasing body of epidemiologic evidence links high intakes of antioxidants with reduced risks of cardiovascular diseases. The evidence is strongest for vitamin E, limited but promising for ß-carotene, and inconsistent for vitamin C.

Vitamin E

The most notable findings on vitamin E come from two large epidemiologic studies conducted in the United States. Key findings of these studies are summarized in Table 2. In an ongoing study of almost 40000 male health professionals, it was found that those in the top fifth of the group in terms of vitamin E intake showed a statistically significant 40% reduction in heart attack risk. In a parallel study of 87000 female nurses, those in the top fifth of vitamin E intake had a 34% reduction in heart attack risk. In both groups, the association was attributable mainly to vitamin E consumed in supplement form. Dietary vitamin E had no significant effect, presumably because vitamin E intakes from food are far lower than those achievable through supplementation. Daily use of single-entity supplements, generally containing at least 100 IU (74 mg ATE) of the vitamin, for 2 or more years was associated with a 37% decrease in heart attack risk among men and a 41% decrease in women. Supplement doses greater than 100–250 IU/day (74–185 mg ATE/day) were not associated with further reductions in risk.

Although these findings are impressive, they do not constitute definitive proof that vitamin E supplementation causes a reduction in heart disease risk. These were not intervention trials; they were observational studies of people who chose for themselves whether or not to use supplements. It is possible that supplement users may have differed from nonusers in other ways that might affect heart disease risk and that these other factors, rather than vitamin intake itself, were responsible for the inverse association between supplement use and coronary risk. However, other aspects of the data argue against this explanation. If supplement use were merely a marker for other aspects of a healthy lifestyle, one would also expect to see effects associated with the use of other types of supplements, including vitamin C. But this was not the case. In both men and women, vitamin C supplementation was not associated with decreased coronary risk. This fact tends to strengthen the case for a specific effect of vitamin E. The idea that vitamin E might protect against heart disease is also supported by other studies.

A cross-cultural study of 16 European population groups (the MONICA study) has shown a strong inverse correlation between blood vitamin E levels and heart disease risk. A study conducted in Scotland showed an inverse correlation between the incidence of angina and blood levels of vitamin E in men. Also, as described below, several trials have shown that high-dose vitamin E supplementation reduces the susceptibility of LDL to oxidation. The EURAMIC study, referred to previously, provides additional evidence about the relationship of vitamin E and other antioxidants to cardiovascular risk. In one arm of this study, vitamin E and ß-carotene concentrations were measured in 683 men who had recently experienced a myocardial infarction and in 727 control men who remained healthy. Coordinated investigations were conducted in nine European centres (Berlin, Germany; Edinburgh, United Kingdom; Granada, Spain; Helsinki, Finland; Malaga, Spain; Moscow, Russia;

Table 2. Some major antioxidant trials for CHD prevention.

Place	Study group	intervention	Risk/ result	References
Harvard school of public health	Middle aged men and women, americans	Vitamin E 100 IU/ D for 10 years	low	Stampfer et al., 1993
-do-	elderly	-do-	low	Losoznsky et al., 1996
United kingdom	Patients with existing heart disease	Vitamin E 400-800 IU/D for 1.5 years	77 percent decrease in heart diseases	Stephens et al., 1996
Finland	Randomly assigned men	Vitamin E (50mg/ D for 12 years on alternate days)	Decrease in heart diseases	Rapola et al., 1996
Harvard school of public health	Nurses health study and Health professionals follow up study	Vitamin E (400 IU or more for 2 years)	20-40 percent reduction in CHD risk	Stampfer et al., 1993 Rimm et al., 1993
GISSI prevention trial	11000 Heart attack survivors	Vitamin E supplements for 3 years	No prevention	Anon, 1999.
HOPE trial	9500 men and women diagnosed with heart diseases or at high risk	Vitamin E supplementation for 3 years	No prevention	Yusuf et al., 2000
Womens health study	40000 women	Vitamin E supplementation (600 IU) for 10 years on alternate days	**(i) Non-significant reduction in major cardiac events** **(ii) 24 percent lower risk of cardiovascular death**	Lee et al., 2005
Physicians health study II	15000 middle aged men, mostly free of heart diseases at start	vitamin E supplements of 400 IU every other day, alone or with vitamin C	No protection against heart attacks, strokes or cardiovascular deaths	Seso et al., 2008

Sarpsborg, Norway; Zeist, The Netherlands; and Zürich, Switzerland) and in Jerusalem, Israel. The researchers found that low adipose tissue vitamin E concentrations were not associated with increased myocardial infarction risk. The lack of effect of vitamin E suggests that protection may be seen only in supplement users. The amounts of vitamin E obtained from foods may be insufficient for protection against myocardial infarction.

ß-Carotene

The EURAMIC study, however, showed that low adipose tissue ß-carotene concentrations were associated with significantly increased risk of myocardial infarction, a risk that was mainly confined to current smokers. The same study of male health professionals that showed an inverse relationship between vitamin E intake and coronary risk also associated high dietary ß-carotene intake with reduced coronary risk in smokers. Current smokers in the top fifth of ß-carotene intake showed a 70% reduction in heart disease risk and former smokers showed a 40% reduction. A preliminary analysis of data from the parallel women's study, reported only in abstract form, also suggests an inverse association between ß-carotene intake and coronary risk. However, these preliminary data show that those in the top fifth of ß-carotene intake (smokers and nonsmokers combined) showed a 22% reduction in heart disease risk.

Vitamin C

As noted above, the U.S. studies of female nurses and male health professionals did not find any association between vitamin C supplementation and coronary risk.

However, several other studies have shown effects of vitamin C. For example, the Basel Prospective Study showed that Swiss men with low blood vitamin C levels had an increased risk of dying from a heart attack during 12 years of follow-up compared with men with normal blood vitamin C levels. Studies comparing different European populations indicate that coronary heart disease mortality is higher in those with blood vitamin C levels that are almost in the deficient range.

An analysis of a 10-year follow-up study of a representative sample of the U.S. population found that men with the highest vitamin C consumption (at least 50 mg/day from diet plus regular use of supplements) had a 42% lower rate of death from cardiovascular diseases and women had a 25% lower rate. Further research is needed to resolve the discrepancy in epidemiologic findings on vitamin C intake and heart disease risk.

Other dietary antioxidants

A small number of studies have suggested that dietary antioxidants other than vitamin E, vitamin C and carotenoids might be protective against heart disease. For example, in a study conducted in The Netherlands, high intakes of flavonoids (found in black tea, onions, and apples) were associated with decreased coronary mortality in a group of elderly men. Garlic also contains antioxidants, and a U.S. study indicated that garlic supplementation can reduce the susceptibility of LDL to oxidation. It has also been suggested that the apparent protective effect of red wine against heart disease, which is believed to contribute to the relatively low cardiovascular death rate in France, may be due not to the alcohol content of the wine but to its antioxidant content. In vitro, antioxidants isolated from red wine have been found to inhibit the oxidation of LDL. The activity of the natural antioxidant ubiquinol is currently under investigation. All of these studies provide intriguing leads for future research.

Intervention trials

As Table 2 indicates three major intervention trials of antioxidants and cardiovascular disease are now in progress, two in the United States and one in the United Kingdom (the MRC/BHF Heart Protection Study, starting date 1994). These studies are likely to provide definitive evidence on the effects of ß-carotene supplementation on cardiovascular risk, and two of the trials should provide valuable information about vitamin E as well. Valuable information has also been obtained from short term trials designed to assess the effect of antioxidant supplementation on the susceptibility of blood lipoproteins to oxidation. In several trials of this type, daily consumption of high- dose vitamin E supplements (800 IU [592 mg ATE] or more), usually for a period of several weeks, has been shown to decrease the susceptibility of LDL from these subjects to in vitro oxidation by copper sulphate. This action appears to be specific to vitamin E; high doses of ß-carotene, another fat-soluble antioxidant, do not have the same effect. Various combinations of antioxidant nutrients have also been shown to be effective in decreasing the oxidation of LDL, but further research is needed to determine whether all of the antioxidants contribute to the effect or whether vitamin E is the only active agent.

Cardiovascular disease: summary

In summary, biochemical studies, epidemiologic investigations and preliminary findings from most intervention trials suggest that antioxidant nutrients may play a role in reducing cardiovascular risk (Losonczy et al., 1996). The evidence is strongest for vitamin E. However, protective effects of vitamin E may be evident only at high doses - much more than can be obtained from a normal diet. Further research is needed to confirm the role of vitamin E and determine the optimal intake. The current evidence on ß-carotene is limited but promising. Clinical trials already in progress should provide definitive answers on the role of this nutrient within the next few years. The evidence on vitamin C and cardiovascular disease is inconsistent, and more study of this nutrient is needed. There is also some limited preliminary evidence suggesting that certain antioxidants other than micronutrient vitamins and minerals in foods may have beneficial effects on cardiovascular health. Results from the Heart Outcomes Prevention Evaluation (HOPE) trial, also showed no benefit of four years worth of vitamin E supplementation among more than 9,500 men and women already diagnosed with heart disease or at high risk for it (Yusuf et al., 2000).

In fact, when the HOPE trial was extended for another four years, researchers found that study volunteers who took vitamin E had a higher risk of heart failure (Lonn et al., 2005). Based on such studies, the American Heart Association has concluded that "the scientific data do not justify the use of antioxidant vitamin supplements [such as vitamin E] for CVD risk reduction" (Kris- etherton et al., 2004).

EYE DISEASES

Age-related diseases of the eye are major health problems around the world. In technologically developed countries, the treatment of cataracts is one of the largest contributors to total health care costs for the elderly. In less developed countries, where treatment is often unavailable, cataracts are a major cause of blindness in older adults. Worldwide, approximately 50 million people are blind from cataracts. In Western countries, the most common cause of new cases of blindness in the elderly is not cataracts but an age-related disorder of the retina called macular degeneration or Maculopathy. Oxidative processes have been implicated in the causation of both cataract and age related maculopathy. Recent evidence suggests that high dietary intakes of antioxidants may help delay or prevent these disorders.

Cataracts

Cataracts occur when transparent material in the lens of the eye becomes opaque. Much of the material in the lens consists of extremely long-lived proteins, which can become damaged over the decades of a human lifetime. Since there is no direct blood supply to the lens, nutrients enter and waste products are removed by a simple diffusion process which is slow and inefficient. Oxidation, induced primarily by exposure to light, is believed to be a major cause of damage to the proteins of the lens. When these proteins become oxidised, they clump together and precipitate, causing portions of the lens to become cloudy. The eye has defence systems which protect the lens from oxidative damage. Antioxidants and antioxidant enzymes inactivate harmful free radicals and proteases (enzymes which break down proteins) selectively remove damaged proteins from the lens. However, these defense systems cannot always keep pace with oxidative damage. As a result, oxidised proteins may accumulate. As people age, the defence systems grow less effective, and damage to lens proteins may become irreversible.

Several recent epidemiologic studies conducted in Western countries have associated high intakes or blood levels of antioxidant nutrients with reduced rates of cataract. All three of the major dietary antioxidants— vitamin C, vitamin E and carotenoids — have been associated with decreased cataract risk. For example, a Finnish study showed that people with low blood levels of vitamin E or ß-carotene had higher risks of developing cataracts during a 15-year follow-up period. The increase in risk was greatest among individuals who had low blood levels of both of these antioxidant nutrients.

In the ongoing study of female U.S. nurses, it was found that intake of carotene was inversely associated with cataract risk. In the same study, the risk of cataract was decreased by 45% among women who had used vitamin C supplements for 10 years or more. A small Canadian epidemiologic study indicated that nonusers of vitamin C supplements were 3.3 times more likely than users to develop cataracts. Similarly, nonusers of vitamin E supplements were 2.3 times more likely than users to develop cataracts. A multicentre U.S. study found that use of multivitamins was associated with a decreased risk of each of four major types of cataract; high intakes of fruits and vegetables also appeared to be protective. In another U.S. study, low blood vitamin C levels were associated with an 11-fold increased risk of one type of cataract, and low blood carotenoid levels were associated with a 7-fold increased risk of another type of cataract. Low intakes of fruits and vegetables were also associated with increased cataract risk. In contrast, 2 studies conducted in the Orient had negative results. An epidemiologic study conducted in fishing communities of Hong Kong found no significant associations between blood levels of antioxidant nutrients and cataract. Also, in the recent nutrition intervention trial in Linxian, China, supplementation with ß-carotene/vitamin E/selenium or with vitamin C/molybdenum did not lead to reductions in cataract risk. However, it should be noted particularly in the Hong Kong study, the subjects were probably exposed to high levels of ultraviolet radiation — a major causative factor in cataract.

Interestingly, a riboflavin/niacin supplement did show a significant protective effect, which may be indirectly related to antioxidant defence mechanisms. Riboflavin is a cofactor for the activity of several essential antioxidant enzymes, and intakes of this vitamin in Linxian are marginal at best. The riboflavin/niacin supplement may have exerted its effect by correcting a subclinical riboflavin deficiency, thus enhancing the activity of antioxidant enzymes. However, a similar benefit of riboflavin supplementation would not be expected in Western populations, where intakes of this vitamin are generally ample. The difference between the Western and Asian findings may reflect differences in overall nutritional status and cataract risk factors in the two types of populations

Age-related macular degeneration

Some scientific evidence suggests that excessive exposure to light and the resulting production of oxidants contribute to the causation of age-related maculopathy. In particular, exposure to blue light has been implicated in this disorder. If blue light is involved, carotenoids might be protective, since they can absorb blue light. An analysis of data from a national U.S. survey showed an inverse correlation between age-related maculopathy and the consumption of "vitamin A-rich" fruits and vegetables. The "vitamin A" found in these foods actually consists of carotenoid precursors of the vitamin rather than preformed vitamin A (retinol). A large multicentre U.S. case-control study that focused on the most severe form of age-related maculopathy found markedly reduced risks in individuals with high blood carotenoid levels. Significant associations were detected for the sum of all carotenoids and for four of five individual carotenoids (ß-carotene, α-carotene, cryptoxanthin and lutein/zeaxanthin, but not lycopene). Vitamin C and vitamin E did not show significant effects. In contrast, a much smaller British study that included all types of age-related macular degeneration found no association between serum carotenoid or vitamin E levels

Table 3. Some major antioxidant trials for eye disease prevention.

Place	Experimental population	intervention	Result	References
USA	US women	Vit C supplement for 10 years or more	45% lower catarct incidence	Hankinson et al., 1992
USA	US physicians	multivitamins	25% lower catarct incidence	Seddon et al., 1994
USA	Elderly men	Multivitamin supplements or vitamin E	Lower risk of opquenes of lens	Leske and Chylack, 1991
Canada	Non-users of vitamin E	-	3.3 times more likely to develop cataract	Robertson et al., 2002
Canada	Non users of vitamin E	-	2.3 times more likely to develop catarct	-do-
Finland	Smokers	Vit E/ β carotene for 6 years	No effect	Tiekari et al., 1997

and the risk of macular degeneration. The difference between these findings and those of the U.S. study may reflect differences in sample size or differences in the pathogenesis of the different types of maculopathy.

In summary, there is evidence suggesting that oxidative processes may play a role in causing age-related disorders of both the lens and the retina of the eye. All of the published epidemiologic studies of antioxidants and cataract in Western populations have shown significant associations for at least one antioxidant nutrient. The overall body of evidence suggests that all three major dietary antioxidants — vitamin C, vitamin E and carotenoids — may be beneficial in reducing cataract risk in Western populations. Less is known about age-related macular degeneration, but some epidemiologic evidence suggests an inverse association with carotenoids. Further research is needed to determine conclusively whether improved nutrition may reduce the risk of these very prevalent eye disorders in older adults (Table 3).

OTHER DISEASES AND PATHOLOGICAL PROCESSES

Although most research on oxidants, antioxidants and disease has focused on cancer, cardiovascular disease and degenerative disorders of the eye, oxidative processes are also implicated in a wide variety of other clinical conditions and pathological processes, as shown in Table 4. A few examples of particular interest are described below.

Neurological disorders

Biochemical studies suggest that oxidation may play a role in the causation of several disorders of the brain and nervous system. Therefore, it has been hypothesized that antioxidants might be helpful in ameliorating the symptoms or in slowing the progression of some neurological disorders. Some studies have shown beneficial effects of vitamin E supplementation in decreasing the severity of Tardive dyskinesia; others have not found an effect. Tardive dyskinesia is a disorder involving involuntary movement. It occurs as a side effect of long-term treatment with certain antipsychotic drugs. An open pilot trial indicated that supplementation with vitamins C and E might be of benefit in slowing the progression of Parkinson's disease. However, a larger placebo-controlled trial found no benefit from vitamin E therapies.

Sperm damage, birth defects and childhood cancer

The children of fathers who smoke cigarettes have increased rates of congenital malformations and childhood cancer. These problems may be related, at least in part, to increased oxidative damage to sperm cells caused by oxidants in cigarette smoke. Cigarette smoking is associated with reduced sperm count and poor sperm quality; it is also associated with decreased blood vitamin C levels. Vitamin C supplementation has been shown to improve sperm quality in heavy smokers. Ample intakes of vitamin C have also been shown to reduce oxidative damage to sperm DNA. Further studies are needed to determine whether improved antioxidant status, particularly with respect to vitamin C, is of benefit in reducing infertility and germ-line mutations in men who smoke cigarettes or who are exposed to oxidative stress from other causes.

Exercise-induced oxidative stress

Because exercise leads to increased oxygen

Table 4. Some major antioxidant trials for neurological disease prevention.

Place	Experimental population	Intervention	Results	References
USA	Patients with moderately severe Alzheimer's disease	2000 IU/ D Vit E or placebo for 2 – 4 years	Treated patients were better (non-significantly)	Sano et al., 1997
USA	Patients with moderately severe parkinsons' disease	Vit E supplementation for 5 years	No effect	Agid et al., 1993

consumption, it causes an increase in the production of oxygen-initiated free radicals. Some studies have reported that antioxidant supplementation reduces evidence of exercise-induced muscle damage. Antioxidants may also play a role in reducing muscle soreness after overexertion. However, it is unclear whether antioxidants can directly enhance physical performance.

Inflammatory disorders

Free radicals and oxidative stress may play a role in inflammatory diseases. Rheumatoid arthritis is one example. The products of free radical reactions have been detected in the blood and joints of patients with this disease. Other lines of evidence also suggest the involvement of oxidative stress in rheumatoid arthritis and in other inflammatory diseases such as glomerulonephritis.

Decreases in immune function

Several aspects of immune function show a marked decline with increasing age. Preliminary studies in elderly people have indicated that this decline can be partly offset by dietary antioxidant supplementation. The age-associated decrease in cell-mediated immunity may be due to a decreased level of small-molecular-weight antioxidants and decreased activity of antioxidant enzymes.

OXIDANTS, ANTIOXIDANTS, AND DISEASE: IMPLICATIONS OF THE SCIENTIFIC EVIDENCE

Although much remains to be learned, there is now convincing evidence for the involvement of free radicals in a number of diseases which constitute major public health problems throughout the world. Free radicals and oxidative processes are believed to play important roles in the pathogenesis of many age-related disorders. The evidence implicating them in the causation of cancer, atherosclerotic cardiovascular disease and cataracts is especially strong. The human body has a complex system of natural enzymatic and nonenzymatic antioxidant defences which counteract the harmful effects of free radicals and other oxidants. Protection against free radicals can be enhanced by ample intakes of dietary antioxidants, of which the best studied are Vitamin E, C and carotenoids. Substantial evidence indicates that foods containing antioxidants and possibly in particular the antioxidant nutrients may be of major importance in disease prevention. Efforts should be made to ensure optimum intakes of foods containing these important molecules.There is a need for improvement in the quality of the diet, especially with respect to increased consumption of fruits and vegetables.

However, other strategies, including optimization of food processing, selective fortification of foods and the use of safe nutritional supplements, may also need to be considered. All three of the major antioxidant nutrients-vitamin E, vitamin C and ß-carotene-are safe even at relatively high levels of intake (Berger et al., 2012). There is, however, a growing consensus among scientists that a combination of antioxidants, rather than single entities, may be more effective over the long term. Antioxidants may be of great benefit in improving the quality of life by preventing or postponing the onset of degenerative diseases.

In addition, there is a potential for substantial savings in the cost of health care delivery. More research is needed to clarify and extend scientific understanding of the health effects of antioxidants. Basic research should continue, and additional large-scale randomised trials and clinical studies should be undertaken. Funding for research in this field is urgently needed. At the same time, efforts should also be made to communicate to the general public on existing information about the importance of protective nutrients in fruits and vegetables. Government agencies, health professionals and the news media should work together to promote the dissemination of scientifically sound information about this aspect of nutrition to all segments of the population.

REFERENCES

Agid Y, Ruberg M, Javoy-Agid F, Hirsch E, Raisman-Vozari R, Vyas S (1993). Are dopaminergic neurons selectively vulnerable to Parkinson's disease? Adv. Neurol. 60:148–64.

Aikens J, Dix TA (1991). Perhydroxyl radical (HOO•) Initiated lipid-peroxidation-The role of fatty-acid hydroperoxides. J. Biol. Chem. 266:15091–15098.

Anon (1999). Dietary supplementation with n-3 polyunsaturated fatty acids and vitamin E after myocardial infarction: Results of the GISSI-Prevenzione trial. Gruppo Italiano per lo Studio della Sopravvivenza nell'Infarto miocardico. Lancet. 354:447-55.

Archer S (1993). Measurement of nitric-oxide in biological models. FASEB J. 7:349-360.

Aslund F, Zheng M, Beckwith J, Storz G (1999). Regulation of the Oxy R transcription factor by hydrogen peroxide and the cellular thioldisulfide status. Proc. Nat. Acad. Sci. U.S.A. 96:6161-6165.

Baeuerle PA, Baltimore D (1996). NF-κB: Ten years after. Cell 87:13-20.

Bergendi L, Benes L, Durackova Z, Ferencik M (1999). Chemistry, physiology and pathology of free radicals. Life Sci. 65:1865-1874.

Berger RG, Lunkenbein S, Strohle A, Hahn A (2012). Antioxidants in food: Mere myth or magic medicine. Critical Rev. Food Sci. Nutr. 52:162-171.

Butler J (2000). Thermodynamic considerations of free radical reactions. In C. J. Rhodes (Ed.), Toxicology of the human environment, Taylor and Francis, London pp. 437–453.

Cadenas E (1997). Basic mechanisms of antioxidant activity. Biofactors 6:391-397.

Cadenas E, Sies H (1998). The lag phase. Free Radic. Res. 28:601-609.

Carr A, McCall, MR, Frei B (2000). Oxidation of LDL by myeloperoxidase and reactive nitrogen species-reaction pathways and antioxidant protection. Arterioscl. Thromb. Vasc. Biol. 20:1716-1723.

Carter BD, Kaltschmidt C, Kaltschmidt B, Offenhauser N, Bohm-Matthei R, Baeuerle PA (1996). Selective activation of NF-κB by nerve growth factor through the neurotrophin receptor p.75. Science 272:542-545.

Chiueh CC (1999). Neuroprotective properties of nitric oxide. Ann. N.Y. Acad Scc. 890:301-311.

Commoner B, Townsend J, Pake GE (1954). Free radicals in biological materials. Nature 174:689-691.

Dalle-Donne I, Scaloni A, Giustarini D, Cavarra E, Tell G, Lungarella G (2005). Proteins as biomarkers of oxidative/ nitrosative stress in diseases: The contribution of redox proteomics. Mass Spectrom. Rev. 24:55-99.

De Grey ADNJ (2002). $HO_2^{\bullet-}$ The forgotten radical. DNA Cell Biol. 21:251-257.

De Coursey TE, Ligeti E (2005). Regulation and termination of NADPH oxidase activity. Cell. Mol. Life Sci. 62:2173-2193.

Droge W (2002). Free radicals in the physiological control of cell function. Physiol. Rev. 82:47-95.

Evens AM (2004). Motexafin gadolinium: A redox-active tumor selective agent for the treatment of cancer. Curr. Opin. Oncol. 16:576-580.

Gerschman R, Gilbert DL, Nye SW, Dwyer P, Fenn WO (1954). Oxygen poisoning and x-irradiation-A mechanism in common. Science 119: 623-626.

Ghafourifar P, Cadenas E (2005). Mitochondrial nitric oxide synthase. Trends Pharmacol. Sci. 26:190-195.

Greenwald P, McDonald SS. (1999). Antioxidants and the prevention of cancer. In: Basu TK, Temple NJ, Garg ML, eds. Antioxidants in Human Health and Disease. Wallingford, Oxon, UK: CAB Int. pp. 217-234.

Guallar E (2005). Meta-analysis: High-dosage Vitamin E supplementation may increase all-cause mortality. Ann. Intern. Med. 142:37-46.

Gutteridge JMC, Halliwell B (2010). Antioxidants: Molecules, medicines, and myths. Biochem. Biophys. Res. Comm. 393:561-564.

Halliwell B, Gutteridge JMC (1999). Free radicals in biology and medicine (3rd ed.). Oxford University Press.

Halliwell BG (2000). The antioxidant paradox. Lancet. 55: 1179-1180.

Halliwell BG, Gutteridge JMC. (2006). Free Radicals in Biology and Medicine, Ed 4. Clarendon Press, Oxford, p.245.

Han D, Canali R, Garcia J, Aguilera R, Gallaher TK, Cadenas E (2005). Sites and mechanisms of aconitase inactivation by peroxynitrite: Modulation by citrate and glutathione. Biochemistry. 44:11986-11996.

Harman D (1956). Aging-A theory based on free-radical and radiation-chemistry. J. Gerontol. 11:298-300.

Heinonen OP, Albanes D, Virtamo J, Taylor PR, Huttenen JK, Hartman AM, Haapakoski J, Malila M, Rautalahti M, Ripatti S, Maenpaa H, Teerenhovi L, Koss L, Virolainen M, Edwards BK (1998). Prostate cancer and supplementation with α-tocopherol and β-carotene: Incidence and mortality in a controlled trial. J. Nat. Cancer Ins. 90:440-446.

Hennekens CH, Buring GE, Manson JE (1996b). Lack of longterm supplementation with beta carotene on the incidenmce of malignant neoplasm and cardio vascular diseases. N. Engl. J. Med. 334:1145-11149.

Hennekens CH, Buring JE, Manson JE, Stampfer M, Rosner B, Cook NR, Belanger C,LaMotte F, Gaziano JM, Ridker PM, Willett W, Peto R (1996a). Lack of effect of long-term supplementation with carotene on the incidence of malignant neoplasms and cardiovascular disease. N. Engl J. Med. 334: 1145-1149.

Jones DP, Carlson JL, Mody VC, Cai JY, Lynn MJ, Sternberg P (2000). Redox state of glutathione in human plasma. Free Radic. Biol. Med., 28:625-635.

Kakhlon O, Cabantchik ZI (2002). The labile iron pool: Characterization, measurement, and participation in cellular processes. Free Radic. Biol. Med. 33:1037-1046.

Kaltschmidt C, Kaltschmidt B, Baeuerle PA (1993). Brain synapses contain inducible forms of the transcription factor NF-κB. Mech. Dev. 43:135–147.

Kern JC, Kehrer JP (2005). Free radicals and apoptosis: Relationships with glutathione, thioredoxin and the Bcl family of proteins. Front Biosci. 10:1727-1738.

Klatt P, Lamas S (2000). Regulation of protein function by *S* glutathiolation in response to oxidative and nitrosative stress. Eur. J. Biochem. 267:4928-4944.

Klein EA, Thompson IM, Jr., Tangen CM (2011). Vitamin E and the risk of prostate cancer: The Selenium and Vitamin E Cancer Prevention Trial (SELECT). JAMA. 306:1549-1556.

Koshland DE (1992). The molecule of the year. Science 258:1861.

Kovacic P, Jacintho JD (2001). Mechanisms of carcinogenesis: Focus on oxidative stress and electron transfer. Curr. Med. Chem. 8:773-796.

Kovacic P, Pozos RS, Somanathan R, Shangari N, O'Brien PJ (2005). Mechanism of mitochondrial uncouplers, inhibitors, and toxins: Focus on electron transfer, free radicals, and structure–activity relationships. Curr. Med. Chem. 12:2601-2623.

Kris-Etherton PM, Lichtenstein AH, Howard BV, Steinberg D, Witztum JL (2004). Antioxidant vitamin supplements and cardiovascular disease. Circulation 110:637-641.

Lee IM, Cook NR, Gaziano JM (2005). Vitamin E in the primary prevention of cardiovascular disease and cancer: The Women's Health Study: A randomized controlled trial. JAMA. 294:56-65.

Leonard SS, Harris GK, Shi X (2004). Metal-induced oxidative stress and signal transduction. Free Radic. Biol. Med. 37:1921-1942.

Leske MC, Chylack LT (1991). The lens opacities caser control study: Risk factors for cataract. Arch opthamology Opthamol. 109:244-251.

Liochev SI, Fridovich I (1994). The role of O_2 in the production of $HO^{\cdot}$ *In vitro* and *in vivo*. Free Radic. Biol. Med. 16:29-33.

Liochev SI, Fridovich I (2002). The Haber–Weiss cycle-70 years later: An alternative view. Redox Rep. 7:55-57.

Liu RH (2004). Potential synergy of chemicals in cancer prevention. The J. Nutr. 22:3479-3485.

Lonn E, Bosch J, Yusuf S (2005). Effects of long-term vitamin E supplementation on cardiovascular events and cancer: a randomized controlled trial. JAMA 293:1338-1347.

Losonczy KG, Harris TB, Havlik FJ (1996). Vitamin E and Vitamin C supplement use and risk of all cause and coronary heart disease mortality in older persons. Am. J. Clinical Nutr. 64:190-196.

Maillard LC (1912). "Action of Amino Acids on Sugars. Formation of Melanoidins in a Methodical Way". Compt. Rend. 154: 66.

Masella R, Di Benedetto R, Vari R, Filesi C, Giovannini C (2005). Novel mechanisms of natural antioxidant compounds in biological systems: Involvement of glutathione and glutathione related enzymes. J. Nutr. Biochem. 16:577-586.

Mattson MP, Culmsee C, Yu ZF, Camaldola S (2000). Roles of nuclear factor-κB in neuronal survival and plasticity. J. Neurochem. 74:443-456.

Mayne ST, Hanleman GJ, Beecher (2000). Beta carotene and lung proportion in heavy smokers: a plausible relationship. J. Nat. Cancer InstInst. 7:440-446.
McCord JM, Fridovich I (1969). Superoxide dismutase an enzymic function for erythrocuprein Hemocuprein. J. Biol. Chem. 44:6049-6055.
McEligot AJ, Yang S, Meyskens FL (2005). Redox regulation by intrinsic species and extrinsic nutrients in normal and cancer cells. Ann. Rev. Nutr. 25:261-295.
Miller DM, Buettner GR, Aust SD (1990). Transition metals as catalysts of “autoxidation” reactions. Free Radic. Biol. Med. 8:95–108.
Miller ER, Pastor-Barriuso R, Dalal D, Riemersma RA, Appel LJ, Guallar E (2005). Meta-analysis: High-dosage Vitamin E supplementation may increase all-cause mortality. Ann. Intern. Med. 142:37–46.
Mittal CK, Murad F (1977). Activation of guanylate cyclase by superoxide-dismutase and hydroxyl radical-Physiological regulator of guanosine 3, 5 monophosphate formation. Proc. Nat. Acad. Sci. U.S.A. 74:4360-4364.
Muller FL, Liu Y, Van Remmen H (2004). Complex III releases superoxide to both sides of the inner mitochondrial membranc. J. Biol. Chem. 279:49064-49073.
Pastor N, Weinstein H, Jamison E, Brenowitz M (2000). A detailed interpretation of OH radical footprints in a TBPDNA complex reveals the role of dynamics in the mechanism of sequence specific binding. J. Mol. Biol. 304:55-68.
Pastore A, Federici G, Bertini E, Piemonte F (2003). Analysis of glutathione: Implication in redox and detoxification. Clin. Chim. Acta., 333:19-39.
Rae TD, Schmidt PJ, Pufahl RA, O’Halloran TV (1999). Undetectable intracellular free copper: The requirement of a copper chaperone for superoxide dismutase. Science. 284:805-808.
Rapola JM, Vitamo J, Hauka JK (1996). Effect of vitamin E and β carotene on nthe incidence od angina pectoris. JAMA 275:693-698.
Ridnour LA, Isenberg JS, Espey MG, Thomas DD, Roberts DD, Wink DA (2005). Nitric oxide regulates angiogenesis through a functional switch involving thrombospondin-1. Proc. Nat. Acad. Sci. U.S.A. 102:13147-–13152.
Rimm EB, Stampfer MJ, Ascherio A, Giovannucci E, Colditz GA, Willett WC (1993). Vitamin E consumption and the risk of coronary heart disease in men. N. Engl. J. Med. 328:1450-1456.
Robertson JM, Donner AP, Trivitchicj JR (2002). Viatamin E intake and risk of cataract in humans. Ann NY Acad. Sci. 570:1392-1394.
Sano M, Ernesto C, Thomas RG, Klauber M, Schafer K, Grundman MA (1997). Controlled trial of selegiline, alpha-tocopherol, or both as treatments for Alzheimer’s disease. N. Engl. J. Med. 336:1216-1222.
Schafer FQ, Buettner GR (2001). Redox environment of the cell as viewed through the redox state of the glutathione disulfide/ glutathione couple. Free Radic. Biol. Med. 30:1191-1212.
Seddon JM, Cristen GJ, Manson JE (1994). The use of vitamin supplements and the risk of cataract among US male physicians. Am..J Public Health 84:788-792.
Shen D, Dalton TP, Nebert DW, Shertzer HG (2005). Glutathione redox state regulates mitochondrial reactive oxygen production. J. Biol. Chem. 280:25305-25312.
Siebenlist U, Franzoso G, Brown K (1994). Structure, regulation and function of NF-κB. Ann. Rev. Cell Biol. 10:405–455.
Siems WG, Grune T, Esterbauer H (1995). 4-Hydroxynonenal formation during ischemia and reperfusion of rat small-intestine. Life Sci. 57:785-789.
Stadtman ER (2004). Role of oxidant species in aging. Curr. Med. Chem. 11:1105-1112.
Stephens NG, Parsons A, Schofold PM (1996). Randomised controlled trial of Vitamin A in patients with coronary diseases. Cambridge heart antioxidant study (CHOS). Lancet 347:781-786.
Temple NJ (2000). Antioxidants and disease: More questions than answers. Nutr. Res. 20:449-459.
Thannickal VJ, Fanburg BL (2000). Reactive oxygen species in cell signaling. Am. J. Physiol. Lung Cell. Mol. Physiol. 279:L1005-L1028.
Tiekari JM, Virtamo J, Rautalahti M (1997). Long term supplementation with α tocopherol, and β carotene and age related cataract. Acta. Opthamol. Scand. 75:734-740.
Valko M, Izakovic M, Mazur M, Rhodes CJ, Telser J (2004). Role of oxygen radicals in DNA damage and cancer incidence. Mol. Cell. Biochem. 266:37-–56.
Valko M, Leibfritz D, Monocol J, Cronin MTD, Mazur M, Telser J (2007). Free radicals and antioxidants in normal physiological functions and human disease. Int. J. Biochem. Cell Biol. 39:44-84.
Valko M, Morris H, Cronin MTD (2005). Metals, toxicity and oxidative stress. Curr. Med. Chem.12:1161-–1208.
Valko M, Morris H, Mazur M, Rapta P, Bilton RF (2001). Oxygen free radical generating mechanisms in the colon: Do the semiquinones of Vitamin K play a role in the aetiology of colon cancer? Biochim. Biophys. Acta. 1527:161-166.
Valko, M, Rhodes CJ, Moncol J, Izakovic M, Mazur M (2006). Free radicals, metals and antioxidants in oxidative stress-induced cancer. Chem. Biol. Interact. 160:1-40.
Van der Vliet A (2000). Cigarettes, cancer, and carotenoids: A continuing, unresolved antioxidant paradox. Am. J. Clin. Nutr. 72:1421-1426.
Voehringer DW, Hirschberg DL, Xiao J, Lu Q, Roederer M (2000). Front. Biosci. 10:1727-1738.
Vorbach C, Harrison R, Capecchi MR (2003). Xanthine oxidoreductase is central to the evolution and function of the innate immune system. Trends Immunol. 24:512–517.
White E, Shannon JS, Patterson RE (1997). Relationship between Vitamin and calcium supplement use and colon cancer. Cancer Epidemiol. Biomark. Prevent. 6:769-774.
Yusuf S, Dagenais G, Pogue J, Bosch J, Sleight P. (2000). Vitamin E supplementation and cardiovascular events in high-risk patients. The Heart Outcomes Prevention Evaluation Study Investigators. N. Engl. J. Med. 342:154-160.

Removal of Erichrome Black T from aqueous solution using low cost waste biomass (cow dung ash) at 303 and 308K

Rashid Mahmood and Afia Naseer

Division of Science and Technology, University of Education, Township Lahore-54000, Pakistan.

Waste biomass cow dung ash was used as an adsorbent for the removal of Erichrome Black T (EBT) from aqueous solution using simple batch technique. The contact time in terms of adsorption capacity was optimized at 303 and 308K, respectively and it was found to be 60 min for both temperatures. All the experiments were carried out at pH = 7. The adsorption capacity of EBT increased from 0.20 to 0.75 and 0.2 to 0.8 mg/g with 1 g of adsorbent at 303 and 308K, respectively by increasing concentration from 2 to 10 mg/L. The data were employed to Langmuir and Freundlich adsorption isotherm models and because of higher regression co-efficient R^2 values at both temperatures the Langmuir adsorption isotherm was found to be well fitted to the data. The adsorption was found to be spontaneous and endothermic at 308K due to negative free energy and positive enthalpy.

Key words: Erichrome Black T, cow dung, adsorption, thermodynamic parameters.

INTRODUCTION

The industries like, textile, paper, printing, leather, food, cosmetics etc, use dyes to color their final products. Discharge of such colored effluents imparts color to the receiving water bodies (River and Lakes) and interferes with its intended beneficial use. Waste water containing even a small amount of dyes can severely affect the aquatic life due to the reduction of light penetration and their toxicity (Mamdouh and Geundi, 1991). Amongst the various techniques such as, coagulation (Weber, 1978), ultra-filtration (Malakootian and Fatehizadeh, 2010), and photo-oxidation (Rajeev and Varshney, 2006), the adsorption is recognized as an efficient, promising and most widely used technique for the treatment of industrial waste water containing color and other organic impurities. The chemical composition of the waste biomass (Stanisiav et al., 2010) is significant for its use as an adsorbent for the purification purpose and the use of waste biomass is highly cost effective as compared to the commonly used adsorbent that is, activated charcoal. Sorbents such as Rosewood Sawdust, Saw dust, Trichoderma Harzianum, Pithophora sp. (fresh water algae), Hydrialla Verticillata and Luffa Cylindrica (Garg et al., 2004; Chakraborty et al., 2005; Sadhasivam et al., 2007; Kumar et al., 2005; Rajeshkhannan et al., 2009, Marchetti et al., 2000) sugar cane dust; algae, red algae (Sari and Tuzen, 2008a), macro fungus (Anayurt et al., 2009), green alga (Sari and Tuzen, 2008b), banana and orange peels (Annadurai et al., 2002) and lichen (Sari et al., 2007) were employed for the removal of dyes, color and trace elements from waste water. Coir pith carbon was found to be useful adsorbent for the removal of Congo red (Namasivayam and Kavitha, 2002). Moreover, many Ligno-cellulosic materials such as banana pith (Namasivayam et al., 1993), coir pith (Namasivayam and Kadirvelu, 1994), eucalyptus bark (Morais et al., 1999), apple pumice and wheat straw (Robinson et al., 2002), rice husk (Vadivelan and Kumar, 2005), pine sawdust (Ozacar and Engil, 2005), peanut hull (Gong et al., 2005), lemon peel (Kumar and Kumaran, 2005), almond shell (Ardejani et al., 2008), neem leaf powder (Tahir et al., 2008) and animal bone meal (Mohammadine, 2012) were also used for the removal of dyes from water.

In the present study the cow dung ash has been used for the removal of EBT from aqueous solution through simple batch technique. The investigations were carried out by varying contact time from 10 to 120 min and concentration of adsorbate from 2 to 10 mgL^{-1} and adsorbent dosage of 1 g at 303 and 308K and pH = 7. The data was employed to the Langmuir and Freundlich adsorption isotherms to determine their constants. Thermodynamic parameters such as free energy, enthalpy and entropy were also calculated.

Molecular structure and molecular formula of Erichrome Black T (EBT)

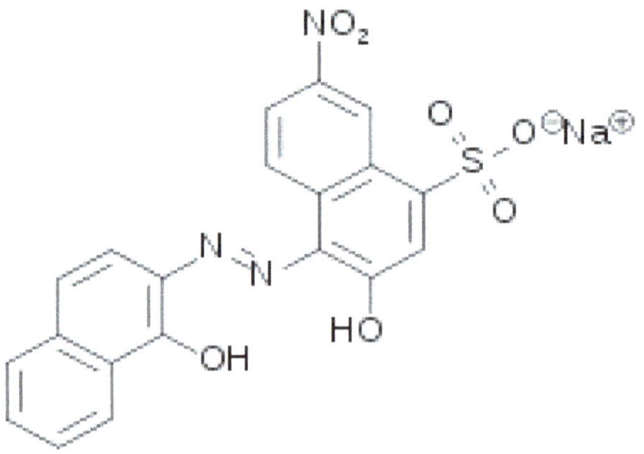

IUPAC names

Sodium (4z)-4-[(1-hydroxynaphthalen-2-yl-hydrazinylidene-7-nitro-3-oxo Y-nephthalen-1-sulfonate.

Systematic names

Sodium 4-[2-(1-hydroxynaphthalen-2-yl) hydrazin-1-ylidene]-7-nitro-3-oxo-3,4 dihydronaphthalene-1-sulfonate.

Molecular formula: $C_{20}H_{12}N_3O_7SNA$

EBT is a complexometric indicator that is part of the complexometric titrations, e.g. In the water hardness determination process. It is an azo dye. In its protonated form, EBT is blue. It turns red when it forms a complex with calcium, magnesium, or other metal ions. Its chemical formula can be written as $HOC_{10}H_6N{=}NC_{10}H_4$ (OH) (NO_2) SO_3Na.

MATERIALS AND METHODS

Adsorption studies were carried out using batch method. EBT (95.5%) supplied by fluka was used without any further purification. The cow dung ash used in the present work was obtained by firing the animal waste in open air and then dried at 105°C for 24 h to remove the moisture and stored in desiccator for further use in the adsorption studies. The stock solution of EBT was prepared by dissolving 10 mg of EBT in distilled water to make 1000 ml of the solution. It was further diluted to the required initial concentrations 8, 6, 4 and 2 mg/L, respectively. The wavelength for maximum absorption (λ_{max}) was determined using spectrophotometer (VIS-1100 (BMS) and it was found to be 623 nm. Absorption of the given solutions was measured at λ_{max}, and calibration curve was drawn to determine the molar absorption co-efficient. The 1 g of adsorbent was added in 100 ml of each dye solution and agitated at 150 rpm for 10 to 120 min in electrical shaking water bath DFD-700 (Glas-Col) at 303 and 308K, respectively. After shaking the samples for given time, their pH was maintained with 0.1 M NaOH and/or 0.1 M HCl using pH-meter pH-720 (Ino LAB). The samples were immediately filtered using Whatmann filter paper (No. 43) and analyzed through spectrophotometer to determine the concentration of each EBT solution after adsorption using Beer's Lambert law. The dye uptake was determined by employing the following mass balance equation.

$$q_e = (C_i - C_e)\ V / M \qquad (1)$$

Where V is the volume of solution (L), M is the mass of the adsorbent used (g) and Ci and C_e are the initial and equilibrium concentrations (mg/L) of the dye, respectively. The removal efficiency or percent adsorption was calculated using the following equation

$$\text{Removal efficiency} = (Ci - Ce/Ci) \times 100 \qquad (2)$$

RESULTS AND DISCUSSION

Effect of contact time

Figures 1 and 2 show the effect of shaking time on the adsorption capacity of EBT at 303 and 308 K, respectively. The results reveal that adsorption increases with increase in agitation time and reaches equilibrium after 60 min for each concentration from (2 to 10 mg/L) for 1 g adsorbent at both temperatures. The results indicate that the removal efficiency for 2 mg/L solution and 1 g adsorbent reaches up to 57.15% and then for 4, 6 and 8 ppm, it reaches to 83.03, 90.40 and 93.73%, respectively after the shaking of initial 10 min. Data also reveal that rate of removal efficiency decreases with increase in shaking time from (10 to 60 min). The increase rate of sorption of the dye in the initial period of the process may be due to higher number of sites available on the surface of the adsorbent and the higher concentration gradient between the adsorbate in the solution and adsorbate on the surface of the adsorbent. As the uptake of the dye progresses the concentration gradient between the dye in the solution and dye on the surface of the adsorbent decreases and the number of vacant adsorption sites on the adsorbent surface also

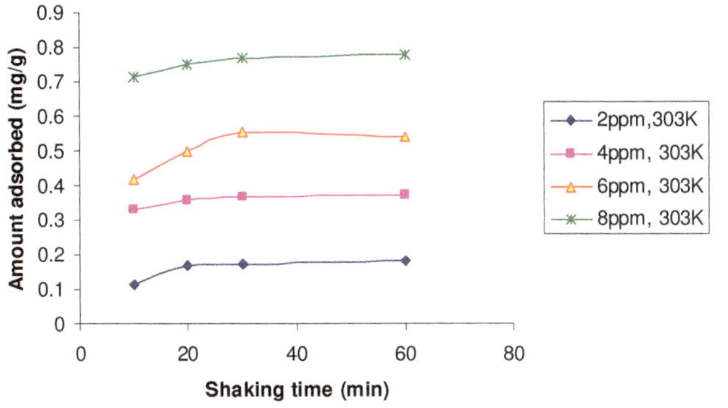

Figure 1. Effect of shaking time on adsorption of EBT onto1 g of cow dung ash at 303 K and pH 7.

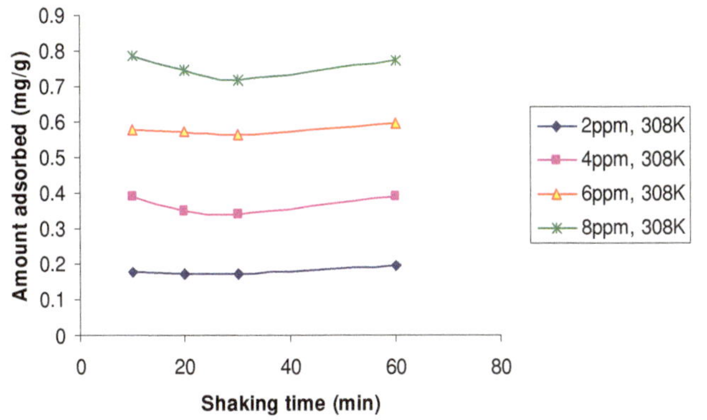

Figure 2. Effect of shaking time on adsorption of EBT onto1 g of cow dung ash at 308 K and pH 7.

decreases leading to a decrease in adsorption rate with time. The maximum removal efficiency was found to be 97.41% for 8 mg/L solution after the shaking of 60 min.

Effect of initial concentration

Figure 3 shows the effect of initial concentration of EBT on its adsorption on cow dung ash from aqueous solution at 303 and 308K. The data reveals that adsorption starts at low concentration and increases with increase in the concentration of adsorbate at both temperatures. The adsorption capacity of EBT was found to increase from 0 .20 to 0.75 and 0.2 to 0.8 mg/g with 1 g of adsorbent at 303 and 308K, respectively by increasing concentration from 2 to 10 mg/L.

Adsorption isotherms

An adsorption isotherm is characterized by certain constants, which express the surface properties and

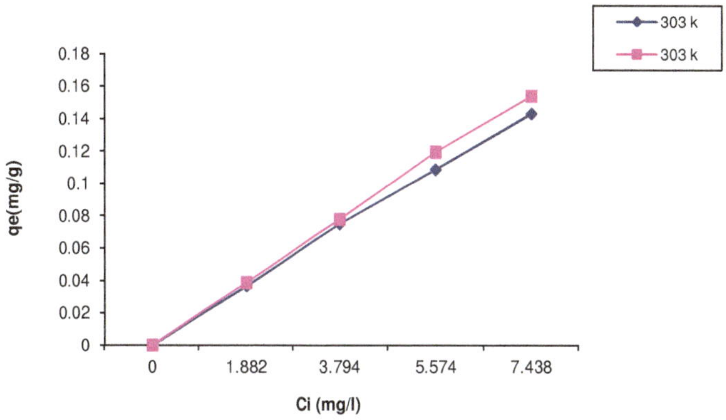

Figure 3. Effect of dye concentration on adsorption over cow dung at 303 and 308K.

affinity of the adsorbent and can also be used to compare the adsorption capacities of the adsorbent for different adsorbate. The equilibrium data owing to the dependence of the adsorption of initial dye concentration is best studied by the Langmuir and Freundlic isotherms. A linearized form of Langmuir equation is as follows

$$1/q_e = 1/Q_o + 1/Q_oK_L . 1/Ce \quad (3)$$

A plot of 1/qe versus 1/Ce was found to be a straight line with $1/Q_o$ as intercept and $1/Q_o K_L$ as slope and hence Q_o and K_L can be calculated. The linearized form of the Freundilch equation can be written as:

$$\ln q_e = \ln K_F + 1/n \ln C_e \quad (4)$$

A plot of $\ln q_e$ versus $\ln C_e$ gives a straight line and K_F and n can be calculated from the intercept and slope, respectively. Figures 4 and 5 show the Langmuir adsorption isotherms and Figures 6 and 7 show the Freundlich adsorption isotherms of EBT at 303 and 308K, respectively. The Langmuir constants K_L and Q_o and Freundlich constants n and K_F and the correlation co-efficient R^2 are given in Table 1. As suggested by some workers (Namasivayam and Yamuna, 1992; Raji and Anirudhan, 1998) if the value of Freundlich constant n is such that $0.1 < n < 1$, the adsorption is favorable. This shows the effectiveness of cow dung ash for the removal of EBT from aqueous solutions. However, the linearized equation did not give a good correlation for the removal of EBT onto cow dung ash, indicating that EBT adsorption by cow dung fits better to the Langmuir model than to the Freundlich model. In the present study the values of n lie in this suggested range. Similarly, a dimensionless constant R_L, called separation factor can be defined from the Langmuir constant K_L (Hall et al., 1966) is as follows:

$$R_L = 1/ (1+ C_i K_L) \quad (5)$$

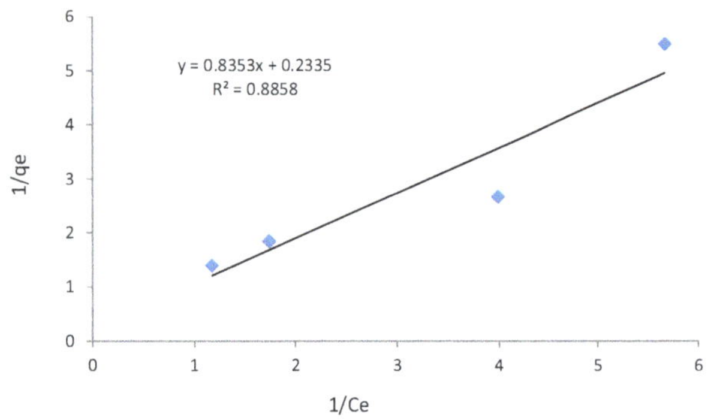

Figure 4. Langmuir adsorption isotherm of adsorption of EBT over cow dung at 303K(adsorbent dose 1g, pH = 7).

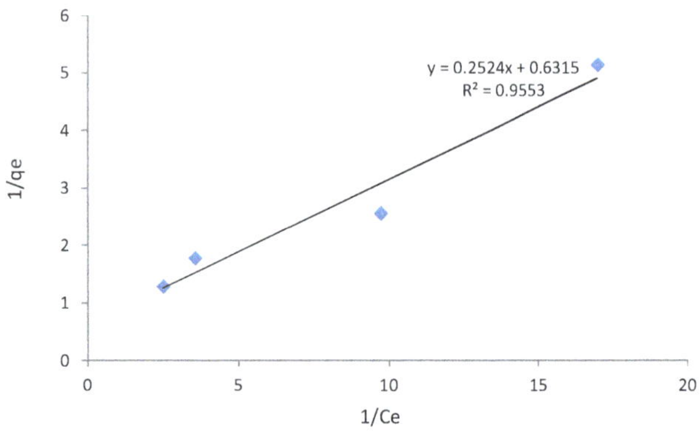

Figure 5. Langmuir adsorption isotherm of adsorption of EBT over cow dung at 308K (adsorbent dose 1 g, pH = 7).

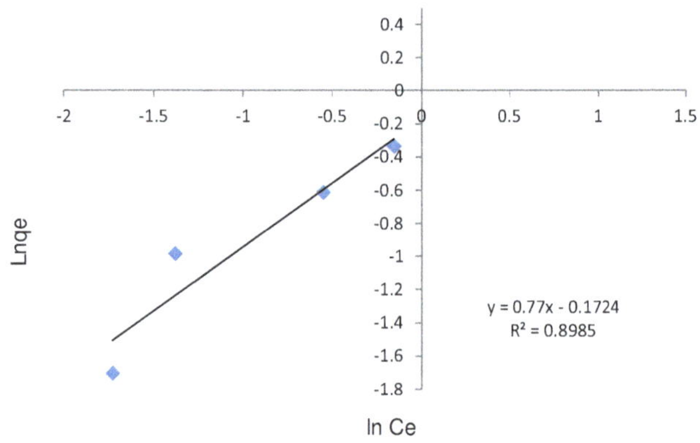

Figure 6. Freundlich adsorption isotherm for adsorption EBT over cow dung at 303K.

Where C_i is the initial concentration of the EBT solution. According to McKay et al. (1982) for favorable adsorption $0 < R_L < 1$, and for unfavorable adsorption $R_L > 1$ and RL = 1 represents linear absorption while $R_L = 0$

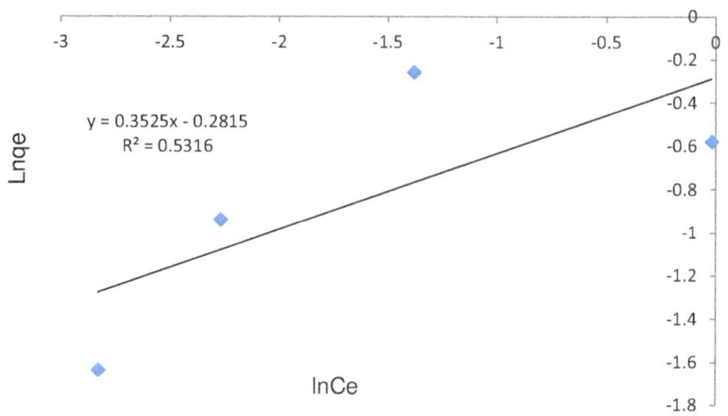

Figure 7. Freundlich adsorption isotherm for adsorption EBT over cow dung at 308K.

represents irreversible adsorption process. As seen in Table 3, for all initial concentrations the value of R_L falls in the range of 0 and 1 showing the favorable adsorption of EBT on cow dung ash from aqueous solution at 303 and 308K, respectively.

Thermodynamic parameters

Thermodynamic parameters such as free energy, enthalpy and entropy changes associated with the adsorption of EBT on the cow dung ash were determined from the binding energy constant K_L obtained from Langmuir equation by using the following equations:

$$\Delta G = - RT \ln K_L$$

$$- RT \ln K_L = \Delta H - T\Delta S$$

$$\ln K_L = (-\Delta H/RT) + (\Delta S/R)$$

$$\ln K_{L1} = (-\Delta H/RT_1) + (\Delta S/R)$$

$$\ln K_{L2} = (-\Delta H/RT_2) + (\Delta S/R)$$

$$\ln K_{L2} - \ln K_{L1} = (\Delta H/RT_1) - (\Delta H/RT_2)$$

$$\ln (K_{L2}/K_{L1}) = \Delta H/R\ (1/T_1 - 1/T_2) \quad (6)$$

$$\Delta H = (RT_1T_2/T_2 - T_1) \ln (K_{L2}/K_{L1}) \quad (7)$$

Where, K_{L1} and K_{L2} are the Langmuir binding constants at T_1 (303K) and T_2 (308K), respectively. By putting the values of ΔG and ΔH into the following equation the entropy of adsorption at T_1 and T_2 can be calculated.

$$(\Delta S) = (\Delta H - (\Delta G)/T)$$

$$(\Delta S)_1 = (\Delta H - (\Delta G)_1/T_1)$$

$$(\Delta S)_2 = (\Delta H - (\Delta G)_2/T_2) \quad (8)$$

Table 1. Freundlic and Langmuir parameters for adsorption of Eriochrome Black-T over cow dung at different temperatures.

Temperature (K)	n	log K_F	R^2	Q_o (mol/g)	K_L(L/mol)	R^2
303	1.299	-0.172	0.898	4.292	0.280	0.885
308	2.841	-0.281	0.531	1.585	2.504	0.955

Table 2. Values of thermodynamic parameters for adsorption of Eriochrome Black-T over cow dung.

Temperature (K)	ΔG^o (kJ mol^{-1})	ΔH^o (kJ mol^{-1})	ΔS^o(k J $K^{-1}mol^{1}$)	Ea (kJ mol^{-1})
303	3.206	326.328	1.063	326.28
308	-2.351	-	1.698	-

Table 3. Separation factor R_L calculated from Langmuir parameter K_L at various initial concentrations of EBT at different temperatures.

Temperature (K)	Mass of adsorbent (g)	K_L	Initial concentration (mg/L)	R_L
			2	0.64
303	1	0.280	4	0.47
			6	0.37
			8	0.31
			2	0.164
308	1	2.504	4	0.091
			6	0.063
			8	0.050

The value of ΔG was calculated using Equation (6), where R is the gas constant (8.314 $JK^{-1}mol^{-1}$). The values of ΔS were calculated using Equation (8). These values are tabulated in Table 2. The value of ΔG was found to be negative at 308K that showed the adsorption was spontaneous at this temperature. The activation energy Ea was calculated using the following equation:

$$Ea = R\ (T_1T_2/T_2\text{-}T_1)\ \ln K_2/K_1 \qquad (9)$$

Conclusions

Cow dung ash was found to a useful waste biomass adsorbent for the removal of EBT from aqueous solution. Adsorption capacity was found to increase with increase in initial concentration. Data employed to both Langmuir and Freundlich adsorption isotherms were found to be well fitted to the Langmuir adsorption isotherm. The adsorption of EBT was found to be endothermic and spontaneous especially at 308K.

ACKNOWLEDGMENT

We are highly thankful to Prof. Dr. Saeed Ahmed UET Lahore Pakistan for his useful help and suggestions during this research work.

REFERENCES

Anayurt RA, Sari A, Tuzen M (2009). Equilibrium, thermodynamic and kinetic studies on bisorption of Pb(II) and Cd(II) from aquous solution by macrofungus (Lactarius Scrobiculatus) biomass. Chem. Eng. J., 151:255-261.

Annadurai G, Juang RS, Lee DJ (2002). Adsorption of heavy metals using banana and orange peels. Water Sci, Technol., 47:185-190.

Ardejani FD, Badii K, Limaee NY, Shafaei SZ, Mirhabibi AR (2008). Adsorption of Direct Red 80 dye from aqueous solution onto almond shells: Effect of pH, initial concentration and shell type. J. Hazard Mater. 151:730-737.

Chakraborty S, De S, Gupta SD, Basu JK (2005) Adsorption study for removal of a basic dye : experimental and modeling. Chemosphere 58:1079-1086.

Garg VK, Amita M, Kumar R, Gupta R (2004). Basic dye (methylene blue) removal from simulated wastewater by adsorption using Indian Rosewood Sawdust a timber industry waste. Dyes Pigm. 63:243-250.

Gong R, Li M, Yang C, Sun Y, Chen J (2005). Removal of cationic dyes from aqueous solution by adsorption on peanut hull. J. Hazard Mater., 121:247-250.

Hall KR, Eagleton LC, Acrivos A, Vermeulen T (1966). Pore and solid-diffusion kinetics in fixed-bed adsorption under constant-pattern conditions. Ind. Eng. Chem. Fund., 5:212-223.

Kumar KV, Kumaran A (2005). Removal of methylene blue by mango seed kernel powder. Biochem. Eng. J., 27:83-93.

Kumar KV, Sivanesan S, Ramamurthi V (2005). Adsorption of malachite green onto pithophora sp., a fresh water algae: Equilibrium and kinetic modeling. Proc. Biochem., 40:2865-2872.

Malakootian M, Fatehizadeh A (2010). Colour Removal From Water By Coagulation/Caustic Soda Lime. Iran J. Environ. Health Sci. Eng., 7:267-272.

Mamdouh NM, Geundi MSE (1991). Comparative cost of color removal from textile effluents using natural adsorbents. J. Chem. Technol. Biotechnol., 50:257-264.

Marchetti V, Gérardin P, Loubinoux B (2000). Preparation and use of sawdusts bearing quaternary ammonium groups for removal of anionic dyes from water. Holz Roh Werkst 58:53-56.

Mckay G, Blair HS, Gardner JR (1982). Adsorption of Dye on Chitin. I. Equilibrium Studies, J. Appl. Polym. Sci., 27:3043

Mohammadine EH (2012). Kinetic and thermodynamic studies on the adsorption behavior of Rhodamine B dye onto animal bone meal. J. Chem. Eng. Mater. Sci., 3:38-44.

Morais LC, Freitas OM, Goncalves EP, Vasconcelos LT, Beca CGG (1999). Reactive dyes removal from wastewaters by adsorption on eucalyptus bark: variables that define the process. Water Res., 33:979-988.

Namasivayam C, Kadirvelu K (1994). Coir pith, an agricultural waste byproduct for the treatment of dyeing wastewater. Bioresour. Technol., 38:79-81.

Namasivayam C, Kanchana N, Yamuna RT (1993). Waste banana pith as adsorbent for the removal of Rhodamine B from aqueous solution. Waste Manage., 13:89-95.

Namasivayam C, Kavitha D (2002). Removal of Congo Red from water by adsorption onto activated carbon prepared from coir pith, an agricultural solid waste. Dyes Pigm., 54:47-58.

Ozacar M, Engil A (2005). Adsorption of metal complex dyes from aqueous solutions by pine sawdust. Bioresour. Technol., 96:791-795.

Rajeev J, Shaily V (2006). Removal Of Reactofix Supra Red HBSL From Industrial Effluent Using Electrochemical and Adsorption Techniques. Int. J. Environ. Poll., 27:75-94.

Rajeshkhannan R, Rajamohan N, Rajasimman M (2009). Removal of malachite green from aqueous solution by sorption in hydrilla verticillata biomass using response surface methodology. Front Chem. Eng. China 3:146-154.

Raji C, Anirudhan TS (1998). Batch Cr(VI) removal by polyacrylamide-grafted Sawdust: Kinetics and Thermodynamics. Water Res., 32, p3772–3780.

Robinson T, Chandran B, Nigam P (2002). Removal of dyes from a synthetic textile dye effluent by biosorption on apple pomace and wheat straw. Water Res., 36:2824-2830.

Sadhasivam S, Savitha S, Swaminathan K (2007). Feasibility of using Trichoderma harzianum biomass for the removal of erioglacine from aqueous solution, World J. Microbiol. Biotechnol., 23:1075-1081.

Sari A, Tuzen M (2008a). Biosorption of cadmium(II) from aquous solution by red algae (ceramium virgatum) : Equilibrium, kinetic and thermodynamic studies. J. Hazard Mater., 157:448-454

Sari A, Tuzen M (2008b). Biosorption of Pb (II) and Cd (II) from aquous solution using green algae (Ulva lactuca) biomass. J. Hazard Mater., 152:302-308.

Sari A, Tuzen M, Oluozlu OD, Soylak M (2007). Biosoption of Pb (II) and Ni (II) from aquous solution by lichen (Cladonia furcota) biomas. Biochem. Eng. J., 37:151-158.

Stanisiav V, Baxter D, Andersen LK, Vassileva CG (2010). An overview of the chemical composition of biomass. Fuel 89:913-933.

Tahir H, Hammed U, Jahanzeb Q, Sultan M (2008). Removal of fast green dye (C.I. 42053) from an aqueous solution using Azadirachtaindicaleaf powder as a low cost adsorbent. Afr. J. Biotechnol., 7:3906-3911.

Vadivelan V, Kumar KV (2005). Equilibrium, kinetics, mechanism and process design for the sorption of methylene blue onto rice husk. J. Colloid Interf. Sci., 286:90-100.

Weber WJ (1978). Physico-chemical Methods of Treatment of Water and Wastewater, Wiley and Sons Inc. 1978.

Comparison of low temperature coal tar solvent extractible and bound biomarker distribution for correlation studies

Ivwurie, W.[1]*, Snape, C. E.[2] and Sun, C.[2]

[1]Department of Chemistry, Federal University of Petroleum Resources, Effurun, Delta State, Nigeria.
[2]Nottingham Fuel and Energy Centre, the University of Nottingham, University Park, Nottingham, United Kingdom.

Solvent extractible and bound biomarker profile for low temperature coal tar were compared using the biomarker ratios obtained from the selective ion resolution for pristane, phytane, n-C_{17} and hopanesratios of GC-MS results, with a view to ascertaining whether or not the data for asphaltene-derived biomarker can be used for effective correlation studies, and if serious alterations arising from biodegradation and other environmental factors could affect the use of the solvent extractible ones. The Pristane/phytane, pr/n-C_{17} ratios and carbon preference index are 4.0, 1.1 and 0.9, respectively for the free aliphatics and 4.2, 1.2 and 1.2 for the bound. These values are quite similar to each other. The hopanes are less mature for the bound than for the free, this is a general trend found for most hydropyrolysis products.

Key words: Biomarker, coal tar, asphaltene.

INTRODUCTION

Biological markers are organic compounds that have an established biological origin, and exhibit little or no change in chemical structure from their parent precursor compound found in living organisms. Their characteristics include source-specificity, restricted occurrence, molecular stability and suitable concentration for analytical detection. They can be used as signatures of organic matter sources (Rogge et al., 1993; Schauer et al., 1996; Ivwurie, 2011a).

For instance, the 17α (H), 21β (H)-hopanes, the extended triterpanes, the steranes and the isoprenoids (pristane and phytane) have been used for this purpose. Pristane and phytane are regular acyclic isoprenoid hydrocarbon biomarkers and are readily present in sedimentary organic rocks, coals and petroleum. For crude oil, transport fuels and vehicular exhaust emissions, Pr/Ph and Pr/n-C_{17} ratios are often ≤ 2, typically ~ 1, with no major variations observed (Schauer et al., 1996; Sun, 2001).

*Corresponding author. E-mail: wivwurie@yahoo.co.uk.

The stereo-chemical changes which take place at the C_{17} and C_{21} chiral positions of hopanes are often used to determine the maturity of geological organic matters, where hopanes with configuration of 17β (H), 21β (H) indicates immaturity, 17β (H), 21α (H), shows moderate maturity; and 17α (H), 21β (H), is an evidence of high maturity. Besides petroleum, however these biomarkers can be found in coal, with their distribution being dependent on coal maturity.

Asphaltenes are of significant interest in geochemical and related studies. Van et al., (1986) investigated biomarker distribution in asphaltenes and kerogens using flash pyrolysis-gas chromatography-mass spectrometry and they observed differences between biomarker distributions of asphaltene and kerogen pyrolysates (Buenrostro-Gonzalez, 2001, Ivwurie et al., 2011b).

This research paper compares the solvent extractible and bound biomarker profile for low temperature coal tar with a view to ascertain whether or not the data used for asphaltene-derived biomarker can be used for effective correlation studies, and if serious alterations, arising from Biodegradation and other environmental factors affect

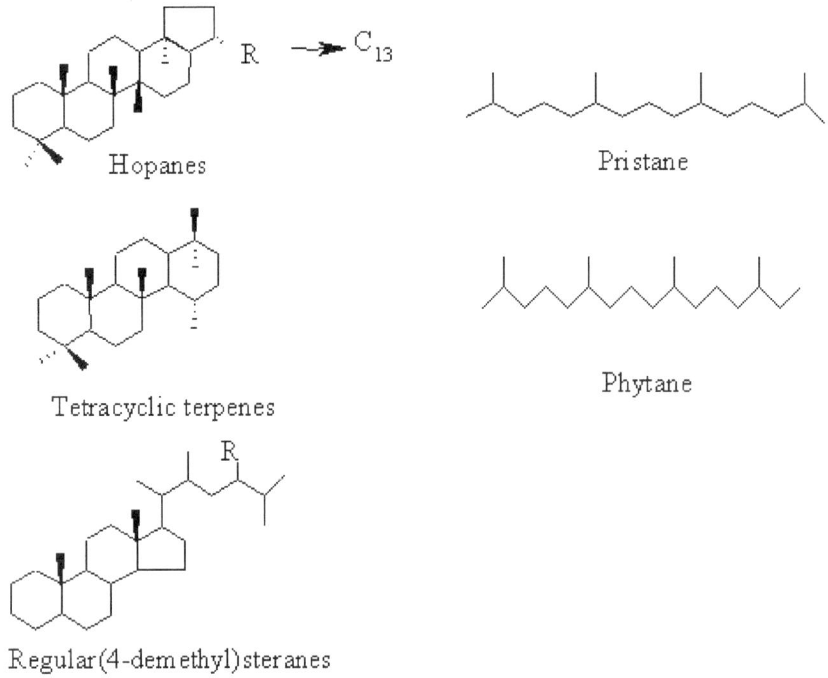

Figure 1. Chemical structures for title molecules/biomarkers.

Table 1. Pristane/phytane, pristane/n-C_{17} ratios and CPI for low temperature coal tar.

Sample name	Pristane / Phytane	Pristane / n-C_{17}	CPI
Low temp coal tar Free aliphatics	4.0	1.1	0.9
Low temp coal tar Hypy aliphatics	4.2	1.2	1.2

the use of the solvent extractible ones.

METHODS

Low temperature coal tar (coal tar processed at a temperature range of 500 to 850°C) sample supplied from the coalite smokeless fuel plant near Chesterfield, United Kingdom, was fractionated using open column chromatography to obtain aliphatic fraction from which different biomarkers were investigated. Asphaltenes were obtained from low temperature coal tar by precipitation with 40-fold excess of n-heptane (Jones et al., 1988). The asphaltenes were mixed with silica in the range of 10 to 50 mg of asphaltene to 1 g of silica. The mixture is then impregnated with an aqueous methanol solution (20%v/v) of ammonium dioxydithiomolybdate catalyst to give a nominal solution of 3-wt %. The catalyst was prepared by slow addition of ammonium heptamolybdate to ammonium sulphide solution. The heptamolybdate salt was then dissolved using a magnetic stirrer and solution stirred until a precipitate of ammonium dioxydithiomolybdate is formed. The precipitate was collected with the aid of a Buchner funnel and dried in a vacuum oven. The sample is then dried in a vacuum oven at 60 to 70°C after which it is placed in a reactor tube. When attached to the hydropyrolysis set-up, it is resistively heated from 50 to 250°C at 300°C min^{-1}, then to 500°C at 8°min^{-1} maintaining a hydrogen pressure of 15 mPa and a flow rate of 10 dm^3min^{-1}. The liquid product is then collected in a trap cooled with dry ice and recovered with dichloromethane (2 to 5 ml) for subsequent separation (Love et al., 1997, 1998; Murray, 2001; Meredith et al., 2004).

The biomarker distribution for the solvent extractible (free) and hydropyrolysis products (bound) were compared using GC-MS on a Fisons instruments 8000 gas chromatograph interfaced to a MD 800 mass spectrometer with a quadrupole mass analyzer (ionizing energy 70 eV, source temperature 280°C). The gas chromatograph capillary column is 50 m long; i.d. is 0.32 mm and a film thickness of 0.25 µm, stationary phase is polyphenylmethylsiloxane.

RESULTS

Chemical structures for title molecules/biomarkers are shown in Figure 1. The pristine/n-C_{17} (Pr/n-C_{17}), pristane/phytane (Pr/Ph) ratios and carbon preference index (CPI) are calculated and listed in the Table 1. Pristane/phytane, pr/n-C_{17} ratios and carbon preference index are 4.0, 1.1 and 0.9, respectively for the free aliphatics and 4.2, 1.2 and 1.2 for the bound.

Figure 2 shows total ion chromatograms for aliphatic hydrocarbons of both free and bound low temperature coal tar. The hopane and sterane maturity parameters for

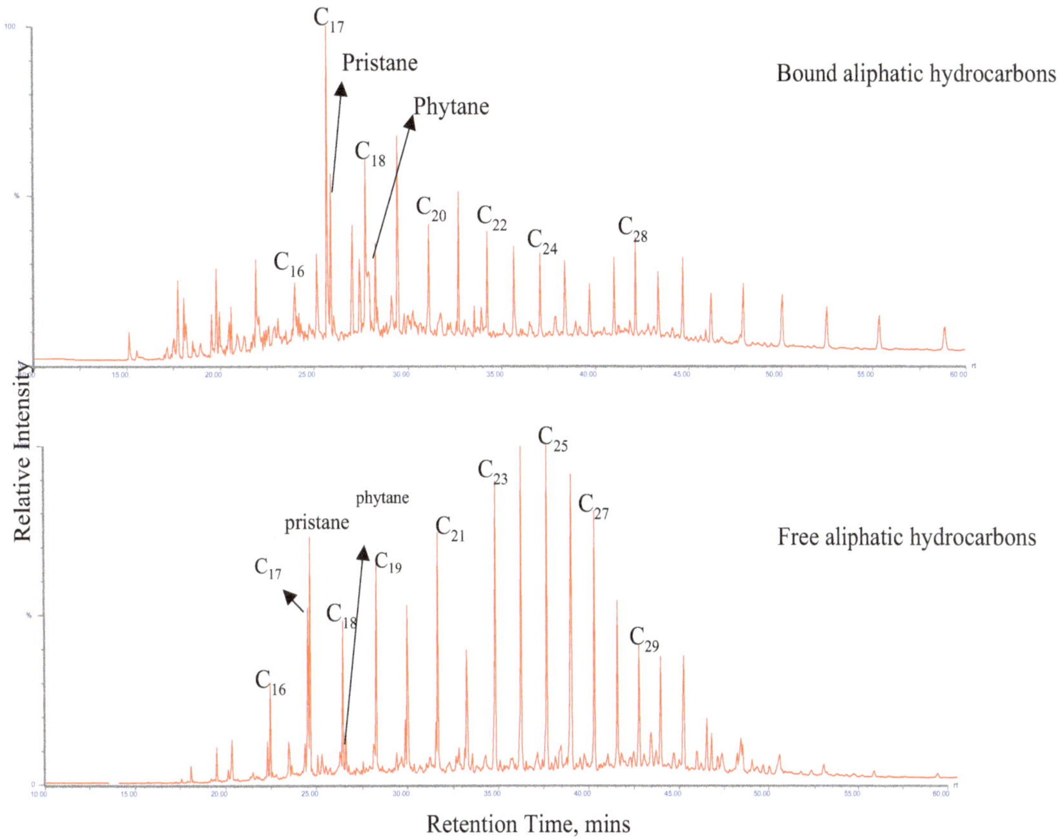

Figure 2. Total ion chromatograms for free and asphaltene-derived aliphatics from low temperature coal tar.

Table 2. Hopane and sterane maturity parameter ratios for free and bound biomarkers for low temperature coal tar.

Biomarker	Low temp. coal tar free biomarker ratio	Low temperature coal tar asphaltene-derived biomarker ratio
29βα/αβ Hopane	0.24	0.28
30βα/αβ Hopane	0.32	0.50
31αβ S/S+R Hopanes	0.55	0.52
29αααS/S+R Sterane	0.53	0.54

Key: ratios were computed using peak areas of m/z 191 and 217 SICs of the biomarkers listed above.

the free and bound aliphatics have been compared in Table 2 as well as in Figures 3 and 4.

DISCUSSION

This study was conducted to be able to generate representative biomarker profiles, in cases of severe biodegradation and environmental alteration that are useful for source apportionment studies, after which a high proportion of the bound biomarkers must be assessed under reaction conditions to minimize the extent of product secondary transformations. Asphaltene hydropyrolysis was used in providing reliable data for correlation studies (Murray, 2001). Low temperature coal tar, a primary source was investigated for this correlation studies to aid source apportionment.

The use of biomarker parameters like pristane/phytane ratios, pristane/nC_{17}, and the maturity of hopanes for correlation studies is well established (Murray, 2001; Meredith et al., 2004).

The pristane/phytane, pristane/nC_{17} values for both free and bound coal tar are quite similar to each other, indicating that source information revealed by the bound constituents are representative of the source as the free constituents (Fu et al., 1992; Hughes et al., 1995; Sun, 2001). Similarly, n-alkanes up to C_{32} and similar values for carbon preference index are visible in both free and

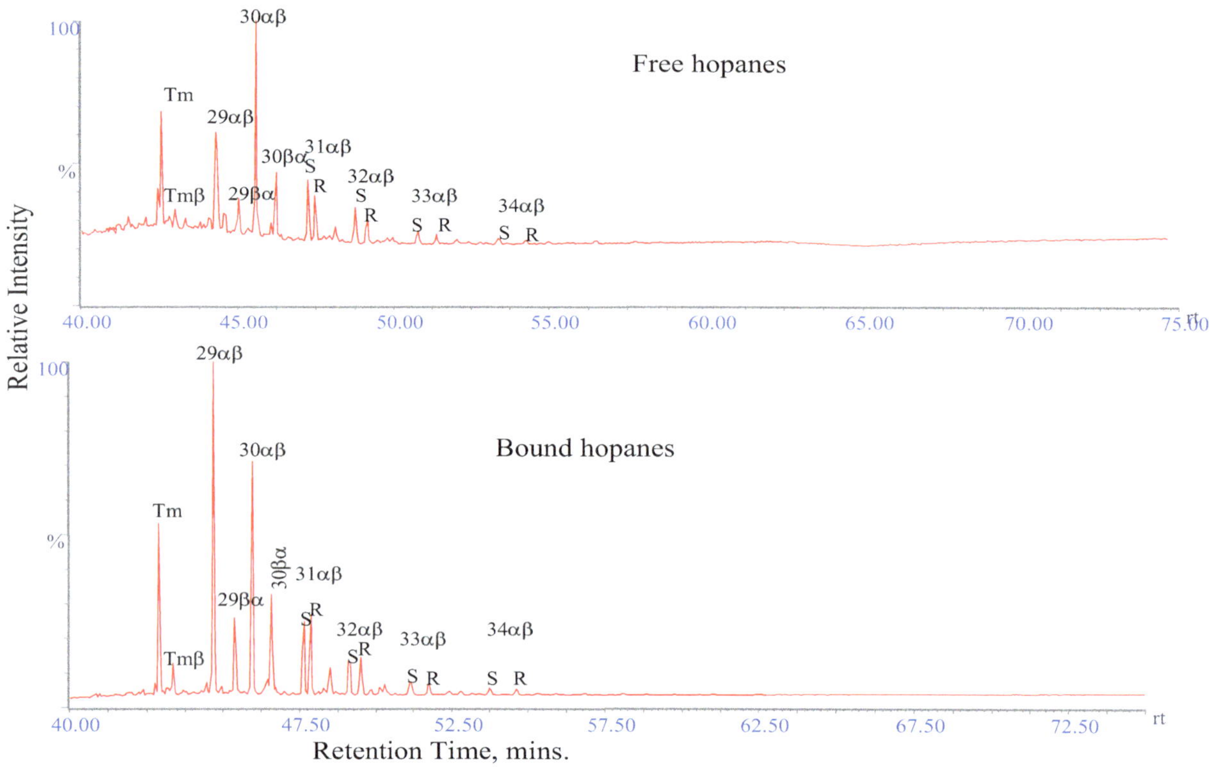

Figure 3. m/z 191 (hopane) fragmentograms for low temperature coal tar free and bound species.

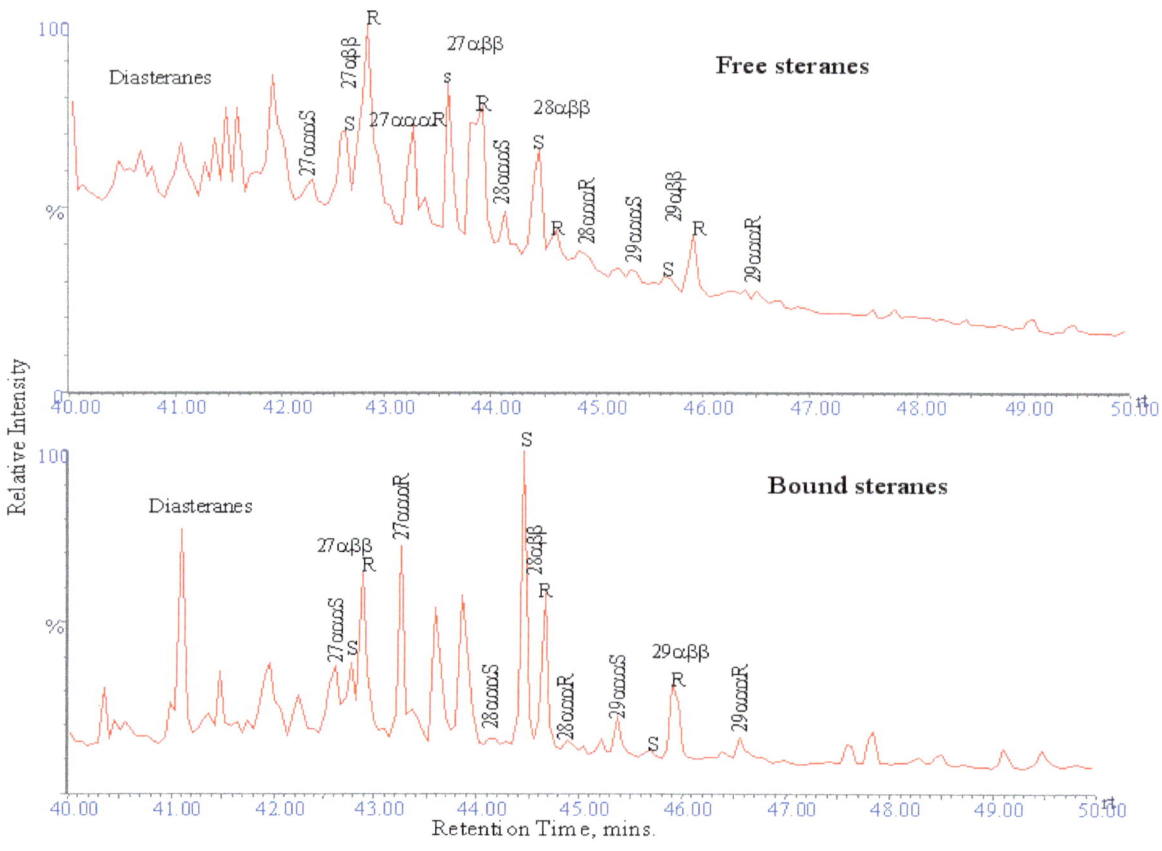

Figure 4. m/z 217 fragmentogram (steranes) for low temperature coal tar free and bound species.

bound aliphatic fractions as shown in Figure 2.

The hopanes are less mature for the bound than for the free, this is a general trend found for most hydropyrolysis products (Figure 3) (Love et al., 1997, 1998; Meredith, 2004). There is also no extended sidechain survival in both. Alkyl- S/S + R ratio is close to thermodynamic equilibrium.

The virtual absence of n-C_{29} sterane suggests that cracking reactions were induced during the carbonization process of the coal tar (Figure 4). Generally, the compositions of alkane fractions are similar to those of the free and bound constituents (Ivwurie et al., 2009).

Conclusion

This paper compared the biomarker (pristane, phytane, hopane and sterane) distribution of free and bound low temperature coal tar as a basis for future source apportionment studies from secondary sources such as pollution sites (example, contaminated soil, sediments and water).

Aliphatics hydrocarbon distribution for both free and bound coal tar shows similar carbon preference index; while pristane/phytane and pristane/C_{17} ratios were within the range reported in previous studies on similar samples (Fu et al., 1992; Hughes et al., 1995; Love et al., 1997, 1998; Sun, 2001). The results obtained, clearly demonstrates that both aliphatic and biomarkers released from asphaltenes via hydropyrolysis could provide useful information such as history of coal formation, environmental influence and processing conditions for source apportionment. This is because molecular profiles of biomarkers for both free and bound coal tar aliphatics show similar profiles, except in the hopane distribution where the free hopanes are more matured than the bound.

REFERENCES

Buenrostro-Gonzalez E, Groenzin H, Lira-Galeana C, Mullins OC (2001).The overriding chemical principles that define Asphaltenes. Energy Fuels, 15: 972-978.

Fu J, Liu D, Sheng G (1992). Geochemistry of coal- generated hydrocarbons; Science Press, Beijing, China, pp. 15-23.

Hughes WB, Holba AG, Dzou LIP (1995). The ratios of dibenzothiophene to phenanthrene and pristane to phytane as indicators of depositional environment and lithology of petroleum source rocks. Geochimica et Cosmochimica Acta, 59: 3581-3598.

Ivwurie W (2011a). The effect of unresolved complex mixtures (UCM) on isotopic profile of aromatic hydrocarbon. Int. J. Chem. Appl., 3(1): 49-54.

Ivwurie W, Snape C E, Sun C (2011b). Comparison of stable carbon isotope ratios for free and bound aromatic in low temperature coal tar. Int. J. Eng. Res. Technol., 4(1): 167-172.

Ivwurie W, Snape CE, Sun C (2009). Comparison of Hydrocarbon yields and molecular distribution of Solvent extractible and Bound Aliphatics of Low Temperature Coal Tar. Abstract, 32nd Int. conf. of the Chemical Society of Nigeria, Bauchi State, Nigeria, p. 110.

Jones DM, Douglas AG, Connan J (1988). Hydrous pyrolysis of asphaltenes and polar fractions of biodegraded oils. *In* L. Matavelli and L. Novelli (eds). Advances in Organic Geochemistry. Pergamon Press, Oxford, pp. 15-40.

Love GD, McAulay A, Snape CE, Bishop AN (1997). Effects of Process variables in catalytic hydropyrolysis on the release of covalently bound aliphatics hydrocarbons from sedimentary organic matter. Energy Fuels, 11: 522-531.

Love GD, Snape CE, Fallick AE (1998). Differences in the mode of incorporation and biogenicity of the principle aliphatic constituents of a Type 1 shale. Org. Geochem., 28: 797-911.

Meredith W, Russell CA, Cooper M, Snape CE, Love GD, Fabbri D, Vane CH (2004). Trapping hydropyrolysates on silica and their subsequent thermal desorption to facilitate rapid fingerprinting by GC-MS. Org. Geochem., 35: 73-89.

Murray IP (2001). Release of covalently bound biological marker molecules from sedimentary organic matter via catalytic hydropyrolysis. Unpublished PhD thesis, University of Strathclyde.

Rogge WF, Hildemann LM, Mazurek MA, Cass GR, Simoneit BRT (1993). Sources of fine organic aerosol 2. Non-catalyst and catalyst equipped automobiles and heavy-duty diesel trucks. Env. Sci. Technol., 27: 636-651.

Schauer JJ, Rogge WF, Hildermann LM, Mazurek MA, Cass GR, Simoneit BRT (1996). Source Apportionment of air-borne particulates matter using organic compounds as tracers. Atmospheric Env., 30: 3837-3855.

Sun C (2001). Sourcing PAH in the environment. Unpublished PhD thesis. University of Strathclyde.

Kinetics of the adsorption of hexavalent Chromium from aqueous solutions on low cost material

Khaldoun Al-Sou'od

Department of Chemistry, Faculty of Science, Al al-Bayt University, P. O. Box 130040, Mafraq 25113, Jordan.
E-mail: khaldoun@aabu.edu.jo.

This study proposes a Jordanian Pottery (Pot-Jo) as a low cost adsorbent for the removal of chromium from aqueous solutions. The effect of contact time, initial concentration of Cr(VI), temperature and ionic strength on the adsorption of Cr(VI) onto Pot-Jo were studied and discussed. The equilibrium (q_e) of adsorption of Cr(VI) onto pottery were found to be 20.287, 17.55, and 13.577 mg Cr(VI) per gram of pottery at temperatures 25, 35, and 45°C, respectively. Five kinetic models, the pseudo first-order equation, the pseudo second-order equation, the Elovich equation, the Ritchie's equation, and the intraparticle diffusion model were tested with respect to their adequacy to describe the adsorption process. The mechanism of adsorption was investigated by using the intraparticle diffusion model. It was found that there was no significant effect of the ionic strength on the removal of Cr(VI) from its solutions. Kinetic parameters, rate constants, equilibrium adsorption capacities and correlation coefficients, for each kinetic equation were calculated and discussed. Activation energy values were found to be between 13.85 and 33.10 kJ mol^{-1}. The Ritchie and pseudo-first order kinetic equation were found to correlate the experimental data well. The rate determining step is well described by intraparticle diffusion process. The k_{id} values ranging from 1.326 to 0.963 mgg^{-1}min$^{-1/2}$.

Key words: Chromium (VI), Sorption kinetics, Pseudo-second order, Lagergren, Tempkin.

INTRODUCTION

Water contamination with heavy metal ions is a very severe environmental problem all over the world. Chromium is a toxic pollutant that causes severe environmental and public health problems. For example, it was reported that $Cr_2O_7^{2-}$ can cause lung cancer (Sikaily et al., 2007; Li et al., 2008). In aqueous phase, Chromium mostly exists in two oxidation states, namely, trivalent Chromium [Cr^{3+}, $Cr(OH)_2^{+}$ or $Cr(OH)^{2+}$] and hexavalent Chromium ($HCrO_4^-$, CrO_4^{2-} or $Cr_2O_7^{2-}$). Moreover, +6 state of Chromium has magnetic properties. Chromium in its compounds has many different colors, so it is widely used in many industrial fields such as plating, alloying, dyeing, tanning, finishing, wood preserving and refractory technologies. The maximum limit for Cr(VI) concentration in drinking water is 0.05 mg/L (Baral et al., 2006).

Conventional techniques for Cr(VI) removal from aqueous solution include precipitation, membrane filtration, solvent extraction with amines, ion exchange, activated carbon adsorption, electro-deposition, and various biological processes (Owlad et al., 2009; Wu et al., 2010; Tiravanti et al., 1997; Rengaraj et al., 2003; Kozlowski and Walkowiak, 2002; Gupta et al., 2001; Bhatti et al., 2010; Zubair et al., 2008; Shafqat et al., 2008; Gupta and Imran, 2004; Gupta and Rastogi, 2009; Gupta et al., 2010; 2012).

A survey of the literature has shown that only very small attempt has been made to use a pottery to remove heavy metals. Jordanian pottery (Pot-Jo) was chosen due to its low cost, its granular structure, insolubility in water, chemical stability, local availability and most importantly being an environment friendly natural material. The objective of this work is to investigate the applicability of Pot-Jo materials to remove Cr(VI) from aqueous solutions. A kinetic study was carried out investigating the pH, Cr(VI) concentration and the adsorbent particle size as parameters, pH, concentration and particle size as

Table 1. Chemical compositions (%wt) of Pot-Jo.

Raw pottery	SiO_2	Al_2O_3	TiO_2	K_2O	MgO	CaO	Fe_2O_3	Others
Pot-Jo	60.41	13.87	1.851	2.35	1.31	10.84	8.67	0.70

parameters. Thermodynamic data have been calculated in some cases to interpret the results. The effects of ionic strength, temperature and initial concentrations on Cr(VI) are particularly investigated. The kinetics on Cr(VI) adsorption on pottery materials was analyzed by various kinetic models. Five models were used in this study: (i) pseudo-first order kinetic model, (ii) pseudo-second order kinetic model, (iii) intraparticle diffusion model, (iv) Elovich equation, and (V) Ritchie's equation. Kinetic parameters such as the half-time, pre-exponential factor and activation energy are evaluated and discussed.

MATERIALS AND METHODS

Sample collection

A pottery used in this study was supplied by a factory near Zarqa (40 km Eastern Amman). All reagents used were of AR grade (Sigma-Aldrich, Germany).

Sample preparation

Raw pottery material (Pot-Jo) was prepared and treated as in the literature (Khazali et al., 2007). The chemical compositions of the Pot-Jo (Table 1) were estimated by XRF (Philips Magix PW 2424) (Khazali et al., 2007).

Qualitative X-ray diffraction (XRD) analysis was performed on the Pot-Jo to determine a mineralogical composition. The XRD pattern was obtained on a Philips, PW 3040/60 X'PertPRO Console using Cu K_α radiator and RTMS detector X'celerator.

The stock solutions of test reagent are prepared by dissolving a known quantity of potassium dichromate ($K_2Cr_2O_7$) in double distilled water to obtain total Chromium concentration levels ranging from 50 to 400 mg/L. The pH values of the initial solutions are adjusted by adding HCl (0.1 M) or NaOH (0.1 M). Ionic strength has been adjusted by adding potassium nitrate. The experimental runs have been carried out in batch mode at the different temperatures.

Experiments for studying adsorption kinetics

For adsorption kinetics, a different set of batch experiments was carried out to establish the effect of time and initial concentration of Cr(VI) on the adsorption and to investigate the parametric effects of initial adsorbate concentration on Cr(VI) adsorption. 100.0 ml Cr(VI) solution with different initial concentrations (50 to 400 ppm) at pH 3 was added to screw-cap conical flask already containing 1.0 g of Pot-Jo (particle size ≤250 µm). [It was found earlier that an initial pH of 3.0 is most favorable for Cr(VI) removal (Al-Sou'od, 2012)].

The mixture was mechanically shaken at a speed of 200 $r \cdot min^{-1}$ with a temperature-controlled water bath at constant pH value of 3. After 300 min, the final pH was measured and samples were filtered using a Whatman No. 1 filter paper. The adsorbents were separated, and the filtrate was analyzed spectrophotometrically (Marczenko, 1976) to determine Cr(VI) content. Cr(VI) concentration was determined using an indirect UV-visible spectrophotometric method based on the reaction of Cr(VI) and diphenylcarbazide, which forms a red-violet colored complex. The absorbance of the colored complex was measured in a double beam spectrophotometer at 540 nm wavelength. All experiments were conducted in three replicates and the relative error between duplicates was less than 5%. The results were reported as mean values.

The controlling mechanism of the adsorption process was investigated by fitting the experimental data with Lagergren pseudo-first-order, pseudo-second-order kinetic models, Elovich equation, Ritchie's equation or the intraparticle diffusion model.

The adsorbed amount of Cr(VI) at equilibrium, q_e [in mg Cr(VI) per g adsorbent g^{-1}] was calculated from the mass balance equation given in Equation 1 (Gode and Pehlivan, 2005):

$$q_e = \frac{C_i - C_e}{m} V \quad (1)$$

Where C_i and C_e are the initial and equilibrium concentrations (mg/L), m is the mass of pottery (g), and V is the volume of the solution (L). When *t* is equal to the equilibrium contact time, then the amount of metal ion sorbed at equilibrium, q_e, is calculated using Equation 1.

The percent Cr(VI) removal (%RE) was calculated for each equilibration by the expression presented as Equation 2.

$$\%\ RE = \frac{C_i - C_e}{C_i} \times 100 \quad (2)$$

Where (%RE) is the percent of Cr(VI) adsorbed or removed.

RESULTS AND DISCUSSION

Effect of contact time and initial concentration

Effect of contact time and initial pottery concentration on adsorption of Cr(VI) by Pot-Jo are shown in Figure 1. The concentrations selected for studies were 50, 100, 150, 200, 300, and 400 ppm. The results obtained have been plotted as %RE versus contact time (Figure 1). The graph shows that for all the concentrations, removal increases in initial stages, goes on increasing and acquires maximum value in 300 min. After 300 min, there is no further increase in removal. As shown in Figure 1, the removal of Cr(VI) increased from 42.2 to 78.1% by increasing the initial concentration of Chromium from 50 to 300 ppm at pH 3, 0.1 M ionic strength, temperature 25°C, 200 rpm and at a particle size of ≤250 µm. When these findings were compared with those reported for other low cost adsorbent materials for the removal of Cr(VI) like silica gel, polymer, carbon nanotubes, fly ash,

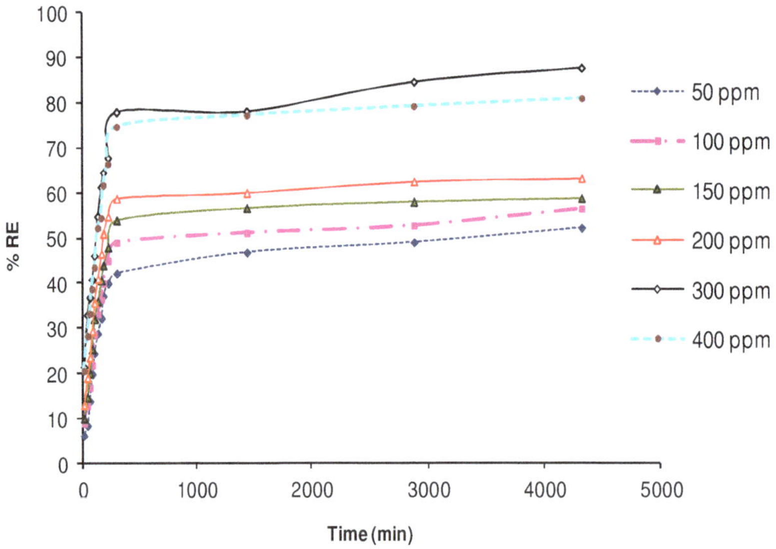

Figure 1. Effect of initial concentration and contact time on removal of Cr(VI) by Pot-Jo at 25°C and pH 3.

Table 2. Comparison between the pottery used and other reported values of %RE in the literature using low cost materials.

Adsorbent	% RE	References
Roots of *Typha latifolia*	45	Barrera-Díaz et al. (2004)
Ashes	60	Barrera-Díaz et al. (2004)
Green Algae spirogyra species	12 - 30	Gupta et al. (2001)
Green alga *Ulva lactuca*	92	El-Sikaily et al. (2007)
Almond Green Hull	94.14	Sahranavard et al. (2011)
Riverbed sand	74.3	Sharma and Weng (2007)
Natural zeolite	93	Campos et al. (2007)
Turkish fly ashes	25.46	Bayat (2002)
Afsin-Elbistan	30.91	Bayat (2002)
Fly ash	89.12	Sharma et al. (2008)
Walnut *(Juglans regia)*	85.32	Pehlivan and Altun (2008)
Hazelnut *(Corylus avellana)*	88.46	Pehlivan and Altun (2008)
Almond *(Prunus dulcis)*	55.00	Pehlivan and Altun (2008)
Coconut shell	80	Nameni et al. (2008)
Wheat bran	87.8	Nameni et al. (2008)
Pre-boiled sunflower stem	81.7	Jain et al. (2009)
Formaldehyde treated sunflower stem	76.5	Jain et al. (2009)
Pine needle	38	Park et al. (2007)
Walnut shell	24.6	Park et al. (2007)
Rice straw	26.3	Park et al. (2007)
Peanut shell	41.0	Park et al. (2007)
Sawdust	19.9	Park et al. (2007)
Jordanian pottery materials	78.1	Present work

clays, zeolites, chitosan, peat moss, biosorbent, food waste etc., it is appeared to act more efficiently than about most of these materials or at least comparable to those classified as very efficient (Table 2).

Effect of ionic strength

Ionic strength is one of the important factors influencing aqueous phase equilibrium. Generally, adsorption

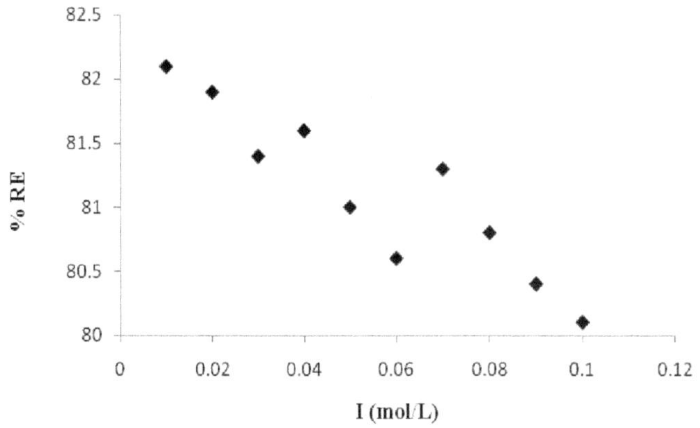

Figure 2. Effect of ionic strength for adsorption of Cr(VI) on Pot-Jo at 25°C and pH 3.

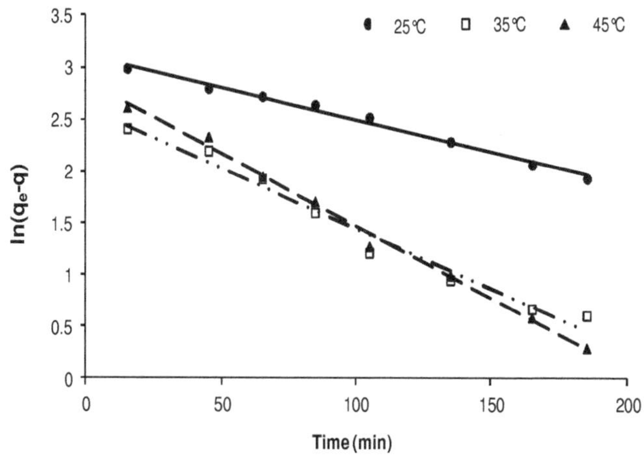

Figure 3. Lagergren plot for adsorption of Cr (VI) on Pot-Jo at 25, 35 and 45°C

decreases with increasing ionic strength of the aqueous solution. The effect of the interfering ion KNO_3 was evaluated. The electrostatic attraction at low ionic strengths applied in this work (0.01 to 0.1 M KNO_3) appears to play a negligible role in the removal of Cr(VI) as indicated by the fact that no significant decrease in the removal of Cr(VI) ion was observed for the interval of (0.01 to 0.1 M) KNO_3 ionic concentration (Figure 2).

Adsorption kinetics

In order to investigate the adsorption mechanism of Cr(VI) onto clay's pottery, the experimental data were modeled using the first-order rate equation of Lagergren (Lagergren, 1898), the pseudo second-order rate equation (Ho and McKay, 1999), Elovich model (Güzel et al., 2008), Ritchie's equation (Viswanathan, 1979) and intraparticle (Weber and Morris, 1963) kinetic models.

In order to understand the kinetics of removal of Cr(VI) by adsorption, many models were tried. The study of adsorption dynamics describes the solute uptake rate and evidently this rate controls the residence time of adsorbate uptake at the solid-solution interface. The kinetics of the interactions were studied by determining the amount adsorbed at different agitation times for various amounts of the adsorbent at constant Cr(VI) concentration. The order of adsorbate-adsorbent interactions has been described by using various kinetic models. Traditionally, the pseudo-first order model derived by Lagergren (1898) has found wide application. On the other hand, several authors (Weber and Morris, 1963; Zacar and Engil, 2004; Zacar, 2003) have shown that second-order kinetics can also very well describe these interactions in certain specific cases.

The pseudo first-order equation

When adsorption is preceded by diffusion through a boundary, the kinetics in most cases follow the pseudo-first-order rate equation of Lagergren (Equation 3).

$$\frac{dq_t}{dt} = k_{ads}(q_e - q_t) \qquad (3)$$

Where q_e and q_t (mg/g) are the amounts of Cr(VI) adsorbed at equilibrium and at any time t (min), and k_{ads} (min^{-1}) is the first-order adsorption rate constant. The integrated rate law, after applying the initial condition of $q_t = 0$ at $t = 0$ is given by Equation 4.

$$\ln(q_e - q_t) = \ln q_e - k_{ads}.t \qquad (4)$$

The linear plot of $\log(q_e\_q_t)$ versus t at three different temperatures (25, 35, and 45°C) shows the applicability of the Lagergren equation (Figure 3).

The values of the Lagergren constants, q_e and k_{ads}, and the correlation coefficients are calculated from linear pseudo first-order equation at all concentration studied and are presented in Table 3. The correlation coefficients for the pseudo first-order kinetic model obtained at all the studied concentrations were high. The R^2 values for the plots were in the range of 0.994 to 0.999 (Table 4).

Numerical values for activation energy (E_a) of adsorption process were determined using the Arrhenius equation (Equation 5) (Srivastava et al., 1995).

$$\ln k_{ads} = \ln A - \frac{E_a}{RT} \qquad (5)$$

The plots of ln k_{ads} versus $1/T$ were found to be linear. Numerical values for activation energy (E_a) were determined from slopes of Arrhenius graphs related to adsorption process of Cr(VI) onto pottery materials.

The activation energy for the adsorption of Cr(VI) was found to be between 13.85 and 33.19 $kJmol^{-1}$. From the value of activation energy, it appears that the adsorption

Table 3. Adsorption kinetic parameters by Lagergren equation at different temperatures and pH 3 [Cr(VI)] = 300 ppm.

Clay's type	T/°C	K_{ads} (min^{-1})	q_e (mgg^{-1})	$t_{1/2}$ (min)	E_a ($kJmol^{-1}$)
	25	0.00602	22.287	115.14	
Pot-Jo	35	0.0116	17.55	59.75	33.19
	45	0.0139	13.577	49.87	

Table 4. Comparison of rate constants calculated based on respective pseudo-first-order, pseudo-second-order, Elovich, Ritchie and Intraparticle diffusion kinetic models.

Temperature 25°C	Pseudo First order			Pseudo-second order			Elovich model			Ritchie model			Intraparticle diffusion		
	K_{ads} (min^{-1})	q_e (mgg^{-1})	R^2	k_2 ($gmg^{-1}min^{-1}$)	q_e (mgg^{-1})	R^2	α ($mgg^{-1}min^{-1}$)	B (gmg^{-1})	R^2	q_∞ (mgg^{-1})	α	R^{Π}	k_{id} ($mgg^{-1}min^{-1/2}$)	C	R^{Π}
25	0.00602	22.29	0.999	0.00058	33.47	0.964	5.398	0.0289	0.923	25.97	0.0147	0.977	1.326	0.792	0.988
35	0.0116	13.58	0.977	0.00056	29.39	0.943	8.295	0.0113	0.955	26.18	0.0451	0.984	1.228	7.373	0.957
45	0.0139	17.55	0.994	0.00800	29.33	0.915	2.987	0.0212	0.911	27.25	0.036	0.969	0.963	10.593	0.947

of Cr(VI) on pottery is a chemical adsorption process. This is confirmed from the fact that the activation energy for chemical adsorption is usually more than 4 to 6 kJ mol^{-1} (Tewaria et al., 2005).

The half-adsorption time ($t_{1/2}$) is the time required to uptake half of the maximal amount of Cr(VI) adsorbed at equilibrium. It characterizes the adsorption rate well. In the case of a pseudo-first-order process, its value is given by the following relationship:

$$t_{1/2} = \frac{\ln 2}{k_1} \qquad (6)$$

Values of $t_{1/2}$ are reported in Table 3.

The pseudo second-order equation

The pseudo second-order adsorption kinetic rate equation is expressed as (Srivastava et al., 1995):

$$\frac{dq_t}{dt} = k_2(q_e - q_t) \qquad (7)$$

Where: k_2 is the rate constant of pseudo second-order adsorption ($g{\cdot}mg^{-1}{\cdot}min^{-1}$).

The linear form of the equation is given by equation:

$$\frac{t}{q_t} = \frac{1}{k_2 q_e^2} + \frac{1}{q_e}(t) \qquad (8)$$

If the initial adsorption rate, h ($mg{\cdot}g^{-1}{\cdot}min^{-1}$) is:

$$h = k_2 q_e^2 \qquad (9)$$

Then Equations 8 and 9 become:

$$\frac{t}{q_t} = \frac{1}{h} + \frac{1}{q_e}(t) \qquad (10)$$

The plot of (t/q_t) and t of Equation 10 should give a linear relationship (Figure 4) from which q_e and k_2 can be determined from the slope and intercept of the plot, respectively.

The applicability of two models (pseudo-first-order model and pseudo-second-order) can be tested by linear fitting of $\ln(q_e - q)$ versus t, and (t/q) versus t, respectively. To quantify the applicability of each model, the correlation coefficient, R^2, was calculated from these plots. A comparison of the correlation coefficients, R^2, showed that the pseudo-first-order model fits better the experimental data ($R^2 \approx 0.999$) than the pseudo-second-order model (R^2 is in the range of 0.964 to 0.915) (Table 4). This confirms that it was not appropriate to use the pseudo-second-order kinetic model to predict the absorption kinetics of Cr(VI) onto clay's pottery for

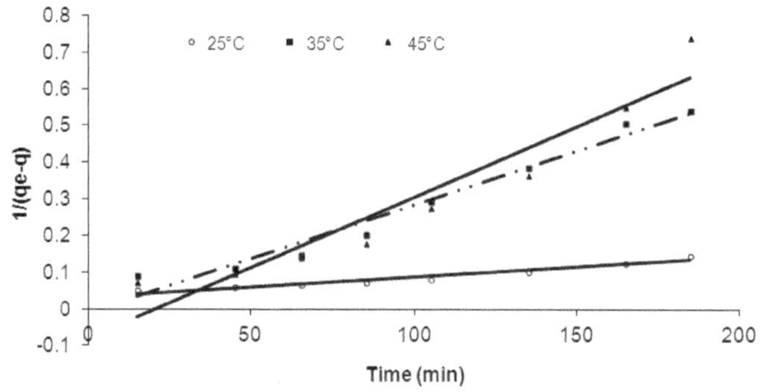

Figure 4. The pseudo second-order adsorption kinetics of Cr (VI) on Pot-Jo at 25, 35 and 45°C.

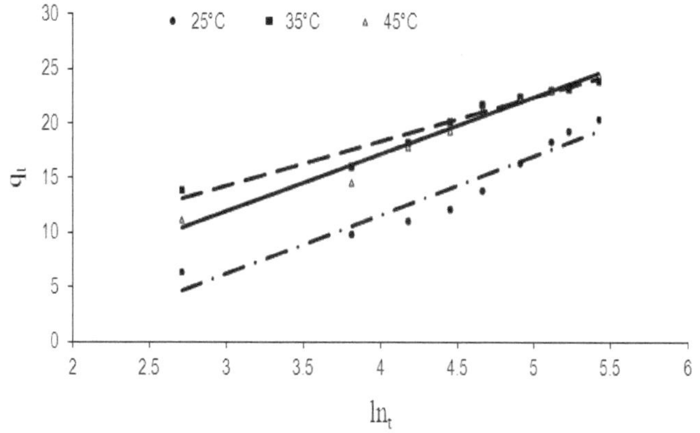

Figure 5. The evaluation of Elovich adsorption kinetics for Cr (VI) adsorption on Pot-Jo at 25, 35 and 45°C.

for the entire adsorption period.

The Elovich equation

Elovich equation is another expression based on the adsorptive capacity of adsorbents and was initially developed to describe the kinetics of gas chemisorption on solids. The Elovich or Roginsky-Zeldovich equation is generally expressed as follows:

$$\frac{dq_t}{dt} = \beta \exp(-\alpha q_t) \tag{11}$$

Where q_t is the amount of Cr(VI) sorbed by pottery at a time t, α is the initial Cr(VI) sorption rate [mmol/(g min)] and β is the desorption constant (g/mmol) during any one experiment.

To simplify the Elovich equation, Chien and Clayton (Zacar and Engil, 2004) assumed $\alpha\beta t >> 1$, and on applying the boundary conditions $q_t = 0$ at $t = 0$ and $q_t = q_t$ at $t = t$, Equation (11) then becomes:

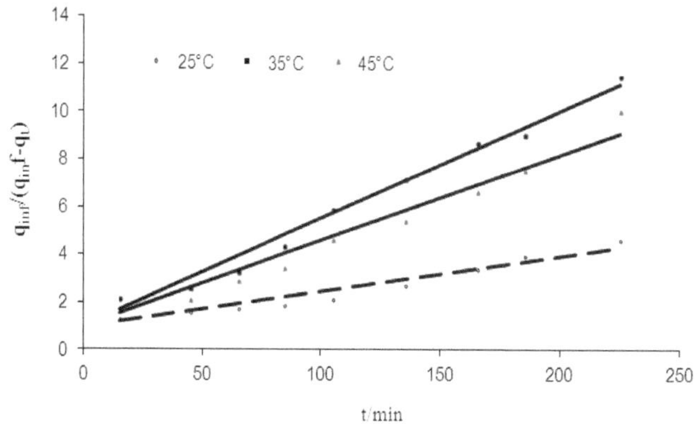

Figure 6. The Ritchie's kinetics of Cr(VI) on Pot-Jo at 25, 35 and 45°C.

$$q_t = \alpha \ln(\alpha\beta) + \alpha \ln t \tag{12}$$

If Cr(VI) adsorption fits the Elovich model, a plot of q_t versus ln(t) should yield a linear relationship with a slope of α and an intercept of αln(αβ) (Figure 5).

Ritchie's equation

Ritchie's equation can be written using the following equation (Ritchie, 1977):

$$\frac{q_\infty}{q_\infty - q_t} = \alpha t + 1 \tag{13}$$

Where q_∞ is the amount of adsorption after an infinite time, *α* the initial adsorption rate.

The value for q_∞ is obtained from the intercept at $(1/t) = 0$ on a plot of $(1/q)$ against $(1/t)$ (Figure 6). Ritchie found a good linear relationship between t and $q_\infty/(q_\infty - q)$, according to Richie's equation as observed by Austin et al. (1971), Bansal et al. (1971), Deitz and Turner (1971), and Samuel and Yeddanapalli (1974). In recent years, the Ritchie equation has also been applied to solution/solid adsorption systems, for example, the adsorption of cadmium ions onto bone char (Inbaraj and Sulochana, 2004), and the adsorption of Cd(II) onto acid-treated jackfruit peel (Cheung et al., 2001).

Intraparticle diffusion

The mechanism of sorption is either film diffusion controlled or particle diffusion controlled. Before adsorption takes place, several diffusion processes known to affect the adsorption process takes place. The sorbate will have to diffuse through the bulk of the

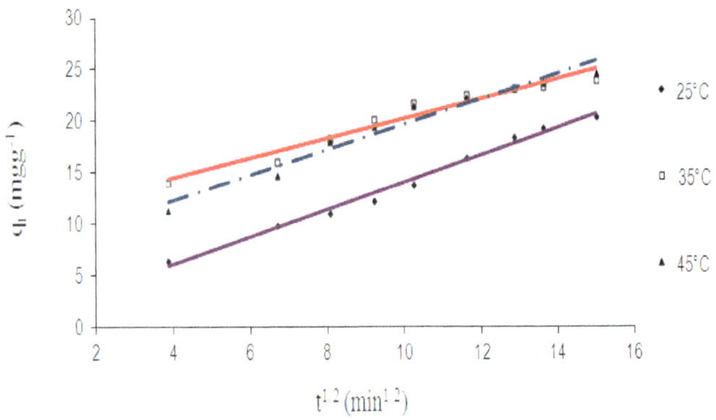

Figure 7. The intraparticle diffusion adsorption kinetics of Cr (VI) on Pot-Jo at 25, 35 and 45°C.

solution to the film surrounding the adsorbent and then into the micropores and macropores of the adsorbent.

The intraparticle diffusion model is expressed (Furusawa and Smith, 1974) as follows:

$$q_t = k_{id}\,(t)^{1/2} + C \quad (14)$$

Where: q_t is the amount of Cr(VI) sorbed by pottery at a time t, C is the intercept of linear plots, k_{id} is the intraparticle diffusion rate constant ($mgg^{-1}min^{-1/2}$). The k_{id} values under different conditions were calculated from the slopes of the straight line portions of the respective plots (Figure 7). Higher values of the intraparticle rate constants (K_{id}) illustrate an enhancement in the rate of adsorption, whereas larger values illustrate a better adsorption mechanism. The values of intercept *C* (Table 4) provides information about the thickness of the boundary layer, the resistance to the external mass transfer increases as the intercept increases. The constant *C* was found to increase with increasing the temperature, corresponding to an increase of thickness in the boundary layer and to a decrease of the chance of external mass transfer to dominate. The high values of R^2 indicate the application of this model.

If intraparticle diffusion is rate-limited, then plots of adsorbate uptake q_t versus the $t^{1/2}$ would result in a linear relationship. Values of k_{id} were found to be between 1.326 and 0.963 mg/g $min^{1/2}$ with a R^2 value of 0.988 to 0.947 (Table 4). If the plot of q_t versus $t^{1/2}$ is linear and passes through the origin, then intraparticle diffusion is the sole rate-limiting step (Ho, 2004). However, the linear plot in our study did not pass through the origin. This indicates some degree of boundary layer control and also that the intraparticle diffusion was not the only rate controlling step (Ho, 2004).

Conclusions

The finding of this study demonstrate the good ability of the naturally occurring Pot-Jo materials in removing heavy metals of interest [Cr(VI) ions in this work]. Therefore, it might be considered as good adsorbent material that could work efficiently to replace the high cost activated carbon. The adsorption kinetics can be well described by the Ritchie and pseudo-first order kinetic equation with intraparticle diffusion as one of the rate determining steps. Both the Ritchie and pseudo-first order models were considered the most appropriate due to their high correlation coefficients relative to other kinetic models. The first-order adsorption rate constant, K_{ads} values were found to be decreased from 0.00602 to 0.0139 with increasing temperature from 25 to 45°C. The results indicate that there was no significant decrease in the removal of Cr(VI) ion for the interval of (0.01 to 0.1 M) KNO_3 ionic concentration. The activation energy of sorption was evaluated for the sorption of Cr(VI) on Pot-Jo. The activation energy (33.19 $kJmol^{-1}$) was determined to be same order of magnitude as the activation energy of chemisorptions.

ACKNOWLEDGEMENT

The author is thankful for the funding provided by Al al-Bayt University.

REFERENCES

Al-Sou'od K (2012). Adsorption Isotherm Studies of Chromium (VI) from Aqueous Solutions Using Jordanian Pottery Materials. APCBEE Procedia 1:116-125.

Austin FE, Brown JG, Dollimore J, Freedman CM, Harrison BH (1971). The use of partial-pressure mass spectrometry in the study of the thermal desorption and oxidation of carbon and graphite. Analyst 96:110-116.

Bansal RC, Vastola FJ, Walker PL (1971). Studies on ultra-clean carbon surfaces—III Kinetics of chemisorption of hydrogen on graphon. Carbon 9:185-192.

Baral SS, Das SN, Rath P (2006). Hexavalent Chromium Removal from aqueous by adsorption on treated Sawdust. Biochem. Eng. J. 13:216-222.

Barrera-Díaz C, Colín-Cruz A, Ureña-Nuñez F, Romero-Romo M, Palomar-Pardavé M (2004). Cr(VI) removal from wastewater using low cost sorbent materials: Roots of *Typha latifolia* and ashes. Environ. Technol. 25(8):907-917.

Bayat B (2002). Comparative study of adsorption properties of Turkish fly ashes II. The case of Chromium (VI) and Cadmium (II). J. Hazard. Mater. B95:275-290.

Bhatti HN, Nasir AW, Hanif MA (2010). Efficacy of Daucus carota L. waste biomass for the removal of Chromium from aqueous solutions. Desalination 253:78-87.

Campos V, Morais LC, Buchler PM (2007). Removal of chromate from aqueous solution using treated natural zeolite. Environ. Geol. 52:1521-1525.

Cheung CW, Porter JF, McKay G (2001). Sorption kinetic analysis for the removal of cadmium ions from effluents using bone char. Water Res. 35:605-612.

Deitz VR, Turner NH (1971). Interaction of ozone with adsorbent charcoals. J. Phys. Chem. 75:2718-2727.

El-Sikaily A, El Nemr A, Khaled A, Abdelwehab O (2007). Removal of toxic Chromium from wastewater using green alga *Ulva lactuca* and its activated carbon. J. Hazard. Mater. 148(1-2):216-228.

Furusawa T, Smith JM (1974). Intraparticle mass transport in slurries by

dynamic adsorption studies. J. AIChE. 20:88-93.
Gode F, Pehlivan E (2005). Adsorption of Cr(III) ions by Turkish brown coals. Fuel Process Technol. 86:875-884.
Gupta VK, Ganjali MR, Nayak A, Bhushan B, Agarwal S (2012). Enhanced heavy metals removal and recovery by mesoporous adsorbent prepared from waste rubber tire. Chem. Eng. J. 197:330-342.
Gupta VK, Imran A (2004). Removal of lead and Chromium from wastewater using bagasse fly ash - A sugar industry waste. J. Colloid Interf. Sci. 271:321-328.
Gupta VK, Rastogi A (2009). Biosorption of hexavalent Chromium by raw and acid-treated green alga Oedogonium hatei from aqueous solutions. J. Hazard. Mater. 163(1):396-402.
Gupta VK, Rastogi A, Nayak A (2010). Adsorption studies on the removal of hexavalent Chromium from aqueous solution using a low cost fertilizer industry waste material. J. Colloid Interf. Sci. 342(1):135-141.
Gupta VK, Shrivastava AK, Jain N (2001). Biosorption of Chromium(VI) from aqueous solutions by green algae spirogyra species. Water Res. 35:4079-4085.
Güzel F, Yakut H, Topal G (2008). Determination of kinetic and equilibrium parameters of the batch adsorption of Mn(II), Co(II), Ni(II) and Cu(II) from aqueous solution by black carrot (*Daucus carota* L.) residues. J. Hazard. Mater. 153:1275-1287
Ho YS (2004). Citation review of Lagergen kinetic rate equation on adsorption reactions. Scientomet 59:171-177.
Ho YS, McKay G (1999). Pseudo-second order model for sorption processes. Biochem. 34:451-465.
Inbaraj BS, Sulochana N (2004). Carbonised jackfruit peel as an adsorbent for the removal of Cd(II) from aqueous solution. Bioresour. Technol. 94:49-52.
Jain M, Garg VK, Kadirvelu K (2009). Chromium(VI) removal from aqueous system using *Helianthus annuus* (sunflower) stem waste. J. Hazard. Mater. 162:365-372.
Khazali O, Abu-El-Halawa R, Al-Sou'od K (2007). Removal of Copper(II) from aqueous solution by Jordanian pottery materials. J. Hazard. Mater. B. 139:67-71.
Kozlowski CA, Walkowiak W (2002). Removal of Chromium(VI) from aqueous solutions by polymer inclusion membranes. Water Res. 36:4870-4876.
Lagergren S (1898). Adsorption geloster stoffe, Kungliga Svenska vetenskapsakademiens. Handlingar 24:1-39.
Li H, Li Z, Liu T, Xiao X Peng Z, Deng L (2008). A novel technology for biosorption and recovery hexavalent Chromium in wastewater by bio-functional magnetic beads A novel technology for biosorption and recovery hexavalent Chromium in wastewater by bio-functional magnetic beads. Bioresour. Technol. 99:6271-6279.
Marczenko Z (1976). Spectrophotometric Determination of Elements, JohnWiley and Sons, New York.
Nameni M, Moghadam MRA, Arami M (2008). Adsorption of hexavalent Chromium from aqueous solutions by wheat bran. Int. J. Environ. Sci. Technol. 5:161-168.
Owlad M, Aroua MK, Daud WAW, Baroutian S (2009). Removal of hexavalent Chromium-contaminated water and wastewater: A review. Water Air Soil Pollut. 200:59-77.
Park D, Lim SR, Yun YS, Park JM (2007). Reliable evidences that the removal mechanism of hexavalent Chromium by natural biomaterials is adsorption-coupled reduction. Chemosphere 70:298-305.
Pehlivan E, Altun T (2008). Biosorption of Chromium (VI) ion from aqueous solutions using walnut, hazelnut and almond shell. J. Hazard. Mater. 155:378-384.
Rengaraj S, Joo CK, Yi Y, Kim J (2003). Kinetics of removal of Chromium from water and electronic process wastewater by ion exchange resins: 1200H, 1500H and IRN97H. J. Hazard. Mater. 102:257-275.
Ritchie AG (1977). Alternative to the Elovich equation for the kinetics of adsorption of gases on solids. J. Chem. Soc. Faraday Trans. I. 73:1650-1653.
Sahranavard M, Ahmadpour A, Doosti MR (2011). Biosorption of Hexavalent Chromium Ions from Aqueous Solutions using Almond Green Hull as a Low-Cost Biosorbent. Eur. J. Sci. Res. 58(3):392-400.
Samuel P, Yeddanapalli LM (1974). Kinetics of hydrogen chemisorption on supported molybdenum disulphide. J. Appl. Chem. Biotechnol. 24:777-786.
Shafqat F, Bhatti HN, Hanif MA, Zubair A (2008). Kinetic and equilibrium studies of Cr(III) and Cr(VI) sorption from aqueous solution using Rosa gruss an teplitz (red rose) waste biomass. J. Chil. Chem. Soc. 54(3):1565-1570.
Sharma YC, Upadhyay SN, Weng CH (2008). Studies on an economically viable remediation of Chromium rich waters and wastewaters by PTPS fly ash Colloids and Surfaces A. Physicochem. Eng. Aspects. 317:222-228.
Sharma YC, Weng CH (2007). Removal of Chromium (VI) from water and wastewater by using riverbed sand: Kinetic and equilibrium studies. J. Hazard. Mater. 142:449-454.
Sikaily AE, Nemr AE, Khaled A, Abdelwehab O (2007). Removal of toxic Chromium from wastewater using green alga Ulva lactuca and its activated carbon. J. Hazard. Mater. 148:216-228.
Srivastava SK, Gupta VK, Anurpam S, Mohan D (1995). Removal of some anionic detergents using an inorganic gel adsorbent. Ind. J. Chem. 34A:342-350.
Tewaria N, Vasudevana P, Guha BK (2005). Study on biosorption of Cr (VI) by *Mucar hiemalis*. Biochem. Eng. J. 23:185-192.
Tiravanti G, Petruzzelli D, Passino R (1997). Pretreatment of tannery wastewaters by an ion exchange process for Cr(III) removal and recovery. Water Sci. Technol. 36:197-207.
Viswanathan B (1979). On Ritchie's equation for the analysis of kinetics of dsorption of gases on solids. J. Chem. Soc. Faraday Trans. 75:477-478.
Weber WJ, Morris JCJ (1963). Kinetics of adsorption on carbon from solution. Sanit. Eng. Div. Am. Soc. Chem. Eng. 89:31-59.
Wu J, Zhang H, He PJ, Yao QL, Shao M (2010). Cr(VI) removal from aqueous solution by dried activated sludge biomass. J. Hazard. Mater. 176:697-703.
Zacar MO (2003). Equilibrium and kinetic modeling of adsorption of phosphorus on calcined alunite. Adsorption. 9:125-132.
Zacar MO, Engil IAS (2004). Application of kinetic models to the sorption of disperse dyes onto alunite. Colloids Surf A. Physicochem. Eng. Aspects. 242:105-113.
Zubair A, Bhatti HN, Hanif MA, Shafqat F (2008). Kinetic and equilibrium modeling for Cr(III) and Cr(VI) removal from aqueous solutions by *Citrus reticulata* waste biomass. Water. Air Soil Pollut. 191:305-318.

19

Decolorization and mineralization of brilliant golden yellow (BGY) by Fenton and photo-Fenton processes

Md. Ahsan Habib[1]*, Iqbal Mohmmad Ibrahim Ismail[2], Abu Jafar Mahmood[1] and Md. Rafique Ullah[3]

[1]Department of Chemistry, Faculty of Science, University of Dhaka, Dhaka 1000, Bangladesh.
[2]Department of Chemistry, Faculty of Science, King Abdulaziz University Jeddah 21589, Saudi Arabia.
[3]Department of Chemistry, Bangladesh University of Engineering and Technology, Dhaka 1000, Bangladesh.

The decolorization and mineralization of an azo dye, brilliant golden yellow (BGY, λ_{max} 416 nm, $\varepsilon = 7.81 \times 10^3$ $Lmol^{-1}$ cm^{-1}) by Fenton and photo-Fenton processes were investigated. The rate of decolorization is strongly dependent on the pH, initial concentrations of the dye, Fe(II) and H_2O_2. The effect of these parameters has been studied and the optimum operational conditions of these two methods were found. The optimum conditions were obtained at pH=3 for both processes. The photo-Fenton method proved to be the most efficient and occurs at a much higher oxidation rate than Fenton process. The results showed that about 99% of the dye decolorized within 20 and 5 min for Fenton and photo-Fenton processes, respectively. Characterization of the products after mineralization is discussed.

Keywords: Dye pollution, Fenton and photo-Fenton processes, brilliant golden yellow, decolorization, mineralization.

INTRODUCTION

Synthetic azo dyes are the major industrial pollutants and water contaminants (Brown et al., 1981; Vaidya and Datye, 1982). Textile wastewater introduces intensive color and toxicity to aquatic systems (Muruganandham and Swaminathan, 2004) which is mostly non-biodegradable and resistant to demolition by physico-chemical treatment methods (Daneshvar et al., 2005). Therefore, it is necessary to find an effective method of wastewater treatment, both in terms of limited watercourses management and the need for nature of preservation. In this direction, Advanced Oxidation Processes (AOPs) have attracted wide interests in wastewater treatment since the 1990s. In principle, AOPs are based on the generation of hydroxyl radicals ($HO^{\bullet} + H^+ + e^- \rightarrow H_2O$, E = 2.8 V vs *NHE*)), which are highly reactive and nonselective oxidants being capable to oxidize organic compounds, in particular, unsaturated organic compounds and azo dyes (Andreozzi et al., 1999). Among them, Fenton's and photo-Fenton's type reactions are very promising since they achieve high reaction yields with a low treatment cost (Bauer and Fallman, 1997). These *AOP*s have been successfully applied to treat reactive dyes (Neamtu et al., 2003; Swaminathan et al., 2003; Torrades et al., 2004) and textile effluents (Perez et al., 2002; Rodrigez et al., 2002; Solmaz et al., 2006).

Fenton's and photo-Fenton's reactions are AOPs where oxidant species are generated from hydrogen peroxide and Fe(II)/Fe(III) as a catalytic couple. In Fenton's reaction (Haber and Weiss, 1934), ferrous salts react with H_2O_2 to generate the $HO^{\bullet}$ radicals as follows (reaction (1)):

$$Fe(II) + H_2O_2 \longrightarrow Fe(III) + H\dot{O} + HO^- \quad k = 76.5\ L\,mol^{-1}\,s^{-1} \tag{1}$$

Then, Fe(III) ion can be reduced by reaction with excess H_2O_2 to form again Fe(II) ions and more $HO^{\bullet}$ radicals. This second process called Fenton's-like (Pignatello, 1992), is slower than Fenton's reaction and allows Fe(II) regeneration giving place to a catalytic mechanism (Reactions (2) – (4)):

*Corresponding author. E-mail: mahabibbit@yahoo.com.

$$Fe(III) + H_2O_2 \rightleftharpoons FeOOH^{2+} + H^+ \quad K_{eq}=3.1 \times 10^{-3} \tag{2}$$

$$FeOOH^{2+} \longrightarrow HO_2^{\bullet} + Fe(II) \quad k=27 \times 10^{3}\ s^{-1} \tag{3}$$

$$Fe(III) + HO_2^{\bullet} \longrightarrow Fe(II) + O_2 + H^+ \quad k < 2 \times 10^{3}\ L\ mol^{-1}\ s^{-1} \tag{4}$$

The perhydroxy radicals ($HO^{\bullet}_2$) are very unstable and converted into hydroxyl free radicals ($HO^{\bullet}$) as shown in Reaction (5):

$$2HO_2^{\bullet} \longrightarrow 2\ HO^{\bullet} + O_2 \tag{5}$$

The rate of contaminant degradation can be considerably increased via photochemical reaction in the photo-Fenton's process (Pignatello, 1992; Pignatello et al., 2006). In this case, the regeneration of Fe(II), with production of new hydroxyl radicals, follows a photo reduction process (Reactions (6) and (7)):

$$Fe(III) + H_2O \longrightarrow Fe(OH)^{2+} + H^+ \tag{6}$$

$$Fe(OH)^{2+} \xrightarrow{h\nu} Fe(II) + HO^{\bullet} \quad \text{wavelength} < 410\ nm \tag{7}$$

Thus, the hydroxyl free radicals are generated from both the Fenton or photo-Fenton processes, and this is a strong oxidizing agent that oxidizes the organic compounds through generation of new organic free radicals as shown in Reaction (8):

$$RH + HO^{\bullet} \longrightarrow R^{\bullet} + H_2O \tag{8}$$

Then the free radicals ($R^{\bullet}$) scavenge oxygen molecules from the environment to form perhydroxy radicals ($RO^{\bullet}$) (Li et al., 1997) which are very reactive and in the presence of Fe(II), they are converted into stable products e.g., H_2O, CO_2 or HNO_3 etc as shown in Reactions ((9) and (10)):

$$R^{\bullet} + O_2 \longrightarrow RO_2^{\bullet} \tag{9}$$

$$RO_2^{\bullet} + Fe(II) \longrightarrow \text{degradated products} \tag{10}$$

It has been reported by Kang et al. (2000) that Fenton process could be adopted rapidly in a textile wastewater treatment system, without the end for reconstructing the existing coagulation unit. The use of $Fe(II)/H_2O_2$ system as an oxidant for wastewater treatment is attractive since iron is highly abundant and non-toxic, and a 30% H_2O_2 aqueous solution is easy to handle and environmentally not harmful. Malik and Saha (2003) reported that the removal rate is strongly dependent on the initial concentration of the dye, Fe(II) and H_2O_2.

Muruganandham and Swaminathan (2004) have carried out studies where similar results were obtained; they suggested a pH of 3 is the optimum for Fenton and photo-Fenton processes.

In the present study, Fenton and photo-Fenton processes have been used to find out the optimal conditions of experimental parameters for the mineralization of brilliant golden yellow (BGY) which is extensively used in textile, dying and in ball-point industries worldwide.

MATERIALS AND METHODS

Chemicals

Chemical structure of brilliant golden yellow (BGY) is shown in Table 1, Mohr's salt, hydrogen peroxide, hydrochloric acid, sodium hydroxide, sulfuric acid, nitric acid and potassium nitrate were purchased from BDH; all the chemicals were AnalaR grade except BGY which was laboratory reagent grade and used without further purification. Double distilled and deionized water was used throughout the experiment.

Experimental procedures

Fenton process's: The experiments were carried out batch wise. Dye solution of 40 mg L^{-1} was adjusted to pH = 3.0 ± 0.1 with NaOH or HCl as required. A known amount of H_2O_2 and Mohr's salt ($FeSO_4.(NH_4)_2SO_4.6H_2O$) were added to 500 ml of the solution. A magnetic stirrer was used for stirring in order to achieve homogeneous mixture.

Photo-Fenton's process: All experiments were carried out in a batch photoreactor. The radiation source was a low-pressure mercury lamp, UV light source (254 nm) (PASCO Scientific) was used for photolysis. The intensity of the light source was measured using spectroradiometer (model IL 1700) with a light source (model XPS 200, Solar light Co.). The lamp was positioned apart from cylindrical Pyrex vessel in such a way that the distance between the light source and the surface of the solution was 10 cm. The vessel was surrounded by circulating water jacket to control the temperature. In each experiment, a known amount of H_2O_2 and Mohr's salt was added to the requisite volume of dye solution and desired pH of the solution was adjusted with NaOH or HCl as required. A magnetic stirrer was used in order to achieve a homogeneous mixture.

Analytical methods

The pH of the solution was measured by using a pH meter (digital, Orion, Japan). Sample solutions were withdrawn at certain time intervals for spectrophotometric analysis. The decolorization of BGY was measured with UV-visible spectrophotometer (Shimadzu 160 A, Japan) at 416 nm. Chemical structure and absorption maxima are shown in Table 1. Calibration plot based on Beer-Lambert's law was established by relating the absorbance to the concentration. The absorbance at 416 nm due to the color of the dye solution is used to monitor the decolorization of the dye. Mineralization was investigated by monitoring the evolving of CO_2 gas (turned transparent lime water to milky white). Moreover, a peak was observed at 224 nm after assuming decolorization when water was reference but this peak disappeared when nitrate solution was reference. All the experiments were performed at 29ºC.

Table 1. Chemical structure and absorption maxima of BGY.

Name	Chemical structure	λmax (nm)
Brilliant golden yellow (BGY)	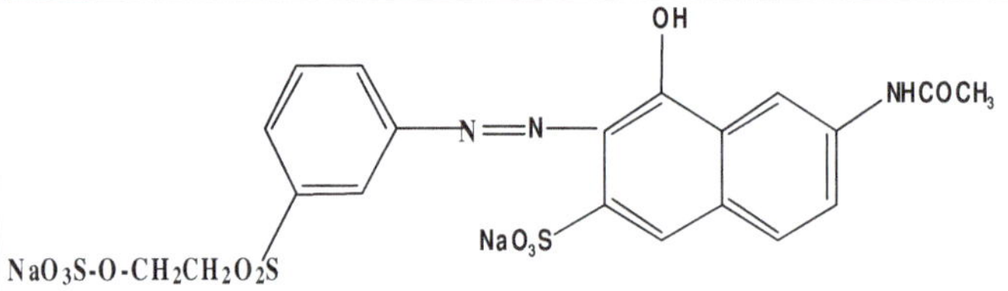	416 224

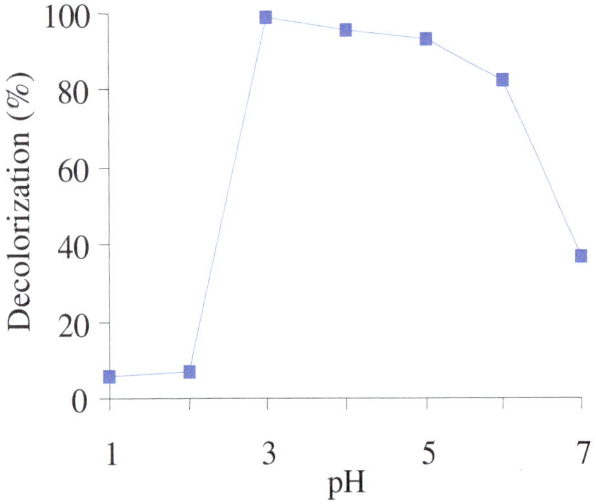

Figure 1. Effect of pH on declorization of BGY in Fenton process (Color online). (BGY)=20 mgL^{-1}, H_2O_2= 400 mgL^{-1}, Fe(II) = 0.05 mmol.

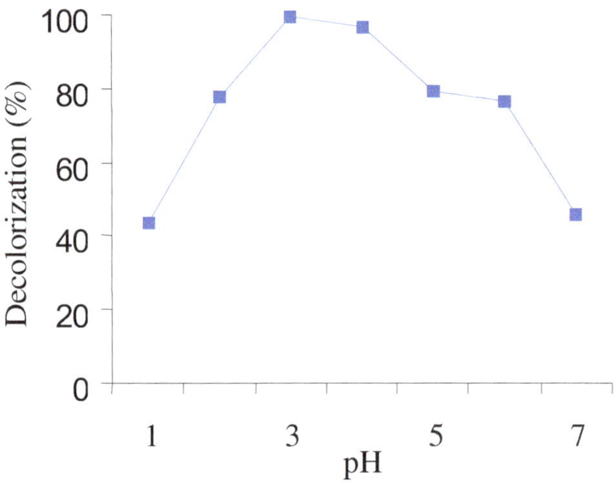

Figure 2. Effect of pH on decolorization of BGY in phot-Fenton process (Color online). BGY = 20 mgL^{-1}, H_2O_2 = 400 mgL^{-1}, Fe(II) = 0.05 mmol, I_0 = 30.3 W m^{-2}

RESULTS AND DISCUSSION

Effect of pH

Solution pH affects the oxidation of organic substances both directly and indirectly. The Fenton and photo-Fenton reactions are strongly pH dependent. The pH values influence the generation of hydroxyl free radicals and thus enhances oxidation efficiency. The effects of pH on Fenton and photo-Fenton processes are shown in Figures 1 and 2, respectively. A maximum decolorization of 99% at 20 min was observed in Fe(II)/H_2O_2 process at pH = 3.0 and decolorization of 99% within 5 min was observed in UV/Fe(II)/H_2O_2 system at the same pH value. As can be seen from Figures 1 and 2, the color removal efficiency increases from 5.6 to 99.0% in Fenton process after 20 min and from 43.6 to 99.2% in photo-Fenton process after 5 min as a function of solution pH from 1 to 3. On the other hand, increasing pH value from 3 to 7 decreases the color removal from 99.0 to 36.45% and 99.2 to 45.8% for Fenton and photo-Fenton processes, respectively. These results are in good agreement with those reported in previous studies (Titus et al., 2004).

At a pH above 4, the decolorization decreases because iron starts to precipitate as hydroxide (Base and Mesmer, 1986). Additionally, the oxidation potential of hydroxyl radical was known to decrease with increasing pH and at a pH below 2, hydrogen peroxide can stay stable, probably solvating a proton to form a hydroxonium ion ($H_3O_2^+$). The hydroxonium ion makes H_2O_2 electrophilic to enhance its stability and presumably to reduce significantly the reactivity with Fe(II) ion (Ghaly et al., 2001; Zhao et al., 2004). Therefore, the amount of hydroxyl radicals would decrease and as a result the removal efficiency decreases. Deng et al. (2000) also found higher degradation rate for dye wastewater by Fenton's and photo-Fenton's processes in a pH range 3.0 - 4.0.

Effect of the initial H_2O_2 concentration

The effect of H_2O_2 concentration on decolorization of BGY by Fenton and photo-Fenton processes are shown in Figures 3 and 4, respectively. The results indicate that

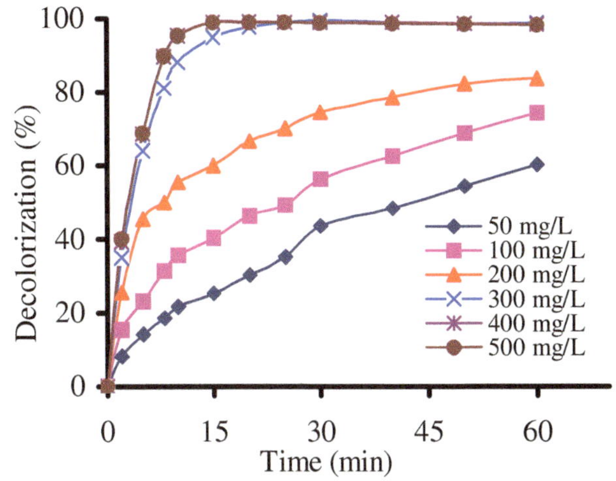

Figure 3. Effect of the initial concentration of H_2O_2 in Fenton process (Color online). [BGY] = 20 mg L^{-1}, [Fe(II)] = 0.05 mmol, pH = 3.

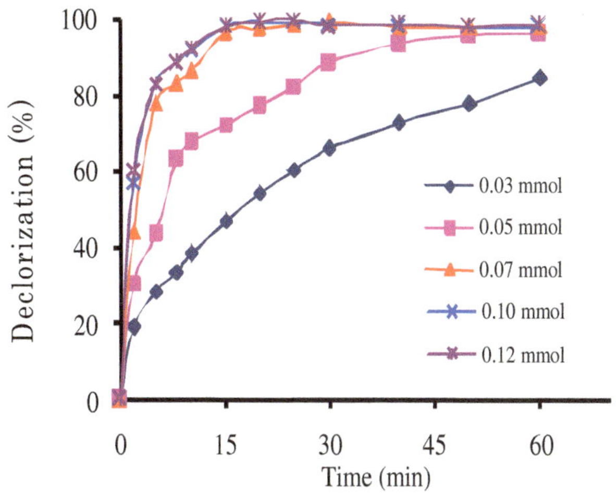

Figure 5. Effect of the initial concentration of Fe(II) in Fenton process (Color online). [BGY] = 20 mg L^{-1}, [H_2O_2] = 400 mg L^{-1}, pH = 3.

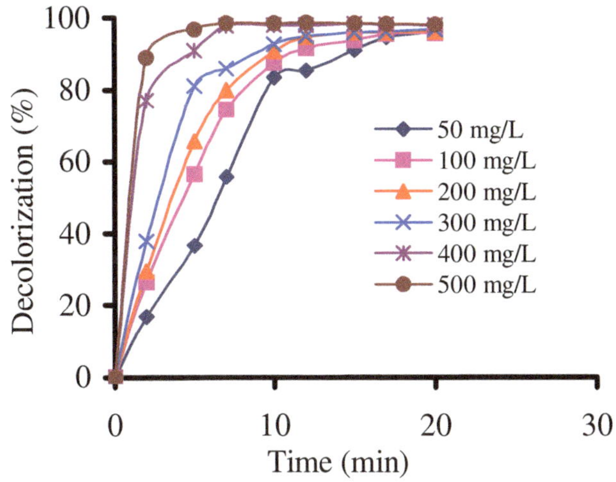

Figure 4. Effect of the initial concentration of H_2O_2 in photo-Fenton process (Color online). [BGY] = 20 mg L^{-1}, [Fe(II)] = 0.05 mmol, pH = 3, I_0 = 30.3 W m^{-2}.

the decolorization of BGY was increased by increasing the concentration of H_2O_2. This is because excess H_2O_2 produce more hydroxyl free radicals resulting in increase in percentage of decolorization. With increasing H_2O_2 concentration from 50 mgL^{-1} to 500 mgL^{-1}, the decolorization percentage increases but above these ranges improvement was not obvious. This may be due to recombination of hydroxyl radicals and also reactions occur between hydroxyl radicals and H_2O_2, contributing to the $HO^{\bullet}$ scavenging capacity (reactions (11)-(13)) (Ghaly et al., 2001).

$$H_2O_2 + HO^{\bullet} \longrightarrow H_2O + HO_2^{\bullet} \quad (11)$$

$$HO_2^{\bullet} + HO^{\bullet} \longrightarrow H_2O + O_2 \quad (12)$$

$$HO^{\bullet} + HO^{\bullet} \longrightarrow H_2O_2 \quad (13)$$

Therefore, it can be concluded that H_2O_2 should be added at an optimum concentration to achieve the highest decolorization rate; hence 400 and 300 mgL^{-1} of H_2O_2 concentrations appear to be an optimum dosage for 40 mg L^{-1} of BGY in Fenton and photo-Fenton processes, respectively.

Effect of Fe(II) concentration

Iron in its ferrous (Fe(II)) or ferric forms Fe(III) acts as photo-catalyst and requires a working pH below 4.0 (Xu et al., 2003). Concentration of Fe(II) ion is one of the main parameters to influence the Fenton's and photo-Fenton's processes. Figures 5 and 6 show the effect of Fe(II) ion concentration on the decolorization of BGY.

As can be seen from Figures 5 and 6, decolorization rate of BGY significantly increased with the increasing of concentration of Fe(II) ion and reaches a maximum at 0.1 mmol and then decreases upon further addition of Fe(II) ion in both the processes. In Fenton's process, generation of hydroxyl free radicals from H_2O_2 depends on the concentration of Fe(II) ion but in photo-Fenton's process, that depends on both the concentration of Fe(II) and intensity of light. Addition of Fe(II) ion above 0.1 mmol in the present study, did not affect the decolorization rate. The decolorization of BGY began to decrease when the concentration of Fe(II) ion was higher than 0.1 mmol.

It is known that Fe(II) had catalytic decomposition effect

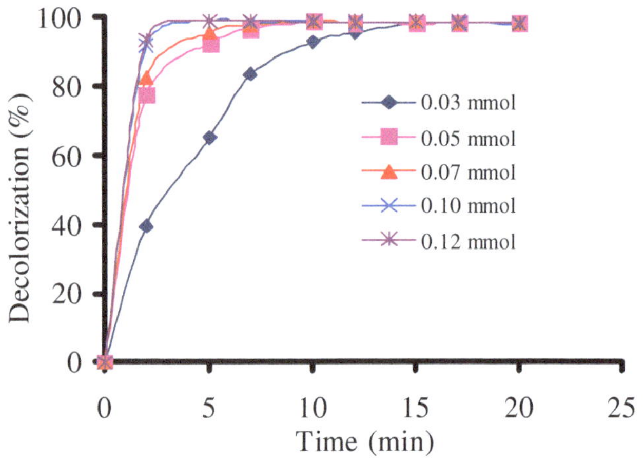

Figure 6. Effect of the initial concentration of H_2O_2 in photo-Fenton process (Color online). [BGY] = 20 mg L^{-1}, [H_2O_2] = 400 mg L^{-1}, pH = 3, I_0 = 30.3 W m^{-2}.

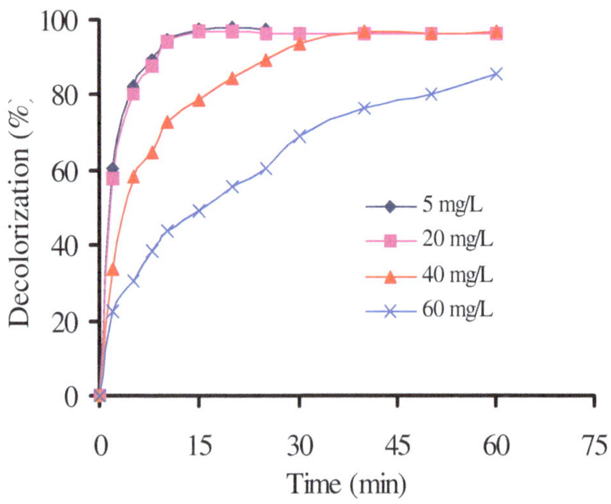

Figure 7. Effect of the initial concentration of BGY in Fenton process (Color online). [Fe(II)] = 0.05 mmol, [H_2O_2] = 400 mg L^{-1}, pH = 3.

on H_2O_2. When Fe(II) concentration is increased, the catalytic effect also accordingly increases and when its concentration is higher, a significant amount of Fe(III) is produced. Then Fe(III) undergoes a reaction with hydroxyl ions to form $Fe(OH)^{2+}$ which has strong absorption tendency and as a result the strength of UV light would decrease (reactions (14)-(16)) (Zhao et al., 2004; Titus et al., 2004).

$$Fe(II) + H_2O_2 \longrightarrow Fe(III) + HO^- + HO^\bullet \quad (14)$$

$$Fe(III) + HO^- \longrightarrow Fe(OH)^{2+} \leftrightarrow Fe(II) + HO^\bullet \quad (15)$$

$$Fe(OH)^{2+} + h\nu \longrightarrow {}^\bullet HO + Fe(II) \quad (16)$$

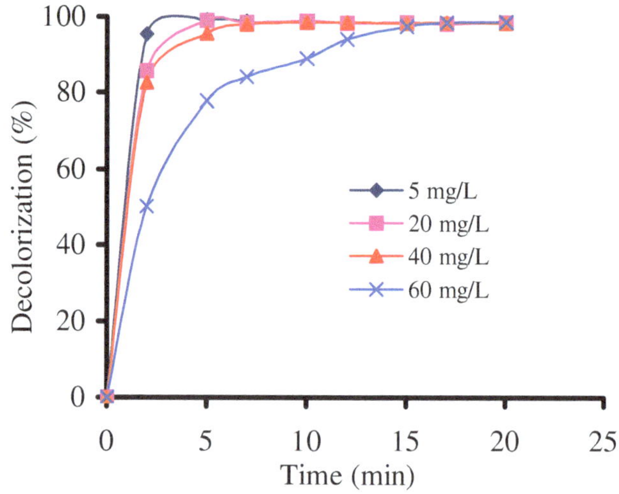

Figure 8. Effect of the initial concentration of BGY in photo-Fenton process (Color online). [Fe(II)] = 0.05 mmol, [H_2O_2] = 400 mg L^{-1}, pH = 3, I_0 = 30.3 W m^{-2}.

Effect of initial BGY concentration

It is important from an application point of view to study the dependence of removal efficiency on the initial concentration of the dye. Therefore, the effect of dye concentration on the decolorization efficiency was investigated at various BGY concentrations. The results are shown in Figures 7 and 8. It was observed that the decolorization decreased with increased in initial concentration of the dye in both the Fenton and photo-Fenton processes. It can be considered that when the initial concentration of the dye is increased, the hydroxyl radical concentrations remain constant for all dye molecules and the removal rate decreases. Once the concentration of the dye increased, it also caused more dye molecules to adsorb light photons and ultimately the efficiency of photodegradation decreases significantly. (Daneshvar et al., 2005).

Effect of salt concentration

Sodium chloride is usually found in wastewater from textile and dying industries. Consequent, a study was carried out to investigate the effects of sodium chloride on degradation of BGY. In the absence of salt, the dye decolorization was 99% in less than 20 and 5 min for Fenton and photo-Fenton processes, respectively. The addition of NaCl (200 mg L^{-1}) to the dye solution caused only a 2% decrease in decolorization. The small decrease in the decolorization is due to the scavenging effect of chloride ion as shown in Reactions (17) and (18) (Muruganandham and Swaminathan, 2004). Further addition of the salt did not reduce the decolorization rate considerably. It is known that the percentage of chloro-/chloro-oxo species of iron ($FeCl_2^+$, $FeOCl_2^-$, $FeCl^{2+}$, $FeCl_3$ etc) in dye solution is much higher compared to hydroxy-species ($Fe(OH)_2^+$, $Fe(OH)^{2+}$, $Fe(OH)_3$ etc) at high chloride concentration and at low pH (e.g., pH = 3) (Base and Mesmer, 1986) which might show increasing trend of decolorization of the dye, on the other hand, chloride ion acts as a scavenger of hydroxyl radicals. Therefore, eventually no effect of the chloride ion on the decolorization was observed upon addition of excess amount of sodium chloride in dye solution.

$$Cl^- + HO^\bullet \longrightarrow HOCl^{\bullet -} \quad (17)$$

$$HOCl^{\bullet -} + H^+ \longrightarrow Cl^\bullet + H_2O \quad (18)$$

The decolorization/degradation results indicate that the dye seems to be completely mineralized after degradation. The aqueous solution after complete mineralization is expected to contain Na^+, NO_3^- and SO_4^{2-} (Li et al., 1997). The carbon dioxide gas, one of the products of mineralization has been detected. Moreover, after mineralization, the aqueous solution showed a peak at about 224 nm as distilled water as reference but no peak was observed at the same wave length (224 nm) when very dilute solution of nitrate salts (e.g., KNO_3) was used as reference.

Conclusions

The results show that Fenton and photo-Fenton are powerful methods for degradation of BGY, but in comparison, photo-Fenton process is more efficient. The results indicated that the degree of decolorization of BGY was strongly affected by the initial concentrations of the dye, Fe(III), H_2O_2 and value of pH. The optimum conditions for mineralization of BGY in Fenton and photo-Fenton processes were observed at pH = 3.0, with an initial concentration of 0.1 mmol Fe(II) and initial H2O2 concentration of 400 mgL^{-1} for Fenton process and 300 mg L-1 for photo-Fenton process with a dye concentration 40 mgL^{-1}.

REFERENCES

Andreozzi R, Caprio V, Insola A, Marotta R (1999). Advanced oxidation processes (AOPs) for water purification and recovery. Catal. Today 53:51-59.

Base Jr CF, Mesmer RE (1986). The Hydrolysis of Cations, Krieger Publishing Company, Florida, USA.

Bauer R, Fallman H (1997). The photo-Fenton oxidation – a cheap and efficient wastewater treatment method. Res. Chem. Intermediat. 23:341-354.

Brown DH, Hitz HR, Schafer L (1981). The assessment of the possible inhibitory effect of dye stuffs on aerobic wastewater: experience with a screening test. Chemosphere 10:245-261.

Daneshvar N, Rabbani M, Modirshahla N, Behnajady MA (2005). Photooxidative degradation of Acid Red 27 in a tubular continuous-flow photoreactor: Influence of operational parameters and mineralization products. J. Hazard. Mater. 118:155-160.

Deng N, Luo F, Xiao M, Wu X (2000). Decoloration of aqueous solution in the UV/Fenton system. Water Res. 34:2408-2411.

Ghaly MY, Hartel G, Mayer R, Haseneder R (2001). Photochemical oxidation of p-chlorophenol by UV/H_2O_2 and photo-Fenton process. A comparative study. Waste Manage. 21:41-47.

Haber F, Weiss J (1934). The catalytic decomposition of hydrogen peroxide by iron salts. Proc. Res. Soc. Series A 147:332-351.

Kang SF, Liao CH, Po ST (2000). Decolorization of textile wastewater by photo-Fenton oxidation technology. Chemosphere 41:1287-1294.

Li ZM, Comfort SD, Shea PJ (1997). Destruction of 2,4,6-Trinitrotoluene by Fenton Oxidation. J. Environ. Qual. 26:480-487.

Malik PK, Saha SK (2003). Oxidation of direct dyes with hydrogen peroxide using ferrous ion catalyst. Sep. Purif. Technol. 31:241-250.

Muruganandham M, Swaminathan M (2004). Decolorization of Reactive Orange 4 by fenton and photo-fenton oxidation technology. Dyes Pigm. 63:315-321.

Neamtu M, Yediler A, Siminiceanu I, Kettrup A (2003). Oxidation of commercial reactive azo dye aqueous solutions by the photo-Fenton and Fenton-like processes. J. Photochem. Photobiol. A Chem. 161:87-93.

Perez M, Torrades F, Domenech X, Peral J (2002). Fenton and photo-Fenton oxidation of textile effluents. Water Res. 36:2703-2710.

Pignatello JJ (1992). Dark and photoassistat Fe^{3+} catalyzed degradation of chlorophenoxy herbicides by hydrogen peroxide. Environ. Sci. Technol. 26:944-951.

Pignatello JJ, Oliveros E, Mackay A (2006). Advanced oxidation processes for organic contaminant destruction based on the Fenton reaction and related chemistry. Crit. Rev. Environ. Sci. Technol. 36:1-84.

Rodrigez M, Sarria V, Esplugas S, Pulgarin C (2002). Photo-Fenton treatment of a biorecalcitrant wastewater generated in textile activities:biodegradibility of the photo-treated solution. J. Photochem. Photobiol. A Chem. 151:129-135.

Solmaz SKA, Birgul A, Ustun GE (2006). Colour and COD removal from textile effluent by coagulation and advanced oxidation processes. Color Technol. 122:102-109.

Swaminathan K, Sandhya S, Carmalin SA (2003). Decolorization and degradation of H-acid and other dyes using ferrous-hydrogen peroxide system. Chemosphere 50:619-625.

Titus MP, Molina VG, Banos MA, Gimenez J, Esplugas S (2004). Degradation of chlorophenols by means of advanced oxidation processes:a general review. Appl. Catal. B:Environ. 47:219-256.

Torrades F, Garcia-Montano J, Garcia-Hortal JA, Nunez L, Domenech X, Peral J (2004). Decolorization and mineralization of homo- and hetro-bireactive dyes under Fenton and photo-Fenton conditions. Color Technol. 120:188-194.

Vaidya AA, Datye KV (1982). Environmental pollution during chemical processing of synthetic fibers. Colourage 14:3-10.

Xu X, Zhao W, Huang Y, Wang D (2003). 2-Chlorophenol oxidation kinetic by photo-assisted Fenton process. J. Environ. Sci. 15:475-481.

Zhao XK, Yang GP, Wang YJ, Gao XC (2004). Photochemical degradation of dimethyl phthalate by Fenton reagent. J. Photochem. Photobiol. A Chem. 161:215-220.

Induced autocatalytic preparation of gold nanostructure

Sameh H. Othman

Nuclear Research Center, Atomic Energy Authority, Cairo, Egypt. E-mail: othman_sameh@yahoo.com.

The present study found that gold nanoparticles with nanosizes below 100 nm can be synthesized by a wet-chemical route, carried out by induced autocatalytic method. The successive liking of the nelghboring nanoparticles seeds by benzenedimethanethiol (BDMT) followed by the induced reduction of Au (III) ions leads to the formation of well defined nanostructure assembly. The size and shape of gold nanoparticles were relatively achieved in a good uniformity. Surface analysis and imaging techniques provided a direct evidence for the successful fabrication of the gold nanostructure. Electrochemical measurements revealed that all of the nanoparticles surfaces are not involved in the adsorption of BDMT.

Key words: Gold nanostructure, benzenedimethanethiol (BDMT), imaging techniques, electrochemical measurements, radiation therapy

INTRODUCTION

Establishing well-defined interconnected arrays of gold nanostructure which can be further addressed is seen to be the key goal for the advancement of nanofabrication methodologies. Gold nanoparticles are amongst the most intensively used form of nanostructure. Their size related properties offer potential applications ranging from electronic to radiation dosimetry and therapy. For example, gold nanoparticles absorb X-rays, γ-radiation and more strongly than most tissues, and causes more dose to be locally deposited (Cho, 2005; Petras et al., 2008; Walker, 2008). In addition, radioactive gold nano-particles present attractive prospects in radiotherapy. They have imageable gamma emissions for dosimetry and pharmacokinetic studies (Cho, 2005; Petras et al., 2008; Walker, 2008). Also, gold nanoparticles are finding their way to neutron materials research facilities (reactors and accelerators) where, in most cases, the materials are not initially radioactive but they become radioactive during the irradiation process (Walker, 2008). Therefore, a controllable preparation of nanostructures with different shapes and exposed surfaces is very important. Gold nanostructures, with unique morphologies, might find use in many areas. Various approaches for fabricating gold nanostructures with various morphologies, such as rods, cubes, prisms, disks and plates have been reported (El-Deab and Ohsaka, 2007).

Two- and three-dimensional arrays of nanoparticles can be readily formed through chemical self-assembly methods (Brust et al., 1995; 1994; Fink et al., 1997; Hostetler and Murray, 1997; Whetten, 1996; Markovich et al., 1997; Dirk, 2005). For instance, alkanethiol capped gold nanoparticles can form well-defined arrays of uniformly spaced nanoparticles with large ordered domain sizes (Brust et al., 1995; 1994; Fink et al., 1997; Hostetler and Murray, 1997). It remains, however, a challenge to 'wire' such assemblies into nanoscale devices (Dirk, 2005). In this context, a promising approach is suggested in this study to fabricate the contacts, followed by self assembly of the nanoparticles networks within the defined contact patterns. The main objectives of the present study are to prepare the gold nanostructures capped by BDMT and to characterize their properties using different electroanalytical as well as microscopic techniques.

MATERIALS AND METHODS

Synthesis of gold nanoparticles

All chemicals used in this study were of analytical grade reagent and were used as received without further purification in laboratory. Gold nanoparticles were prepared by induced autocatalytic method in presence of dithiol. Different volumes of 10 mM aqueous solution of $NaAuCl_4$ (200, 150 and 100 μl) were added dropwisely with continuous vigorous stirring to 5 ml of different samples of ultra super saturated aqueous solution of BDMT. BDMT aqueous solution was prepared by the vigorous stirring addition of 5 ml of 50 mM of ethanol solution of BDMT to deionized water to complete 20 ml. To induce the formation of gold nanoparticles enough amount of

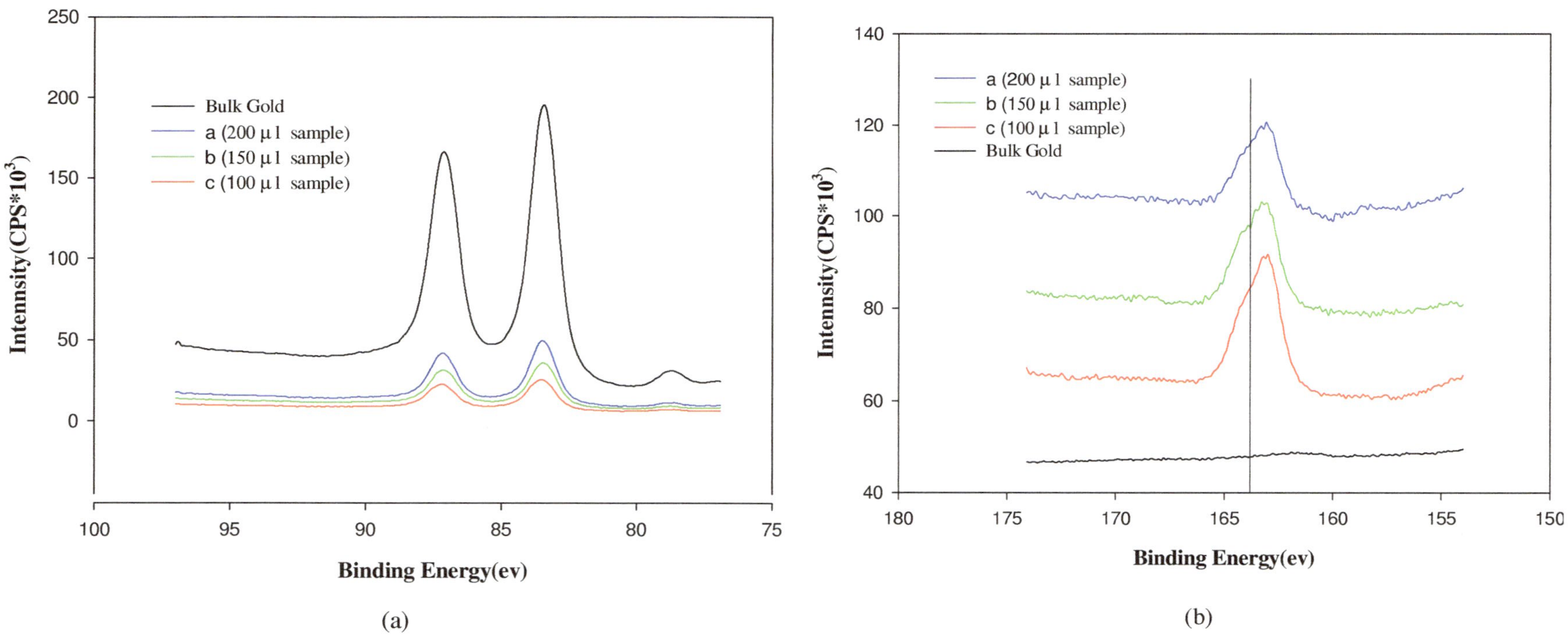

Figure 1. The X-ray photoelectron spectra of BDMT capped Au nanoparticle structures for (a) Au 4f and (b) S 2p regions.

a freshly prepared 1 mM aqueous sodium borohydride solution was added to each one of the previous samples with contentious vigorous stirring.

Characterization

After the fabrication of the gold nanostructures, it was immobilized on different metal substrates (copper and gold sheets or electrode) by immersion of the metal substrate for 30 min in the freshly prepared gold nanoparticles solution. The immobilized gold nanostructure was characterized by atomic force microscope (AFM), scanning electron microscope (SEM), X-ray photoelectron spectroscopy (XPS) and electrochemical measurements.

X-ray photoelectron spectroscopy (XPS) analysis was performed by ESCA3400 electron spectrometer (SHIMADZU) using monochromatized X-ray source with Mg Kα (hv = 1253.6 eV) radiation. Atomic force microscopic (AFM) images were obtained using an AFM Nanoscale Hybrid Microscope VN–8000–KEYENCE in direct contact mode. The SEM images of the electrode surface were observed using a JSM-T220 scanning electron microscope (JEOL, Japan) at an acceleration voltage of 15 kV and a working distance of 4 to 5 mm.

The gold electrode was mechanically polished with aqueous slurries of successively fine alumina powders (down to 0.06 µm) and was sonicated for 10 min in Milli-Q water. The poly-gold electrode was further electrochemically pretreated in N_2-saturated 0.5 M H_2SO_4 solution by cycling the potential scan in the potential range of -0.2 to 1.5 V vs. Ag/AgCl/KCl (sat.) at 100 mV s^{-1} for 10 min or until the cyclic voltammetry (CV) characteristic for a clean poly-gold electrode was obtained prior each measurement. The CV measurement was performed using a computer controlled BAS 100 B/W electrochemical analyzer at room temperature.

RESULTS

Preparation and chemical characterization

XPS spectra of BDMT capped gold nanostructures are shown in Figure 1. Figure 1(a) shows two peaks for gold $4f_{5/2}$ and $4f_{7/2}$ electronic states. The sulfur 2p peaks are shown in Figure 1(b). AFM image for gold sheet and SEM image for copper sheet after the assembly of the gold

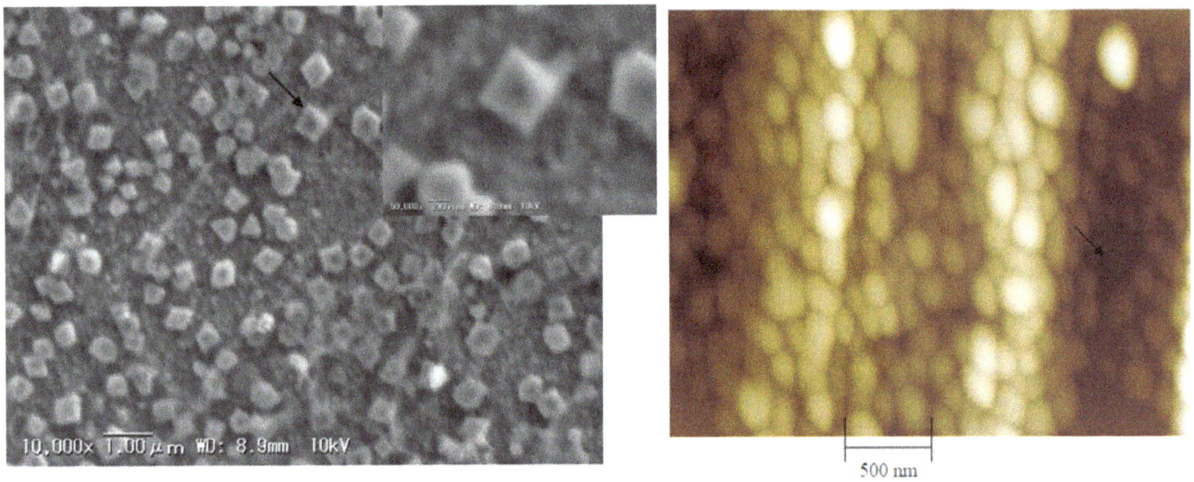

(a) (b)

Figure (2). (a) SEM image for copper sheet and (b) AFM image for gold sheet after the assembly of the gold nanoparticle on their surfaces after immersion for 30 min in the gold nanoparticles solution.

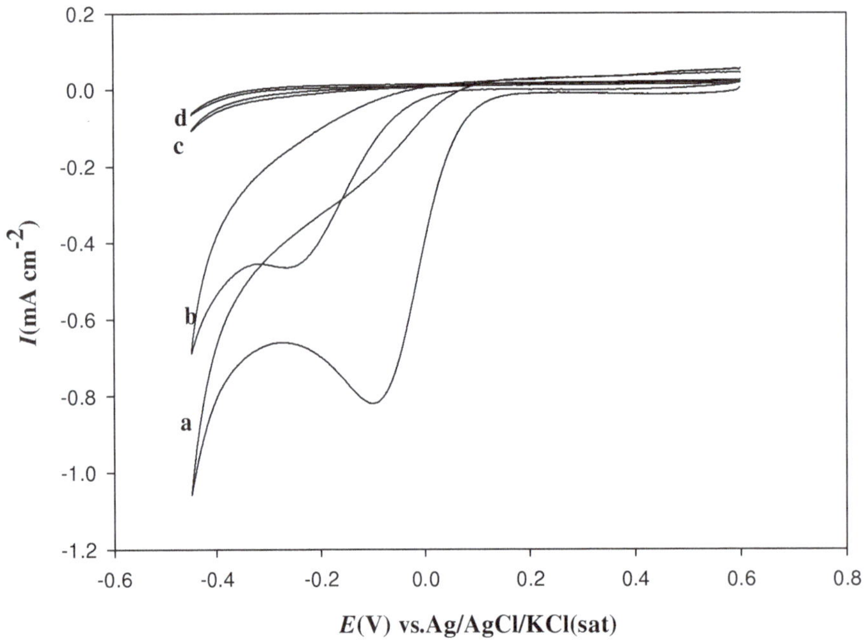

Figure (3). CVs for the ORR at 0.1 Vs^{-1} of ORR in 0.1 M H_2SO_4 for (a) bare gold electrode, (b) gold electrode after immersion for 30 min in gold nanoparticle three-dimensional network solution, (c) gold electrode after immersion for 30 min in gold nanoparticles three-dimensional network solution and then in ultra super saturated aqueous solution of BDMT for each immersion and (d) gold electrode after immersion for 30 min in ultra super saturated aqueous solution of BDMT.

nanoparticle on their surfaces by the immersion for 30 min in the gold nanoparticles solution are shown in Figures 2(a) and 2(b), respectively. The average size of nanoparticle was about of 75 nm, although a range of sizes from 50 to 100 nm were observed.

Figures (3) and (4) show two electrochemical sensitive experiments to characterize self assembled molecules (SAMs) of BDMT capped gold nanostructures on ploy-crystalline gold electrode. Figure 3 shows CVs for oxygen reduction reaction (ORR) at 0.1 Vs^{-1} of ORR in 0.1 M H_2SO_4 for (a) bare gold electrode, (b) gold electrode after immersion for 30 min in gold nanoparticle solution, (c)

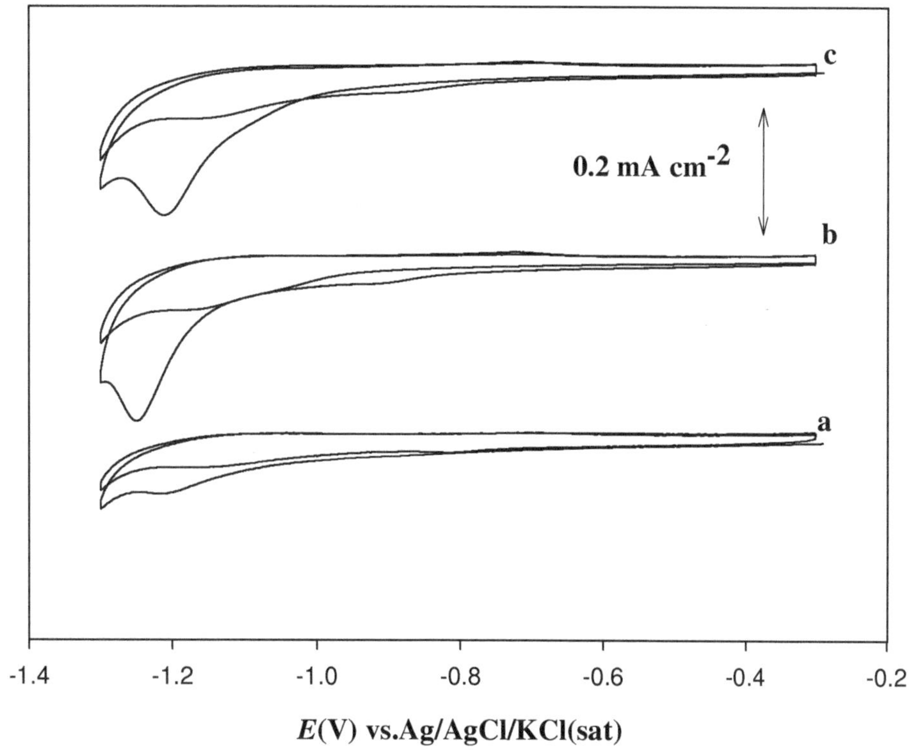

Figures (4). CVs at 50 mVs^{-1} for the reductive desorption in N_2-saturated 0.5 M KOH for (a) gold electrode after immersion for 30 min in gold nanoparticle three-dimensional network solution, (b) gold electrode after immersion in gold nanoparticle three-dimensional network solution and then in ultra super saturated aqueous solution of BDMT for 30 min for each immersion and (c) gold electrode after immersion for 30 min in ultra super saturated aqueous solution of BDMT

gold electrode after immersion in gold nanoparticles solution as well as in ultra super saturated aqueous solution of BDMT for 30 min for each immersion, and (d) gold electrode after immersion for 30 min in ultra super saturated aqueous solution of BDMT. Figure (4) shows the CVs at the sweep rate of 50 mVs^{-1} for the reductive desorption of SAMs in N_2-saturated 0.5 M KOH for (a) gold electrode after immersion for 30 min in gold nanoparticle three-dimensional network solution, (b) gold electrode after immersion in gold nanoparticle three-dimensional network solution and then in ultra super saturated aqueous solution of BDMT for 30 min for each immersion and (c) gold electrode after immersion for 30 min in ultra super saturated aqueous solution of BDMT.

DISCUSSION

The gold particles are grown through two steps for the growth of nanoparticles. In the first step, small gold seeds are generated by the pre-addition of aqueous sodium borohydride. In the second step, the induced formed nanoparticle seeds were bridged with the partially aromatic molecule of benzenedimethanethiol (BDMT). The partially aromatic BDMT is chosen as the bridging molecule since it exhibits higher conductance than a comparable length saturated dithiol molecule. It stabilizes and links the neighboring gold particle seeds which formed by the addition of the one drop of 1 mM aqueous solution of sodium borohydride. Consequently, the liking of the neighboring particles seeds induces further reduction of another Au (III) ion and facilitates the bridging of the gold. The successive liking of the neighboring particles by BDMT followed by the induced reduction of other Au (III) ions leads to the formation of gold nanoparticle having 3-D network as described in the following scheme:

$$n\text{Au (III)} + 3n\, e^- \rightarrow \text{Au}_n \quad (1)$$

$$\text{Au}_n + \text{ArS}_2\text{H}_2 \rightarrow \text{Au}_{n-1}\text{Au}^+(\text{ArS}_2\text{H})^- + \text{H}^+ + e^- \quad (2)$$

$$\text{Au}_n + \text{Au}_{n-1}\text{Au}^+(\text{ArS}_2\text{H})^- \rightarrow \text{Au}_{n-1}\text{Au}^+\text{S}^-\text{-Ar- S}^-\text{Au}^+\text{Au}_{n-1} + \text{H}^+ + e^- \quad (3)$$

$$\text{Au}_{n-1}\text{Au}^+\text{S}^-\text{-Ar-S}^-\text{Au}^+\text{Au}_{n-1} + \text{ArS}_2\text{H}_2 \rightarrow \text{Au}_{n-1}\text{Au}^+\text{S}^-\text{-Ar-S}^-\text{Au}^+\text{Au}_{n-2}\,\text{Au}^+(\text{ArS}_2\text{H})^- + \text{H}^+ + e^- \quad (4)$$

$$\text{Au}_n + \text{Au}_{n-1}\text{Au}^+\text{S}^-\text{–Ar–S}^-\text{Au}^+\text{Au}_{n-2}\ \text{Au}^+(\text{ArS}_2\text{H})^- \rightarrow \ldots\ldots\ldots\ldots \text{(Like reaction (3))} \quad (5)$$

mAu (III) +....... →......... (Like reaction (1)) (6)

Au_m +... →... (Like reaction (2), (3), (4) and (5)) (7)

The formed gold nanostructure should have a uniform size through the uniform growth of the seeds. The gold nanoparticle growth is driven by the diffusion and dramatic effect on surface energy of the seeds. XPS spectra and SEM as well as AFM images provided a confirmation for the successful fabrication of the gold nanostructures according to the suggested scheme in Reactions 1 to 7. Figure 1(a) shows the XPS spectra of BDMT capped gold. The Au $4f_{5/2}$ and $4f_{7/2}$ peaks at binding energy (BE) 87.7 and 84.0 eV, respectively, are characteristic of bulk metallic state of gold. Absence of any observable shift of peak binding energy indicates that the fraction of atoms involved in bonding is low in the clusters (Markovich et al., 1997).

Even for a cluster of 2.8 nm core diameters, only 10% of the total numbers of atoms are bound to thiolate groups (Brust et al., 1994, Fink et al. 1997; Venkataramanan et al., 1999; Jorg et al., 2003) and the absence of Au (I) and Au (II) in the XPS spectrum is not unexpected. Final state effects do not seem to be shifted the peak position greatly in this size range as seen earlier (Brust et al., 1994, Fink et al., 1997; Venkataramanan et al., 1999; Jorg et al., 2003). The increase of the height of the XPS spectra of Au 4f is qualitatively correlated to the amount added of $NaAuCl_4$ solution which can be further confirmed from the sulfur 2p regions shown in Figure 1(b).

The thiolate peak in the clusters, as well as planar monolayer, is observed in the range 162 to 164 eV binding energy (Brust et al., 1994, Fink et al., 1997; Venkataramanan et al., 1999; Jorg et al., 2003). The presence of shoulders within the sulfur peaks indicates the presence of more than one kind of sulfur (Venkataramanan et al., 1999). As shown in Figure, the peak binding energy is shifted from the characteristic thiolate position. The intensity in the high binding energy side of the XPS spectrum to the X-ray beam is attributed to induced damage of the monolayer. Although the nature of the surface sulfur group is unclear, it may be noted that even in relatively clean UHV conditions, photochemical degradation is observed (Fink et al., 1997). X-ray induced disulfide formation could also contribute to the beak width, but it has not been confirmed independently.

Figures 2(a) and 2(b) show the SEM and AFM images, respectively, for copper and gold sheets after the assembly of the gold nanoparticle on their surfaces by the immersion for 30 min in the gold nanoparticles solution. The assembly of the gold nanostructure on the metal substrate can be explained by a reaction similar to Reaction (7). The average size of nanoparticle is about of 75 nm although a range of sizes from 50 to 100 nm are observed. A decrease in the overall surface energy of the of 75 nm although a range of sizes from 50 to 100 nm are observed. A decrease in the overall surface energy of the electrodeposited gold nanostuctures might reasonably account for the observed aggregation of gold nanoparticles. No superlattice formation is evident, which would have been the case of each functional groups of the BDMT molecule. Geometrically this is not possible, as the size of the clusters is much larger than the dimensions of BDMT, and therefore, not all the molecules can be involved in intercluster bonding. However, it is possible for a considerable fraction of the adsorbates to do so, and this should be the reason for the agglomeration of the particles [Figure 1(b)]. In addition, the morphology structure of the gold nanoparticles reflects the geometry of the –S– groups to the underlying substrate and the partial coagulation of the nearby gold nanoclusters during their stages of assembly. The mechanistic details of reaction and the partial coagulation reasons are not well known. It is reasonable to assume, however, that the attachment of the –S– groups results in the inhibition of the growth of the metal clusters in some directions and allows to growth in other directions in a manner similar to the presence of stabilizer around the gold atom. That is, the metal core growth in some directions is prohibited by the binding of the stabilizing ligand –S– group. It was proposed that the presence of a large ligand on the surface of the nanoparticles would result in the retarding of the metal cores growth, as suggested by many authors (Brust et al., 1995; 1994; Fink et al., 1997; Hostetler and Murray, 1997). Using different bulky stabilizer in the media throughout the steps of synthesis may control the size and the crystallographic orientation of the nanoparticle if this would be a consequence of steric crowding of the nanoparticle surface.

The oxygen reduction reaction (ORR) at gold electrodes in alkaline solutions is well known to be very dependent on the crystallographic orientation of the gold surface (El-Deab, 2005). Figure 3 reflects three important catalytic features of the Au electrode loaded with the Au nanoparticles, that is, (i) a significant negative shift of the O_2 reduction peak from -80 mV (in the case of electrode 3(a)) to -250 mV upon loading of the Au particles (electrode 3(b)), (ii) a considerable decrease in the peak current (in the case of electrode 3(b)), and (iii) no catalytic effect for O_2 reduction is observed for electrodes 3(c) and 3(d). At the bulk Au electrode, the reduction peak at -80 mV vs. Ag/AgCl/KCl (sat.) is due to the 2-electron reduction of O_2 to H_2O_2 and the further reduction of H_2O_2 to H_2O was not observed under the present experimental conditions, since the H_2 evolution occurred predominantly and in addition the reduction of H_2O_2 on Au is very slow.

These results suggest that not all the surfaces of the assembled gold nanoparticles and of the gold electrode are involved in the intercluster bonding and that should be the reason for the catalytic effect of the assembled gold nanoparticles on the gold electrode [Figure 3 (b)]. That is, the most important characteristic for the 3-dimontional SAMs is that the surface on which adsorption occurs in the latter is three dimensional as the name itself indicates. This would make the molecules at the surface diverging as they move away from the gold surface and this is not the case of the electrode in Figures 3 (c) and 3 (d). Due to this, there is a definite possibility of the interpenetration of the BDMT molecules to the electrode surface through the assembled gold molecules as occurred to the electrode in Figure 3(c).

In another trial to probe the change of the specific activity of the bulk Au electrode upon loading of the Au nanoparticles, the cyclic voltammograms for the reductive desorption of the SAMs of BDMT were measured. Figure 4(c) shows the existence of one characteristic reductive desorption peaks for the bulk polycrystalline Au/BDMT electrode at ca. -1210 mV and two shoulders appeared in the second cycle of desorption at ca. -1150 and -890 mV versus Ag/AgCl/KCl (sat.). The origin of the peak and shoulders is attributed to the existence of different crystal faces of Au on the polycrystalline Au surface. Similar peaks with lower currents appeared for electrode as shown in Figure. 3(a). As obviously can be observed, there is a difference between the amounts adsorbed of BDMT molecules on the three different electrodes. This can be concluded from the increase in the peak current from electrode 3(a) to electrode 3(b). It is also reflected on the total charge consumed during the reductive desorption of BDMT of Au nanoparticles. The total charge consumed is proportional to the total amount of BDMT adsorbed on the Au electrode. These results confirm the previous discussions about not all of the surfaces are involved in the adsorption of BDMT. In a recent investigation of the adsorption of BDMT on planar surfaces, BDMT is adsorbed with one thiol group in the case of gold, both the thiol groups are bonded with. This difference in the amount adsorbed can be attributed to completely different adsorbate geometries. Consequently, it is possible for a fraction of the adsorbates to do so, and that should be the reason for the catalytic effect of the assembled gold nanoparticles on the gold electrode.

Conclusion

The study suggests that the role of BDMT is key factor for producing the nanostructure gold in certain morphologies. The successful fabrication of polyhedral gold exemplifies the exquisite shape control that can be achieved through careful growth-rate regulation along different crystallographic directions. The gold nanoparticles may be generally applicable to other material systems.

REFERENCES

Brust M Bethell D, Schiffrin J, Kiely J (1995). Novel gold-dithiol nano-networks with nonmetallic electronic-properties. Adv. Mater., 7: 795–7.

Brust M Walker M, Bethell D, Schiffrin J, Whyman R, (1994). Synthesis of thiol-derivatized gold nanoparticles in a 2-phase liquid-liquid system. J. Chem. Soc.-Chem. Commun., 7: 801–2.

Cho S (2005). Estimation of tumor dose enhancement due to gold nanoparticles during typical radiation treatments: a preliminary Monte Carlo study. Phys. Med. Biol., 50: 163-173.

Dirk M (2005). Novel one-dimensional nanogap created with standard optical lithography and evaporation procedures. Nanotechnol., 16: 1983–5.

El-Deab M, Ohsaka T, (2007). Electrocatalysis by nanoparticle: fabrication and electrochemical applications of tailor-design nanoparticles-based electrocatalyst. Electrochem., 75 (11): 858-866.

El-Deab M, Sotomura T, Ohsaka T, (2005). Oxygen reduction at electrochemically deposited crystallographically oriented Au(1 0 0)-like gold nanoparticles. Elec. Comm., 7: 29–34.

Fink J, Kiely J, Bethell D, Schiffrin J (1997). Self assembly of nanosized gold clusters into regular arrays. Inst. Phys. Conf. Ser., 153: 601–604.

Hostetler J, Murray W, (1997). Colloids and self-assembled monolayers. Colloid Interface Sci., 2: 42–50.

Jorg R, Christian M, Peter C, (2003). On the origin of binding energy shifts of core levels of supported gold nanoparticles and dependence of pretreatment and material synthesis. Phys. Chem. Chem. Phys., 5: 172–177.

Markovich G, Leff V, Chung W, Soyez M, Dunn B, Heath R (1997). Parallel fabrication and single-electron charging of devices based on ordered, two-dimensional phases of organically functionalized metal nanocrystals. Appl. Phys. Lett., 70: 3107–309.

Petras J, Wei Chen, Ya-Ping Sun, Manuel C, Roman G, Natalia G, Ingeborg C, (2008). Quantum dots and nanoparticles for photodynamic and radiation therapies of cancer. Adv. Drug Delivery Rev., 60(15): 1600-1400.

Venkataramanan M, Shuguang M, Pradeep T (1999). 3D Monolayers of 1,4-Benzenedimethanethiol on Au and Ag Clusters:Distinct Difference in Adsorption Geometry with the Corresponding 2D Monolayers. J. Coll. Intl. Soc., 216: 134–142.

Walker L (2008). The 53rd Annual Meeting of the Health Physics Society, Pittsburgh, PA.

Whetten L (1996). Nanocrystal gold molecules. Adv. Mater., 8: 428.

New compounds based on anthracene as a good candidate for organic dye-sensitized solar cells: Theoretical investigations

N. Belghiti[1], M. Bennani[1], M. Hamidi[2], S. M. Bouzzine[2] and M. Bouachrine[3*]

[1]Laboratoire de Recherche «Chimie-Biologie appliquées à l'environnement», Faculté des Sciences, Université Moulay Ismail Meknès, Marocco.
[2]URMM/UCTA, Faculté des Sciences et Techniques d'Errachidia, Université Moulay Ismaïl, Maroc
[3]ESTM, Université Moulay Ismail, Meknes, Marocco.

The specific properties of organic-conjugated molecules and polymers are of great importance since they have become the most promising materials for the optoelectronic device technology such as solar cells. The use of low band gap materials is a viable method for better harvesting of the solar spectrum and increasing its efficiency. The control of the parameter of these materials is a research issue of ongoing interest. In this work, quantum chemical investigation was performed to explore the optical and electronic properties of a series of different compounds based on anthracene. Different electron side groups were introduced to investigate their effects on the electronic structure. The theoretical knowledge of the highest occupied molecular orbital (HOMO) and lowest unoccupied molecular orbital (LUMO) energy levels of the components is basic in studying organic solar cells; so the HOMO, LUMO, Gap energy and open circuit voltage (Voc) of the studied compounds have been calculated and reported. These properties suggest these materials as good candidate for organic dye-sensitized solar cells.

Key words: π-Conjugated molecules, anthracene, organic solar cells, density function theory (DFT), low band-gap, electronic properties.

INTRODUCTION

The π-conjugated molecules and polymers based on heterocylcic ring and characterized by an extended π-conjugation is of broad interest in many areas of research, for their applications in organic electronics (Nalwa, 1999; Lim et al., 2006). Particularly in the last decade, optoelectronic device technology (Nalwa, 1999; Lim et al., 2006), such as light-emitting diode (LEDS) (Dimitrakopolous and Malenfrant, 2002), thin film transistors (TFTs) (Kraft et al., 1998; Kim et al., 2006) and low-cost solar cells (Manoj and Narayan, 2003; Wienk et al., 2003) are among these applications. Obtained polymers are highly amorphous, while oligomers are not amorphous and can be synthesized as well defined compounds. Recently, many researchers have become interested in synthesizing short-chain organic OLED compounds based on conjugated oligomers (Müllen and Wegner, 1998; Cornil et al., 1998).

Considerable efforts have been directed towards the study of the electro-optic properties of these conjugated systems, with the aim to exploit their potential as materials for stable and easy processable devices (Sariciftci et al., 1992; Brabec et al., 2001; Roquet et al., 2006). These materials offer advantages over polymeric systems in terms of easy synthesis and purification, and generally exhibit high charge carrier mobility. Therefore, designing and synthesizing molecules with interesting properties play a crucial role in technology (Bundgaard and Krebs, 2007; Brocks and Tol, 1996). At the same time, it is important to understand the nature of the relationship between the molecular structure and the

*Corresponding author. E-mail: bouachrine@gmail.com.

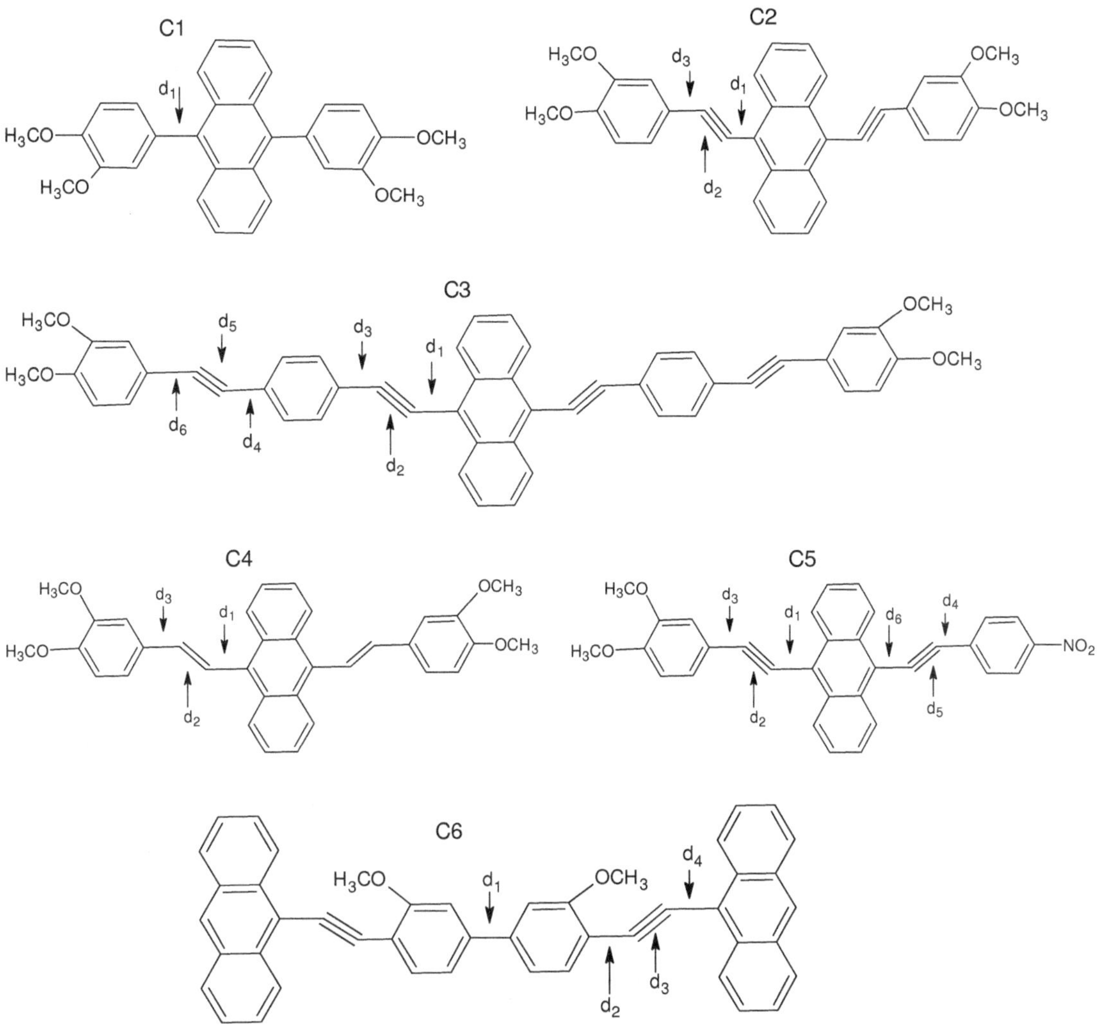

Figure 1. The sketch map of studying structures (C1, C2, C3, C4 C5 and C6).

electronic properties to provide guidelines for the development of new materials. Recent work in this area has been focused on the synthesis and design of new molecules combining donor and acceptor blocks, or conjugated systems with narrow band gaps (Cravino et al., 2002). Theoretical analysis of the electronic structure of conjugated systems can establish the relationships between molecular structure and electronic properties.

Theoretical studies on the electronic structures of π-conjugated compounds have given great contributions to the rationalization of the properties of known materials and to the properties prediction of those yet unknown. In this context, quantum-chemical methods have been increasingly applied to predict the band gap of conjugated systems (Bouzakraoui et al., 2006; Bouzakraoui et al., 2005; Zgou et al., 2008; Bouzzine et al., 2008; Mondal et al., 2010). We note that theoretical knowledge of the highest occupied molecular orbital (HOMO) and lowest unoccupied molecular orbital (LUMO) energy levels of the components is crucial in studying organic solar cells. So, we can save time and money by choosing the adequate organic materials to optimize photovoltaic device's properties. The HOMO and LUMO energy levels of the donor and of the acceptor components for photovoltaic devices are very important factors to determine whether the effective charge transfer will happen between donor and acceptor. The offset of band edges of the HOMO and LUMO levels will prove responsible for the improvement of all photovoltaic properties of the organic solar cells.

Recently, Marrocchi et al. (2010) described the synthesis of a series of anthracene-containing semiconductors (Figure 1). Anthracene derivatives may exhibit large charge carrier mobility and excellent stability. To the best of our knowledge, a systematic theoretical study of such compounds has not been reported.

This work shall further elucidate geometric and electronic properties with the aim to evidence the relationship between molecular structure and electronic properties, thus, driving towards the next syntheses compounds useful as active

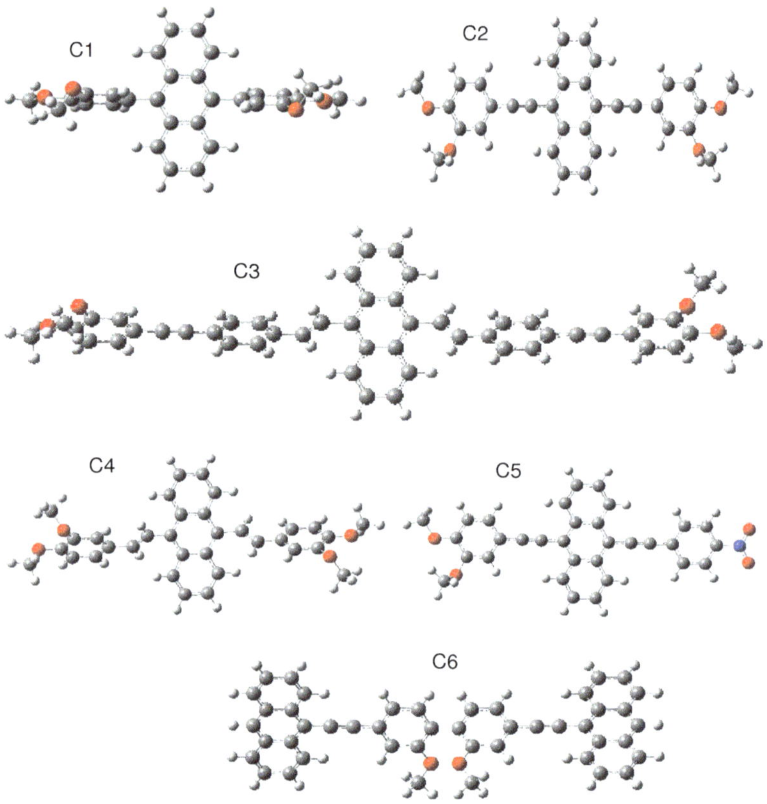

Figure 2. Optimized structure of C1, C2, C3, C4, C5 and C6 obtained by B3LYP/6-31(d) level.

materials in optoelectronic; the quantum chemical investigation has been performed to explore the optical and electronic properties of a series of different compounds based on anthracene. Different electron side groups were introduced to investigate their effects on the electronic structure. The theoretical knowledge of the HOMO and LUMO energy levels of the components is a basis in studying organic solar cells as the HOMO, LUMO and gap energy of the studied compounds have been calculated and reported. Their properties suggest they are good candidates for organic solar cells

THEORETICAL METHODOLOGY

Density function theory (DFT) method of three-parameter compound of Becke (1993) (B3LYP) was used in all the study of the neutral and polaronic compounds. The 6-31G(d) basis set was used for all calculations (Ditchfield et al., 1971; Hehre et al., 1972; Hariharan and Pople, 1974; Gordon, 1980). To obtain the charged structures, we start from the optimized structures of the neutral form. The calculations were carried out using the GAUSSIAN 03 program (Frisch et al., 2003). The geometry structures of neutral and doped molecules were optimized under no constraint. We have also examined HOMO and LUMO levels; the energy gap is evaluated as calculated by the difference between the HOMO and LUMO energies. The ground state energies and oscillator strengths were investigated using the ZINDO/s, calculations on the fully optimized geometries. In fact, these calculation methods have been successfully applied to other conjugated polymers (Meille et al., 1994).

RESULTS AND DISCUSSION

Molecular design and geometry structures

The optimized structures of all studied compounds are shown in Figure 2. It reveals that the π-electron delocalization between the different aromatic units is clear. For the effect of increasing additional conjugated, π-bridge are studied. The addition of end alkoxy chains to backbone might increase the materiel solubility. In addition, molecules C2 and C5 are designed in order to examine the effect of asymmetry in comparison with the corresponding ones and also the effect of the electro accepting group (NO_2) and finally in order to examine the effect of the position of anthracene compounds, molecule C6 and C7 are designed. Finally, in order to study the effect of terminating anthracene moieties on the electronic properties, the doubled end-capped dianthryl compound C6 has also been considered. All the molecular geometries have been calculated with the Hybrid B3LYP function combined with 6-31G (d) basis

Table 1. Geometrical parameters of study compounds C1 to C6 obtained by B3LYP/6-31G(d) in their neutral (N) and doped (D) states.

Compound state	d_1(Å)	d_2(Å)	d_3(Å)	d_4(Å)	d_5(Å)	d_6(Å)
C1(N)	1.497					
C1(D)	1.474					
C2(N)	1.421	1.219	1.417			
C2(D)	1.404	1.227	1.394			
C3(N)	1.422	1.217	1.421	1.464	1.349	1.475
C3(D)	1.415	1.220	1.412	1.464	1.356	1.464
C4(N)	1.467	1.384	1.467			
C4(D)	1.447	1.367	1.445			
C5(N)	1.420	1.219	1.426	1.415	1.220	1.417
C5(D)	1.400	1.229	1.391	1.402	1.223	1.416
C6(N)	1.420	1.218	1.423	1.486		
C6(D)	1.405	1.220	1.411	1.472		

Table 2. Values of HOMO (eV), LUMO (eV) and Gap energy (eV) energies calculated for, C1 to C6 obtained by B3LYP/6-31G(d).

Compound	C1	C2	C3	C4	C5	C6	C7
HOMO-1	-6.75	-6.59	-6.66	-6.71	-6.85	-6.45	-6.61
HOMO	-4.95	-4.67	-4.80	-4.77	-5.11	-4.93	-4.86
GAP	3.47	2.57	2.78	2.95	2.35	2.90	2.20
LUMO	-1.48	-2.09	-2.02	-1.82	-2.75	-2.03	-2.65
LUMO+1	0.57	0.53	-0.002	0.35	-0.25	0.17	0.09

sets using GAUSSIAN 03 program. It was found in other works (Bouachrine et al., 2011) that the DFT-optimized geometries were in excellent agreement with the data obtained from X-ray analysis.

The results of the optimized structures for all studied compounds show that they have similar conformation (quasi planar conformation) (Figure 2). We found that the consecutive units have similar dihedral angles (180°) and inter-ring distances (d_1=1.424 Å, d_2=1.259 Å, d_3=1.387 Å, d_7=1.440 Å, d_8=1.3689 Å, d_9=1.482 Å) (Table 1). The incorporation of several groups does not change these parameters.

On the other hand, it is interesting to study how the p-doped π-conjugated molecule becomes the ultimate responsible of charge transport. As earlier said, to obtain oxidized optimized structure, we started from the optimized structure of the neutral form. We can conclude that during the doping process and for all studied compounds the simple bonds become shorter, while the double ones become longer. The inter-rings bonds are longer than normal double bonds. A quinoid-like distortion emerges as a result of the oxidation. These results are consistent with the ab-initio HF and DFT calculations performed by Casado et al. (1997) and Bouzzine et al. (2005) for substituting oligothiophenes. The optimized geometry of the cationic compound indicates the formation of the positive polaron defect localized in the middle of the molecule and extending over the adjacent repeat units. The charged species are characterized by a reversal of the single double C-C bond pattern; the geometry process thus induces the appearance of a strong quinoid character within the molecule.

Electronic parameters

Electronic structures are fundamental to the interpretation and understanding of the absorption spectra. The calculated frontier orbital energies (four occupied orbital and four unoccupied orbital) and energy gaps between HOMO and LUMO are shown in Table 2. As shown in Table 2, one can notice that all studied molecules (C2 to C6) exhibit destabilization of the LUMO and stabilization HOMO levels in comparison with those of compound C1. The HOMO and LUMO energies of C1 to C6 change significantly, (-4.96 and -1.48 eV, -4.67 and -2.09 eV, -4.80 and -2.02eV, -4.77 and -1.82eV, -5.11 and -2.76eV, and -4.93 and -2.03eV, respectively). It can also be found that, the HOMO and LUMO energies of the studied compound are slightly different. This implies that different structures play key roles on electronic properties and enhancing the electron-accepting ability and the effect of

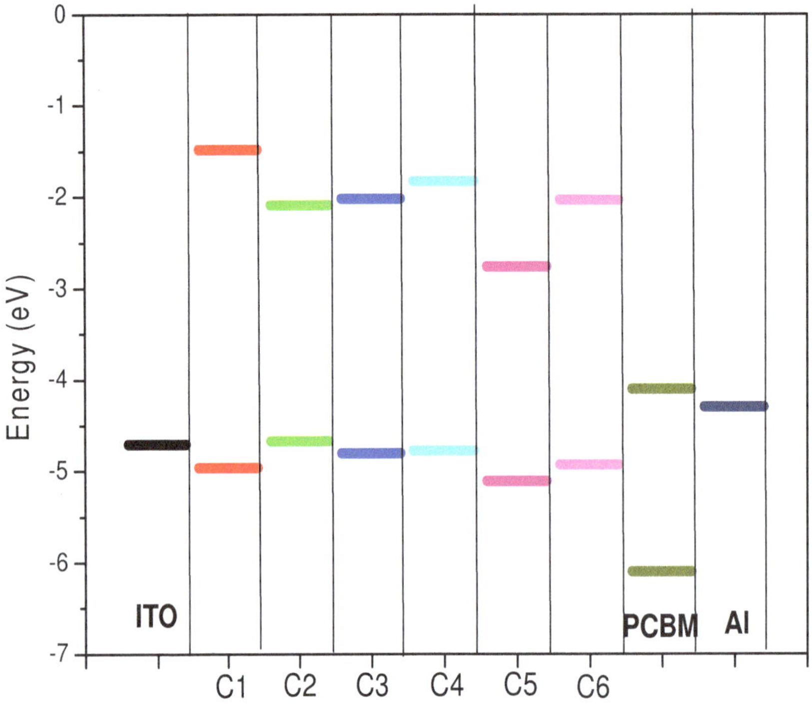

Figure 3. Data of the absolute energy of the frontier orbitals HOMO and LUMO for the studied molecules and ITO, PCBM and the aluminum (Al).

the symmetry of ethylene-thiophene branch on the HOMO and LUMO energies cannot be neglected. In addition, the energies of E_{gap} C1 to C6 differ slightly from 2.35 to 3.47eV depending on the different structures. They are studied in the following order C1 > C4 > C6 > C3 > C2 > C5.

It should be emphasized that C2 and C5 are asymmetric, one can see that the energies of HOMO and LUMO for the C5 molecule (HOMO: -5.11 eV, LUMO: -2.76 eV) are obviously lower than that of the molecule C2 (HOMO: -4.67 eV, LUMO: -2.09 eV) and the energy gap between HOMO and LUMO of C5 is also lower than that of C2 with a lower energy gap (2.35 eV). This may be attributed to the presence of electron-attracting NO_2 side groups. On the other hand, the comparison between C2 and C4 show that the replacement of two triple with two double bonds causes a decrease of band gap and a net destabilization of both HOMO and LUMO levels. This is in agreement with what was found in experimental results (Marrocchi et al., 2010).

The calculated band gap E_{gap} of the studied compound increases in the following order: C1 > C4 > C6 > C3 > C2 > C5. Figure 3 shows detailed data of absolute energy of the frontier orbitals for studying compounds, Indium tin oxide (ITO), [6.6]-phenyl-C61-butyric acid methyl ester (PCBM) (substituted C_{60}) and aluminum (Al) is included for comparison purposes. It was deduced that substitution pushes up/down the HOMO/LUMO energies in agreement with their electron acceptor character. To evaluate the possibilities of electron transfer from the excited studied molecules to the conductive band of PCBM, we have studied the photovoltaic properties of the compounds C1 as donor blended with PCBM, which is the most broadly used as an acceptor in solar cell devices. As shown in Table 3, the change of the electron-donor shows a great effect on the HOMO and on the LUMO levels. The experiment phenomenon was quite consistent with previous literature (Derouiche and Djara, 2007; Zhang et al., 2008), which reported that the increase of the HOMO levels may suggest a negative effect on organic solar cell performance due to the broader gap between the HOMO level of the organic molecules and the LUMO level of PCBM (open circuit voltage, Voc). The theoretical values of Voc were calculated from the following expression:

$$Voc = E_{HOMO}\,(\text{Donor}) - E_{LUMO}\,(\text{Acceptor}) - 0.3$$

As shown in Figure 3 and knowing that in organic solar cells, the Voc is found to be linearly dependent on the HOMO level of the donor and the LUMO level of the acceptor (Gunes et al., 2007). The calculated Voc of the studied molecules range from 1.14 to 1.58 eV; these values are sufficient for a possible efficient electron injection. Therefore, all the studied molecules can be used as sensitizers because the electron injection process from the excited molecule to the conduction band of PCBM and the subsequent regeneration is

Table 3. Energy values of E_{LUMO} (ev), E_{HOMO} (ev) and the Voc (ev).

Compound	E_{LUMO} (ev)	E_{HOMO} (ev)	V_{oc} (ev)
C1	-1.488	-4.960	1.434
C2	-2.096	-4.672	1.146
C3	-2.025	-4.806	1.280
C4	-1.820	-4.774	1.248
C5	-2.760	-5.112	1.586
C6	-2.033	-4.939	1.413
PCBM	-3.226	-5.985	-

Table 4. Values of Gap energy (eV) of the several studied compounds in their neutral and doped forms.

Compound	E_{gap} (eV) neutral	E_{gap} (eV) polaronic
C1	3.47	2.87
C2	2.57	2.26
C3	2.78	2.60
C4	2.95	2.35
C5	2.35	2.32
C6	2.90	2.74
C7	2.20	2.01

possible in an organic sensitized solar cell.

On the other hand, and since the p-doped π-conjugated molecule has becomes the ultimate responsible of charged transport, the values of the energy of the HOMO, LUMO, E_{gap} (E_{HOMO} - E_{LUMO}) for several compounds in their neutral and polaronic forms were determinate and shown in Table 4. Available values of the gap energy in neural state are presented also in this table for comparison. Traditionally and until now, charged state in conjugated molecules has been discussed in terms of one-electron band model. We note that the energy gap decreases when passing from the neutral to the doped form for all oligomers.

Finally, it is important to examine the HOMO and the LUMO for these compounds because the relative ordering of occupied and virtual orbital provides a reasonable qualitative indication of excitations properties (Yang et al., 2005). In general, as shown in Figure 4 (LUMO, HOMO), the HOMOs of these oligomers in the neutral form possess a π-bonding character within subunit and a π-anti-bonding character between the consecutive subunits; while the LUMOs possess a π-anti-bonding character within subunit and a π-bonding character between the subunits, whereas it is the opposite in the case of doped forms.

Absorption and electronic properties

Based on the optimized molecular structures with B3LYP/6-31G(d) method, we have calculated the ultraviolet-visible (UV-Vis) spectra of the studied compounds C1, C2, C3, C4, C5 and C6 using ZINDO/s method. As shown in Table 5, we can find the values of calculated wavelength λ_{max} and for comparison experimental ones excitation to the S_1 state corresponds almost exclusively to the promotion of an electron from the HOMO to the LUMO orbital. The absorption wavelengths arising from $S_0 \rightarrow S_1$ electronic transition increase progressively with the increasing of conjugation lengths. It is reasonable, since HOMO→LUMO transition is predominant in $S_0 \rightarrow S_1$ electronic transition; the results are a decrease of the LUMO and an increase of the HOMO energy.

These values are calculated by ZINDO method starting with optimized geometry obtained at B3LYP/6-31G(d) level. However, we believe that the bulk of intermolecular effect must be taken into account. This effect is the source of the deviation between the calculation and the experimental results.

We can remark for comparing calculated and experimental results (Marrocchi et al., 2010) a linear relationship between calculated and experiment results. Therefore, the DFT theoretical calculations give a good description of the absorption properties of the studied compound and can be employed to predict the electronic characteristics of other materials. It should be noted that the difference between theoretical and experimental values can be explained by the fact that the calculations assume that the molecules are in the vapour state.

Another considered point is that the position of λmax shows a bathochromic shift when passing from C1 to C6,

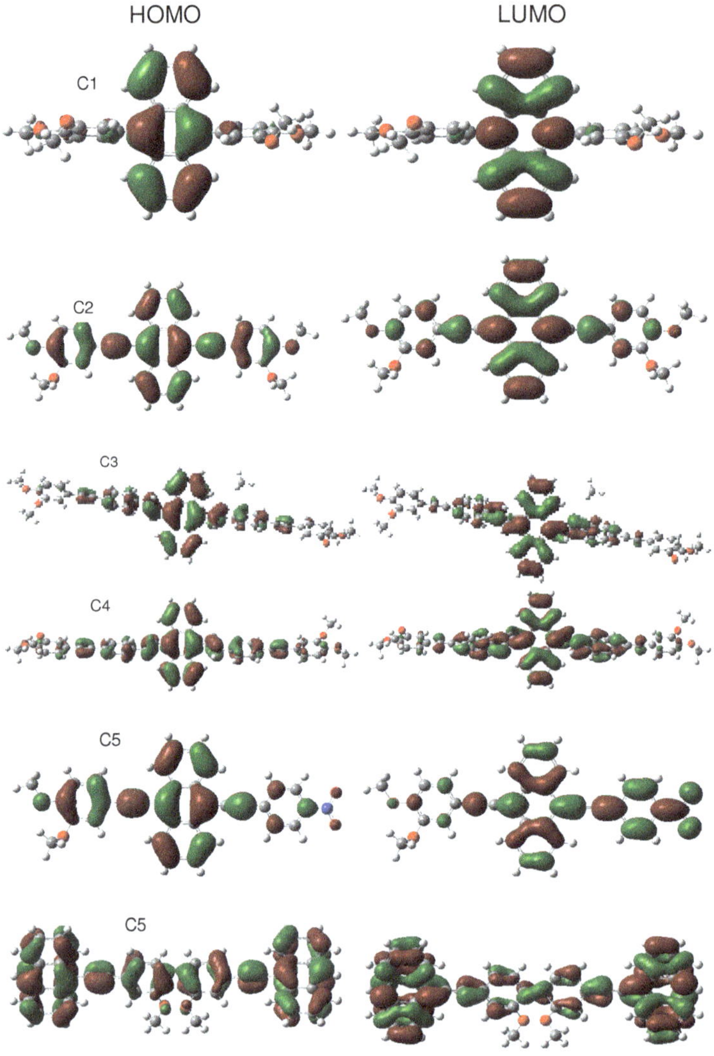

Figure 4. The contour plots of HOMO and LUMO orbitals of study compounds C1 to C6 in neutral form.

which also can be seen, respectively in C1 (397.54 nm), C2 (494.19 nm), C3 (460.17 nm), C4 (447.35 nm), C5 (514.20 nm) and C6 (442.76 nm) due to the increasing of the extended conjugation through the system of aryl groups and multiple bonds. The asymmetrically substituted compounds C5, containing a nitro-group, show absorption spectra red-shifted of about 20 nm with respect to the symmetric analogous to C2 and finally the presence of two anthracene units in C6 does not cause a significant red shift of the absorption. Those interesting points are seen both in the theoretical and experimental results.

Conclusion

In this study, theoretical analysis of the geometries and electronic properties of several various compounds based on the anthrance was performed using the DFT/B3LYP method. The effect of substituted groups on the structural and opto-electronic properties of these materials and the possibility of their application as organic solar cells are discussed. The concluding remarks are:

1) The results of the optimized structures for all studied compounds so that they have similar conformations (quasi planar conformation). We found that the incorporation of several groups does not change the geometric parameters.
2) During the doping process and for all studied compounds the simple bonds become shorter, while the multiple ones become longer. The inter-rings bonds are longer than normal double bonds. A quinoid-like distortion emerges as a result of the oxidation
3) The calculated frontier orbital energies HOMO and

Table 5. Absorption λ_{abs} (nm) obtained by the ZINDO/DFT method.

Compound	λ_{max} (nm) ZINDO	OS	λ_{max} exp (nm)	
			Chloroform	Cyclohexane
C1	397.54	0.4317	398	375, 395
C2	494.49	1.1089	475	478
C3	460.17	1.5308	464	460
C4	447.35	0.9057	418	413
C5	514.20	1.3002	494	492
C6	442.76	1.5183	444	439

LUMO and energy gaps showed that the energy gaps of the studied molecules differ slightly from 2.35 to 3.47 eV depending on the different structures. The calculated band gap E_{gap} of the studied compound increases in the following order C1 > C4 > C6 > C3 > C2 > C5.

4) The replacement of two triple with two double bonds causes a decrease of band gap and a net destabilization of both HOMO and LUMO levels

5) All the studied molecules can be used as sensitizers because the electron injection process from the excited molecule to the conduction band of PCBM and the subsequent regeneration are feasible in the organic sensitized solar cell. This calculation procedure can be used as a model system for understanding the relationships between electronic properties and molecular structure and also can be employed to explore their suitability in electroluminescent devices and in related application. Presumably, the procedures of theoretical calculations can be employed to predict and assume the electronic properties on yet prepared and efficiency proved other materials, and to further to novel materials for organic solar cells.

ACKNOWLEDGEMENTS

This work was supported by Volubilis Program (N° MA/11/248), and the convention CNRST/CNRS (Project chimie1009). We are grateful to the "Association Marocaine des Chimistes Théoriciens" (AMCT) for its pertinent help concerning the programs

REFERENCES

Becke AD (1993). Density-functional thermochemistry. III. The role of exact exchange J. Chem. Phys. 98:5648-5653.

Bouachrine M, Echchary A, Bouzzine SM, Amine M, Hamidi M, Amine A, Zair T (2011). Oligothiophenes bridged by silicon groups, DFT study of structural and electronic properties. Phys. Chem. News. 58: 61-66.

Bouzakraoui S, Bouzzine SM, Bouachrine M, Hamidi M (2005). Density functional theory study of conformational and opto-electronic properties of oligo-para-phenylenes. J. Mol. Struct. (Theochem). 725:39-44.

Bouzakraoui S, Bouzzine SM, Bouachrine M, Hamidi M (2006). Theoretical investigation of electroluminescent alkoxy substituted 4,4′-bis(2-phenylethenyl)biphenyls as guest in blue OLEDs; Sol. Energy Mater Sol. Cells. 90:1393-1402.

Bouzzine SM, Bouzakraoui S, Bouachrine M, Hamidi M (2005). Density functional theory (B3LYP/6-31G*) study of oligothiophenes in their aromatic and polaronic states. J. Mol. Struct. (Theochem). 726:271-276.

Bouzzine SM, Makayssi A, Hamidi M, Bouachrine M (2008). Bridging effect on structural and optoelectronic properties of oligothiophene. J. Mol. Struct. (Theochem). 851:254-262.

Brabec CJ, Sariciftci NS, Hummelen JC (2001). Plastic Solar Cells. Adv. Funct. Mater. 11: 15-26. Zhou Y, Peng P, Han L, Tian W (2007). Novel donor–acceptor molecules as donors for bulk heterojunction solar cells. Synth. Met. 157: 502-507.

Brocks G, Tol A (1996). Small Band Gap Semiconducting Polymers Made from Dye Molecules:∞ Polysquaraines. J. Phys. Chem. 100(5):1838-1846.

Bundgaard E, Krebs FC (2007). Low band gap polymers for organic photovoltaics. Sol. Energy Mater. Sol. Cells 91:954-985.

Casado J, Ramirez FJ, Hotta S, Lopez Navarrete JT, Hernandez V (1997).Vibrational Spectroscopy study of doping induced charged defects in a series of α,α'-Dimethyl end-capped Oligothiophenes. Synth. Met. 84:571-572.

Cornil J, Beljonne D, Brédas JL, Mûllen K, Wegner G (Eds.) (1998). Electronic Materials: The oligomers Approach, Wiley-VCH, Weinheim, New York. pp. 432-447.

Cravino A, Zerza G, Neugebauer H, Maggini M, Bucella S, Menna E, Svensson M, Andersson MR, Brabec CJ, Sariciftci NS (2002). Electrochemical and Photophysical Properties of a Novel Polythiophene with Pendant Fulleropyrrolidine Moieties: Toward "Double Cable" Polymers for Optoelectronic Devices. J. Mater. Chem. 12:1931-1943.

Derouiche H, Djara V (2007). Impact of the energy difference in LUMO and HOMO of the bulk heterojunctions components on the efficiency of organic solar cells. Sol. Energy Mater. Sol. Cells. 91:1163-1167.

Dimitrakopolous CD, Malenfrant PRL (2002). Organic Thin Film Transistors for Large Area Electronics. Adv. Mater. 14:99-117.

Ditchfield R, Hehre WJ, Pople JA (1971). Self-Consistent Molecular-Orbital Methods. IX. An Extended Gaussian-Type Basis for Molecular-Orbital Studies of Organic Molecules J. Chem. Phys. 54:724-728.

Frisch MJ, Trucks GW, Schlegel HB, Scuseria GE, Robb MA, Cheeseman JR, Montgomery JA, Vreven T, Kudin KNJr, Burant JC, Millam JM, Iyengar SS, Tomasi J, Barone V, Mennucci B, Cossi M, Scalmani G, Rega N, Petersson GA, Nakatsuji H, Hada M, Ehara M, Toyota K, Fukuda R, Hasegawa J, Ishida M, Nakajima T, Honda Y, Kitao O, Nakai H, Klene M, Li X, Knox JE, Hratchian HP, Cross JB, Adamo C, Jaramillo J, Gomperts R, Stratmann RE, Yazyev O, Austin AJ, Cammi R, Pomelli C, Ochterski JW, Ayala PY, Morokuma K, Voth GA, Salvador P, Dannenberg JJ, Zakrzewski VG, Dapprich S, Daniels AD, Strain MC, Farkas O, Malick DK, Rabuck AD, Raghavachari K, Foresman JB, Ortiz JV, Cui Q, Baboul AG, Clifford S, Cioslowski J, Stefanov BB, Liu G, Liashenko A, Piskorz P, Komaromi I, Martin RL, Fox DJ, Keith T, Al-Laham MA, Peng CY,

Nanayakkara A, Challacombe M, Gill PMW, Johnson B, Chen W; Wong MW, Gonzalez C, Pople JA (2003). GAUSSIAN 03, Revision B. 04; Gaussian: Pittsburgh, PA.
Gordon MS (1980). The isomers of silacyclopropane. Chem. Phys. Lett. 76:163-168.
Gunes S, Neugebauer H, Sariciftci NS (2007). Conjugated Polymer-Based Organic Solar Cells. Chem. Rev. 107:1324-1338.
Hariharan PC, Pople JA (1974). Accuracy of AH n equilibrium geometries by single determinant molecular orbital theory. Mol. Phys. 27:209-214.
Hehre WJ, Ditchfiesld R, Pople JA (1972). Self-Consistent Molecular Orbital Methods. XII. Further Extensions of Gaussian-Type Basis Sets for Use in Molecular Orbital Studies of Organic Molecules. J. Chem. Phys. 56:2257-2261.
Kim DH, Ohshita J, Lee KH, Kunugi Y, Kunai A (2006). Synthesis of π-Conjugated Oligomers Containing Dithienosilole Units. Organometallics 25:1511-1516.
Kraft A, Grimsdale AC, Holmes AB (1998). Electroluminescent Conjugated Polymers-Seeing Polymers in a New Light. Angew. Chem. Int. Ed. 37:402-428;
Lim E, Jung BJ, Shim HK, Taguchi T, Noda B, Kambayashi T, Mori T, Ishikawa K, Takezoe H, Do LM (2006). Nanoscale thin-film morphologies and field-effect transistor behavior of oligothiophene derivatives. Org. Electron. 7:121-31.
Manoj AG, Narayan KS (2003). Photovoltaic properties of polymer p-n junctions made with P3OT/BBL bilayers. Opt. Mater. 21:417-420.
Marrocchi A, Spalletti A, Ciorba S, Seri M, Elisei F, Taticchi A (2010). Synthesis and photophysical properties of conjugated anthracene-based compounds. J. Photochem. Photobiol. A: Chem. 211:162-169.
Meille SV, Farina A, Bezziccheri F and Gallazi MC (1994). Green Electroluminescence in Poly-(3-cyclohexylthiophene) light-emitting diodes. Adv. Mater. 6:488-490.
Mondal R, Becerril HA, Verploegen E, Kim D, Norton JE, Miyaki SKo, Lee S, Toney MF, Bredas JL, McGehee MD, Bao Z (2010). Thiophene-rich fused-aromatic thienopyrazine acceptor for donor–acceptor low band-gap polymers for OTFT and polymer solar cell applications. J. Mater. Chem. 20:5823-5834.
Müllen K, Wegner G (Eds) (1998). Electronic Materials: The oligomers Approach, Wiley-VCH, Weinheim, New York. pp. 105-197.
Nalwa HS, Ed. John Wiley and Sons (1999). Handbook of Organic Conductive Molecules and Polymers.
Roquet S, Cravino A, Leriche P, Aleveque O, Frere P, Roncali J (2006). Triphenylamine–Thienylenevinylene Hybrid Systems with Internal Charge Transfer as Donor Materials for Heterojunction Solar Cells. J. Am. Chem. Soc. 128:3459-3466.
Sariciftci NS, Smilowitz L, Heeger AJ, Wudl F (1992). Photoinduced electron-transfer from a conducting polymer to Buckminster fullerene. Science. 258:1474-1476.
Wienk MM, Kroon JM, Verhees WJH, Knol J, Hummelen JC, Van Hal PA (2003). Janssen RA. Angew. Chem. Int. Ed. 42:3371-3375.
Yang L, Feng J, Ren A (2005). Theoretical studies on the electronic and optical properties of two thiophene–fluorene based π-conjugated copolymers. Polymer. 46:10970-10981.
Zgou H, Bouzzine SM, Bouzakraoui S, Hamidi M, Bouachrine M (2008). Theoretical study of structural and electronic properties of oligo(thiophene-phenylene)s in comparison with oligothiophenes and oligophenylenes. Chin. Chem. Lett. 19:123-126.
Zhang L, Zhang Q, Ren H, Yan H, Zhang J, Zhang H, Gu J (2008). Calculation of band gap in long alkyl-substituted heterocyclic-thiophene-conjugated polymers with electron donor–acceptor fragment. Sol. Energy Mater. Sol. Cells. 92:581-587.

22

Microstructure contradistinction for indirect/direct desulfurization product layer of limestone particle in coal-fired flue gas

Jian-yu Shang[1] , Zhong-liang Liu[1] and Chun-bo Wang[2]

[1]College of Environmental and Energy Engineering, Beijing University of Technology, Beijing City, Chaoyang, 100124, China.
[2]School of Energy and Power Engineering, North China Electric Power University, Hebei Province, Baoding 071003, China.

The micro morphology of limestone sorbent is a key for improving the SO_2 removal efficiency in the coal-fired power plant. The tube furnace system was built for imitating the indirect desulfurization reaction in traditional boiler atmosphere (65%N_2, 15%CO_2, 4.7%O_2, 15%H_2O, 0.3%SO_2) and the direct desulfurization reaction in oxygen-fired atmosphere (80%CO_2, 4.7%O_2, 15%H_2O, 0.3%SO_2). The reacted limestone particle samples were observed by SEM and are analyzed by Fourier transformation 3-D reconstruction technique. It is shown that the sizes of internal pore within sorbent particle in two kind atmospheres differ by an order of magnitude, one is about 200 *nm* for indirect desulfurization, and the other is about 20 *nm* for direct desulfurization. The $CaSO_4$ products of indirect desulfurization are chaotic derivatives with serious blockage, and the direct desulfurization product layer belongs to uniform honeycombed micro-structure without blockage. Those numerous micro-porous passages can guarantee continuous reactive substance transmission.

Key words: Limestone powder, SO_2 removal, particle surface, 3-D reconstruction, product layer.

INTRODUCTION

The hazards to human survival from free SO_2 gas molecules in the atmosphere continue to get the attentions of scientists from various countries (Sipila et al., 2010; Mihaela et al., 2010; Ramírez et al., 2008). The limestone powder sorbent is widely used in the dry SO_2 removal system in many large coal-fired power plants. Especially in recent years, the circulating fluidized bed (CFB) boiler oxy-fuel combustion technology is emerging rapidly in many countries including china, which can effectively control the emissions of three gaseous pollutions (NO_x, SO_2, CO_2) at O_2/CO_2 atmosphere (Ramon and Carmen, 2008; Krzywanski et al., 2010). The SO_2 removal efficiency of this technology, which still takes the limestone as sorbent, should be higher than that of the dry limestone injection pattern in the traditional coal-fired boiler, and the latter is 40% (Winaya and Basu, 2001; Avila et al., 2012). The reason for this difference in the desulfurization efficiency is that the researchers tried to find, especially its microscopic mechanism (Sınırkaya et al., 2012). In the traditional coal-fired boiler, air is the fuel accelerant, and the partial pressure of CO_2 in its flue gas is low about 14 to 16%. So from the chemical point of view under such a low partial pressure of CO_2, the limestone will be firstly decomposed thermally at very high temperature, and then take part in the desulfurization reaction (Ramón and Carmen, 2009).

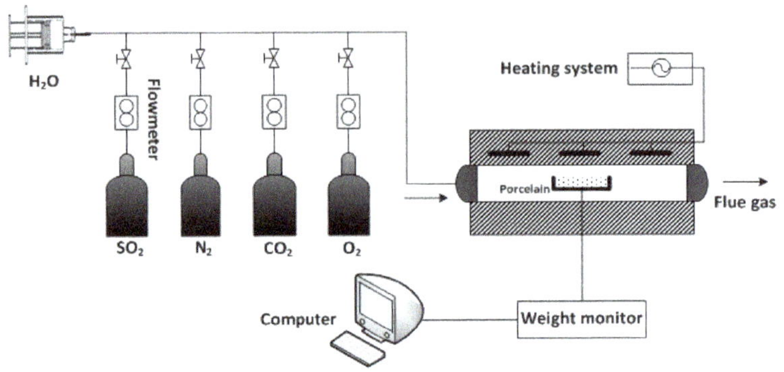

Figure 1. Schematic diagram of tube furnace system.

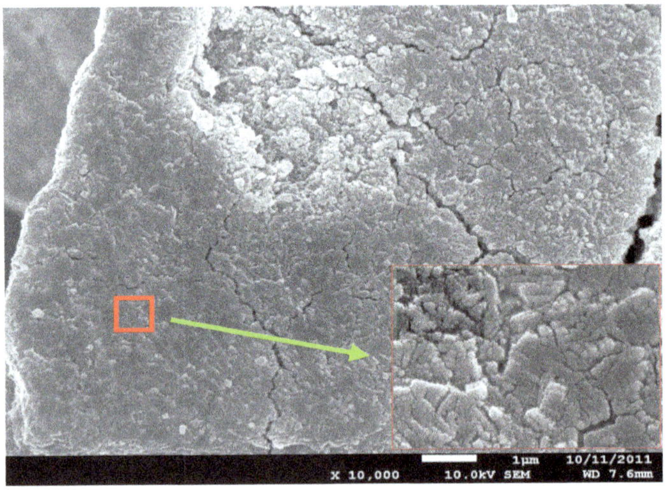

Figure 2. The indirect desulfurization particle morphology in the reaction atmosphere of 65%N_2, 15%CO_2, 4.7%O_2, 15%H_2O, 0.3%SO_2.

The chemical reaction equation is as follows:

$$CaCO_3(s) \rightarrow CaO(s) + CO_2(g) \quad (1)$$

$$CaO(s) + SO_2(g) + \tfrac{1}{2}O_2(g) \rightarrow CaSO_4(s) \quad (2)$$

According to above equations, the limestone does not directly involve in the desulfurization reaction, and the desulfurization reaction cannot happen without intermediate CaO, so it is called indirect desulfurization (De Diego et al., 2011; Pisani De Moraes 2004). However in the CFB oxy-fuel boiler, pure oxygen (O_2) is the fuel accelerant, and the partial pressure of CO_2 in its flue gas is high about >80% (Wu et al., 2011; De Joannon et al., 2012). Under such a high partial pressure of CO_2, the thermal decomposition of limestone particle like Equation (1) cannot occur anyhow. So recently, some researchers include us consider that its reaction in this case should be as follows (Scala et al., 2010; Stewart et al., 2012).

$$CaCO_3(s) + SO_2(g) \rightarrow CaSO_4(s) + CO(g) \quad (3)$$

$$CO(g) + \tfrac{1}{2}O_2(g) \rightarrow CO_2(g) \quad (4)$$

Because CaO does not appear in this reaction, it is also called direct desulfurization. The reports on the direct desulfurization are rarely, especially for the evidence on CO. As we all know, any reaction mechanism is closely related to its reactant microstructure, so those limestone powder's surface micro-structure properties may be different between two cases, and some clues were found by us as following.

EXPERIMENTAL SETUP

In order to verify that indeed some differences exist, the tube furnace system was built for indirect/direct desulfurization reaction experiments as shown in Figure 1, by which the aforementioned two reactions could be simulated for getting the limestone desulfurization samples. It includes: (1) In first test, the indirect desulfurization atmosphere was make up by N_2(65%), CO_2(15%), O_2(4.7%), H_2O(15%), SO_2(0.3%). (2) In second test, the direct desulfurization atmosphere was make up by CO_2(80%), O_2(4.7%), H_2O(15%), SO_2(0.3%). At above two tests, the working temperatures were both 850°C, the gas flows were both 60 ml/min, the limestone particles were both sieved by 15.6 um, the limestone powder were put into the porcelain and sent into the furnace by a track, the reaction times were both 40 min. After tests, the samples were removed and kept under seal. Both of above samples were observed by scanning electron microscope (SEM; KYKY3800, China).

RESULTS AND DISCUSSION

Figure 2 is the sample particle from the first test in the flue gas of traditional coal-fired boiler, and its magnification is 10000 times. It is shown that the surface

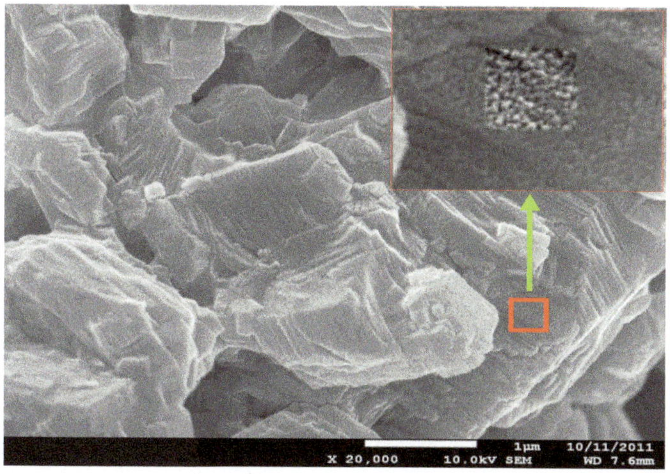

Figure 3. The direct desulfurization particle morphology in the reaction atmosphere of 80%CO2, 4.7%O2, 15%H2O, 0.3%SO2.

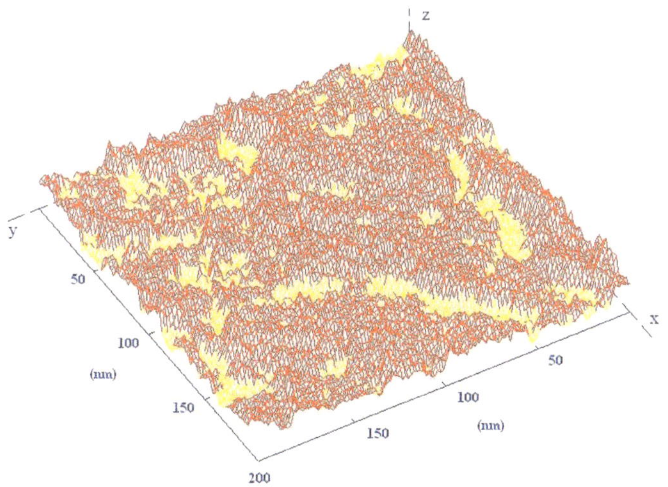

Figure 4. The indirect desulfurization micro-morphology reconstruction by Fourier transformation.

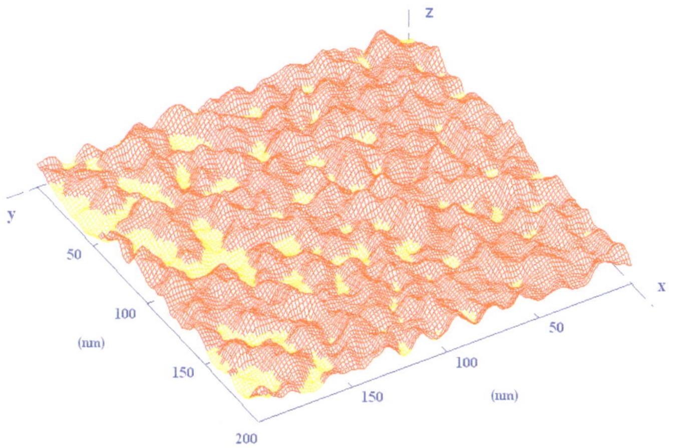

Figure 5. The direct desulfurization micro-morphology reconstruction by Fourier transformation.

of sample particle is covered by dark desulfurization product layer, on which there is a breakpoint. At the breakpoint, a large number of smaller un-reacted grains appear to agglomerate together, the grain sizes are about 200 nm. Pores between those grains should be same size order of magnitude with those grains. The origin for those pores is that CO_2 gas fully precipitates according to Equation 1 during the pre-calcination process before the beginning of indirect desulfurization reaction. Obviously, the blockage phenomenon is serious on the surface of sulfide layer. In the enlarged image as shown in the red-region, the product of $CaSO_4$ is cotton-shaped and this shape may be caused by the outward diffusion (Hsia et al., 1995) of Ca^{2+}. Figure 3 is the particle morphology in the second test simulated the flue gas of CFB oxy-fuel boiler. There is an obvious structural difference with Figure 2 that CaO grains do not appear anywhere and those larger pores like Figure 2 inside particle do not exist yet. However we have found that there are a lot of extremely smaller micro-pores on the surface of product layer in Figure 3. For example in the red region, those pore's quantity is great, and their size and their growth law are that we care about. In the enlarged region in Figure 3; the surface of product layer is honeycombed with extremely smaller pores, which should also be a large number below the surface. For the above two experimental samples, their morphologies are different. In order to deeply and microscopically get their structural properties, the mathematical analysis is necessary by Fourier transformation (Solookinejad et al., 2011; Szeloch et al., 2012) which can transform the image pixel into frequency height at 3-D space as shown in Figures 4 and 5.

In Figure 4, the 3-D morphology is reconstructed from the red region in Figure 2. The sulfide products are chaotic derivatives like numerous cluster mountains, in which outward growth heights are very uneven. Any valuable pore cannot be found; it may be two reasons, one is that CO_2 does not continue to separate out from the dense internal of grains, the other is that many pores between grains are blocked by those outward growth sulfide products (Vieille et al., 2012; Adnadjevic and Popovic., 2008). From the point of gas diffusion theory, such topography is not conductive to the reaction mass diffusion, and therefore it will result in lower desulfurization efficiency. In Figure 5, the 3-D morphology is reconstructed from the red region in Figure 3. The product growth heights are uniform; the distribution of those pores is also uniform and pore sizes are about 20 nm, any blockage cannot be found, the active sites for adsorption are very rich. So, we may reasonably conclude that gaseous reactive substances can constantly and uniformly precipitate through those micro pores during the reaction process of desulfurization. Because the partial pressure of CO_2 in this reaction atmosphere is high about >80%, the gaseous precipitated product should not be CO_2 (Yu et al., 2013). During the process of desulfurization reaction as shown in Figure 6, Ca^{2+} combines SO_2 molecule into

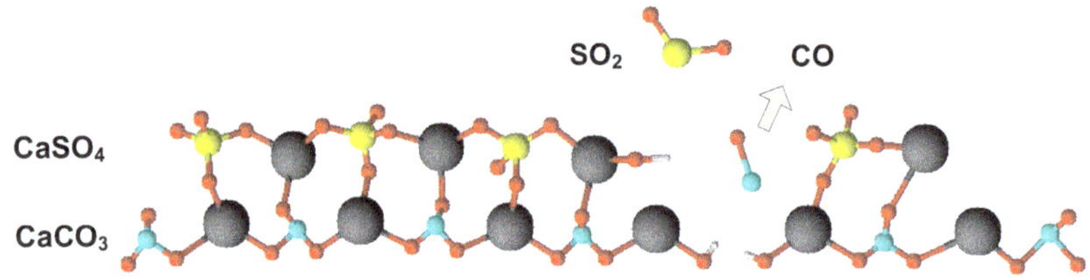

Figure 6. The precipitation of CO from $CaSO_4$ product layer leads to micro-porous passages.

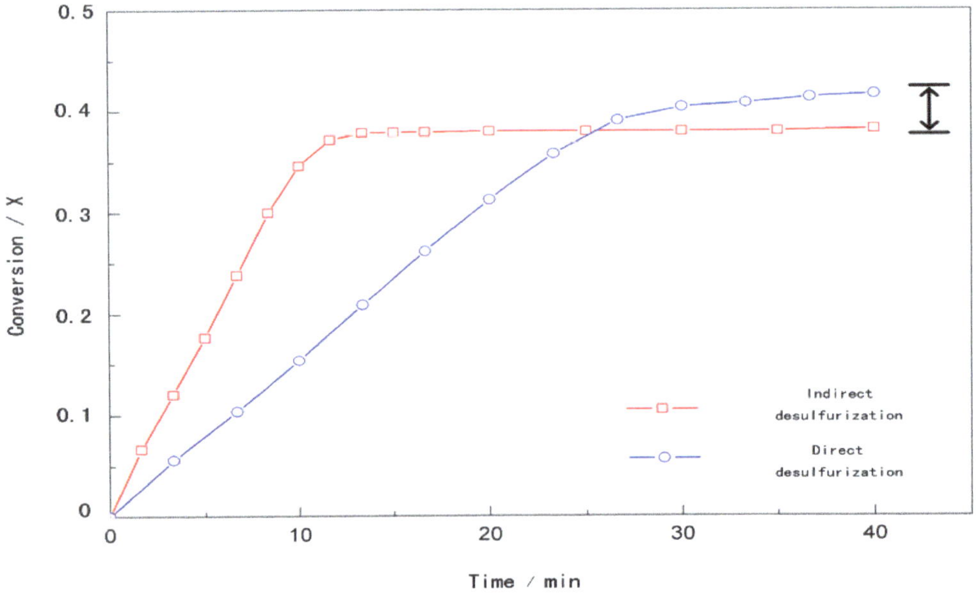

Figure 7. SO_2 removal Conversion of indirect/direct desulfurization.

calcium sulfate, at the same time CO molecule is precipitated and combined with free O_2 into stable CO_2 molecule at high temperatures. The precipitation of CO is continuous and is synchronized to the desulfurization reaction, so some channels will occur for CO emissions.

The type of chemical mechanism leads to its type of microscopic structure and this supports Equation (3) from a new perspective which considers that the gas is CO. This reaction pattern is highly efficient because those micro-porous passages can guarantee continuous reactive substance transmission, and it will result in high calcium utilization.

Indirect desulfurization rate is calculated as:

$$X = \frac{M_{CaO}}{M_{CaSO_4} - M_{CaO}} \times \frac{\Delta m}{m_0} \qquad (5)$$

Direct desulfurization rate is calculated as:

$$X = \frac{M_{CaCO_3}}{M_{CaSO_4} - M_{CaCO_3}} \times \frac{\Delta m}{m_0} \qquad (6)$$

Where X is conversion rate, M is molecular weight, m_0 means inlet gas quality, Δm means quality change. To observe the TGA curve as shown in Figure 7, we can find that the final conversion of direct desulfurization is really higher than that of indirect desulfurization; this is just a proof of the above discussion. However, these micro-pores are about one magnitude smaller than those in indirect desulfurization, so in the internal of sulfide product layer, it is easy to form many closed micro-cavities. Theoretically, for the outward growth solid-state Ca^{2+}, whether each closed cavity in the internal maybe play an impediment role or play a short-circuit role, which has not been reported anywhere, will also be that we are very interested in.

Conclusions

In summary, for the limestone particle in coal-fired flue gas, its microstructure is closely related to its desulfurization reaction mechanism. The sizes of internal pore within sorbent particles differ by an order of magnitude in two kind reaction atmospheres, one is about 200 nm for indirect desulfurization, and the other is about 20 nm for direct desulfurization. For the indirect desulfurization reaction, the sulfide products are chaotic derivatives like numerous cluster mountains, and the blocking phenomenon occurs seriously. For the direct desulfurization reaction, the product layer belongs to honeycombed micro structure without blockage, and a lot of smaller micro-porous passages can guarantee continuous reactive substance transmission, which may be a meaningful enlightenment to further improve the efficiency of SO_2 removal in industry.

ACKNOWLEDGMENTS

This work was supported by the National Key Technology R&D Program of China (2012BAC24B01), the National Natural Science Foundation of China (51276064), the Beijing Nature Science Foundation (3132028), and the Beijing Postdoctoral Research Foundation (2012ZZ-19).

REFERENCES

Adnadjevic B, Popovic A (2008). Influence of crystal form and morphological characteristics of $CaCO_3$ particles on kinetic of combustion gases desulfurization. Fuel Proces. Technol. 89:773-776.

Avila I, Crnkovic PM, Milioli FE, Luo KH (2012). Investigation of the pore blockage of a Brazilian dolomite during the sulfation reaction. Appl. Surf. Sci. 258:3532–3539.

De Diego LF, De las Obras-Loscertales M, Garcia-Labiano F (2011). Characterization of a limestone in a batch fluidized bed reactor for sulfur retention under oxy-fuel operating conditions. Int. J. Greenho. Gas Control. 5:1190-1198.

De Joannon M, Chinnici A (2012). Optimal post-combustion conditions for the purification of CO2-rich exhaust streams from non-condensable reactive species. Chem. Eng. J. 211:318-326.

Hsia C, Pierre GRS (1995). Isotope study on diffusion in $CaSO_4$ formed during sorbent-flue- gas reaction. AIChE J. 41:2337-2340.

Krzywanski J, Czakiert T, Muskala W (2010). Modeling of solid fuels combustion in oxygen-enriched atmosphere in circulating fluidized bed boiler: Part 1. The mathematical model of fuel combustion in oxygen-enriched CFB environment. Fuel Proces. Technol. 91:290–295.

Mihaela O, Magdalena A, Bogdana S (2010). Effect of SO_2 Dry Deposition on Porous Dolomitic Limestones. Materials. 3:216-231.

Pisani R, De Moraes D (2004). Removal of sulfur dioxide from a continuously operated binary fluidized bed reactor using inert solids and hydrated lime. J. Hazard. Mater. 109:183-189.

Ramon AR, Carmen CJ (2008). Oxidation of the sulphurised dolomite produced in the desulphurization of the gasification gases. Fuel. 11:1011-1036.

Ramírez JP, Hevia MAG, Abelló S (2008). SO_2-promoted catalytic N_2O removal over iron zeolites. Chem. Commun. 0:5351-535.

Ramón ÁR, Carmen CJ (2009). Oxidation of the sulphurised dolomite produced in the desulphurisation of the gasification gases. Fuel. 88:2507-2519.

Scala F, Salatino P (2010). Flue gas desulfurization under simulated oxyfiring fluidized bed combustion conditions: The influence of limestone attrition and fragmentation. Chme. Eng. Sci. 65:556-561.

Sipila M, Berndt T, Petaja T (2010). The role of sulfuric acid in atmospheric nucleation. Science 327:1243-1246.

Solookinejad G, Rozatian ASH, Habibi MH (2011). ZnO thin film characterization by X-ray reflectivity optimization using genetic algorithm and Fourier transformation. Appl. Surf. Sci. 258:260-264.

Stewart MC, Symonds RT, Manovic V (2012). Effects of steam on the sulfation of limestone and NO_x formation in an air- and oxy-fired pilot-scale circulating fluidized bed combustor. Fuel. 92:107-115.

Szeloch RF, Janus P, Serafińczuk J (2012). Characterization of fatigued Al lines by means of SThM and XRD: Analysis using fast Fourier Transform. Microelectron. Reliab. 52:711-717.

Sınırkaya M, Kadir Özer A, Gülaboglu MS (2012). Investigation of the changes of P_2O_5 content of phosphate rock during simultaneous calcination/sulfation. Powder Technol. 211:72-76.

Vieille L, Govin A, Grosseau P (2012). CO_2 capture enhancement in a fluidized bed of a modified Geldart C powder. Powder Technol. 228:319-323.

Winaya NS, Basu P (2001). Effect of pressure and carbon dioxide concentration on heat transfer at high temperature in a Pressurized Circulating Fluidized Bed (PCFB) combustor. Int. J. Heat Mass Tran. 44:2965-2971.

Wu WF, Zhao CS, Li QZ (2011). Experimental investigation on calcination/sulphation characteristics of limestone modified by acetic acid solution in O_2/CO_2 atmosphere. P. Combust. Inst. 33:3455-3462.

Yu J, Zhou K, Ou W (2013). Effects of Stefan flow and CO oxidation on char particle combustion in O_2/CO_2 atmosphere. Fuel. 106:576-585.

23

Chemical constituents and biological activities of the genus *Zanthoxylum*: A review

J. S. Negi[1]*, V. K. Bisht[1], A. K. Bhandari[1], P. Singh[2] and R. C. Sundriyal[1]

[1]Herbal Research and Development Institute, Mandal, Gopeshwar-246401, Uttarakhand, India.
[2]Department of Chemistry, HNB Garhwal University, Srinagar-246174, Uttarakhand, India.

Zanthoxylum **known as "Timoor" is used as mouth fresh, tooth care, spice and in intoxicating of the fishes. The present study deals with documentation of isolated chemical constituents and biological activities of the genus *Zanthoxylum*. The chemical constituents and essential oil have been isolated and characterized by several analytical methods. Alkaloids and cumarins are the main constituents of the genus. The presence of alkaloids and essential oil, *Zanthoxylum* possessed several types of biological activities such as larvicidal, anti-inflammatory, analgesic, antinociceptive, antioxidant, antibiotic, hepatoprotective, antiplasmodial, cytotoxic, antiproliferative, anthelminthic, antiviral and antifungal.**

Key words: *Zanthoxylum*, alkaloids, essential oil, biological activities.

INTRODUCTION

Zanthoxylum are deciduous shrubs and trees of the family Rutaceae which comprise 250 species that are native to warm temperate and subtropical region of the world. The genus has much ethanobotanical importance and is used as sources of pharmaceutical and cosmetics raw material. Traditionally, leaves and fruits are used for mouth fresh and tooth care while bark is used for intoxicating the fishes (Gaur, 1999). Leaves, fruits and barks are used as spice (Samant and Dhar, 1997). Alkaloids were isolated from *Zanthoxylum* (Rodriguez-Guzman et al., 2010; Yang et al, 2008; Liang et al., 2006; Martin et al., 2005). Twenty microelements were also determined in the root, stem and leaf of traditional Chinese herbs *Zanthoxylum nitidum* by ICP-AES (Qin et al., 2009). Due to presence of nitrogen containing compound, *Zanthoxylum* has wide spectrum of biological activities. The purpose of this review to collect all the possible information regarding the chemical constituents and biological effects of the genus *Zanthoxylum*, thus this will help to researchers and pharmaceutical companies to take action in this discipline.

*Corresponding authour. E-mail: negi_js1981@yahoo.co.in.

CHEMICAL CONSTITUENTS

2,4-dihydroxypyrimidine, syringic acid, 2,6-dimethoxy-1,4-benzoquinone, 4-hydroxybenzoic acid, ethylparaben, (Z)-3-(2,3,4-trimethoxyphenyl) acrylic acid, 5,6,7-trimethoxycoumarin, stigmast-9(11)-en-3-ol, daucosterol and beta-sitosterol were isolated and characterized from *Z. nitidum* (Hu et al., 2006). 6-(2',3'-dihydroxy-3'-methyl-butyl)-7-hydroxy-5-methoxy-2H-1-benzopyran-2-one,6-(2',3'-dihydroxy-3'-methyl-butyl)-7-methoxy-8-(3'-methyl-but-2'-enyl)-2H-1-benzopyran-2-one, 6-(2',3'-dihydroxy-3'-methyl-butyl)-7-hydroxy-2H-1-benzopyran-2-one, 6-(2',3'-oxiranyl-3'-methyl-butyl)-7-methoxy-8-(3-methyl-but-2-enyl)-2H-1-benzopyran-2-one, 7-methoxy-8-(3'-methyl-but-2'-enyl)-2H-1-benzopyran-2-one were also isolated from *Zanthoxylum* (Tao et al., 2001, 2003, 2005).

Coumarins and alkaloids such as bergarpten, umbelliferone, skimmianine and schinifoline were isolated from *Zanthoxylum* (Liu et al., 1991, Liu et al., 2008). A Novel amide N-(4'-methoxyphenyl ethyl)-3,4-methylenedioxy cinnamoyl amide was isolated and characterized from *Z. armatum* (Kalia et al., 1999) and norchelerythrine, magnoflorine and (-)(S)-O-methylbalfourodinium cation from *Z. scandens* bark (Nguyen et al., 2002). Aliphatic acid amides have been isolated from the pericarp of *Z. piperitum* fruits (Hatano et

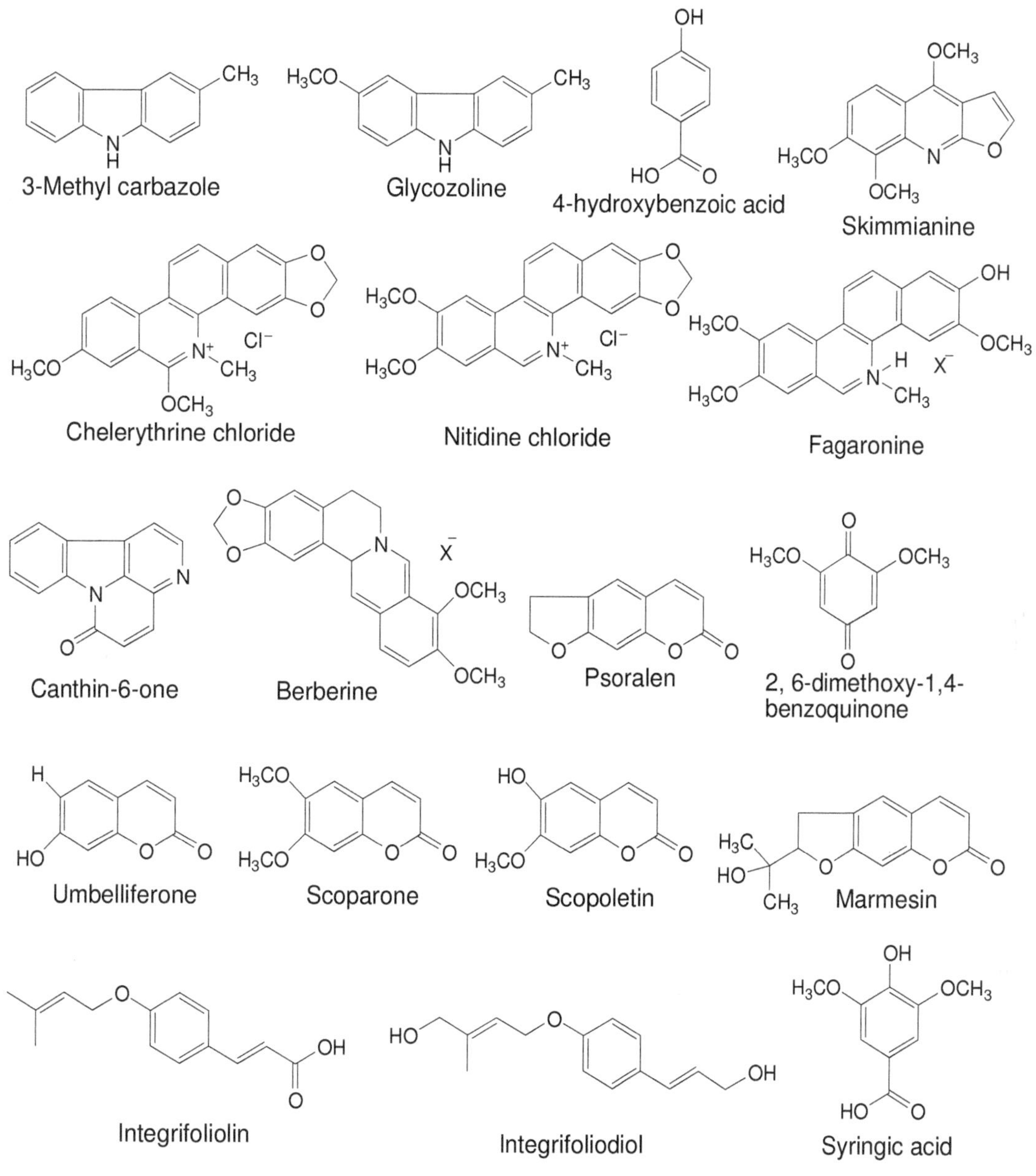

Figure 1. Structure of some isolated bioactives of *Zanthoxylum*.

al., 2004). Ahmad et al. (2003) isolated pseudophrynamine, lunacridine and 2-(2',4',6'-trimethyl-heptenyl)-4-quinozolone from the leaves of *Z. budrunga*. Quinoline alkaloids, (-)-R-geilbalansine and hyemaline were isolated from the stem barks of *Z. hyemale*. The antispasmodic activity of the crude ethanolic extract of *Z. hyemale* was studied in two different antispasmodic test models (De Moura et al., 2002). Canthin (fungicide) and pellitorine (insecticide) have been isolated from dichloromethane extracts of the roots and the bark of *Z. usambarense* (He et al., 2002) and 3-methoxy-3',4'-methylenedioxylignan-4,8,9,9'-tetraol from *Z. monophyllum*. Lupeol, ß-sitosterol, (+)-sesamin, *trans*-dimethylmatairesinol, hesperidin, (-)-cis-N-methylcanadine and sucrose were isolated from *Z. spruce* (Binutu and Cordell, 2000). L-sesamin and L-asarinin were determined by HPLC from different parts of *Zanthoxylum* (Roxb.) DC (Zhang et al., 2002). All the information regarding chemical and pharmaceutical aspect of Nigerian *Zanthoxylum* have been documented by Adesina (2005). The structures of some isolated bioactives of *Zanthoxylum* are depicted in Figure 1.

ESSENTIAL OIL CONTENT

The main components of *Zanthoxylum* oil are 38.91 oleic acid, 29.45 palmitic acid, 10.80 linoleic acid methyl ester and 8.64% linolenic acid methyl ester (Li et al., 2001). Furthermore, Yang (2008) reported 15 linalyl acetate, 13 linalool and 12% limonene as major components. GC-FID and GC-MS analyses of the essential oil of *Z. schinifolium* and *Z. bungeanum* showed that limonene was the main component (21%), followed by 4-terpineol, gamma-terpinene, alpha-terpineol acetate, beta-pinene, alpha-terpineol and beta-linalool (Iseli et al., 2007). Itthipanichpong et al. (2002) had reported 31.09 limonene, 13.94 terpin-4-ol and 9.13% sabinene as major components of the essential oil of *Z. limonella* Alston. Chemical constituents of essential oil of *Z. bungeanum* were also determined by Guo et al. (2001). Bisht and Chanotiya (2011) analyzed the essential oil of *Z. armatum* DC leaf by capillary gas chromatography (GC-FID) and gas chromatography-mass spectrometry (GC-MS) and found 2-undecanone as dominate for the first time, although, linalool was reported as dominant content in several previous studies on *Z. armatum* oil (Yoshitito et al., 2000). The essential oil of *Z. gillettii* leaves were analyzed by GC-MS and olfactoric methods. Monoterpenes such as myrcene, trans-beta-ocimene, limonene, camphene, alpha- and beta-pinene are main constituents of the oil. Sesquiterpenes such as beta-caryophyllene, alpha- and beta-farnesene, beta-bisabolol, gamma-cadinene and nerolidol and alphatic components such as 2-undecanone and dodecanal were also identified from the essential oil of this species (Jirovetz et al., 1999). Bhattacharya and Zaman (2009) identified essential oil composition from the fruits and leaves of *Z. nitidum* by GC-MS and found monoterpenes in the fruit oil as linalool (23.3%), limonene (12.9%), aterpineot (8.3%), α-pinene (7.9%), γ-terpinene (6.6%), cis-β-ocimene (6.2%), terpinen-4-ol (4.7%) and isomenthone (2.0%), while, among the monoterpenes of leaf oil, the predominant was limonene (33.1%), followed by geraniol (10.6%), carvone (9.6%), 7-hydroxy-3,7-dimethyloctanal (4.3%), geranial (3.9%) and nerol (2.2%).

The essential oils from aerial parts of juvenile leaves, mature leaves, fruit and flowers of *Z. hyemale* were isolated by hydrodistillation and analyzed by GC, GC-MS and chiral phase gas chromatography (CPGC). The major constituent of the juvenile leaf essential oil was the sesquiterpene trans-nerolidol (51%), while the main constituent of mature leaf and flower oils was 31 and 22% hyemalol, respectively. In the fruit, the most abundant components were the monoterpenes β-pinene (25%) and α-pinene (10%). The essential oils of the leaves and the fruits of *Z. leprieurii*, *Z. macrophylla* and *Z. xanthoxyloides* of Cameroon were extracted by hydrodistillation. Extraction yields of 0.42 and 7.22%, 0.14 and 0.11%, 0.12 and 3.88% were obtained for leaves and the fruits of *Z. leprieurii*, *Z. macrophylla* and *Z. xanthoxyloides*, respectively. Essential oil from green fruit, ripe fruit and dried pericarp of Z. *piperitum* were also determined by using GC-MS (Jiang and Kubota, 2004). Zhang and Jiang (2008) developed a technique to produce biodiesel from *Z. bungeanum* seed oil.

BIOLOGICAL ACTIVITIES

The fruits are used as digestive appetizer, to cure asthma and bronchitis, eliminate pain, use to treat heart diseases, piles, diseases of mouth, teeth and throat disorder, also prescribed in dyspepsia and diarrhea (Kritikar and Basu, 1983). The roots bark and leaves of many species are used in various medicinal preparations for curing stomachache, tooth-ache, coughs, urinary and veneral diseases, leprous ulcerations, rheumatism and lumbago (Oliver-Bever, 1982). *Zanthoxylum* has been studied for several types of biological activities such as larvicidal, anti-inflammatory, analgesic, antinociceptive, antioxidant, antibiotic, hepatoprotective, antiplasmodial, cytotoxic, antiproliferative, anthelminthic, antiviral, anticonvulsant and antifungal (Tiwary et al., 2007; Lima et al., 2007; Chen et al., 2007, 2008; Guo et al., 2011; Rodrigures et al., 2010; Lee and Lim, 2008; Ross et al., 2004; Chou et al., 2011; Barnabas et al., 2010; Song et al., 2010; Yang and Chen, 2008; Islam et al., 2001a, b; Gansane et al., 2010; Amabeoku and Kinyua, 2010; Seal, 2011; Peneluc et al., 2009).

Essential oil of *Z. armatum* showed strong inhibition of mycelial growth against the test fungus *Bipolaris sorokiniana* (Manandhar and Tiwari, 2005). The anti-tumor activity of the volatile oil of *Z. rhoifolium* Lam. leaves were investigated *in vitro* and *in vivo* using the *Ehrlich ascites* tumor model (Da Silva et al., 2007a; Da Silva et al., 2007b). Methyl-N-methylanthranilate and alpha-bisabolol represented the main compounds of essential oil of *Z. tingoassuiba* which showed antimicrobial activity (Detoni et al., 2009). Toothpaste containing *Z. nitidum* extract decreased the incidence of dental plaque and enhanced gingival health (Wan et al., 2005). An alkaloidal extract of the stem barks of *Z. chiloperone* exhibited antifungal activity against *Candida albicans*, *Aspergillus fumigatus* and *Trichophyton mentagrophytes* (Thouvenel et al., 2003). Bafi-Yeboa et al. (2005) investigated *Z. americanum* leaf, fruit, stem, bark and root for antifungal activity with 11 strains of fungi. All extracts exhibited a broad spectrum of antifungal activity. Alkamides isolated from the leaves of *Z. syncarpum* showed moderate antiplasmodial activity, with IC50 values of 4.2 and 6.1 mM against *Plasmodium falciparum* D6 and W2 clone (Ross et al., 2005). Ethanolic extracts of the trunk bark of *Z. fagara*, *Z. elephantiasis* and *Z. martinicense* showed antifungal activity (Dieguez-Hurtado et al., 2003). The petroleum ether, chloroform and methanol extracts of the leaves and barks of *Z. budrunga* have been evaluated for their antibacterial, antifungal and cytotoxic properties (Islam et al., 2001a). Benzophenanthrene alkaloids, 8-acetonyldihydronitidine and 8-acetonyldihydroavicine were isolated from *Z. tetraspermum* stem bark which showed significant antibacterial activity (Nissanka et al., 2001). *Z. limonella* showed mosquito repellent activity (Trongtokit et al., 2005). Canthin isolated from *Z. chiloperone* possessed a broad spectrum of antifungal and leishmanicidal activities (Ferreira et al., 2006). The hepatoprotective activity of the ethanolic extract of *Z. armatum* leaves have been evaluated in CCl_4-induced hepatotoxicity in rats (Verma and Khosa, 2010). The antimicrobial activity of the essential oils of *Z. hyemale* was also reported (Simionatto et al., 2005). The antioxidant properties of the fruit samples of *Z. leprieurii* and *Z. xanthoxyloides* demonstrated an SC_{50} of 3.72 and 4.05 g/l, respectively, while the butyl hydroxy toluene reference gave an SC_{50} of 7.10 mg/l. The fruit essential oils of *Z. leprieurii* and *Z. xanthoxyloides* could be used as food supplements to protect against emergent diseases such as cardiovascular problems, cancer and diabetes (Dongmo et al., 2008).

Cytotoxic activity of essential oil of *Z. rhoifolium* was evaluated against HeLa (human cervical carcinoma), A-549 (human lung carcinoma), HT-29 (human colon adenocarcinoma), Vero (monkey kidney) cell lines and mice macrophages by Da Silva et al. (2007a). They observed that the essential oil is cytotoxic against tumoral cells (CD50 = 82.3, 90.7 and 113.6 µg/ml for A-549, HeLa, HT-29 cell lines, respectively), while noncytotoxic against non-tumoral cells (Vero and mice macrophages). Essential oil and methanolic extract of *Z. budrunga* showed significant hypnotic activity (Bhadoriya et al., 2009). Kamsuk et al. (2007) investigated protective effects of *Z. piperitum* for repellency against mosquitoes under laboratory and field conditions using synthetic repellent standard, N,N-diethyl-3-methylbenzamide.

CONCLUSION

The genus *Zanthoxylum* posses several types of biological activities such as larvicidal, anti-inflammatory, analgesic, antinociceptive, antioxidant, antibiotic, hepato-protective, antiplasmodial, cytotoxic, antiproliferative, anthelminthic, antiviral and antifungal, mainly due to the presence of alkaloids and essential oil.

REFERENCES

Adesina SK (2005). The Nigerian *Zanthoxylum*; Chemical and Biological Values. Afr. J. Trad. C. A. M., 2(3): 282-301.

Ahmad MU, Rahman MA, Huq E, Chowdhury R (2003). Alkaloids of *Zanthoxylum budrunga*. Fitoterapia, 74: 191-193.

Amabeoku GJ, Kinyua CG (2010). Evaluation of the anticonvulsant activity of *Zanthoxylum capense* (Thunb.) Harv. (Rutaceae) in mice. Int. J. Pharmacol., 6: 844-853.

Bafi-Yeboa NF, Arnason JT, Baker J, Smith ML (2005). Antifungal constituents of Northern prickly ash, *Zanthoxylum americanum* Mill. Phytomed., 12: 370-377.

Barnabas BB, Mann A, Ogunrinola TS, Anyanwu PE (2010). Screening for anthelminthic activities from extracts of *Zanthoxylum zanthoxyloides*, *Neocarya macrophylla* and *Celosia laxa* against ascaris infection in Rabbits. Int. J. Appl. Res. Natural Prod., 3: 1-4.

Bhadoriya U, Yadav A, Aggarwal N, Jaiswal D, Yadav IK, Singh HP, Chandra D, Jain DA (2009). Hypnotic effect of essential oil and methanolic extract of fruits of *Zanthoxylum budrunga* W. Int. J. Pharmtech. Res., 1: 1494-1498.

Bhattacharya S, Zaman K (2009). Essential oil composition of fruits and leaves of *Zanthoxylum nitidum* grown in upper Assam region of India. Phcog. Res., 1: 148-151.

Binutu OA, Cordell GA (2000). Constituents of *Zanthoxylum sprucei*. Pharm. Biol., 38: 210-213.

Bisht D, Chanotiya CS (2011). 2-Undecanone rich leaf essential oil from *Zanthoxylum armatum*. Nat. Prod. Commun., 6: 111-114.

Chen JJ, Chen PH, Liao CH, Huang SY, Chen IS (2007). New phenylpropenoids, bis(1-phenylethyl)phenols, bisquinolinone alkaloid, and anti-inflammatory constituents from *Zanthoxylum integrifoliolum*. J. Nat. Prod., 70: 1444-1448.

Chen JJ, Wang TY, Hwang TL (2008). Neolignans, a coumarinolignan, lignan derivatives, and a chromene: Anti-inflammatory constituents from *Zanthoxylum avicennae*. J. Nat. Prod., 71: 212-217.

Chou ST, Chan HH, Peng HY, Liou MJ, Wu TS (2011). Isolation of substances with antiproliferative and apoptosis-inducing activities against leukemia cells from the leaves of *Zanthoxylum ailanthoides* Sieb. Zucc. Phytomed., 18(5): 344-348.

Da Silva SL, Figueiredo PM, Yano T (2007a). Cytotoxic evaluation of essential oil from *Zanthoxylum rhoifolium* Lam. Leaves. Acta Amazonica, 37: 281-286.

Da Silva SL, Figueiredo PM, Yano T (2007b). Chemotherapeutic potential of the volatile oils from *Zanthoxylum rhoifolium* Lam leaves. Eur. J. Pharmacol., 576: 180-8.

De Moura NF, Morel AF, Dessoy EC, Zanatta N, Burger MM, Ahlert N, Porto G, Baldisserotto B (2002). Alkaloids, amides and antispasmodic activity of *Zanthoxylum hyemale*. Planta Med., 68: 534-538.

Detoni CB, Cabral-Albuquerque EC, Hohlemweger SV, Sampaio C, Barros TF, Velozo ES (2009). Essential oil from *Zanthoxylum tingoassuiba* loaded into multilamellar liposomes useful as antimicrobial agents. J. Microencapsul., 26: 684-691.

Dieguez-Hurtado R, Garrido-Garrido G, Prieto-Gonzalez S, Iznaga Y, Gonzalez L, Molina-Torres J, Curini M, Epifano F, Marcotullio MC (2003). Antifungal activity of some Cuban *Zanthoxylum* species. Fitoterapia, 74: 384-386.

Dongmo PMJ, Tchoumbougnang F, Tchinda SE, Manedong KS, Amvam ZPH, Menut C (2008). Antioxidant and anti-inflammatory potential of essential oils of some *Zanthoxylum* (Rutaceae) of Cameroon. Int. J. Essent. Oil Therap., 2: 82-88.

Ferreira ME, Nakayama H, de Arias AR, Schinini A, de Bilbao NV, Serna E, Lagoutte D, Soriano-Agaton F, Poupon E, Hocquemiller R, Fournet A (2006). Effects of canthin-6-one alkaloids from *Zanthoxylum chiloperone* on *Trypanosoma cruzi*-infected mice. J. Ethnopharmacol., 109: 258-263.

Gansane A, Sanon S, Ouattara PL, Hutter S, Ollivier E, Azas N, Traore A, Traore AS, Guissou IP, Nebie I, Sirima BS (2010). Antiplasmodial activity and cytotoxicity of semi purified fractions from *Zanthoxylum zanthoxyloides* Lam. Bark of Trunk. Int. J. Pharmacol., 6: 921-925.

Gaur RD (1999). Flora of the District Garhwal North West Himalaya: With Ethnobotanical Notes. TransMedia, Srinagar, p. 811.

Guo T, Deng YX, Xie H, Yao CY, Cai CC, Pan SL, Wang YL (2011). Antinociceptive and anti-inflammatory activities of ethyl acetate fraction from *Zanthoxylum armatum* in mice. Fitoterapia, 82: 347-351.

Guo ZA, Zhao JC, Xie ZH (2001). Study on chemical constituents of the essential oil from *Zanthoxylum bungeanum* Maxim by gas chromatography-mass spectrometry. Se Pu., 19: 567-568.

Hatano T, Inada K, Ogawa TO, Ito H, Yoshida T (2004). Aliphatic acid amides of the fruits of *Zanthoxylum piperitum*. Phytochemistry, 65: 2599-2604.

He W, Van Puyvelde L, De Kimpe N, Verbruggen L, Anthonissen K, Van der Flaas M, Bosselaers J, Mathenge SG, Mudida FP (2002). Chemical constituents and biological activities of *Zanthoxylum usambarense*. Phytother. Res., 16: 66-70.

Hu J, Zhang WD, Liu RH, Zhang C, ShenYH, Xu XK, Liang MJ, Li HL (2006). Chemical constituents in root of *Zanthoxylum nitidum*. Zhongguo Zhong Yao Za Zhi., 31: 1689-1691.

Iseli V, Potterat O, Hagmann L, Egli J, Hamburger M (2007). Characterization of the pungent principles and the essential oil of *Zanthoxylum schinifolium* pericarp. Pharmazie, 62: 396-400.

Islam A, Sayeed A, Anwar-Ul Islam M, Astaq Mohal Khan GRM, Biswas MHU, Shah Alam Bhuiyan M (2001b). A flavonone from leaves of *Zanthoxylum budrunga*: Its *In vitro* antimicrobial activity and cytotoxic evaluation. J. Med. Sci., 1: 209-213.

Islam A, Sayeed A, Bhuiyan MSA, Mosaddik MAG (2001a). *In vitro* Antimicrobial effect of three Terpenes, Isolated from the Bark of *Zanthoxylum budrunga*. Pak. J. Biol. Sci., 4: 711-713.

Itthipanichpong C, Ruangrungsi N, Pattanaautsahakit C (2002). Chemical compositions and pharmacological effects of essential oil from the fruit of *Zanthoxylum limonella*. J. Med. Assoc. Thai., 1: S344-S354.

Jiang L, Kubota K (2004). Differences in the volatile components and their odor characteristics of green and ripe fruits and dried pericarp of Japanese pepper (*Zanthoxylum piperitum* DC.). J. Agric. Food Chem., 52: 4197-4203.

Jirovetz L, Buchbauer G, Fleischhacker W, Ngassoum MB (1999). Analysis of leaf volatiles of *Zanthoxylum gillettii* used in folk medicine of Cameroon. Planta Med., 65: 181-183.

Kalia NK, Singh B, Sood RP (1999). A new amide from *Zanthoxylum armatum*. J. Nat. Prod., 62: 311-312.

Kamsuk K, Choochote W, Chaithong U, Jitpakdi A, Tippawangkosol P, Riyong D, Pitasawat B (2007). Effectiveness of *Zanthoxylum piperitum* derived essentialnoil as an alternative repellent under laboratory and field applications. Parasitol. Res., 100: 339-345.

Kritikar KR, Basu BD (1983). Indian Medicinal Plants Text, Latit Mohan Basu, Leader Road, Allahabad, 1: 464-465.

Lee SJ, Lim KT (2008). Glycoprotein of *Zanthoxylum piperitum* DC has a hepatoprotective effect via anti-oxidative character *In vivo* and *In vitro*. Toxicol *In vitro*, 22: 376-85.

Li Y, Zeng J, Liu L, Jin X (2001). GC-MS analysis of supercritical carbon dioxide extract from seeds of *Zanthoxylum bungeanun* Maxim. Zhong Yao Cai., 24: 493-494.

Liang M, Zhang W, Hu J, Liu R, Zhang C (2006). Simultaneous analysis of alkaloids from *Zanthoxylum nitidum* by high performance liquid chromatography-diode array detector-electrospray tandem mass spectrometry. J. Pharm. Biomed. Anal., 42: 178-183.

Lima LM, Perazzo FF, Carvalho JCT, Bastos JK (2007). Anti-inflammatory and analgesic activities of the ethanolic extracts from *Zanthoxylum riedelianum* (Rutaceae) leaves and stem bark. J. Pharm. Pharmacol., 59: 1151-1158.

Liu S, Zhang W, Lei P, Li XZ, Liang YZ (2008). Separation and purification of alkaloids from radix of *Zanthoxylum nitidum* by macroporous adsorption resin. Zhongguo Zhong Yao Za Zhi, 33: 377-379.

Liu SL, Wei LX, Wang D, Gao CY (1991). Studies on the chemical constituents from the peel of *Zanthoxylum schinifolium* Sieb et Zucc. Yao Xue Xue Bao., 26: 836-840.

Manandhar A, Tiwari RD (2005). Antifungal efficacy of *Zanthoxylum* oil against *Bipolaris sorokiniana* (Sacc.) Shoem. Int. J. Ecol., 12: 91-93.

Martin MT, Rasoanaivo LH, Raharisololalao A (2005). Phenanthridine alkaloids from *Zanthoxylum madagascariense*. Fitoterapia, 76: 590-593.

Nguyen QA, Van-Dufat HT, Michel S, Tillequin F, Dumontet V, Sevenet T (2002). A new phenylpropanoid ester from the bark of *Zanthoxylum scandens* (Rutaceae). Z Naturforsch C., 57: 986-989.

Nissanka AP, Karunaratne V, Bandara BM, Kumar V, Nakanishi T, Nishi M, Inada A, Tillekeratne LM, Wijesundara DS, Gunatilaka AA (2001). Antimicrobial alkaloids from *Zanthoxylum tetraspermum* and *caudatum*. Phytochemistry, 56: 857-861.

Oliver-Bever B (1982). Medicinal plants in tropical West Africa, plants acting on cardiovascular system. J. Ethnopharmacol., 5: 1-72.

Peneluc T, Domingues LF, de Almeida GN, Ayres MC, Moreira EL, da Cruz AC, dos Santos Calmon de Bittencourt TC, de Almeida MA, Batatinha MJ (2009). Anthelmintic activity of aqueous extract of *Zanthoxylum rhoifolium* Lam. leaves (Rutaceae). Rev. Bras. Parasitol. Vet., 18: 43-48.

Qin JL, Chen ZF, Liu YC, Liang H (2009). Determination of trace metal elements in *Zanthoxylum nitidum* by ICP-AES. Guang Pu Xue Yu Guang Pu Fen Xi., 29: 2851-2854.

Rodriguez-Guzman R, Radwan MM, Burandt CL, Williamson JS, Ross SA (2010). Xenobiotic biotransformation of 4-methoxy-N-methyl-2-quinolone, isolated from *Zanthoxylum monophyllum*. Nat. Prod. Commun., 5: 1463-1464.

Rodrigures FF, Costa JG, Coutinho HD (2010). Enhancement of the antibiotic activity of gentamicin by volatile compounds of *Zanthoxylum articulatum*. Indian J. Med. Res., 131: 833-835.

Ross SA, Sultana GN, Burandt CL, ElSohly MA, Marais JP, Ferreira D (2004). Syncarpamide, a new antiplasmodial (+)-norepinephrine derivative from *Zanthoxylum syncarpum*. J. Nat. Prod., 67: 88-90.

Ross SA, Al-Azeib MA, Krishnaveni KS, Fronczek FR, Burandt CL (2005). Alkamides from the leaves of *Zanthoxylum syncarpum*. J. Nat. Prod., 68: 1297-1299.

Samant SS, Dhar U (1997). Diversity, endemism, and economic potential of wild edible plants of the Indian Himalayas. Int. J. Sustain. Dev. World Ecol., 4: 179-191.

Seal T (2011). Determination of nutritive value, mineral contents and antioxidant activity of some wild edible plants from Meghalaya State, India. Asian J. Appl. Sci., 4: 238-246.

Simionatto E, Porto C, Dalcol II, Da Silva UF, Morel AF (2005). Essential oil from *Zanthoxylum hyemale*. Planta Med., 71: 759-763.

Song JH, Chae SW, Yoon KA, Park JS, Choi HJ (2010). Antiviral Activity of *Zanthoxylum* Species against Porcine Epidemic Diarrhea Virus. J. Cosmetics Pub. Health, 6: 42-44.

Tao ZY, Chen WS, Zheng SQ, Zhang WD, Qiao CZ (2001). Studies on the chemical constituents of *Zanthoxylum dimorphophyllum* Hemsl. var spinifolium Rhed. et Wils. Yao Xue Xue Bao., 36: 511-513.

Tao ZY, Chen WS, Zhang WD, Sun LN, Zheng SQ, Qiao CZ (2005). Coumarins from root of *Zanthoxylum dimorphophyllum* var. spinifolium. Zhongguo Zhong Yao Za Zhi, 30: 832-834.

Tao ZY, Chen WS, Zhang WD, Sun LN, Zheng SQ, You L, Qiao CZ (2003). Studies on the coumarins in the root of *Zanthoxylum dimorphophyllum*. Zhongguo Zhong Yao Za Zhi, 28: 344-346.

Thouvenel C, Gantier JC, Duret P, Fourneau C, Hocquemiller R, Ferreira ME, de Arias AR, Fournet A (2003). Antifungal compounds from *Zanthoxylum chiloperone* var. angustifolium. Phytother. Res., 17: 678-680.

Tiwary M, Naik SN, Tewary DK, Mittal PK, Yadav S (2007). Chemical composition and larvicidal activities of the essential oil of *Zanthoxylum armatum* DC (Rutaceae) against three mosquito vectors. J. Vect. Borne. Dis., 44: 198-204.

Trongtokit Y, Rongsriyam Y, Komalamisra N, Apiwathnasorn C (2005). Comparative repellency of 38 essential oils against mosquito bites. Phytother. Res., 19: 303-309.

Verma N, Khosa RL (2010). Hepatoprotective activity of leaves of *Zanthoxylum armatumn* DC in CCl4 induced hepatotoxicity in rats. Indian J. Biochem. Biophys., 47: 124-127.

Wan HC, Hu DY, Liu HC (2005). Clinical observation of toothpaste containing *Zanthoxylum nitidum* extract on dental plaque and gingivitis. Zhongguo Zhong Xi Yi Jie He Za Zhi, 25: 1024-1026.

Yang G, Chen D (2008). Alkaloids from the roots of *Zanthoxylum nitidum* and their antiviral and antifungal effects. Chem. Biodivers., 5(9): 1718-1722.

Yang CH, Cheng MJ, Chiang MY, Kuo YH, Wang CJ, Chen IS (2008). Dihydrobenzo[c]phenanthridine alkaloids from stem bark of *Zanthoxylum nitidum*. J. Nat. Prod., 71: 669-673.

Yang X (2008). Aroma constituents and alkylamides of red and green huajiao (*Zanthoxylum bungeanum* and *Zanthoxylum schinifolium*). J. Agric. Food Chem., 56(5): 1689-1696.

Yoshitito U, Yuriko N, Masayoshi H, Shuichi H, Seiji H (2000). Essential oil constituents of Fuya-sanshoo (*Zanthoxylum armatum* DC) in Nepal. Koryo Terupen Oyobi Seiyu Kagaku ni Kansuru Toronkai Koen Yoshishu, 44: 59-61.

Zhang J, Jiang L (2008). Acid-catalyzed esterification of *Zanthoxylum bungeanum* seed oil with high free fatty acids for biodiesel production. Bioresour Tech., 99: 8995-8998.

Zhang SY, Zhou BJ, Wang Y (2002). Determination of L-sesamin and L-asarinin in *Zanthoxylum* (Roxb.) DC. by high performance liquid chromatography. Di Yi Jun Yi Da Xue Xue Bao, 22: 654-655.

Polycyclic aromatic hydrocarbons in smoked Lates niloticus from selected markets, Gulu District, Uganda

Acaye Ongwech[1], George William Nyakairu[2] , Jolocam Mbabazi[2] Justus Kwetegyeka[3] and Margaret Masette[4]

[1]Department of Chemistry Gulu University, P. O. Box 166, Gulu, Uganda.
[2]Department of Chemistry, School of Physical Sciences, College of Natural Sciences, Makerere University, P. O. Box 7062, Kampala, Uganda.
[3]Department of Chemistry, Kyambogo University, P. O. Box 1, Kyambogo, Uganda.
[4]Food Bioscience Research Centre – National Agriculture Research Organisation Kawanda, P. O. Box 7852, Kampala Uganda.

Hot smoking is among the oldest methods of preservation which mankind has used in food processing. Potential health hazards associated with smoked foods may be caused by carcinogenic components of wood smoke – mainly polycyclic aromatic hydrocarbons (PAHs) and their derivatives. This paper presents results based on the determination of PAHs in smoked Lates niloticus from three markets in Gulu district, northern Uganda. The samples were analysed using gas chromatography-mass spectrometry (GC-MS). The PAHs detected in the fish samples were acenaphthylene, fluorene, phenanthrene, anthracene, fluoranthene, pyrene, chrysene, benzo[b]fluoranthene and indeno [1,2,3-cd]pyrene. The analysed samples showed PAH levels ranging from non-detectable (n.d) levels to 53.23 µg/kg of smoked fish. High molecular weight (HMW) PAHs constituted 71.1% by mass of the total PAHs detected and quantified in the samples, with indeno [1,2,3-cd]pyrene and benzo[b]fluoranthene featuring substantially in 95.8% of all the samples analysed. However, seven of the nine (77.8%) compounds detected in the samples were low molecular weight (LMW) PAHs. The higher concentrations of HMW PAHs suggest that the fish could have been smoked using soft wood or smoked for longer time and, also, may be due to the resistance of these PAHs to environmental degradation. Generally most of the samples analysed had ∑BaPeq and ∑PAH4 levels within the maximum acceptable risk limits of 5 and 30 µg/kg respectively, as recommended by European Commission Regulations for muscle meat of smoked fish. Hence the fish could therefore be deemed fit for human consumption.

Key words: Lates niloticus, polycyclic aromatic hydrocarbons; smoked fish, gas chromatography-mass spectrometry (GC-MS), toxic equivalency.

INTRODUCTION

Polycyclic aromatic hydrocarbons (PAHs) are a large class of organic compounds containing two or more fused aromatic rings without heteroatoms (Anyakora et al., 2005; Doris and Ken, 2009). The compounds are lipophilic, chemically stable and poorly degraded by hydrolysis. The majority of PAHs are readily metabolised and broken down in mammals though, some are bioaccumulated especially in organisms higher up the food chain (Atlas, 1991; WHO, 2000). They occur in every type of environment as complex mixtures that originate from environmental sources, industrial food processing (e.g. drying and smoking processes), packaging materials and certain alimentary practices (Agerstad and Skog, 2005; Duedahl-Olesen et al., 2006; EFSA, 2008). Pyrosynthesis and pyrolysis are two processes that may explain formation of PAHs during combustion. Lower hydrocarbons form PAHs by pyrosynthesis (Chen and Chen, 2001). Higher alkanes present in fuels and plant materials form PAHs by pyrolysis (Ravindraa et al., 2008).

Many of the heavy PAHs reported to be highly mutagenic and carcinogenic in laboratory animals have also been implicated in breast, lung and colon cancers in humans (Alonge, 1988; Simko, 2002). Some PAHs that do not exhibit carcinogenicity have been reported to act as synergists (Simko, 2002). PAHs have also been reported to cause hemato, cardio, renal, neuro, immuno, reproductive and developmental toxicities in humans and laboratory animals (Ramesh et al., 2004). Prenatal exposures of humans to PAHs have been reported to affect IQ at the age of five (Edwards et al., 2010). In 2001, PAHs were ranked the ninth most life threatening compounds to human health (King et al., 2002). More than 100 PAHs have been characterized in nature, 16 of which were classified by United States Environmental Protection Agency (USEPA), as priority pollutants. They include: naphthalene, acenaphthylene, benzo[b]fluoranthene, phenanthrene, dibenzo[a,h]anthracene, chrysene, benzo[a]pyrene, acenaphthene, benzo[k]fluoranthene, fluorene, pyrene, benzo[a]anthracene, anthracene, fluoranthene, indeno[1,2,3-cd]pyrene, and benzo[g,h,i]perylene (Chen and Chen, 2001; EPA, 1993). Benzo[a]pyrene (BaP), the most carcinogenic PAH in soot, has for long been used as a marker for the occurrence and carcinogenicity of PAHs (Simko, 2002). However, the total carcinogenic potency of all PAHs contained in a food product is 10 times higher than the content of BaP alone (SCF, 2002). European Food Safety Authority (EFSA, 2008) suggested the use of PAH8 (benzo[a]anthracene, chrysene, benzo[a]pyrene, benzo[b]fluoranthene, benzo[k]fluoranthene, dibenzo[a,h]anthracene, indeno[1,2,3-cp]pyrene, and benzo[g,h,i]perylene) or a subgroup of these, PAH4 (the sum of the first four in PAH8) as more suitable indicators for the occurrence and toxicity of PAHs. European Commission Regulation (EC) No 1881/2006 was amended in August 2011 and the use of PAH4 as marker was proposed to take effect on the 1st September 2012 (European Commission, 2011).

Food products can be contaminated through various ways including hot smoking, contaminated soil and packaging materials, polluted air and water sources (Chung et al., 2011). Although high levels of PAHs are not usually observed in raw foods (WHO, 1998), with uncontrolled smoking procedures, PAH levels up to 200 µg/kg of smoked fish have been reported (Bonny, 1983). During smoking, heating and drying processes, combustion products come into direct contact with foods and PAH contamination can occur. The compounds have been detected in various food and non-food substances and beverages including duck meat (Chen and Lin, 1997), smoked cheese (Pagliuca et al., 2003), fish (Anyakora and Coker, 2007; Linda et al., 2011; Moret et al., 1999; Serden et al., 2010; Wretling et al., 2010) and pork (Chung et al., 2011) among others.

Gulu district, where the study was conducted, experienced over two decades of insurgency since 1986. This decimated domestic animals and smoked fish become the dominant source of animal proteins. *Lates niloticus*, the dominant fish species on the markets, is transported from the regions around Lake Victoria, over 300 km away. The catch is normally smoke cured, since consignments must be bulked and stored to await transportation to the markets. The hot smoking procedure to which the fish are subjected has been reported to generate a lot of PAHs (Larsson et al., 1983) that get diffused to the fish muscles. Very little, if any, information exists on the occurrence and levels of PAHs in the smoked fish consumed in Gulu district in particular and Uganda in general. This study aimed at establishing the presence and amounts of PAHs in smoked *L. niloticus* sold in the selected markets in Gulu district, Uganda.

MATERIALS AND METHODS

Reagents and standards

Analytical grade reagents and chemicals used, including dichloromethane, anhydrous sodium sulphate, silica gel, potassium hydroxide, cyclohexane and methanol were purchased from British Drug House (BDH, England) store in Kampala. The 16 USEPA's priority PAH standards were obtained from Supelco Inc. (Bellefonte, PA), Sigma Aldrich Co., Germany.

10 ml of each stock of PAH standard solution was prepared at a concentration of 100 mg/L in cyclohexane. The stock solutions were transferred into amber glass tubes with Teflon lined closures. The tubes, appropriately labeled with the name of the standard, and date of preparation were kept at 4°C. These stock solutions were later diluted to the appropriate concentrations to get the working standards for analysis after extraction and purification of the samples.

Sampling

Smoked fish samples were collected from three markets, G (Bobi, from Gulu municipality), R (Omoro, from Omoro county) and S (Unyama, from Aswa county) which lie 32°18'03.31"- 32°21'25.94"E and 2°33'23.84"- 2°48'59.90" N. From each of these markets, four fish samples each weighing about 500 g were purchased once a week for four weeks. The samples were collected, wrapped in aluminium foils and transported to Makerere University, Department of Chemistry, for laboratory analysis.

Extraction and purification

The method employed (with minor modifications), were described

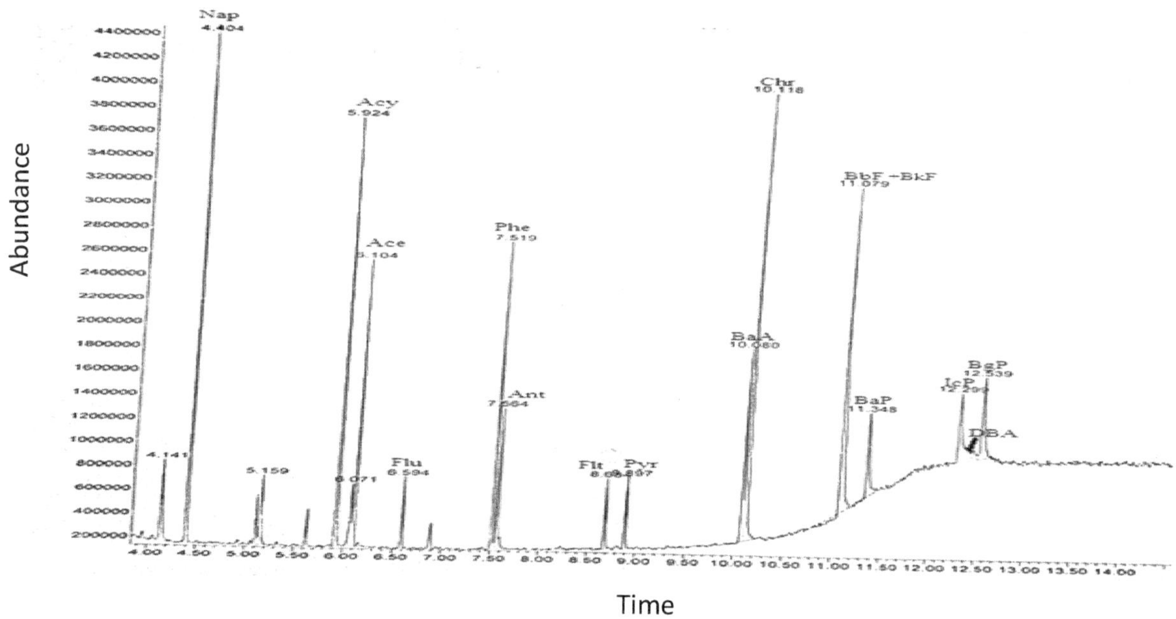

Figure 1. GC-MS chromatogram of PAHs standard mixture.

and used by Wretling et al. (2010) and Anyakora and Coker (2007). Homogenized fish muscles (30 g) were weighed in a 250 ml Erlenmeyer flask. The weighed amount was saponified by adding methanolic potassium hydroxide solution (methanol/water, 9:1v/v; 90 ml, 3.5 M), thoroughly sealing the flask and keeping it in a drying cabinet at 70°C for two hours. The flask was shaken after 1 h. Cyclohexane (50 ml) was added to the flask; the mixture was shaken and allowed to cool at ambient temperature. The contents in the flask were transferred to a 250 ml separation funnel, the flask rinsed with methanol/water (30 ml; 4:1 v/v) and the washing added to the funnel. The mixture was then shaken vigorously for 3 min and allowed to stand so that the layers separate. After separation, the aqueous layer was transferred to a second 250 ml separation funnel and washed with cyclohexane (30 ml). This second cyclohexane phase was combined with the first and the mixture washed with methanol/water (30 ml; 4:1 v/v), then with methanol/water (30 ml; 1:1 v/v) and three times with water (30 ml). The cleaned cyclohexane phase was then dried by passing it through anhydrous sodium sulphate, collected in a round bottomed flask and concentrated to 1 ml using a rotary evaporator. This was kept for purification.

The extracts were purified as described by Mottier et al. (2000); Wang et al. (1999). Chromatographic column (1 cm i.d. × 15 cm) was plugged with glass wool at the base. Activated silica gel was loaded in the column to a height of 5 cm. An additional 1 cm of anhydrous sodium sulphate was added to the column. The packed column was then conditioned with dichloromethane (15 ml), the 1 ml concentrate (extract) loaded onto it and eluted with dichloromethane (20 ml). The eluate was dried by passing it through anhydrous sodium sulphate, concentrated to 1 ml using a rotary evaporator and transferred to an amber glass sample vial for gas chromatography-mass spectrometry (GC-MS) analysis. The vials were kept at 4°C prior to the analysis.

Gas chromatographic analysis

An Agilent 6890N network GC system equipped with a 5975 MSD, 7683B auto sampler and a Zebron, ZB-5MSi column was used to separate the compounds. The system was operated in a constant flow mode with an initial rate of 4.9 ml/min, nominal initial pressure of 35.53 psi and average velocity of 81 cm/s. The oven was initially set at 55°C for 1 min, ramped to 320°C at 25°C/min and held for 3 min.

The mode of injection used was pulsed (1.0 µl) splitless at an initial temperature of 300°C and pressure 35.53 psi, until 0.2 min. The purge flow rate was 30 ml/min with a purge time 0.75 min. The carrier gas was helium (99.999%) at a constant flow rate of 45 cm/s. The detector source was set at 300°C, quadrupole at 180°C, transfer line at 280°C; scan range at 45 to 450 amu and electronic impact at 70eV with a solvent delay of 3.75 min. PAHs were identified based on the match in the retention times of the compounds in the samples against those of the PAH standards (Figure 1). A retention time match of ± 1% was considered for confirmation (Samuel et al., 2010). Mass spectral data base library search was also performed for retention time match outside this window. A total of nine PAHs were identified in the samples analysed (Figure 2).

Quantification of the PAHs

This was based on absolute calibration methods. A minimum of four concentration levels of the standards ranging from 0.1 to 15 ppm were injected into the GC-MS and calibration curve for each standard was obtained by plotting peak area against concentration of the standards. The regression equation and coefficient of determination (r^2) were determined. Each PAH in the sample was then quantified using the formula;

$$As\,(ppb) = \frac{Xs \times 100}{Ws \times R}$$

where, *As* is the concentration of PAH in the sample in µg/kg, *Xs* is the concentration (ng/ml) relative to the peak area in the injection volume (µl), *Ws* is the mass (in grams) of the sample extracted and *R* is the recovery.

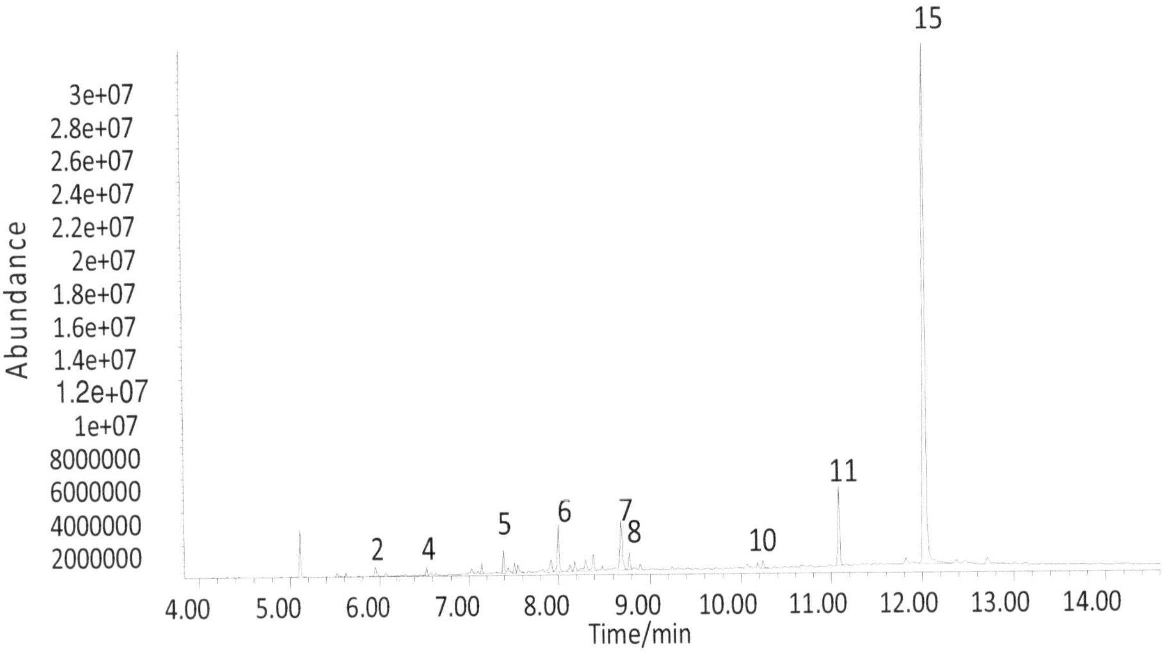

Figure 2. The 9 PAHs identified in the samples analysed: 2- acenaphthylene, 4- fluorene, 5- phenanthrene, 6- anthracene, 7- fluoranthene, 8- pyrene, 10- chrysene, 11- benzo[b]fluoranthene, 15- indeno[1,2,3-cd]pyrene.

Quality control

Both method's (prepared at a frequency of 12.5%), and solvent blanks were included in every run during the analysis. Individual reference standards were used to identify and quantify the PAHs. Arithmetic means and standard deviations were calculated from positive quantifiable samples only, and in all cases, the differences were considered significant, if the exact *p* value was $\alpha \leq 0.05$. A computer programme XLSTAT (version 7.5.2) was used for computation. Linearity was established by computing the relative standard deviations (RSDs) and observing the coefficient of determination (r^2) values. The RSDs obtained were all below 10% while the r^2 values ranged from 0.9930 to 0.9999.

To establish the mean percent recoveries, a set of three aliquots (30 g) of the homogenized fish muscles were spiked with 50 µl mixture of the 16 PAH standards of concentration ranging from 2 to 10 µg/L. Another set of three aliquots (30 g) were set as controls. Both the spiked and the control aliquots were allowed to stay overnight to allow for absorption of the added PAHs into the matrix. The two sets of the aliquots were then taken through the same analytical procedures. The percent recoveries were then computed from the formula

$$\mathrm{Recovery} = \%R = \frac{Cs - Cu}{Cn} \times 100$$

Where, *Cs* is the concentration of PAHs in the spiked sample, *Cu* is the concentration of PAHs in the unspiked (control) sample, *Cn* is the (nominal) theoretical concentration increase that results from spiking the samples.

The mean percent recoveries obtained were within the range (70 to 130%) with a mean of 80.23%. These values were relatively quantitative and were therefore used without any correction. Limits of detection (LODs) and limits of quantification (LOQs) ranged from 0.01 to 0.08 µg/kg and 0.04 to 0.24 µg/kg respectively.

RESULTS

PAHs in smoked *Lates niloticus*

In the samples from three sampled markets, 9 PAHs were detected in the smoked fish (*L. niloticus*) Figure 2, and the total concentrations were 58.10, 23.40 and 43.20 µg/kg at sites G, R, S, respectively, as shown in Table 1. Five, nine and eight PAHs were identified and quantified in samples from sites R, G and S, respectively. Acenaphthylene was present in samples from both sites G and S while indeno[1,2,3-cd]pyrene featured prominently in all except in samples from site S. Fluorene was in all samples collected from all the sites. Most of the PAHs (77.8%) identified in the samples were low molecular weight (LMW, 3-4 ringed) compounds though there were high concentrations of the HMW PAHs (indeno [1,2,3-cd]pyrene and benzo[b]fluoranthene).

Carcinogenity of PAHs

Toxic equivalency as a measure of carcinogenicity of PAHs

The carcinogenic potency of each PAH identified in the samples collected was also determined in terms of BaP

Table 1. Levels of PAHs (µg/kg) in smoked *L. niloticus* from three markets in Gulu district.

PAH	Sample collection site n = 16		
	G (Bobi)	R (Omoro)	S (Unyama)
Acenaphthylene	2.5 ± 0.5	n.d	2.50
Fluorene	0.9 ± 0.2	0.7 ± 0.1	0.85
Phenanthrene	5.5 ± 3.3	1.5 ± 0.1	5.54
Anthracene	3.7 ± 2.6	n.d	3.67
Fluoranthene	4.9 ± 1.3	n.d	4.85
Pyrene	3.4 ± 1.0	n.d	3.40
Chrysene	1.5 ± 1.0	4.4 ± 0.1	1.54
Benzo[b]fluoranthene (HMW)	9.5 ± 3.8	3.8 ± 2.7	9.46
Indeno[1,2,3-cd]pyrene (HMW)	26.2 ± 8.6	13.0 ± 5.6	26.18

Results are presented as mean ± standard deviation for samples from the three sites, n.d, not detected (p value 0.001).

Table 2. Toxic equivalency factors (TEF) for the 16 priority PAHs (Nisbet and LaGoy, 1992).

PAH	TEF	PAH	TEF
Naphthalene	0.001	Benzo[a]anthracene	0.1
Acenaphthylene	0.001	Chrysene	0.01
Acenaphthene	0.001	Benzo[b]fluoranthene	0.1
Fluorene	0.001	Benzo[k]fluoranthene	0.1
Phenanthrene	0.001	Benzo[a]pyrene	1.0
Anthracene	0.01	Indeno[1,2,3-cd]pyrene	0.1
Fluoranthene	0.001	Dibenzo[a,h]pyrene	1.0
Pyrene	0.001	Benzo[g,h,i]perylene	0.01

equivalent concentration, commonly referred to as the toxic equivalency (TEQ). A list of toxic equivalency factors (TEFs), completed by Nisbet and LaGoy (1992) (Table 2), was used to estimate the carcinogenic potency of total PAHs (that is, total BaP equivalent concentration) using the formula:

$$TEQ = (PAHi \times TEFi)$$

Where TEQ is the toxic equivalency of the reference compound, PAHi and TEFi the concentration and toxic equivalent factor of congener *i*.

The computed BaPeq (TEQ) values for PAHs detected and identified in the smoked fish sampled are shown in Table 3. Indeno[1,2,3-cd]pyrene had the highest mean TEQ, while Fluorene had the lowest. The sum TEQ of the PAHs in sampled area is G> S >R.

PAH4 as a measure of carcinogenicity of PAHs

Based on the conclusions of the Scientific Panel on Contaminants in the ood Chain (CONTAM Panel), new maximum levels for PAH4 were introduced whilst maintaining a separate one for BaP. The use of PAH4 ensure that PAH levels in foods are kept at low levels and that the amount of PAH can also be controlled in those samples in which BaP is not detectable, but where other PAHs are present. European Food Safety Authority (2008) concluded that the use of PAH4 should be preferred to PAH8 since the latter would not provide much added value.

Furthermore, both Scientific Committee on Foods (SCF, 2002) and Joint Expert Committee on Food Additives (JECFA, 2005), concluded that a TEQ approach to assessment of carcinogenic potency of PAHs was not very appropriate due to limitations in the available data and because of the different modes of action amongst different PAHs (EFSA, 2008). The use of PAH4 was also employed to estimate the carcinogenic potency of the PAHs detected and quantified in the samples (Table 4). The PAH4 values were 11.00, 8.14, and 10.98 µg/kg for sites G, R and S, respectively. The PAH4 values for each sample and for each of the sites were below the maximum level (30 µg/kg) set by Regulation (EU) 835/2011, for muscle meat of smoked fish and smoked fishery products.

Table 3. BaPeq PAH concentrations detected in the *L. niloticus* samples.

PAH	Site of sample collection				
	G	R	S	∑TEQ	Mean TEQ
Acenaphthylene	0.003	n.d	0.002	0.005	0.002
Fluorene	0.001	0.001	0.001	0.003	0.001
Phenanthrene	0.003	0.002	0.002	0.007	0.002
Anthracene	0.037	n.d	n.d	0.037	0.012
Fluoranthene	0.005	n.d	0.007	0.012	0.004
Pyrene	0.003	n.d	0.003	0.006	0.002
Chrysene	0.015	0.043	0.04	0.098	0.033
Benzo[b]fluoranthene	1.175	0.382	0.701	2.258	0.753
Indeno[1,2,3-cd]pyrene	2.618	1.303	1.767	5.688	1.896
∑TEQ	3.860	1.731	2.523	8.114	2.705

n.d = not detected, BaPeq = BaP equivalent concentration. TEQs were calculated on the basis of the content and TEFs for the 16 PAHs as given by Nisbet and LaGoy (1992).

Table 4. Levels of PAHs, PAH4 and PAH8 detected and quantified in the samples.

PAH	Levels of PAHs, PAH4 and PAH8 detected and quantified in the samples site of sample collection		
	G	R	S
Acenaphthylene	2.50	n.d	1.59
Fluorene	0.85	0.74	0.92
Phenanthrene	5.54	1.51	1.52
Anthracene	3.67	n.d	n.d
Fluoranthene	4.85	n.d	7.13
Pyrene	3.40	n.d	2.92
Chrysene	1.54	4.32	3.97
Benzo[b]fluoranthene	9.46	3.82	7.01
Indeno[1,2,3-cd]pyrene	26.18	13.03	17.67
∑PAH	56.14	23.41	42.73
∑PAH8	37.18	21.17	28.65
∑PAH4	11.00	8.14	10.98

n.d = not detected, PAH4 is the sum of benzo[a]anthracene, chrysene, benzo[b]fluoranthene and benzo[a]pyrene.

DISCUSSION

The relatively high levels of PAHs could be attributed to smoking of the fish. In a similar study by Olabemiwo et al. (2011), acenaphthylene, fluorene and indeno [1,2,3-cd] pyrene were not detected in non-smoked fish (used as control) but were present in all the smoked samples. Smoking as a method of preserving fish could have contributed to PAH content and amount in the smoked fish since they were not detected in the control samples by Olabemiwo et al. (2011).

In most of the samples (77.8%), low molecular weight (LMW, 3 to 4 ringed) compounds were detected. This could possibly be traced to the type of wood used during the smoking process. Studies by Pagliuca et al. (2003) showed that smoke produced by woods of deciduous trees (hard woods) show high concentrations of low molecular weight PAHs. However, although more of the LMW PAHs were detected, their concentrations were much lower than those of the high molecular weight (HMW, 5-6 ringed) PAHs that constituted up to 71.1% of the total mass. This can be attributed to the fact that the HMW PAHs are more resistant to degradation both in the fish and the environment. A similar profile was noted by Anyakora and Coker (2007), when they assessed PAH content in four species of fish from the Niger Delta. Linda et al. (2011) also noted higher concentrations of HMW compared to LMW PAHs when they characterized PAHs in smoked fish from Ghana. They attributed that difference to residues of previous pyrolytic processes that occurred in the smoking chamber.

The high concentrations of indeno [1,2,3-cd]pyrene and benzo[b]fluoranthene, Table 1, could also be explained from the mechanism of formation of the PAHs. Pagliuca et al. (2003) noted that vendors sometimes re-smoke the fish in order to increase their shelf life. Linda et al. (2011) noted that during re-smoking, it is possible that the pyrolytic products from the wood combustion add to the intact PAH molecules forming HMW PAHs. Similarly, during prolonged smoking, chances are that the LMW PAHs formed are subsequently converted to the HMW compounds through addition of the pyrolytic products from the continued wood combustion. Furthermore, pyrolysis of aromatic hydrocarbon residues leads to the formation of additional HMW PAHs and consequently increases their concentrations in the samples (Guillen and Sopelana, 2004).

The difference in the number of PAHs from the sites could be due to the difference in the setup of the markets. Sites G and S are enclosed markets (with lockup shops) while site R is an open space close to the busy Gulu-Kampala highway. It is therefore possible that the intense heat and free wind movement at site R could have facilitated the volatilization and loss of especially the LMW PAHs from the fish. Furthermore, such open space could have allowed atmospheric fallouts of such compounds as sulphurdioxide, ozone and oxides of nitrogen with which PAHs are known to react readily forming sulphonic acid, diones, nitro and dinitro PAHs respectively (ATSDR, 1995). It is also important to note that the commercial samples used in the study were smoked and handled differently and that could result in varying PAH load. That could explain the slight variation in the PAH profile noted from the three sites particularly if these fish were not obtained from the same smoking kiln.

BaP, the marker for the occurrence and carcinogenicity of PAHs was not detected in any of the samples analysed. Similarly, a study by Olabemiwo et al. (2011), reported that BaP was conspicuously absent in both the control and smoked fish samples. In another study, Anyakora and Coker (2007), reported BaP levels as high as 2.32 μg/kg in *Clarias garieppinus* which they attributed to the fact that the fish were caught in highly polluted rivers of the Niger Delta region.

The reason for the absence of BaP in the samples could therefore be that the fish were not caught from aquatic environments polluted by petroleum. However, according to Lawrence and Weber (1984), the ability of fish to metabolise PAHs may also explain why BaP is not often detected or found only at very low levels in fish from environments heavily contaminated with PAHs. On the other hand Mottier et al. (2000), noted that a principal loss normally occurs due to partitioning of BaP into the alkaline methanolic saponification solvent. Takatsuki et al. (1985) added that such partitioning leads to formation of emulsions that are difficult to breakdown hence lowering the concentration of free BaP. A study by Larsen and Poulsen (1987) as cited in Moret et al. (1999), concluded that smoked fish normally contains 0.1-0.5ppb of BaP but added that the level could be exceeded in heavily smoked products.

The total toxic equivalency (TEQ), (Table 3), for all samples considered at the three sites was 8.114 μg/kg, with a mean value of 2.705 μg/kg. This total was above the maximum risk limit for BaP level set by European Commission Regulation (EU) 835/2011 but the mean was below the limit for muscle meat of smoked fish. The sum of TEQ for each of the sites considered was 3.860, 1.731 and 2.523 μg/kg for sites G, R, S, respectively. All these values were within the maximum tolerable risk limit for BaP in muscle meat of smoked fish. Values for site G and S were however above that for muscle meat of fish other than smoked fish, set at 2 μg/kg (European Commission, 2011). The total TEQ for indeno[1,2,3-cd]pyrene(IcP), (5.688 μg/kg) was just above the limit while the mean (1.896 μg/kg) was well within the limit for BaP in muscle meat of smoked fish. According to earlier studies (Mottier et al., 2000), BaP could have simply been degraded during saponification and could have actually added to these TEQ values. There is therefore need to monitor the levels of PAHs especially the HMW components in the smoked fish sold in these markets. However this value is the total for all the samples from the three sites and would only raise serious health concern if an individual consumes all the samples and accumulate these PAHs without any breakdown or metabolism. However, studies have shown that PAHs are readily metabolised or broken down even in humans (WHO, 2000). This could therefore allay the fears of eating smoked *L. niloticus* from markets in Gulu.

Based on the use of PAH4 as indicator for the carcinogenicity of PAHs, (Table 4), smoked *L. niloticus* from each of these three markets could be deemed safe for human consumption. However, there is need for regular monitoring of the concentrations of especially the HMW PAHs in the smoked fish from these markets.

Conclusion

Traditionally smoked *L. niloticus* from selected three markets from Gulu district were found to contain BaP at undetectable levels as in none of the analysed samples was detected. Highest single concentration of PAH was noted for indeno[1,2,3-cd]pyrene from a smoked fish sampled from site R (53.23 μg/kg), while the lowest single quantifiable amount was for Fluorene (0.69 μg/kg) sourced from site G. Five PAHs were detected in samples from R, eight from S and nine from G. Generally the samples from the different sampling sites exhibited fairly uniform profiles, suggesting similarity in source of contamination. Benzo[b]fluoranthene and indeno[1,2,3-cd]pyrene were most representative, each being found in 95.8% of all samples analysed. The samples contained varying amounts of both the LMW and HMW PAHs that could be attributed to the smoking process. BaPeq concentrations commonly referred to as TEQ values were within tolerable risk levels (5 µg/kg), set by European Commission regulations. The use of PAH4 as marker for the occurrence and carcinogenicity of PAHs was also employed and mean values obtained for individual samples (n = 3), and individual sites were all within the maximum tolerable risk limits (30 µg/kg), set by European Commission Regulation (EU) 835/2011, for muscle meat of smoked fish.

Generally, the fish could be deemed fit for human consumption. However, the higher levels obtained for the HMW PAHs and the inclusion of exposures from other dietary sources can augment the risk levels hence there is need for regular monitoring of the levels of PAHs in the smoked fish.

ACKNOWLEDGEMENTS

Authors acknowledge the Editor, AJAPC, the anonymous reviewers whose comments and criticisms have greatly enriched and improved this manuscript. The Department of Chemistry, Makerere University is thanked for housing this research. This study was partially supported financially by "Encouraging Research Capacity in Gulu University (ENRECA-GU)" project to Acaye Ongwech. We would like to thank Jane Kayanja and Christopher Bitainensha for the technical support during GC-MS analysis and NARO-Kawanda for financial and technical cooperation.

REFERENCES

Agerstad MJ, Skog K (2005). Review genotoxicity of heat-processed foods. Mutat. Res. Fundam. Mol. Mech. Mutagen. 574:156-172.

Alonge DO (1988). Carcinogenic polycyclic aromatic hydrocarbons (PAH) determined in Nigerian kundi (smoke-dried meat). J. Sci. Food Agric. 43(2):167-172.

Anyakora C, Coker H (2007). Assessment of polynuclear aromatic hydrocarbon content in four species of fish in the Niger Delta by gas chromatography/mass spectrometry. Afr. J. Biotechnol. 6:737-743.

Anyakora C, Ogbeche A, Palmer P, Coker H, Ukpo G, Ogah C (2005). GC/MS analysis of polynuclear aromatic hydrocarbons in sediment samples from the Niger Delta region. Chemosphere 60(7):990-997.

Atlas RM (1991). Microbial hydrocarbon degradation: Bioremediation of oil spills. Biotechnology 52:149-156.

ATSDR (1995). Toxicological profile for polycyclic aromatic hydrocarbons (PAHs) (Final Report*)*. Atlanta, GA: Agency for Toxic Substances and Disease Registry.

Chen BH, Chen YC (2001). Formation of Polycyclic Aromatic Hydrocarbons in the Smoke from Heated Model Lipids and Food Lipids. J. Agric. Food Chem. 49(11):5238-5243.

Chen BH, Lin YS (1997). Formation of Polycyclic Aromatic Hydrocarbons during Processing of Duck Meat. J. Agric. Food Chem. 45(4):1394-1403.

Chung SY, Ramesh RY, Kim JS, Kwon K, Kim MC, Min BD (2011). Effects of grilling and roasting on the levels of polycyclic aromatic hydrocarbons in beef and pork. Food Chem. 129(4):1420-1426.

Doris S, Ken L (2009). GC/MS Analysis of European Union Priority Polycyclic Aromatic Hydrocarbons using an Agilent J&W DB-EUPAH GC Column with a Column Performance Comparison. Agilent Technologies, USA, 6 pp, pub 5990-4883EN, Oct 2009.

Duedahl-Olesen L, White S, Binderup ML (2006). Polycyclic aromatic hydrocarbons (PAHs) in Danish smoked fish and meat products. Polycycl. Aromat. Compd. 26(3):163-184.

Edwards SC, Jedrychowski W, Butscher M, Camann D, Kieltyka A, Mroz E, Flak E, Li Z, Wang S, Rauh V, Perera F (2010). Prenatal exposure to airborne polycyclic aromatic hydrocarbons and children's intelligence at 5 years of age in a prospective cohort study in Poland. Environ. Health Perspect. 118(9):1326-1331.

EFSA (2008). Scientific opinion of the Panel on contaminants in the food chain on a request from the European Commission on Polycyclic Aromatic Hydrocarbons in Food. The EFSA J. 724:65-114.

EPA (1993). Provisional guidance for the quantitative risk assessment of polycyclic aromatic hydrocarbons. US environmental protection agency report (No. EPA/600/R-93/089): US Environmental Protection Agency.

European Commission (2011). Commission Regulation (EU) No 835/2011 of 19 August 2011 amending Regulation (EC) No 1881/2006 as regards maximum levels for polycyclic aromatic hydrocarbons in foodstuffs. Official J. Eur. Union.

Guillen MD, Sopelana P (2004). Occurrence of polycyclic aromatic hydrocarbons in smoked cheese. J. Diary Sci. 87:556-564.

JECFA (2005). Joint FAO/WHO Expert Committee on Food Additives. Sixty-fourth meeting. Rome, 8-17 February 2005. Summary and conclusions. Rome: FAO/WHO. http://www.who.int/ipcs/food/jecfa/summaries/summary_report_64_final.pdf.

King S, Meyer JS, Andrews ARJ (2002). Screening method for polycyclic aromatic hydrocarbons in soil using hollow fibre membrane solvent microextraction. J. Chromatogr. A 982:201-208.

Larsson BK, Sahlberg GP, Eriksson AT, Busk LA (1983). Polycyclic aromatic hydrocarbons in grilled food. J. Agric. Food Chem. 31(4):867-873.

Lawrence JF, Weber DF (1984). Determination of polycyclic aromatic hydrocarbons in some Canadian commercial fish, shellfish, and meat products by liquid chromatography with confirmation by capillary gas chromatography-mass spectrometry. J. Agric. Food Chem. 32(4):789-794.

Linda MNP, Carboo D, Yeboah OP, Quasie JW, Gorleku OM, Darko A (2011). Characterization of Polycyclic Aromatic Hydrocarbons (PAHs) present in smoked fish from Ghana. Adv. J. Sci. Technol. 3(5):332-338.

Moret S, Conte L, Dean D (1999). Assessment of Polycyclic Aromatic Hydrocarbon Content of Smoked Fish by Means of a Fast HPLC/HPLC Method. J. Agric. Food Chem. 47(4):1367-1371.

Mottier P, Parisod V, Turesky RJ (2000). Quantitative Determination of Polycyclic Aromatic Hydrocarbons in Barbecued Meat Sausages by Gas Chromatography Coupled to Mass Spectrometry. J. Agric. Food Chem. 48(4):1160-1166.

Olabemiwo OM, Alade AO, Tella AC, Adediran GO (2011). Assassment of polycyclic aromatic hydrocarbons content in smoked C. gariepinnus and T. guineensis fish species available in Western Nigeria. Int. J. Basic Appl. Sci. 11(2):135-150.

Pagliuca G, Gazzotti T, Zironi E, Serrazanetti GP, Mollica D, Rosmini R (2003). Determination of High Molecular Mass Polycyclic Aromatic Hydrocarbons in a Typical Italian Smoked Cheese by HPLC-FL. J. Agric. Food Chem. 51(17):5111-5115.

Ramesh A, Walker SA, Hood DB, Guillén MD, Schneier K, Weyand EH (2004). Bioavailability and risk assessment of orally ingested polycyclic aromatic hydrocarbons. Int. J. Toxicol. *23*:301-333.

Ravindraa K, Ranjeet S, Rene VG (2008). Atmospheric polycyclic aromatic hydrocarbons: Source attribution, emission factors and regulation. Atmos. Environ. 42(13):2895-2921.

Samuel G, Angela M, Bryan G, Jill G, David J, John R, Laura C, Heather M, Gerry , David C, Walter K, Terri A, Jason W, Michele F, Yoko J, Douglas H, Fred F (2010). Screen for the presence of Polycyclic aromatic hydrocarbons in select seafoods using LC-Fluorescence. Minnesota: 1Forensic Chemistry Center (FCC), Food and Drug Administration. 4475:1-39.

SCF (2002). Opinion of the Scientific Committee on Food (SCF) in the

risks to human health of Polycyclic Aromatic Hydrocarbons in Food. Brussels: European Commission, Health and Consumer Protection Directorate General.
Serden B, Gülgün FŞ, Fatma TK (2010). The Detection of Potential Carcinogenic PAH Using HPLC Procedure in Two Different Smoked Fish, Case Study: Istanbul/Turkey. Turk. J. Fish. Aquat. Sci. 10:351-355.
Simko P (2002). Determination of polycyclic aromatic hydrocarbons in smoked meat products and smoke flavouring food additives. Journal of Chromatography B: Analytical Technologies in the Biomedical and Life Sciences 770(1-2):3-18.
Takatsuki K, Suzuki S, Sato N, Usizawa I (1985). Liquid chromatographic determination of PAHs in Fish and Shell fish. J. Assoc. Official Anal. Chem. 68(5):945-949.
Wang G, Lee AS, Lewis M, Kamath B, Archer RK (1999). Accelerated Solvent Extraction and Gas Chromatography/Mass Spectrometry for Determination of Polycyclic Aromatic Hydrocarbons in Smoked Food Samples. J. Agric. Food Chem. 47(3):1062-1066.
WHO (1998). World Health Organisation (WHO). Environmental Health Criteria 202, Selected Non-heterocyclic PAHs. Geneva.
WHO (2000). Polycyclic Aromatic Hydrocarbons. In Air Quality Guidelines (2 ed.). Copenhagen: World Health Organization.
Wretling S, Eriksson A, Eskhult GA, Larsson B (2010). Polycyclic aromatic hydrocarbons (PAHs) in Swedish smoked meat and fish. J. Food Compos. Anal. 23(3):264-272.

Potentials of Nigerian calcined kaolin as paint pigment

Alabi F. M. and Omojola. M. O.

Raw Materials Research and Development Council PMB 232, Garki Abuja, Nigeria.

Pigment is one of the main components in paint formulation which is dominated with the use of titanium dioxide, due primarily to three important physical characteristics; which are high refractive index, high opacity and chemical inertness. However, due to increasing demand of this important pigment and the attendant increase in price, it has become imperative to develop cheaper alternative source with similar properties. One such alternative is calcined kaolin. Knowing fully that not all kaolin deposits can be calcined to meet the peculiar requirements, this paper therefore highlights the potentials of some Nigeria kaolin deposits that can be calcined and utilized as alternatives to titanium dioxide in paint formulation. The paper also highlights the other pigments used in the coating industry, their advantages, disadvantages and their national and international market trends.

Key words: Pigments, calcined kaolin, titanium dioxide, refractive index, opacity.

INTRODUCTION

Pigment is one of the essential components of paint and it serves three main functions which include the optical function of providing colour, opacity and gloss; a protective function with regard to the surface underneath the paint and the binder which can be destroyed by UV; and a reinforcing function for the paint itself helping the binder to stick. Pigments usually have a particle size of about 0.5 to 5 µms in standard paints but can be as small as 0.01 µms in the case of some bright organic pigment, a size that enables them to refract light (Ghosh et al., 1990; Murray, 2007; Clark, 2013). A variety of natural and synthetic pigments are used in paints, providing a complete spectrum of colours and finishes.

Classification of pigments

Pigments can be classified into three main groups:

(1) Inorganic e.g Titanium dioxide, red and yellow iron oxides, mixed metal oxide and chrome oxide or organic e.g Black, Bright yellow, Green, Orange, Violet, Magenta;
(2) Extender pigments, and
(3) Anti-corrosive and special function pigments.

Inorganic pigments are of mineral origin and are usually metal oxides. Many metal oxides are found naturally such as red and yellow iron oxides. However, most are now made synthetically to get the cleanest colour. The advantages and disadvantages of some common inorganic pigments used in paint formulation are presented in Table 1.

Reaction 1. $4ZnO + 2CrO_3 + K2Cr_2O_7 + H_2O \longrightarrow K_2CrO_4 + 3ZnCrO_4Zn(OH)_2$

Reaction 2. $4Fe + 2H_2O + 3O_2 \longrightarrow 2Fe_2O_3H_2O$ (spongy mass)

Reaction 3. $2Cr_6+ + 2Fe + 6OH- \longrightarrow 2Cr_3^+ + Fe_2O_3$ (hard, protective coating) $+ 3H_2O$

Reaction 4. $ZnSO_4 + BaS \longrightarrow BaSO_4(s) + ZnS(s)$

Organic pigments are based on the chemical element

Table 1. Some common inorganic pigments.

Type	Production	Advantages	Disadvantages	Colour(s)
Carbon black	Decomposition of carbonaceous matter	High strength, good colour, light and weather resistant	Thickens paint	Black
Titanium dioxide	Synthesised	High strength, high opacity, cheap, good UV resistance	Forms radicals that degrade the binder	White
Iron oxides	Usually mined although can be synthesised	Light and weather resistant, unreactive	Cannot produce clean shades	Yellow, red, brown, black
Zinc chromates	Synthesised according to reaction (1)	Corrosion formed by reaction (2) is inhibited according to reaction (3)	Thickens paint	Yellow
Azurite $Na_7Al_6Si_4O_{24}S_2$	Kaolin, Na_2CO_3, sulphur and carbon heated together above 800°C	Rich colours	Fades on contact with acid	Blue
Chromium oxides		Light, weather, alkali and acid resistant; thermally stable	Doesn't give clear colours	Green, blue
Cadmium sulphides	Reacting cadmium salts with sodium sulphide	Heat and light resistant, clear pigment, high opacity	Expensive, poor weather resistance	Greenish yellow to red to Bordeaux
Lithopone ZnS mixed with $BaSO_4$	Synthesised according to reaction (4)	Pure tints when mixed with organics, unreactive, easily wetted	Poor weather resistance	white

Source: (Clark, 2013).

carbon. Most organic pigments are not naturally occurring and are made from petroleum by-products. Organic pigments are the brightly coloured ones used for clean colours red, yellow, bright green and blue, purple and magenta. They sprung into prominence in the last 10 to 20 years due to the awareness of the dangers of lead in the older bright red and yellow lead chromate type pigments and the advancing technology in manufacturing and chemical engineering in the organic pigment industry. Some advantages and disadvantages of organic pigments are highlighted in Table 2. Extender pigments are usually naturally occurring minerals such as clay, mica, talc and limestone. They are often used as a cheap way of reducing the cost of a can of paint. These mineral are usually produced bymining. Table 3 shows the advantages and disadvantages of some common extender pigments. Anti-corrosive pigments are a very important area of study. Steel requires the greatest protection against corrosion and the development of pigments to inhibit the corrosion of steel has occupied chemists for many years. The anti-corrosive pigment which has been used for many years is red lead, Pb_3O_4. In combination with linseed oil, red lead has in the past been the standard anti-corrosive primer for iron and steel. Unfortunately, red lead is toxic and much research has been carried out and is still being carried out worldwide to find anti-corrosive pigments which are as effective as red lead. Anti-corrosive pigments for iron and steel commonly in use include zinc chromate, zinc phosphate and zinc molybdate and barium metaborate, although alternatives to chromates are being sought as they are toxic and environmentally hazardous (Resene, 2013). Special purpose pigments such as copper can be used in anti-fouling paints for ships. Another area for specialty pigments is in intumescent coatings. These pigments break down when heated and initiate chemical reactions that eventually results in the formation of a thick skeleton of carbon about 50 to 100 times the original paint thickness (Resene, 2013). This

Table 2. Advantages and disadvantages of organic pigments.

Advantages	Disadvantages
They are available in bright vivid colours to match all shades of the rainbow	They are very expensive
They have greater tinting strength and cleaned toned except black	Very difficult to get clean reds and yellows with good opacity and good exterior light fastness
Pigments are generally low in density and not prone to settling out	They are very difficult to grind into paint millbases
Some organic pigment absorbs UV light thereby preventing it from damaging the binder	Organic pigments often have a very high oil absorption or demand for binder which results in paint with poor opacity
	They have low density and may tend to float to top of paint
	Very small particle size of 0.01 to 0.1 μms. This makes handling in the factory difficult as spilling just a few grams can contaminate a large area
	They are often not heat stable

Source: (Clark, 2013).

Table 3. Some common extender pigments.

Type	Advantage	Disadvantage	Colour(s)
Mica	Chemical and solar resistant	Poor adherent to surface and rapid settling leading to inhomogeneity in aqueous paint	Transparent
Barytes	Wear resistant	It appears darker when wet with less covering power than after drying	Colourless, white, light shades of blue, yellow, grey and brown
Calcium carbonate	Helps flatten the gloss of semi-gloss and low sheen paints. Usually a very cheap pigment	Does not contribute much in film integrity and its chemistry makes it subject to acid attack. High ionic content that adversely affects paint stability	White
Talc	Assists titanium dioxide dispersion and improves sanding	It leads to high viscosity	Pale green, white, grey and colourless
Silica	Good flattening agent and wear resistant. It gives extra hardness to paint	High viscosity and less adherent to surface	White
Clay	Helps to prevent settling and increases bulk density of paint	Usually has coarse particle and low hiding power	white

Source: (Clark, 2013).

skeleton forms a physical barrier, which save the substrate from burning for long enough for people to escape from a fire. Other special pigments are the non-leafing aluminium flake pigments and the special effect, iridescent, pearl afflair pigments used in automotive finishes.

Of all the various pigments, titanium dioxide is regarded as the most important. Therefore, as a result of increasing demand for titanium dioxide, its price has been increasing. Titanium dioxide price rose about 35% in 2011 and is expected to increase at an average annual rate of about 7% for the near future (Huntsman, 2013). For this reason, the search for an alternative to substitute titanium dioxide as paint pigment is desirable. One such alternative is calcined kaolin with its physical and chemical properties compare favourably with those of titanium dioxide.

With regard to the environmental impact, the process for production of calcined kaolin does not have any negative effect on the environment and requires less capital investment. On the other hand, the production process for titanium dioxide using the sulphate method has great negative impact on the environment. It also requires huge capital investment which probably translates to its high cost. Also, the eco-friendly chloride process is associated with negative economic factors and

Table 4. Trade summary of various pigments imported into Nigeria between 2008 and 2011.

Type of pigment	Value (₦)			
	2008	**2009**	**2010**	**2011**
Dye	130,311,757	215,000,783	394,199,941	183,461,121
Organic pigments	185,679,432	237,346,081	487,522,555	458,957,593
Metallic pigments	229,026,522	422,056,834	520,309,509	406,210,647
Inorganic pigment (others)	2,222,759,520	6,818,715,214	2,783,398,429	4,525,305,180
Titanium dioxide	822,075,763	1,883,243,736	4,141,382,689	3,388,338,441
Inorganic carbon	6,988,250	246,858,511	2,947,283	1,942,292
Organic black	49,917,883	78,000,447	43,559,436	47,202,583
Total	**3,646,759,127**	**9,901,221,606**	**8,373,319,842**	**9,011,417,857**

Source: (NFTS, 2008, 2009, 2010 and 2011).

requires much more capital investment than the sulphate process.

INTERNATIONAL MARKET TREND OF PIGMENTS

The global market for pigment has been forecast to reach 9.0million MT valued at $24.2billion by 2015 (Joses, 2011). This increase in market growth is expected to be fuelled by growth in the paints, coatings, ink and construction industries. Asian pacific represents the largest regional market for pigments worldwide both in terms of volume and value. Europe trails behind Asia while US constitute the next largest market. China is the largest and fastest growing market for pigment across the globe.

Segment wise, inorganic pigments represent the largest segment in volume and value terms, closely followed by organic pigment and then the specialty pigment representing a promising market due to their eco-friendly characteristics. Amongst the inorganic pigment, titanium dioxide is the most important pigment in the world accounting for approximately 70% of total volume (IHS Chemical, 2013). About 96% of titanium produced is used as feedstock for titanium dioxide pigment (Huntsman, 2013). Currently, the major producing feedstock regions are Australia, South Africa, Canada and India, with Australia being the largest producer.

The major consumers of titanium dioxide for paint production is China where consumption doubled from 2004 to 2011 and is still expected to grow at an average annual rate of 7% in the next decade (IHS Chemical, 2013). Western European and Japanese consumption of titanium dioxide has remained stagnant whereas North American consumption declined in the last decade due to decrease in coatings consumption caused by poor construction and manufacturing markets. There has also been a steady price increase of titanium dioxide. The price of titanium dioxide was about $500/MT in 2010 and rose to $2,500/MT in 2012. The price rose about 11% in 2011(IHS Chemical, 2013).

Similarly, China is also the largest single consumer of organic pigment in the World and the fastest growing national market. The specialty pigment market is also growing at approximately 7% per year and is estimated to be worth $2.1billion in 2013 (IHS Chemical, 2013).

MARKET TREND OF PIGMENTS IN NIGERIA

The paint industry in Nigeria is highly import dependant as over 70% of the raw materials for the sector are imported. The major reason for non-availability of local raw materials for the sector is absence of the Petrochemical industry in the country which is expected to produce most of the chemicals used in the industry. Also, it is because most beneficiating plants that process locally available non-metallic raw materials for the industry are operating below installed capacity due to the obsolete technology and equipment still being used. Furthermore, there is a low level of R&D in the processing of metallic, non-metallic and agro-based raw materials needed in the sector (PMA, 2010). According to National Bureau of Statistics, Nigeria spent about ₦3.6b on the importation of paint pigment in 2008. This figure rose to ₦ 9.9b (NFTS, 2008, 2009, 2010 and 2011). Table 4 shows a breakdown of various paint pigment imported for 2008, 2009, 2010 and 2011.

POTENTIALS OF USING CALCINED KAOLIN AS PIGMENT

Pigments for paint formulations are submicron particles which must possess certain properties such as good opacifying strength (commonly called hiding power) and brightness. Other important features are excellent resistance to chemical attack, good thermal stability and resistance to UV degradation.

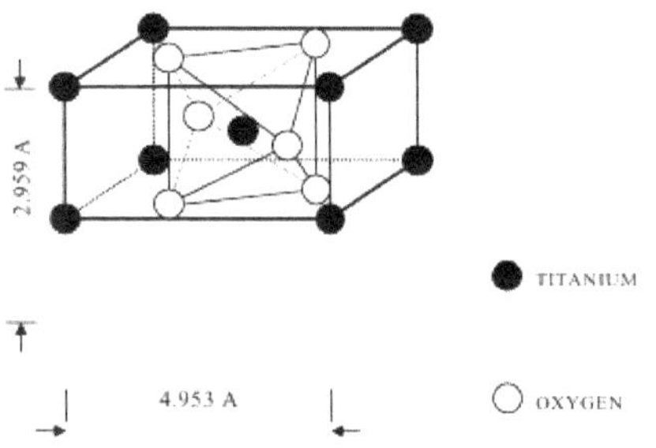

Unit cell of Rutile TiO_2

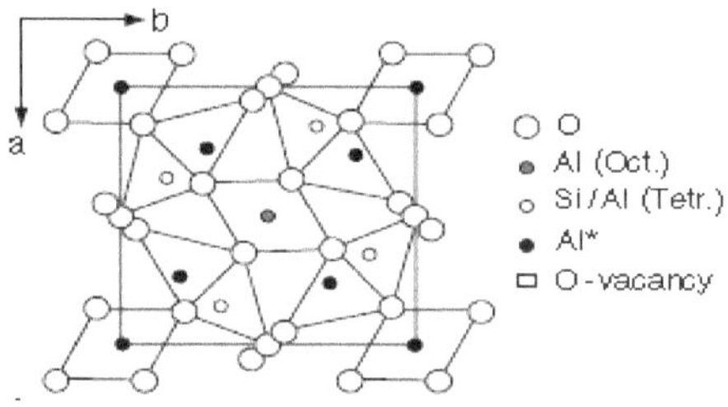

unit cell of calcined kaolin (mullite)

Figure 1. Chemical structures of TiO_2 and calcined kaolin.

Pigments from mineral sources offer opacity and brightness because they have the ability to scatter light. The degree to which minerals scatter light depends on their intrinsic index of refraction and also on the structure they form in the coating (Imerys, 2013). Titanium dioxide has high refractive index, strong absorption in the UV region of light spectrum and strong reflectance in the visible spectrum thereby giving it a very high light scattering properties which, when combined with its particle size creates an effective pigment for brightness and opacity (Imerys, 2013) (Figure 1).

Calcined kaolin also exhibit properties similar to titanium dioxide and therefore can be used as an alternative pigment to titanium dioxide. These properties include brightness and opacity. Calcined kaolin effectively scatters light due to its porous structure, hence when light encounters it, some of the light gets reflected off the planar surfaces, a portion of the light enters the structure and these rays upon reflection encounter other crystals in close proximity. This light then undergoes multiple internal reflections and diffractions giving excellent optical scatter (Imerys, 2013). The similarities in properties can be attributed to the similarity in their chemical structure as shown in Figure 1. According to the structure, the rutile TiO_2 form has TiO_6 octahedrons that form columns by sharing two edges. Titanium has six oxygen nearest neighbours and oxygen has three titanium nearest neighbours. Adjacent columns are linked through sharing corners of other octahedra. Likewise, the mullite structure consists of chains of distorted edge-sharing Al-O octahedral at the corners and centre of each unit cell. These chains are further cross-linked by Si-O and Al-O corner-sharing tetrahedra.

EFFECT OF CALCINATION ON THE PHYSICOCHEMICAL PROPERTIES OF KAOLIN

For kaolin to be suitable for pigment, several empirical tests are run on the kaolin to determine its suitability. The mineralogical composition and fundamental structure of the kaolin are very important since they control the resultant physical and chemical properties that determine its industrial application (Murray and Lyon, 1956). Calcination increases the silica and alumina content of kaolin. The silica content determines the texture of the paint while aluminium content and particle size the brightness and opacity of the paint. Absence of iron oxide improves the brightness (Duval et al., 2008; Raw Materials Research and Development Council (RMRDC) 2004). The ideal composition of silica and alumina in calcined kaolin for paint pigment is of the range of 45 to 47% and 37 to 39% respectively, while that of iron oxide content should be less than 2% (American, 2013). Silica content higher than the above range increases the abrasion of the calcined kaolin pigment. Reduction of the silica content to the required level can be carried out by controlled leaching of the silica from the calcined kaolin using caustic solution which dissolves the silica out of the calcined kaolin (Engelhard, 2013). The removal of the iron oxide can be carried out using hydrosulphite, high intensity magnetic separation and the froth flotation technique. The brightness, opacity or both of calcined kaolin can be increased by increasing the fineness of the kaolin feed supplied to the calciner (Engelhard, 2013). Also there is a correlation between brightness and degree of crystal perfection of kaolin. Poorly crystalline kaolin is expected to contain more impurities on their surfaces because of more available valence charges (Murray and Lyon, 1956). Tables 5 and 6 show the chemical composition and physical properties of kaolin suitable for pigment, while Table 7 is a comparison of the properties of titanium dioxide and calcined kaolin useful in paint formulation. Table 7 shows a comparison of the properties of titanium dioxide and calcined kaolin useful in paint formulation.

NIGERIAN KAOLIN DEPOSITS WITH POTENTIALS AS PAINT PIGMENT

Nigeria has a huge reserve of kaolin with an estimated reserve of about 3 billion metric tonnes scattered in different parts of the country (Edomwonyi-Out et al.,

Table 5. Chemical composition of kaolin suitable for calcined kaolin as pigment.

Component	Weight percent
SiO_2	45-47
Al_2O_3	37-39
Fe_2O_3	0.3-1.9
TiO_2	0.4-0.8
CaO	Trace
MgO	Trace
Na_2O	0.2-0.3
K_2O	0.1-0.3
Loss on ignition	14.45-14.50

Source: (Glass, 1957; Ghosh et al., 1990; Mesbah et al., 2011; Murray, 2007).

Table 6. Physical properties of kaolin suitable for paint formulation.

Property	Raw kaolin	Calcined kaolin
Oil absorption	44	43
pH	5.6	6.95
Conductivity ($\Omega^{-1}cm^{-1}$)	111.3	34.32
Bulk density (gcm-1)	0.64	0.48
Relative density	2.55	2.71
Refractive index	1.56	1.55
Reflectance	82.7	90.1

Source: (Ghosh et al., 1990).

2013). There are at least 45 known deposits of kaolin in Nigeria. There is no state in the country without at least one deposit (Badmus and Olatinsu, 2009).

The market for kaolin in Nigeria is large, sustainable and expanding because of the numerous applications of the product. Also there is an outright ban on the importation of kaolin into the country in order to stimulate the growth of the local industry. The major kaolin deposit in the country and their chemical composition are shown in Table 8. It can be observed from the table that Nahuta, Kankara, Onibode, Ifon and Major porter deposits have the potentials to meet industrial quality as pigment because their Alumina, Silica and Iron contents almost fall within the required range. The chemical composition of these deposits can be improved to achieve the ideal if the kaolin undergoes the process of beneficiation and calcination.

PHASE CHANGES OF KAOLINITE DURING HEAT TREATMENT

Kaolinite, $Al_2O_32SiO_2.2H_2O$, undergoes some significant changes in its structure when heated to higher temperatures. It loses its (OH) lattice water between 500 to 600°C (Quinfu et al., 2002; Schieltz and Soliman, 1964). The loss of lattice water breaks up the regular periodicity in the kaolinite structure, along the *c*-axis producing a dehydrated phase known as metakaolin (Schieltz and Soliman, 1964). By further heating, metakaolin transforms to a crystalline phase, indicated by the first exothermic peak at about 980°C to form primary mullite crystalites. Further heating to approximately 1200°C produces secondary mullite crystalites and this is followed at approximately 1350°C by a third exothermic reaction to the formation of beta-cristobalite (Wahl and Grim, 2012) (Figure 2).

$$\underset{Al_2O_3.2SiO_2.2H_2O}{\text{Kaolinite}} \xrightarrow[\Delta]{550\text{-}600^\circ C} \underset{Al_2O3.2SiO_2}{\text{Metakaolin}} \xrightarrow[\Delta]{980-1050^\circ C} \underset{Al_2O_3.SiO_2}{\text{Mullite}} \xrightarrow[\Delta]{1100\text{-}1350^\circ C} \underset{SiO_2}{\text{Cristoballite}}$$

Mullite being the only stable intermediate phase in the Al_2O_3-SiO_2 system, is an important material for both conventional and high-technology applications (Temuujin et al., 2000; Titanium and titanium dioxide, 2013). Most pigment grade calcined kaolin is heated to temperature between 1000 and 1050°C which indicates the mullite

Table 7. Comparative properties of Titanium dioxide and calcined kaolin.

Property	Titanium dioxide	Calcined kaolin
Chemical stability	Stable	Stable
Crystal structure	Tetragonal	Triclinic with hexagonal flaky layers
Durability	Good	Good
Particle size (μm)	0.2-0.3	0.1-1.0 (thickness)
Brightness	95%	92%
Refractive index	2.76	1.56

Source: (Ghosh et al., 1990; Murray, 2007).

Table 8. Chemical composition of major Nigerian kaolin deposits.

Deposit	Chemical composition (%)										Colour
	SiO_2	Al_2O_3	Fe_2O_3	TiO_2	CaO	MgO	Na_2O	K_2O	SO_2	L.O. I	
Ozubulu	60	26	5	5	Trace	Trace	Trace	Trace	-	-	White, light grey and dark brown
Nahuta	47	32	1.3	2.4	Trace	Trace	Trace	Trace	0.55	12-18	White yellow, pink, grey and brown
Jagalwa River	50	30	1.4	2.4	Trace	Trace	Trace	trace	-	-	-
Darazo	51.9	32.9	2.9	-	0.39	0.29	Trace	0.89	0.13	-	Orange, pink, brown, white and grey
Kankara	43.3	36.7	0.21	Nil	Trace	1.49	Trace	0.7	-	11.10	Off-white with pink and brown patches
Onibode	43.1	36.1	3.1	1.9	Trace	Trace	Trace	Trace	-	12.65	-
Ifon	48	33.2	0.06	1.72	Trace	Trace	Trace	Trace	-	-	Pink
Okitipupa	50	29	-	-	-	-	-	-	-	10	-
Major porter	45.6	35.2	2.7	-	Trace	Trace	Trace	0.97	-	-	-

Source: (RMRDC, 2004).

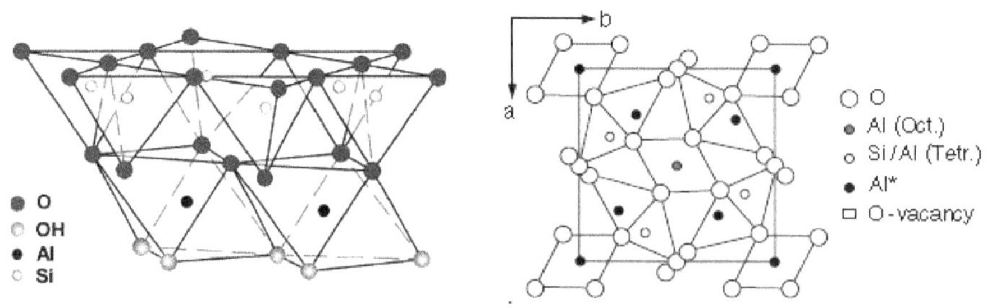

The Kaolinite structure   Average structure of mullite

Figure 2. Structural transformation of kaolinite to mullite.

phase (Murray, 2007) (Figure 3).

CONCLUSION

In any production process, cost reduction is an important consideration. The cost of titanium dioxide is about six times the cost of calcined kaolin products (Murray, 2007). The similarities in both the physical and chemical properties of calcined kaolin and titanium dioxide indicate that calcined kaolin could effectively be used as pigment

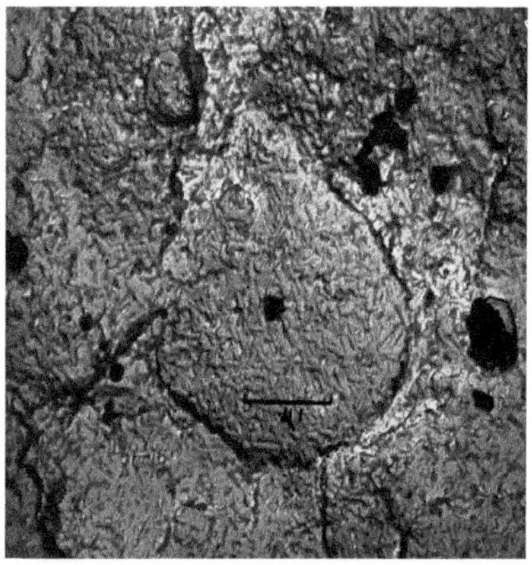

Surface structure of calcined kaolin

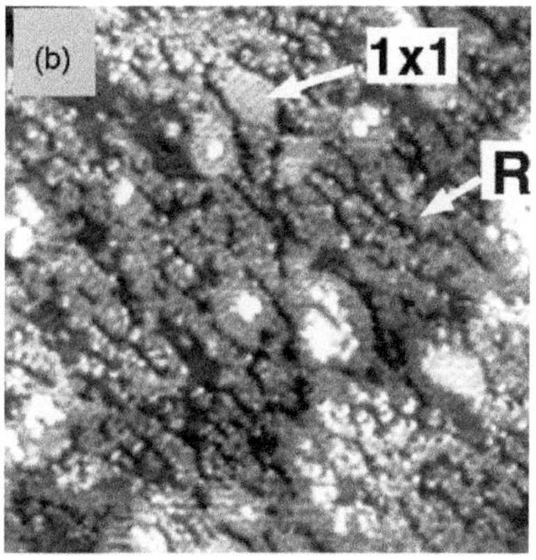

Surface structure of TiO_2

Figure 3. Structure of Calcined kaolin and TiO_2.

in paint formulation and not just as a partial substitute to titanium dioxide. Since the import value of titanium dioxide into Nigeria has been on the increase which invariably translates to huge foreign exchange being expended on its importation. Therefore with the estimated 30 million MT reserve of Kaolin from the five major deposits with good chemical composition, a vast investment opportunity exists in the production of calcined kaolin as paint pigment.

REFERENCES

American E (2013). Calcined Aluminium Silicate Powder review. http://www.americanelements.com/kaolin.html .

Badmus BS, Olatinsu OB (2009). Geophysical evaluation and chemical analysis of kaolin clay deposit of Lakiri village, southwestern Nigeria. Intl. J. Phys. Sci. 4 (10):592-606.

Clark MDT (2013). Paints and pigments review. http://www.nzic.org.nz/ChemProcess/polymer/10D.pdf

Duval JD, Risbud SH, Shackelford JF (2008). Ceramic and glass materials: Structure, properties and processing, pp 27-39. Retrieved August 14, 2013, from http;//www.springer.com/cda/content/9780387733616-c1.pdf

Edomwonyi-Otu LC, Aderemi BO, Ahmed AS, Coville NJ, Maaza M (2013). Influence of thermal treatment on kankara kaolinite. *Opticon* 1 826(15): 5, DOI:http://dx.doi.org/10.5334/opt.bc.

Engelhard C (2013). Low abrasion calcined kaolin pigments and enhanced filtration method. Retrieved September 16, 2013, from http://www.freepatentsonline.com/5997626.html

Ghosh S, Warrier KGK, Damodaran AD (1990). Thermally treated kaolin as an extender pigment. J. Mater. Sci. Letts. 9(1):1046-1048.

Huntsman C (2013). Titanium and Titanium dioxide Industry review. Retrieved August 22, 2013, from http://www.rsc.org/images/PDF1_tcm18-40441.pdf

IHS Chemical. (2013). Colour pigment, Inorganic review. Retrieved August 27, 2013, from http://www.ihs.com/products/chemical/planning/ceh/inorganic-colour.aspx.

Imerys P (2013). Calcined clays in light weight coatings technical guides. Retrieved September 19, 2013, from http://www.imerys-paper.com/Technical_Guide_Calcined_Clay_in_LWC_pdf

Jose S (2011). Pigments and dyes: A global strategic business report. Retrieved September 18, 2013, from http://www.strategicr.com/pigmentsanddyesmarketreport.asp

Mesbah H, Wilson MA, Carter MA (2011). The role of the kaolinite-mullite reaction sequence in moisture mass gain in fired kaolinite. Retrieved August 13, 2013, from http://www.scientific.net/ATS.68.38.pdf

Murray HH (2007). Developments in Clay Science, First Edition, Elsevier. Radarweg 29, PO Box 211, 1000 AE Amsterdam, The Netherlands Pp.85-94.

Murray HH, Lyons SC (1956). Correlation of paper-coating quality with degree of crystal perfection of kaolinite. Retrieved September 18, 2013, from http://www.clays.org/journal/archive/volume%204/4-1-31.pdf

NFTS (2008, 2009, 2010 and 2011). Nigerian Foreign Trade Summary Report. A publication of the National Bureau of Statistics, Plot 762 Independence Avenue, Central Business Area, Abuja-Nigeria.

Paints Manufacturers Association (2010). Paints Manufacturers Association of Nigeria Annual Report. A publication of the Paints Manufacturers Association of Nigeria, P. O. Box 67, Ikeja, Lagos, Nigeria.

Quinfu L, Shuli D, Tongjun Z, Ping L (2002). The effect of kaolinite structure changes on its transformation temperature of phases. J. China Univ. Min. Technol. 12(2):155-162.

Resene P (2013). What's in a can of Paint review. Retrieved August 20, 2013, from http://www.resene.com/homeown/prosolv/whatsinacanofpaint.pdf

RMRDC (2004). Technical brief on Minerals in Nigeria: Kaolin, pp. 8-14. A publication of Raw Materials Research and Development Council, P. M. B. 232, Garki-Abuja, Nigeria.

Schieltz NC, Soliman MR (1964). Thermodynamics of the various high temperature transformations of kaolinite. Clays Clay Miner. 13(1):419-428.

Temuujin J, Mackenzie KJD, Schmucker M, Schneider H, McManus J, Wimperis S (2000). Phase evolution in mechanically treated mixtures of kaolinite and alumina hydrates (gibbsite and boehmite). J. Euro. Ceram. Soc. 20(1):413-421

Wahl FM, Grim RE (2012). High-Temperature D.T.A and X-ray diffraction studies of reactions, pp 69-81. Retrieved September 18, 2013, from http://www.clays.org/journal/archive/volume%2012/12-1-69.pdf

Didecanoate compound: Isolated from *Momordica charantia* Linn. seeds from Nigeria

Oragwa Leonard N.[1] , Olajide Olutayo O.[2], Efiom Otu O.[1] and Okwute Simon K.[1]

[1]Chemistry Department, University of Abuja – Nigeria.
[2]Chemistry Advanced Laboratory, Sheda Science and Technology Complex, Abuja- Nigeria.

The seeds of *Momordica charantia* were extracted with hexane, ethyl acetate and 95% ethanol successively by percolation and concentrated at 37°C. Fractions 19-21 and 247 gave two compounds. However, only one has been elucidated to be 4a-phorbol-12, 13-didecanoate using the state- of –art-tools of spectrometry. The phytochemical screening of seeds of *M. charantia* showed the presence of flavonoids, glycosides, sterols, fat and oil in hexane, ethyl acetate and ethanol extracts. Anthraquinone is only present in hexane and ethyl acetate while alkaloid is in hexane and ethanol. And also phyto-medicine should be integrated into the health system of Nigeria and developed in such a way to bring harmony between the traditional and modern system of health care with minimum threat to each other.

Key words: *Momordica charantia,* Cucurbitaceae, gas chromatography, thin layer chromatography, phytochemicals.

INTRODUCTION

Momordica charantia Linn. (bitter melon) belongs to the family of Cucurbitaceae, a climbing vine which is commonly seen growing on walls and shrubs in the tropics. The textured leaves look as a bite has been taken from them giving the plant its Latin name Momordica which means to bite. The orange fruits are soft when ripe and inside black seeds have a red covering. It is used as food, bitter flavouring and medicine (Basch et al; 2003). Earlier claims showed that its bitter fruits have carminative, aphrodisiac, anthelmintic properties and are used in syphilis, rheumatism, and troubles of spleen and ophthalmic. It is also useful in piles, leprosy and jaundice and also used as a vermifuge (Kirtikar and Basu 2006). The ethanolic extract showed an analgesic and antipyretic effect which was significant higher than that in the control rats. The observed pharmacological activities provide the scientific basis to support traditional claims as well as explore some new and promising leads (Roshan et al., 2010). The mineral and amino acid analysis showed that the bitter melon contained nutritionally useful quantities of most of the essential minerals and amino acids. The predominant fatty acid was α- eleostearic acid in non-polar lipids, linolenic acid in glycolipids and palmitic acid in phospholipids (Kuri et al., 1991). The plant inhibits primary human adipocyte differentiation by modulating adipogenic genes and it was effective in reducing lipid accumulating in primary human adipocytes by regulating adipogenic transcription factors and adipocytokine gene expression (Pratibha et al., 2010). Many pharmacological properties have been reported including antioxidant, anti-diabetic, anticancer, anti-fertility, antimicrobial, antiviral and hepatoprotective activity. Popularity of the plant in various systems of traditional medium for several ailments such as anti-diabetic, contraceptive, dysmenorrhoea, eczema has been reported. Traditionally, it has also been used in treating peptic ulcers, interestingly in a recent experimental studies have exhibited its potential against *Helicobacter pylori* (Grover, 2004). The plant contains several biologically active compounds chiefly momordicin, cucurbitacin and glycosides such as momordin, charantin, momordicosides and other terpenoids compounds such as momordenol, momordol, momordicin-28 and momordicilin (Fatope et al., 1990; Ortigao and Better, 1792). However, there are reports to our knowledge on isolation of some bioactive compounds from the leaves of *M. charantia* which are Asian variety, hence the present study was undertaken to carry out phytochemicals and to isolate newly identified compounds from the seeds of the bitter melon for the first time with African specie using the state-of -art –tools spectrometry.

MATERIALS AND METHODS

Plant materials

The matured gourds of *M. charantia* were harvested at Chika, along Airport Road, Abuja and identified at the Herbarium Unit of Pharmaceutical Research and Development, Idu -Abuja, where voucher specimens were deposited.

Extraction/ partitioning procedure

Blended 500 g of the seeds was extracted by percolation as described by (Singh et al., 2006) using hexane, ethyl acetate and 95% ethanol as extracting solvents which were in increasing order of polarity and they were concentrated using rotary evaporator at 37°C, at reduced pressure. Thereafter, phytochemical screening was carried out and ethyl acetate extract was subjected to flash column chromatography based on the results of the screening. Using Chemo-glass 500 ml of column diameter 27 mm and optimal flow range of 10-30 ml/min, 10 g Silica gel powder of 150 angstrom spore, 35-75 micron particle and 239-400 mesh size and 5 g of the crude extract dissolved in 15 ml ethyl acetate. The gradient system, in which the solvent composition changes during the course of elution, was used. The solvent system consisted of hexane (Hex), ethyl acetate (Ea), methanol (Meth) in different ratios. 287 fractions of 40 mls were collected and evaporated to dryness. Thin layer chromatography was also used for fingerprint profiling of the fractions collected using solvent mixture of hexane and ethyl acetate (9:2). Those with the same retention values (Rf.) were combined and two major fractions namely samples 19-21 which is denoted as compound PS1 and sample 247denoted as compound PS2 were obtained.

Apparatus, reagents and equipment

Solvents used were of analytical grade manufactured by Sigma-Aldrich Laboratory, Germany but were redistilled for certainty of quality. Thin layer chromatography (TLC) analysis was carried out on pre-coated Merck-Kiesegel 60F 254 with 0.25 mm thickness. The TLC plates were visualized by exposure to iodine vapor and further clarified by UV light at 254/336 nm. The GC-MS analysis was also carried out. All these were done at the Sheda Science and Technology Complex, Abuja.

Phytochemical screening of extract

For the purpose of this study, phytochemical screenings were carried out on the extracts to confirm the presence or absence of the following plant secondary metabolites: alkaloids, phenols, sterols, terpenes, tannins, flavonoids, anthraquinones, cardiac glycosides, saponins, fats and oil (Harborne, 1973; Trease and Evans, 1989).

Phenols: Equal volumes of each extract and ferric chloride solution are added together. A deep bluish green precipitate indicates the presence of phenol.

Alkaloids: Each extract was added to 1% aqueous HCl over water bath and filtered. The filtrate was treated with iodine in potassium iodide. Formation brown or reddish brown precipitate indicates presence of alkaloids.

Steroids: Each extract was added to 2 ml acetic anhydride and 2 ml H_2SO_4. Colour change from violet to blue or green indicates the presence of steroids.

Terpenes: Each extract was added to 0.5 ml acetic anhydride and few drops of concentrated H_2SO_4. A bluish green precipitate indicates the presence of terpenes.

Cardiac glycosides: Extract was treated with 2 ml glacial acetic acid with a drop of ferric chloride solution and underplayed with 1 ml H_2SO_4. A browning at the interface indicates the presence of cardiac glycosides.

Tannins: Each extract was boiled in 20 ml water and filtered. A few drops of 0.1% ferric chloride solution were added. Brownish green or blue-black color indicates the presence of tannins.

Flavonoids: 5 ml ammonium solution was added to aqueous filtrate of each extract and then few drops of concentrated H_2SO_4. Yellow coloration indicates the presence of flavonoids.

Anthraquinones: 10 ml benzene was added to each extract and filtered. 0.5 ml of 1% Ammonium solution was added and shaken. Pink, red, or violet color in the ammoniacal lower phase indicates the presence of Anthraquinones.

Saponins: 1 g each extract was boiled with 5 ml distilled water and filtered. 3 ml distilled water was added to the filtrate and shaken vigorously for 5 min. Persistent frothing on warming indicates the presence of Saponins.

Fats and Oils: Small quantity of each extract was pressed between two filter papers. Oily stains indicate the presence of fats and oils.

Spectral analysis

The fraction with higher yield-Compound PS1 was used for this work. It was elucidated and characterised by H-NMR, GC-MS and FTIR and discussed. The other compound PS2 was only analysed by H-NMR, GC-MS and FTIR and Mass Spectrometer search gave a complex compound and was not discussed in this report. Quantitative and qualitative data were determined by GC-MS respectively. The samples were injected into a thermo-scientific trace GC ultra system coupled to DSQ II mass spectrometer, and equipped with an AS 3000 auto sampler and a split/split-less injector. The column used was an TR-5MS, 30 m × 0.25 mm i. d., 0.25 µm d. f., coated with 5% diphenyl-95% polydimethyl siloxane, operated with the following oven temperature programmed: 140°C, held for 1 min, rising at 8°C/min to 300°C, injection temperature and volume, 250° C and 1.0 µl respectively; injection mode, split, split ratio, 15:1; carrier gas, helium at 30 cm/s linear velocity and inlet pressure 99.8KPa; detector temperature, 280°C. The components of the sample were identified based on the basis of their retention indices. Identification confirmation was by comparison of their mass spectra with published spectra (Adams,

Table 1. Result for phytochemical screening of *M. charantia* seed extracts

Phytochemical	Result		
	Hexane	**Ethyl Acetate**	**95% Ethanol**
Alkaloids	+	-	+
Flavonoids	+	+	+
Glycosides	+	+	+
Saponins	-	-	+
Taninns	-	+	-
Sterols	+	+	+
Terpenes	+	+	-
Phenols	-	-	-
Anthraquinones	+	+	-
Fats and Oils	+	+	+

Key: (+) = Present; (-) = Absent.

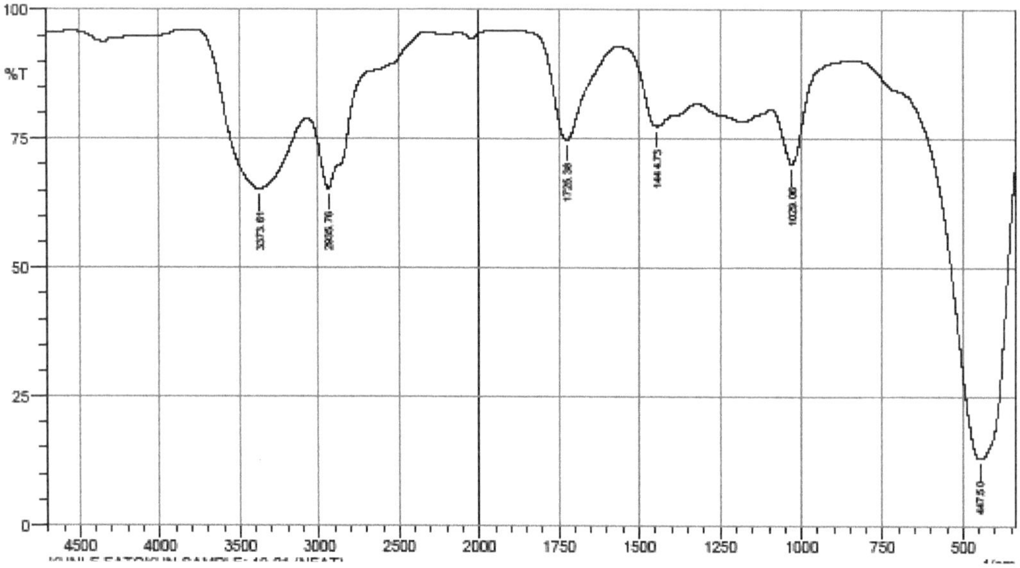

Figure 1. Infra red spectrum of compound PS1.

1989) and those of reference compounds from Library of National Institute of Standard and Technology (NIST) database.

RESULTS AND DISCUSSION

The phytochemical screening of seeds of *M. charantia* showed the presence of flavonoids, glycosides, sterols, fat and oil in hexane, ethyl acetate and ethanol extracts in Table 1. Only hexane and ethanol extracts contain alkaloids while phenols were not detected in all the extracts. These chemical constituents present in the extracts have many therapeutic values. Flavonoids have both antifungal and antibacterial activity. They possess anti-inflammatory properties (Iwu et al., 1999). GC-MS analysis and FTIR, HNMR Spectral analyses of Compound PS1 gave the following results. From Figure 1 of the FTIR analysis, the intense and broad absorption band at 3373.61 cm^{-1} is associated with both the O-H and the C-O stretching vibrations; 1444.73 cm^{-1} is suggestive of C–C stretch (in-ring); 1725.38 cm^{-1} is suggestive of esters. This is confirmed by the C–O stretches which appear as two or more bands in the region 1029.06 cm^{-1} typical of esters. The band at 2935.76 cm^{-1} is characteristic of spectrum of simple alkanes due to C–H stretching and bending. This is confirmed by the C–H bend or scissoring of 1444.73 cm^{-1} for alkanes. Thus from these results the compound is an alcoholic ester with extended and saturated alkyl chains. From Figure 5 of the H-NMR analysis, it showed chemical shifts at 0.86 ppm suggestive of terminal methyl $-CH_3$ in alkyl; 1.2-1.49 ppm

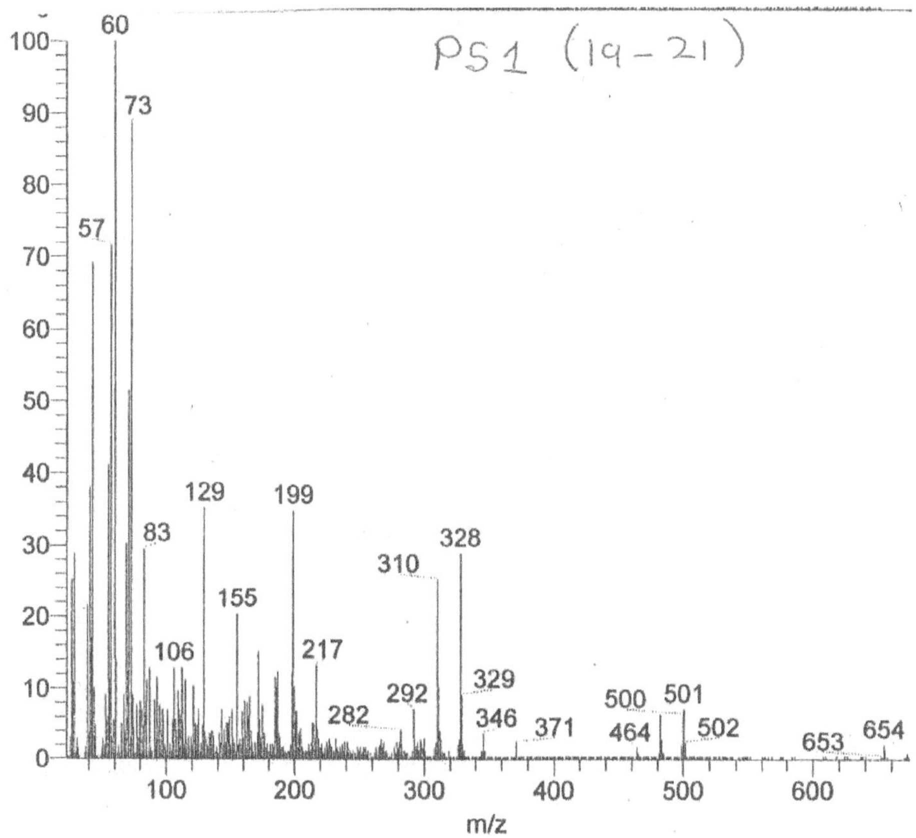

Figure 2. Mass spectrum of compound PS1.

is suggestive of saturated alkyl chain $-CH_2$; 3.5 ppm is suggestive of hydroxyl proton R-OH; 4.25-4.4 ppm is suggestive of esterified glycerol $-CH_2$-O-CO-R or esterified phorbol. From Figure 2 of the GC-MS, the gas chromatography showed one major peak at with retention time of 2.95 min. Thus it has one component.

The mass spectral showed a molecular ion at m/e 654 with a base peak at m/e 60. Other major peaks include 501, 328, 310, 199, 129, 83 and 73. From the computer MS Library in Figure 3, phytochemical is suggestive of 4a-Phorbol, 12, 13-didecanoate and its fragmentation pattern was showed in Figure 4. Also, from Figure 5 of the H-NMR analysis, it showed chemical shifts at 0.86ppm suggestive of terminal methyl —CH3 in alkyl; 1.2-1.49ppm is suggestive of saturated alkyl chain -CH2; 3.5ppm is suggestive of hydroxyl proton R—OH; 4.25-4.4ppm is suggestive of esterified glycerol —CH2—O—CO—R or esterified phorbol. From the ethno medical uses and biological studies the seeds were reported to be anti diabetic. This agrees with report that showed that decanoic acid found attached to phorbol as obtained above had the same modulating effect on the sub-family of nuclear receptors called Perixisome Ploriferator Activited receptors (PPAR) as the anti-diabetic drug-Thiazolidinedione (TDZ). These receptors are responsible for glucose metabolism. This may be responsible for its reported anti-diabetic therapeutic properties. They concluded that this acid can serve as a regulator of blood sugar levels in cells and may have important application in designing better and safer drugs for diabetes treatment since TDZ has negative side effects which include weight gain, fluid retention and increased risk for cardiovascular disease. Based on the foregoing it appears that in *M. charantia*, 4a-phorbol-12, 13-didecanoate is hydrolyzed in situ to release the active decanoic acid probably by enzymatic process (es). Thus this plant is a store for decanoic acid. It is also reported that decanoic acid is among the three most valuable and potent antimicrobial medium chain fatty acids (MCFAs) that display antibacterial, antiviral, antifungal, anti parasitic and anti protozoal properties. The others are lauric acid and caprylic acid (Kabara, 1978). A study conducted in the Philippines showed that decanoic and lauric acids were efficient in destroying the AIDS virus (HIV) in laboratory cultures thus revealing a possible treatment for AIDS that was a lot safer and cheaper than the antiviral drugs being used at the time (Thormar, 1987). This decanoic acid component is present in 'all important' mothers' milk and Coconut oil (Isaacs, 1990). Pharmaceutically, some drugs called depot injections

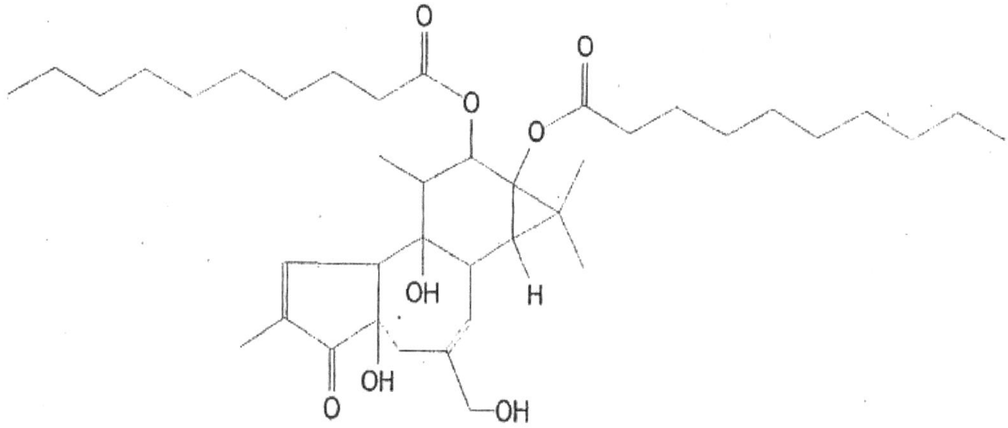

Figure 3. MS search result of Compound PS1- 4a-phorbol-12, 13-didecanoate.

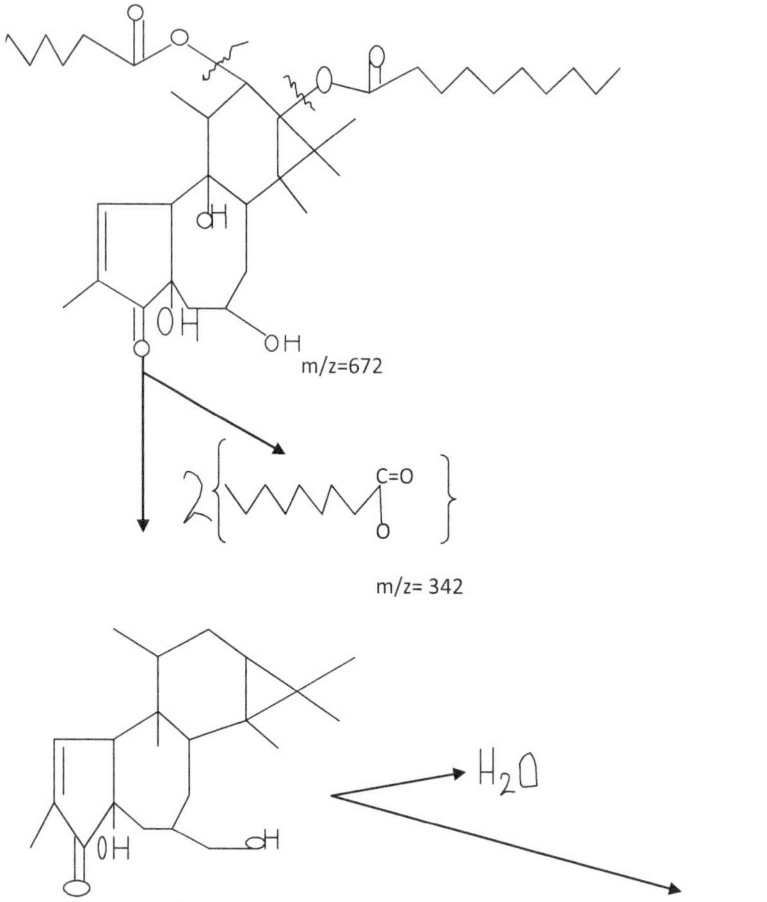

Figure 4. Fragmentation pattern of 4a- Phorbol-12,13-didecanoate.

contain decanoic acid in its ester form. This increases drug lipophilicity and its affinity for fatty acids because distribution of drug in fatty tissues is slow. These drugs include nandrolone used in the treatment of osteoporosis

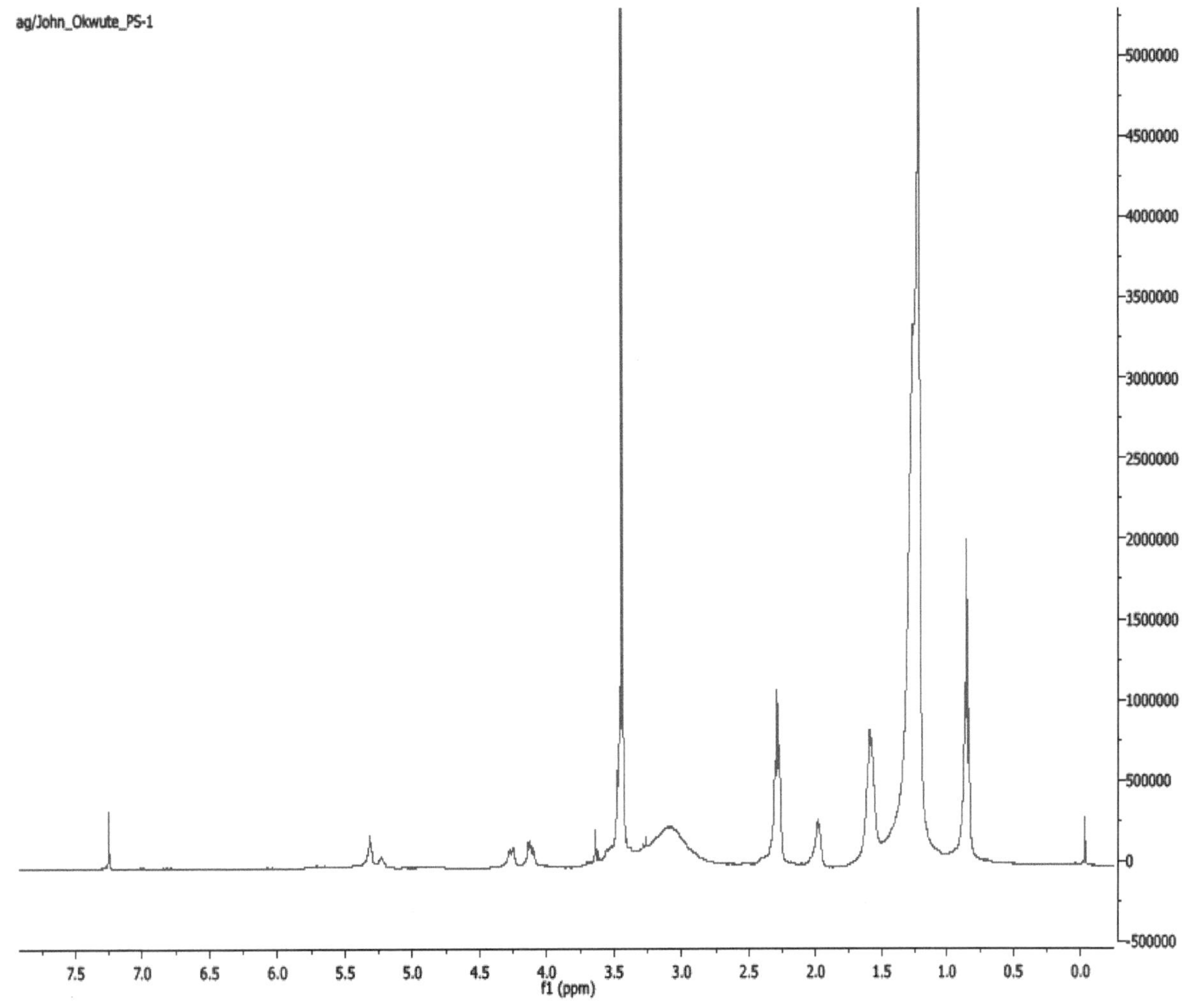

Figure 5. H-NMR spectrum of compound PS1.

in menopausal women and fluphenazine used for the treatment of psychoses e.g Schizophrenia. Finally, there is no past record on the isolation and characterization of 4a-Phorbol, 12, 13-didecanoate from this plant which is suggestive of this being the first. Therefore this study has contributed to the body of knowledge on phytochemicals, antimicrobials and characterization and identification of compounds in *M. charantia* seeds.

Conclusion

We recommend that if capric or decanoic acid possesses strong antimicrobial properties by killing or inhibiting the growth of microbes such as bacteria, fungi, or viruses, the possibility of the hydrolysis of 4a-Phorbol, 12, 13-didecanoate in plant cell to release decanoic acids should be studied further for use in designing of better and safer drugs for the prevention and treatment of infections and formulation of potential disinfectants, preservatives and pharmaceuticals in food products, and in cosmetics. And also phyto-medicine should be integrated into the health system of Nigeria and developed in such a way to bring harmony between the traditional and modern system of health care with minimum threat to each other.

ACKNOWLEGDEMENT

The authors would like to thank Dr. Abayomi Orishadipe and Mr. 'Kunle Fatokun for their technical assistance during the preparation of this manuscript.

REFERENCES

Adams RP (1989). Identification of essential oils by ion trap mass spectroscopy, Academic Press. Inc. USA.

Basch E, Gabardi S, Ulbricht C (2003). Bitter melon *(Momordica charantia* Linn.): A review of efficacy and safety. Am. J. Health System Pharm. 60:356-359.

Fatope M, Takeda Y, Yamashita H, Okabe H, Yamauchi T (1990). New Cucurbitane triterpenoids from *Momordica charantia*. J Nat. Prod. 53(6):1491-1497.

Grover JK, Yadav SP (2004). Pharmacological Actions and potential uses of *Momordica charantia*, A review. J. Ethno. Pharmacol. 93(1):123-132.

Harborne JB (1973). Phytochemical Methods, a Guild to Modern Techniques of Plant Analysis. Chapman and Hall, London pp.182-201.

Isaacs CE (1990). Antiviral and antibacterial lipids in human milk and infant formula feeds. Archives of Disease in Childhood. 65:861-864.

Iwu MM, Angela RD, Chris O (1999). New microbial of plant origin in Janick (ed) perspective on crops and their uses. ASHS press Mexandrria pp.457-462.

Kabara JJ (1978). The Pharmacological Effect of Lipids. Champaign, Ill: The American Oil Chemists' Society, pp.1-14.

Kirtikar KR, Basu BD (2006). Indian Medicinal Plants, 2nd edition, vol.2 Dehradum. Intl. Book Distributors, P. 1130.

Kuri EY, Koyyalamudi SR, Chalapan K, Gwyu PJ, Donald ER (1991). Chemical Composition of *Momordica charantia* fruits. J. Agric. Food Chem. 39:1762-1763.

Ortigao M, Better M (1792). Momordin II a ribosome inactivating protein from *Momordica balsamina* is a homologous to other plant protein. Nucleic Acids Res. 20(17):4662.

Pratibha V Nerurkar, Yun-keng L, Vivek RN (2010). Momordica charantia Inhibits Primary Human Adipocyte Differentiation By modulating Adipogenic genes. BMC Complement. Altern. Med. 10(34):1-10.

Roshan P, Naveen M, Nitin U, Naheed W, Hetal T, Zalak P (2010). Analgesic and Antipyretic Activities of *Momordical charantia Linn* fruits. J. Adv. Pharm Technol. Res. 1(14):415-418.

Singh RK, Dhiman RC, Mittal PK (2006). Mosquito larvicidal properties of *Momordica charantia* Linn (Cucurbitaceae) J .Vect. Borne Dis. 43:88-91.

Thormar H (1987). Inactivation enveloped viruses and killing of cells by fatty acids and monoglycerides. Antimicrob. Agents Chemother. 31:27-31.

Trease CE, Evans WC (1989). A Textbook of Pharmacognosy (13th ed.) Bailliere, Tindal Ltd, London 40-58:224-233.

Oxidation of acha (*Digitaria exilis*) starch using 3.5% active chlorine sodium hypochlorite

S. Isah, A. A. Oshodi and V. N. Atasie

Bells University of Technology, Nigeria.

Acha (*Digitaria exilis*) starch was chemically modified using sodium hypochlorite (3.5% active chlorine) as oxidative agent. The physicochemical properties of the native acha starch and oxidized derivative were investigated. Oxidation improved the water and oil absorption capacity of the native acha starch and also the emulsion capacity was significantly improved. The bulk density, foam capacity and solubility reduced upon oxidation. The granule morphologies investigated using scanning electron microscopy showed significant difference in the ratio of crystalline and amorphous regions. However, oxidation of native acha starch did not affect the shape, appearance and structural arrangement of the starch granules. The granules were polygonal in shape with size range of 6 to 8.57 µm. The infra spectra showed additional band at 3600 cm^{-1} for the oxidized derivative indicating hydroxyl group stretching vibration of carboxylic acid. This indicates that the oxidation of native acha starch was successfully carried out. Improved physicochemical properties impacted upon oxidative modification is quite desirable in impacting greater stability and less retrogradation and seneresis of the native acha starch. Thus, oxidized acha starch will find suitable applications in food, pharmaceutical, paper and textile industries as good dispersants, emulsifying agent, surface sizing, adhensive, disintegrants, excipients and preparation of biopolymer based flocculants.

Key words: Modified starch, starch oxidation, physicochemical properties, scanning electron microscopy, starch granules.

INTRODUCTION

The genus *Digitaria* referred to as fonio, family Poaceae, is one of the smallest cereal grains indigenous to most West African countries. It is one of the primary cereals of southern Sudan and Ethiopia. It is classified as millet but unlike other millets, it is low in protein (Jideani and Akingbala, 1993). They are consumed whole or milled into flour and can be processed into a variety of preparations such as gruels, porridges, beverages etc. (Coda et al., 2010).

To obtain oxidized starch, the most common approach is treating the native starch with a variety of oxidizing agents such as alkali metal hypochlorite. Oxidation improves whiteness and reduces microbiological content. In addition, the hydrogen bonding reduces the tendency to retro-gradation producing soft- bodied gels of high clarity.

The oxidation reaction are controlled by many factors such as the amount of alkali metal hypochlorite used, the pH, the temperature, and the use of metal and/or bromide ions as catalyst. The oxidative reactions lead to the introduction of carboxyl and carbonyl groups, and to the degradation of the starch molecule.

The degradation of the starch molecule during oxidation leads to a lower viscosity of a solution (or dispersion) of the oxidized starch, which is usually desired of an oxidized starch. It has been found that the degradation occurs to a farther extent at neutral pH of about 7 to 7.5 than at alkaline pH such as pH of 9 or higher. In other words, in order to obtain an optimal yield of oxidized starch providing a dispersion of low viscosity, the oxidation reaction should preferably be carried out at neutral pH.

In practice, the pH during oxidation of starch using an alkali metal hypochlorite is chosen at 8.5 or higher, dependent mostly on the desired viscosity of the oxidized starch. Alkali metal hypochlorite often used as oxidizing agent is relatively cheap and large oxidizing power. The reaction temperature at which the starch is treated with an oxidizing agent is preferably chosen between 20 and 50°C.

The duration of alkaline treatment lasts at least 30, more preferably at least 60 min. Oxidized starch is used in the paper industry as coating binders, surface sizing and adhesives. It is also as emulsifying agent, protective colloid for providing desired stability and also to improve weaving operation in the textile industry by improving abrasion resistance of the warp yarn sizing. Arabic gum may be substituted in confectioneries by an oxidized starch to provide excellent stability of the food product, leading to a more clear food product.

MATERIALS AND METHODS

Oxidation of acha starch

25 g of starch was dissolved into 100 ml of distilled water and the pH of starch solution adjusted to 10 to 11 with sodium hydroxide solution. The starch slurry was heated to a temperature of 30°C and 20 ml of sodium hypochlorite solution (3.5% active chlorine) added to the starch solution dropwise over a period of 20 min with stirring. During the addition of the reagent and the course of reaction the pH of the slurry was maintained at the desired value with NaOH or HCl solution. The mixture was stirred under the defined conditions above (temperature of 30°C) and then terminated after 2 h by addition of sodium metabisulphite. Filtered and suspended in distilled water and the pH adjusted to 6.5 to 7.0 and oven dried at 50°C.

Determination of physicochemical properties

Solubility

The native starch and modified starch samples (2 g each) were suspended in 20 ml of distilled water. Then heated to 70°C for 30 min with continuous shaking. The mixture was then centrifuged at 4000 rpm for 15 min. An aliquot of supernatant (5 ml) was evaporated at 105°C and weighed. The solubility of starch is the ratio in mass (g) of the dried supernatant to the initial mass (g) of dried starch.

Water and Oil absorption capacity

1 g of native and modified starch was weighed into test tubes. 10 ml of distilled water (and 10 ml of groundnut oil in the second test tube) were added, and then heated in a water bath at 60°C for 30 min. The starch slurry was centrifuged at 1000 rpm for 15 min and the supernatant carefully decanted and the weight of the starch paste taken.

WAC/OAC= weight of starch paste/weight of dry starch sample.

Bulk densities of native and modified starch

2 g each of native Acha starch and the modified starch were placed in a 10 ml measuring cylinder and the volume occupied by the sample without tapping recorded. The bulk density is the ratio of the weight to volume occupied.

The pH of starch

The pH of 1% w/v slurry of both the native starch and modified starch were determined using a pH meter.

Least gelation concentration of starch

8 samples each for native and modified starches (1 to 16% w/v) were prepared in test tubes with 5 ml of distilled water. The starch solutions were mixed using magnetic stirrer for 5 min and heated for 30 min at 80°C in a water bath followed by rapid cooling under running cold water. Further cool at 4°C for 2 h. Least gelation concentration was determined as that conc. when the samples from the inverted test tube did not fall down or slip.

Pasting properties of the starch

The pasting property of native starch and modified sample was carried out using Brookfield viscometer.

Foam capacity of starch

2 g of native Acha starch and each of modified starch were homogenized in 100 ml of distilled water using a magnetic stirrer for 5 min. The homogenate was poured into a 250 ml measuring cylinder and the volume occupied was recorded after 30 s. The foam capacity is expressed as the percent increase in volume.

Emulsion capacity of starch

2 g of native and modified acha starch were dispersed in 25 ml of distilled water using a magnetic stirrer for 30 s. After complete dispersion, 25 ml of vegetable oil (groundnut oil) was added gradually and the mixing continued for another 30 s. Then centrifuged at 1600 rpm for 5 min. The volume of oil separated from the sample was read directly from the tube. Emulsion capacity is the amount of oil emulsified and held per gram of sample.

Starch granules morphology

The starch granule morphology of both the native starch and modified starch were obtained using scanning electron microscopy (sem).

Table 1. Physicochemical properties.

Properties	NAS	OAS
Solubility (g)	10.24±0.02	9.32±0.04
Water absorption capacity (%)	488	499
Oil absorption capacity (%)	122	115
Bulk density (g/ml)	0.41±0.03	0.44±0.01
pH	6.85	6.40
Foam capacity (%)	4.0±0.00	3.2±0.02
Emulsion capacity (%)	36±0.02	39±0.03
Viscosity (cP)	31.5	20.2

NAS= Native acha starch ; **OAS**= Oxidized acha starch.

RESULTS AND DISCUSSION

Solubility

The results of solubility of native acha starch (NAS) and oxidized starch (OAS) are shown on Table 1. The solubility expressed as gram per 100 g of starch (g/100 g) reduced from 10.24 value observed with the native acha starch to 9.32 for oxidized starch. The decrease in solubility of oxidized samples is probably due to introductions of bulky functional groups reducing the mobility of starch molecules. This pattern agreed with the results of Olu-owolabi et al. (2014). Thus, modifications altered this physical property (that is, solubility) of native acha starch.

Water absorption capacity

The water absorption capacity of native acha starch and chemically modified sample is presented in Table 1. Oxidation improved water absorption capacity of native acha starch. This may be attributed to the incorporation of carboxyl functional groups on the starch molecules which enhanced binding capacity more than the native starch. The increase in water absorption capacity following oxidation is very important especially in the application of this starch either as a drug carrier or disintegrant in tablets and capsule formulation (Emeje et al., 2012).

Oil absorption capacity

The result of oil absorption capacity of native acha starch and modified sample is presented in Table 1. The values expressed as percentage varied from 122 observed with native acha starch to 115 observed for oxidized sample. This may be attributed to the functional groups incorporated onto the starch molecule following chemical modification. This agrees with the report of Sathe and Salunkhe (1981) that acetylation and oxidation do not improve oil absorption capacity of great northern bean. This increase in starch crystallinity restricted access of oil into the granule of the starch. This is because chemical modification is thought to occur in the amorphous region of the starch molecules leading to increase in starch crystallinity.

pH of starch slurry

The result of pH of starch slurries of native and modified acha starch are presented in Table 1. The pH values reduced from 6.85 to 6.40. A reduction on pH value was observed following modification by oxidation. Reduction in pH values of oxidized samples may be attributed to the incorporation of acetyl functional group to the starch molecule thereby increasing the acidity of starch molecules.

Bulk density

The bulk density of native acha starch and modified sample are presented on Table 1. The bulk density values were in the range of 0.39 to 0.50. A reduction of bulk density was observed with all chemical modifications of the native acha starch. The least bulk density was observed with the acid thinned sample (0.39). The reduction in bulk density might be attributed to increased crystallinity following chemical modification. Increase crystallinity is characteristic of more ordered state and this might impact greater stability on the modified samples. Thus retrogradation of native acha starch as well as seneresis may be improved upon modification. This reduction in bulk density is in agreement with Emeji et al. (2012). This improved physical property following chemical modification of native acha starch is desirable in food and pharmaceutical applications as good dispersant and preparation of biopolymer based flocculants.

Foam capacity

The result of foam capacity of native acha starch and

Table 2. Least gelation concentration (%).

Concentration (%)	NAS	OAS
2	Viscous	viscous
4	Viscous	Viscous
6	Gel	Gel
8	Gel	Gel
10	Gel	Gel
12	Gel	Gel
14	Gel	Gel
16	Gel	Gel

NAS= Native acha starch; **OAS** = Oxidized acha starch.

modified samples are presented in Table 1. The foam capacity of oxidized sample was reduced to 3.2% from 4% observed with native acha starch. Reduction in foam capacity following oxidation could find application as an emulsifier in the food industries (Ihegwuagu et al., 2009).

Emulsion capacity

The emulsion capacity of native acha starch and oxidized samples is presented on Table 1. The emulsion capacity increased in the modified samples from 36 to 39%. This suggests that chemically modified acha starch are better emulsifying agent due to the introduction of functional groups in the starch molecules increasing the binding force of the starch granules.

Least gelation concentration

The gelation properties of native acha starch (NAS) and oxidized samples are presented in Table 2. The least gelation concentration remained at 6%.

Starch pasting properties

The pasting properties of NAS and modified samples are presented in Table 1. These are shown as viscosities (Cp) at room temperature using Brookfield Viscometer. The pasting viscosities of modified samples were reduced from 31.5 centipoise for native acha starch to 20.2 centipoise for oxidized sample. The pasting viscosity reduced by about 30% on modification. In such instances where high content of starch is desirable in order for the starch to be able to form a gel as in wine gums and liquorice the pasting property of native starch would be too viscous during heating. In such applications, we can take advantage of oxidized starch. Also in instant soups compositions, thin boiling starches are often used as filler.

Starch granule morphology

Starch is stored in most green plants as minute granules in the leaves, stems, roots, fruits and seeds. Such starch is stored traditionally for the future use of the plant. The granule morphology of NAS and modified samples are shown in Figure 1. These morphologies were investigated using scanning electron microscopy at 15kv accelerating voltage and 2500 magnification each. Though increased crystalline region is observed on modifications, it is obvious that modifications may not have destroyed the shape, appearance and structural arrangements of the starch. The starch granules retain its polygonal shape with sizes ranging from 6 to 8.57 µm.

Starch utilization as a food or feed relate to its physical and chemical properties. The solubility of starch starts with swelling in the gel or amorphous region leaving the crystalline structure unaffected. But with higher temperature, the crystalline region melts or dissolves and the entire crystalline structure are destroyed. With increased crystallinity from cross-linked and acid treated modifications, less solubility and improved stability are expected from these modified samples.

Infra red (IR) spectra

The infra red (IR) spectra of NAS and modified derivative are presented on Figures 2 and 3 respectively. The infra red spectra have similar peaks, except for the additional peak at 3600 cm^{-1} on Figure 3 which is ascribed to the – OH stretching vibration of the carboxylic acid groups. These bands confirm that modifications of NAS were successfully carried out.

Conclusion

Oxidation of NAS was successfully carried out. The physicochemical properties of NAS and modified derivative were determined. Chemical modifications enhanced emulsion capacity, water absorption capacities

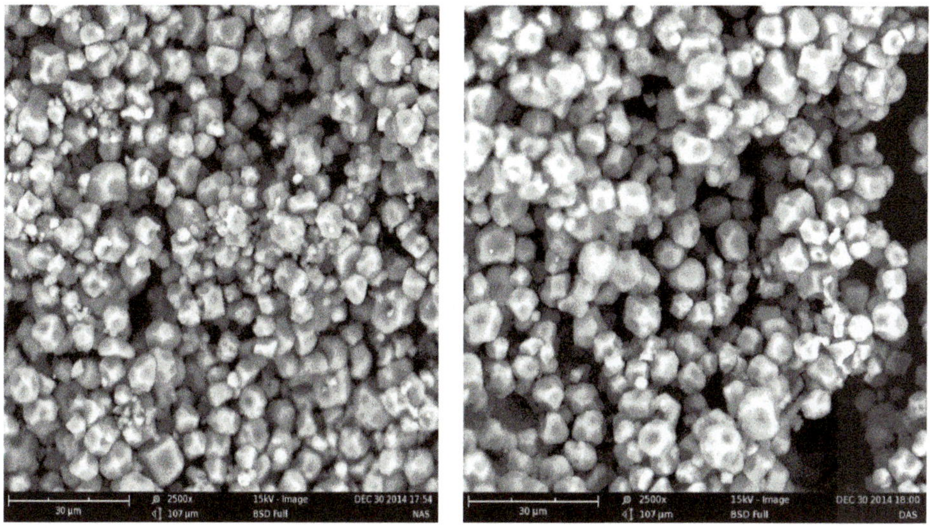

Figure 1. Comparison of Granule morphology NAS vs OAS.

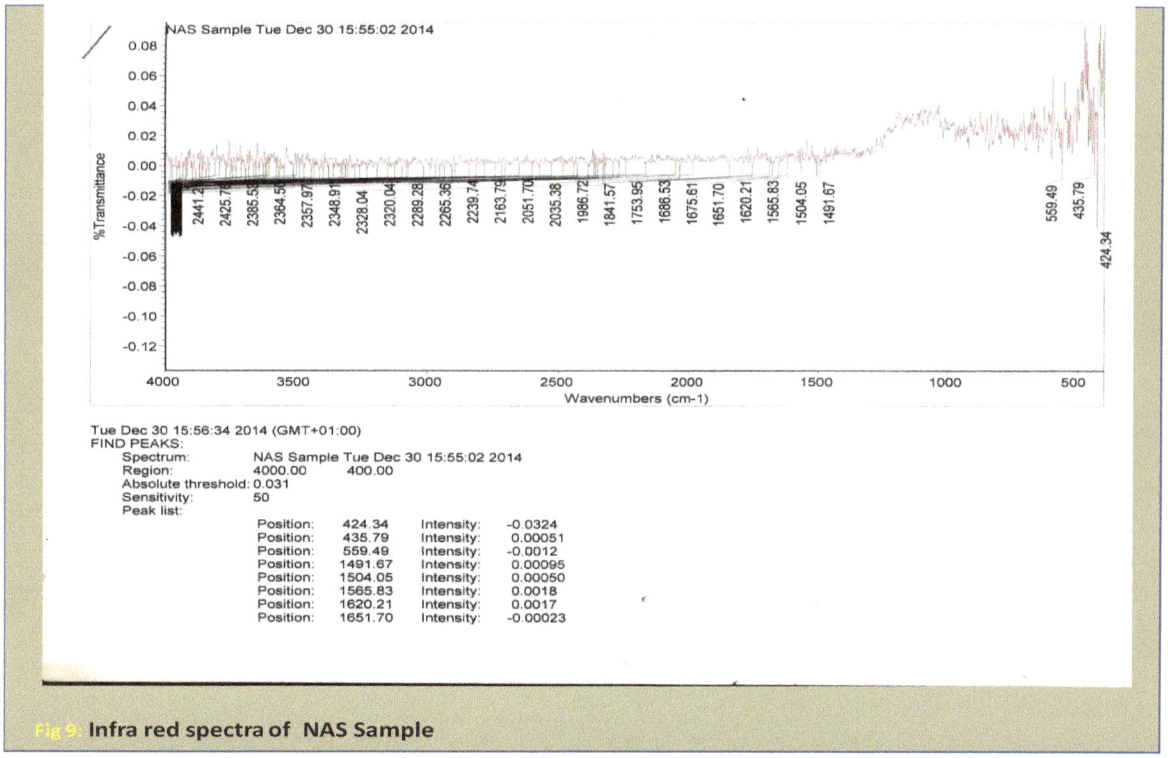

Figure 2. Infra red spectra of NAS Sample.

but reduced the paste viscosity, oil absorption and foam capacity. Oxidation improved NAS stability and less retrogradation. Potential applications of modified acha starch include good emulsifying agent, starch thickened sauces, soups, paper binding and pharmaceutical drug carriers and disintegrants including paper surface sizing and adhensives.

Conflict of Interest

The authors have not declared any conflict of interest.

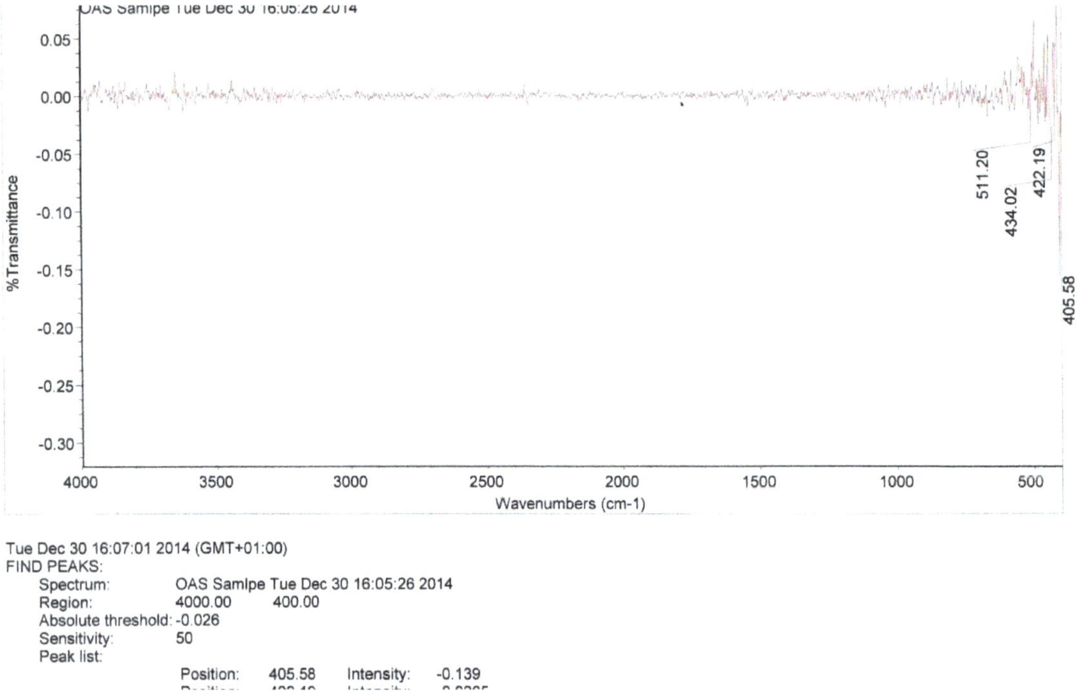

Figure 3. Infra red spectra of OAS sample.

REFERENCES

Coda R, Cagno RD, Edema MO, Nioeli L, Gobbetti M (2010). Exploitation Of Acha (*Digitaria exillis*) And Iburu (*Digitaria Iburua*) Flours: Chemical characterization and their use for sourdough fermentation. Food microbial J. 27:1043-1053.

Emeje M, Kaita R, Isimi C, Buragohain A, Kunle O, Ofoefule S (2012). Synthesis, physicochemical characterization, and functional properties of an esterified starch. Afr. J. Food Agric. Nutr. Dev. 65:369-374.

Ihegwuagu EN, Omojola OM, Emeje MO, Kunle OO (2009). Isolation and evaluation of some physicochemical properties of Parkia Biglobosa starch. Pure Appl. Chem. 81(1):97-104.

Jideani AI, Akingbala JO (1993). Some physicochemical properties of acha (*Digitaria exilis* Stapf) and iburu (*Digitaria iburua* Stapf) grains. J. Sci. Food Agric. 65(4):465-476.

Olu-Owolabi BI, Olayinka OO, Adegbemile AA, Adebowale KO (2014). Comparison of functional properties between native and chemically modified starches from acha (*Digitaria* stapf) grains. J. Food Nutr. Sci. 5:222-230

Sathe SK, Salunkhe DK (1981). Isolation, partial characterization and modification of the great Northern bean (*Phaseolus vulgaris* L.) Starch. J. Food Sci. 46:617-621.

Studies on urea treated rice milling waste and its application as animal feed

Simon Terver Ubwa, James Abah, Barnabas Atsinafe Oshido and Esther Otokpa

Department of Chemistry, Benue State University, P. M. B 102119, Makurdi, Nigeria.

The composition of urea treated rice milling waste and its application as animal feed was studied. The proximate analysis of the urea treated rice milling waste showed that it contained 94.90% dry matter, 10.38% crude protein, 5.89% crude fibre, 0.16% ether extract, 7.47% ash, and 54.81% nitrogen-free extracts. The untreated rice milling waste contained 94.34% dry matter, 9.11% crude protein, 6.37% crude fibre, 0.18% ether extract, 8.11% ash, and 54.69% nitrogen-free extracts. Four experimental diets were prepared containing two different levels (30 and 35%) each of untreated and urea treated rice milling waste mixed with commercial pelletized chick mash to assess their effects on weaner rabbits. Data on the feed intake, growth rate, and feed conversion ratio (FCR) of weaner rabbits fed diets containing two different levels each of untreated and urea treated rice milling waste were compared using two-way analysis of variance (ANOVA). There were no significant effects ($P > 0.05$) of dietary treatment and level of inclusion for average feed intake, average body weight gain, and the FCR. The study indicated that rabbits can be successfully raised on a commercial chick mash mixed with 35% rice milling waste treated or untreated without any adverse effect on growth.

Key words: Rice milling waste, weaner rabbit, proximate analysis, analysis of variance, animal feed, urea.

INTRODUCTION

Rice milling waste is one of the commonest agro-industrial wastes generated in large quantities in most parts of Nigeria. Rice processing generates a great volume of by-products that constitute a large proportion of agro-industrial waste in many parts of the world (NAERLS and PCU, 2004). Although, one can hardly classify rice milling waste among hazardous wastes, its treatment is very important in view of the great volume of waste materials involved. Waste treatment techniques are normally employed to alter the physical, chemical, or biological characteristics of waste and make it safer for disposal. These include composting, pyrolysis, gasification, and combustion. In Nigeria and many other developing countries, where the bulk of rice produced is for consumption, the most common waste treatment technique employed is combustion which has several disadvantages including environmental pollution (Thipwimon et al., 2004; Bhattacharya et al., 1999). Moreover, some countries under the environmental protection legislation now strongly oppose and even prohibit this practice.

Rice milling waste is believed to contain various nutrients that would enable it to serve as animal feed. The major challenges are however, its high level of fibre and low protein and energy. Studies have shown that the nutritional value of rice milling waste can be significantly improved by processing/treatment techniques such as mechanical treatment, ensilage, biological treatment, and chemical treatment with alkalis and urea (Belewu and Babalola, 2009; MacDonald et al., 1987a) since urea treatment increases rice milling waste utilization and fibre fraction degradation .In developed countries, the by-products of the rice industry are separated into different components and converted into value-added products. In Nigeria, however, these by-products are largely dumped together. The quantity of rice milling by-products generated in Nigeria annually was estimated at about 1,032,993.6 metric tons (NAERLS and PCU, 2004). A large amount of these by-products is dumped as waste thereby posing disposal problems and bringing about methane emissions (Thipwimon et al., 2004; Bhattacharya et al., 1999). The disposal problems posed by rice milling waste have led to indiscriminate burning of the waste and subsequent accumulation of ash in rice producing areas resulting in environmental pollution and loss of land. Rice milling waste can also cause respiratory problems due to its characteristics (Thipwimon et al., 2004; Beagle, 1978).

The problems associated with rice milling waste can be greatly reduced if the waste can be effectively utilized as animal feed. Effective utilization of rice milling waste as animal feed has not been possible and reports have shown that the major constraints to this utilization include low crude protein, energy and mineral content (Belewu and Babalola, 2009; MacDonald et al., 1987b; Maikano, 2007; Yakubu et al., 2007). However, some research findings indicated that rice milling wastes contain moderate level of crude protein, and also have low crude fibre and high metabolizable energy (Crampton and Harris, 1969; Ambasankar and Chandrasekan, 2002; Singh and Marwaha, 1968). Crude fibre consists of cellulose, hemicelluloses and lignin (Yakubu et al., 2007) which are not well utilized by monogastric animals (Amdt et al., 1980). Lignin, which envelopes some nutrients, is highly resistant to chemical and enzymatic degradation and is poorly degraded by rumen microbes (Belewu and Babalola, 2009; MacDonald et al., 1987a). Strong chemical bonds exist between lignin and many plant polysaccharides and cell wall proteins, which render these compounds unavailable during digestion. These bonds are however, broken by chemical treatment thereby increasing the digestibility of fibrous feeds (Belewu and Babalola, 2009; MacDonald et al., 1987b; Rexen and Vestergaard, 1976). Among the chemicals that have been utilized, sodium hydroxide has proven to be the most effective in improving digestibility but lacks nitrogen. Furthermore, there is increased sodium load in animals fed with diets treated with sodium hydroxide (Adeniji, 2010). Another effective chemical that has been used successfully in achieving this is ammonia, which weakens the hard cell walls, allowing better penetration by rumen microorganisms to produce more effective fermentation and liberation of nutrients (Chenost, 1995). In developing countries like Nigeria, one of the more successful procedures available to improve the digestibility and therefore nutritional value of fibrous feeds is urea treatment since this requires little equipment or expenses, even subsistence farmers can apply urea treatment. Chemical treatment of rice milling waste with urea can lead to significant improvement in nutritional quality and therefore greater utilization (Taiwo et al., 1992). The effective utilization of rice milling waste as animal feed will greatly reduce its disposal problems and contributes towards value addition in the rice sector. This research was therefore designed to study the urea treated rice milling waste and its application as animal feed.

MATERIALS AND METHODS

Rice milling waste collection

The sample was obtained from Naka rice processing industry waste dump 25 km from Makurdi, Benue State, Nigeria. Random sampling technique was employed to collect samples of the waste taken at depths of 10 cm from different points using a clean plastic container and shovel. The sub-samples were pooled together, mixed thoroughly and packed into an empty 10 kg polyethene bags which were conveyed to animal laboratory.

Sources of feed and experimental animals

The rice milling waste was obtained from Naka rice processing industry dumpsite along Makurdi road. 5.0kg of urea (fertilizer grade) was purchased from a certified dealer. The experimental animals (16 weaned rabbits) were purchased at the University of Agriculture Quest House, Makurdi. A small quantity (2 kg) of commercial chick mash was obtained from a certified dealer at Wurukum, Makurdi and fed to the rabbits before the experiment began. 25 kg of the commercial feed was later obtained at Asaba, Delta State and used for the experimental feed preparation. The experiment was conducted at the Federal College of Education (Technical) Asaba, in the animal unit using a metabolic cage constructed under the supervision of certified animal scientist and biochemist.

Sample treatment

The rice milling waste was treated by dissolving 2.5 kg of urea in 50 dm^3 of water and applied to 50 kg of rice milling waste. The treated rice milling waste was placed in a plastic container and made airtight by covering it with polyethene sheets and the container's lid. It was then allowed to ferment for 3 weeks after which it was removed and sun dried for several days (Adeniji, 2010; Sundstol and Coxworth, 1984). A portion of the rice milling waste was left untreated and used for the control experiment.

Feed preparation

Four diets (Diets 1, 2, 3, and 4) were prepared by thoroughly mixing the rice milling waste samples and the commercial pelletized chick mash in different ratios. Diet 1 contained 30% urea treated rice milling waste and 70% commercial diet, Diet 2 contained 30% untreated rice milling waste and 70% commercial diet, Diet 3 contained 35% untreated rice milling waste and 65% commercial diet while Diet 4 contained 35% urea treated rice milling waste and

Table 1. Chemical composition of urea treated and untreated rice milling waste.

Parameter	Components (%)					
	DM	CP	CF	EE	Ash	NFE
UTRMW	94·90	10.38	5·89	0.16	7.47	54.81
UNTRMW	94·34	9.11	6·37	0.18	8.11	54.69

UTRMW = Urea treated rice milling waste; UNTRMW = untreated rice milling waste; DM = dry matter; CP = crude protein; CF = crude fibre, EE = ether extract; NFE = nitrogen-free extracts.

65% commercial diet. Diets 2 and 3 served as the first and second control (2 for Diet 1 and 3 for Diet 4), respectively.

Experimental animals design and diets

In this experiment, 16, weaned, mixed breeds male rabbits of age 28 days were used. The rabbits were kept for a week and fed with commercial diet and fresh potato (*Ipomoea batatas)* leaves under laboratory conditions. They were then fed with the experimental diets for an adaptation period of 2 weeks. During this period, no data on feed intake and body weight was recorded. The rabbits weighed between 530 and 560 g when the experiment started. The rabbits were randomly assigned to four treatment diets with four rabbits per treatment after balancing for body weight. The rabbits were housed in a metabolic cage constructed and partitioned with wood and wire gauze.

The cage was raised above the ground to allow for separation of urine and faeces. The animals were fed twice daily (morning and evening). Clean water was provided *ad libitum* and Anupco Vitalyte Extra (a combination of vitamins, electrolyte, and amino acids), Embazin-forte (anticoccidial with vitamin K) and Neotreat WSP (antibiotics and vitamins) were added to the water at intervals and according to the manufacturer's prescription throughout the experimental period of 9 weeks.

Data collection

The animals were weighed when the experiment started and then once weekly during the growth period. The weekly body weights were determined by weighing the animals individually. This was usually done early in the morning prior to feeding. The quantity of feed offered as well as the leftover was weighed each day to determine the daily intake. The daily feed consumption was recorded as the quantity of feed served minus the leftover. Data on feed intake, weight gain, and feed conversion ratio (FCR) were computed. Nutrient digestibility study was done on the 9th week of the experiment.

Digestibility study

The faeces were collected individually during 4 consecutive days. The faecal samples from each experimental unit were bulked, sun dried, and analysed for proximate composition.

Proximate analysis

The proximate compositions of the treated and untreated rice waste samples as well as the faecal samples were determined according to official methods of analysis (AOAC, 1980). All samples were ground to pass a 1-mm sieve prior to analysis.

RESULTS AND DISCUSSION

Chemical composition of urea treated and untreated rice milling waste

The results of the chemical compositions of the urea treated and the untreated rice milling waste are presented in Table 1.

Chemical composition of diets

The result of the chemical compositions of the experimental diets is presented in Table 2.

Feed intake of weaner rabbits

The results of the weekly average feed intake (g) of the weaner rabbits fed with different levels of untreated and urea treated rice milling waste based diets are presented in Table 3. The average feed intakes were comparable among the four groups of rabbits.

Weekly weight of weaner rabbits

The result of the average weekly live weight (g) of weaner rabbits fed with the experimental diets is presented in Table 4.

Performance of the weaner rabbits fed with the experimental diets

The performance of the weaner rabbits fed with varying levels of rice milling waste based diets is shown in Table 5.

The feed conversion ratio was calculated as: $\dfrac{\text{Feed intake}}{\text{Weight gain}}$

Coefficients of digestibility of nutrients were determined according to the formula:

$$\frac{[(\text{quantity of feed} \times \text{nutrients of feed}) - (\text{quantity of faeces} \times \text{nutrients of faeces})] \times 100}{(\text{Quantity of feed} \times \text{nutrients of feed})}$$

Table 2. Chemical composition of the experimental diets.

Component (%)	Diet 1	Diet 2 (T1)	Diet 3 (T2)	Diet 4
Dry matter	96.20	96.07	96.37	96.27
Crude protein	15.71	15.33	14.89	15.33
Crude fibre	5·97	6·26	6·13	6·11
Ether extract	6.00	6.00	5.59	5.59
Ash	3.85	4.11	4.33	4.04
NFE	59.28	59.25	58.92	58.96

T1 = Control 1; T2 = Control 2; NFE = Nitrogen-free extracts.

The nutritive compositions of the urea treated and untreated rice milling waste are presented in Table 1. The results showed that the both feedstocks have comparable nutrient profile although the urea treated rice milling waste had a higher crude protein value (10.38%) than the untreated rice milling waste (9.11%). This suggests that urea treatment increased the crude protein content of rice milling waste. The treatment of rice milling waste with urea increased its nitrogen content due to the addition of non-protein nitrogen. This collaborate the reports of other studies that urea ammoniation increases the crude protein content of feed materials (Yakubu et al., 2007; Ambaye, 2009; Amaefule et al., 2003, 2006; Oluokun, 2005). The percentage increase in crude protein (13.94%) due to urea treatment in this study is lower than the values reported by Yakubu et al. (2007) and Ambaye (2009). This may be attributed to the higher content of crude protein in the original material. The crude protein value of 9.11% for the untreated rice milling waste is within the range, 7.8 and 9.4% reported by Ambasankar and Chandrasekan (2002) and Crampton and Harris (1969), respectively. It is however, higher than the 5.03, 5.83, and 6.00% previously reported by Maikano (2007), Yakubu et al. (2007), and Aduku (2004). The crude fibre content of the urea treated rice milling waste was slightly lower than that of the untreated rice milling waste. The reduction in crude fibre content of the urea treated rice milling waste is in agreement with the report of Yakubu et al. (2007). The crude fibre value of 6.37% obtained in this present study is higher than the 0.96% reported by Ambasankar and Chandrasekan (2002) and 3.14% reported by Singh and Marwaha (1968) but much lower than the 30.39, 42.15, and 33.00% obtained by Maikano (2007), Yakubu et al. (2007), and Aduku (2004), respectively. The dry matter, ether extract, ash, and nitrogen-free extract values are 94.90, 0.16, 7.47, and 54.81% for the urea treated rice milling waste and 94.34, 0.18, 8.11 and 54.69% for the untreated rice milling waste, respectively. The urea treated rice milling waste had slightly higher dry matter and nitrogen-free extract contents but lower ether extract and ash contents than the untreated rice milling waste. The dry matter content of the untreated rice milling waste in this study was quite similar to that (94.42%) published by Maikano (2007) but higher than the 89.50 and 90.53% previously reported by Yakubu et al. (2007) and Ambasankar and Chandrasekan (2002), respectively. Maikano (2007), Yakubu et al. (2007), Ambasankar and Chandrasekan (2002), and Aduku (2004) reported ether extract values of rice milling waste of 3.40, 6.45, 3.20, and 5.60% and Maikano (2007), Yakubu et al (2007), and Aduku (2004) reported ash content of 16.67, 21.76 and 19.10%, respectively. The rice milling waste used in the present study contained considerably less ether extract and ash. The nitrogen-free extract (54.69%) obtained in this experiment was higher than the 46.10 and 30.05% reported by Maikano (2007) and Yakubu et al. (2007), respectively but considerably lower than the value (85.26%) reported by Aduku (2004). The variations between the proximate compositions of different rice milling waste have been attributed to the rice variety, growing conditions, pre-treatment before milling, milling system, degree of milling, and the proportion of husk (Saunders, 1990; Farell, 1994).

The proximate compositions of the experimental diets (Table 2) showed that the four dietary treatments have comparable nutrient compositions. The dry matter content of the four dietary treatments ranged from 96.07 to 96.37%, crude protein ranged from 14.89 to 15.71%, crude fibre ranged from 5.97 to 6.26%, ether extract ranged from 5.59 to 6.00%, ash ranged from 3.85 to 4.33%, and nitrogen-free extract ranged from 58.92 to 59.28%. The comparable nutrient compositions of the experimental feed may be attributed to the initial quality of the treated material. It has been reported that the effect of urea treatment is more pronounced for materials whose initial quality is very poor compared to those with better original quality (Chenost, 1995).

The result of the average daily feed intake of the weaner rabbits from Week 1 to 8 is shown in Table 3. The average daily intake did not follow any defined pattern. The least average daily intake (61.65 ± 11.34 g, 57.29 ± 10.32 g, 55.65 ± 3.76, and 58.57 ± 13.45 g for Diets 1, 2, 3, and 4, respectively) was recorded in Week 1. The highest average daily intake recorded (79.68 ± 14.10 g, 76.29 ± 22.36 g, 67.35 ± 12.01 g, and 75.45 ±

Table 3. Weekly average feed intake (g) of the weaner rabbits fed with different levels of untreated and urea treated rice milling waste based diets.

Week	Diet 1	Diet 2	Diet 3	Diet 4
1	61.65 ± 11.34	57.29 ± 10.32	55.65 ± 3.76	58.57 ± 13.45
2	65.15 ± 6.07	61.96 ± 16.01	58.24 ± 6.73	60.75 ± 12.49
3	79.68 ± 14.10	76.29 ± 22.36	65.71 ± 10.18	75.45 ± 9.32
4	66.86 ± 12.20	63.57 ± 13.45	60.35 ± 5.59	62.14 ± 12.20
5	64.65 ± 9.51	62.05 ± 11.67	64.09 ± 8.70	65.10 ± 8.00
6	70.45 ± 11.10	68.55 ± 6.55	63.70 ± 11.32	69.15 ± 13.25
7	75.39 ± 7.69	71.93 ± 9.65	67.35 ± 12.01	74.65 ± 11.30
8	67.26 ± 10.12	63.28 ± 13.00	61.30 ± 7.07	63.45 ± 9.78
Total	551.29 ± 22.65	524.92 ± 18.56	496.39 ± 13.17	529.26 ± 23.50

Values measured as mean ± standard deviation, n = 7.

Table 4. Average weekly weight per diet (g) of the weaner rabbits fed with treated diets.

Average weight (g)	Diet 1	Diet 2	Diet 3	Diet 4
Initial weight	550.00	555.00	540.00	540.00
Week 1	665.00	662.00	638.00	641.00
Week 2	776.00	798.00	760.00	761.00
Week 3	895.00	900.00	875.00	867.00
Week 4	1021.00	1020.00	1000.00	1002.00
Week 5	1161.00	1151.00	1137.00	1145.00
Week 6	1303.00	1295.00	1284.00	1284.00
Week 7	1456.00	1444.00	1434.00	1429.00
Week 8	1616.00	1601.00	1580.00	1584.00
Weekly average	133.25 ± 18.63	130.75 ± 19.75	130.00 ± 18.25	130.50 ± 19.42

Values measured as mean weight of four animals, except for weekly average, which equals Mean ± standard deviation of eight measurements.

9.32 g for animals on Diets 1, 2, 3, and 4, respectively was on different weeks.

The average daily feed intake (Table 5) of the rabbits fed with the four treatment diets, D1, D2, D3, and D4 (in grams) for the experimental period of eight weeks are 68.89, 65.62, 62.05, and 66.16, respectively. The rabbits fed with diets containing urea treated rice milling waste (D1 and D4) consumed more feed than those fed with diets containing untreated rice milling waste (D2 and D3). The trend of the average daily feed intake showed that D1 > D4 > D2 > D3. The higher feed intake of animals on D1 and D4 may be due to urea treatment, which is believed to increase intake (Preston and Leng, 1987). The higher intake however, did not result in better growth performance of rabbits. The 2-way analysis of variance (ANOVA) result of Tests of Between-Subjects Effects [TBSE (Table 6)] showed no significant main effect ($P > 0.05$) of treatment and level of inclusion. There was also no significant interaction effect between the treatment and level of inclusion ($P > 0.05$). The parallel lines of the marginal means in the profile plot (Figure 1) further illustrate the non-significant interaction effect.

The average weekly weight gains of the experimental rabbits per treatment diet are recorded in Table 4. The TBSE result (Table 7) indicated that average weekly weight gain was not significantly affected ($P > 0.05$) by the dietary treatments and the level of inclusion of rice milling waste. The interaction effect was also not significant as revealed by the profile plots of the marginal means (Figure 2). The average weekly weight gain per diet (g) over the eight weeks are 133.25, 130. 75, 130.00, and 130.50 (Tables 4) for D1, D2, D3, and D4, respectively. The pattern for the weight gain per diet revealed that D1 > D2 > D4 > D3. Generally, the animals in this study irrespective of the dietary treatment and the level of inclusion of the rice milling waste did not record significant ($P > 0.05$) difference in the average weekly weight gain. There were however, significant effects of initial weight on performance of animals during the first 2 weeks with the heavier animals having a higher rate of gain. Similar findings were reported for growing rabbits fed low protein diets supplemented with or without urea

Table 5. Performance of the weaner rabbits fed with different levels of untreated and urea treated rice milling waste based diets.

Parameter	Diet 1	Diet 2	Diet 3	Diet 4	SEM	SL
No. of days of trial	56	56	56	56		
No. of animals	4	4	4	4		
Average initial live weight (g)	550.00	555.00	540.00	540.00		
Average final live weight (g)	1616.00	1601.00	1580.00	1584.00		
Total weight gain (g)	1066.00	1046.00	1040.00	1044.00		
Average weekly weight gain (g)	133.25	130.75	130.00	130.50	3.36	NS
Average daily weight gain (g)	19.04	18.67	18·57	18.64		
Total feed intake (g)	3859.03	3674.44	3474.73	3704.82		
Average daily feed intake (g)	68.89	65.62	62.05	66.16	1.01	NS
FCR	3.62	3.51	3.34	3.55	0.11	NS

FCR = Feed conversion ratio; SEM = Standard error of the mean; SL = Significance Level, NS = not significant.

Table 6. ANOVA tests of between-subjects effects

Source	Type iii sum of squares	df	Mean square	F	P/Sig.
DT	108.892	1	108.892	3.367	0.077
LI	79.286	1	79.286	2.452	0.129
DT * LI	1.399	1	1.399	0.043	0.837
Error	905.472	28	32.338		

Dependent Variable: Feed Intake, (α = 0.05); DT = Dietary treatment; LI = Level of inclusion; df = Degrees of freedom.

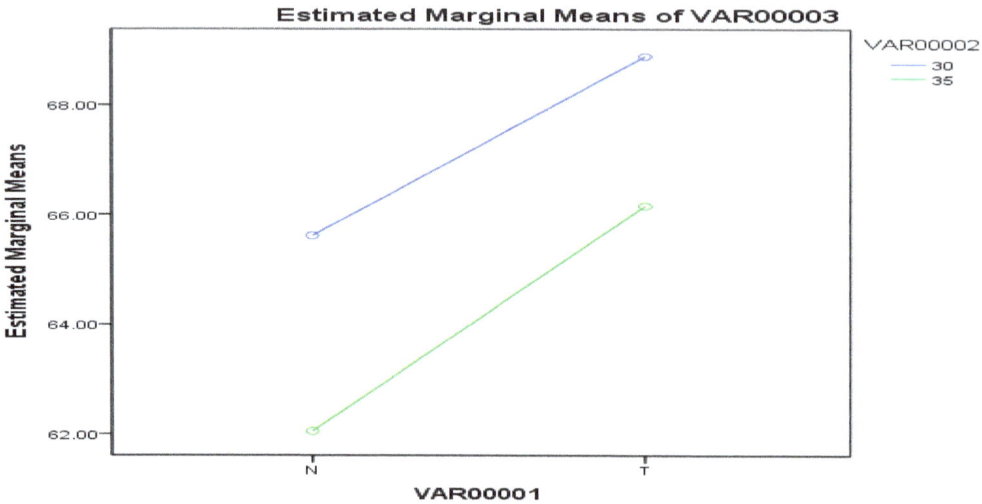

Figure 1. Profile plots of marginal means of feed intake. N = Untreated diets, T = Treated diets, VAR00001 = Dietary treatment, VAR00002 = Level of inclusion, VAR00003 = Dependent variable (Feed Intake).

or methionine (Raharjo, 1986).

The FCR obtained (Table 5) showed the same trend as the average daily feed intake: D1 > D4 > D2 > D3. The FCR of 3.67, 3.59, 3.39, and 3.62 for the rabbits on D1, D2, D3, and D4, respectively did not show significant (P > 0.05) (Table 8) interaction within the dietary treatments

Table 7. ANOVA tests of between-subjects effects.

Source	Type III sum square	df	Mean square	F	P/Sig
DT	18.000	1	18.000	0.050	0.825
LI	24.500	1	24.500	0.068	0.797
DT * L	I 8.000	1	8.000	0.022	0.883
Error	10135.000	28	361.964		

Dependent variable: Weight Gain, (α = 0.05); TM = Dietary treatment, LI = Level of inclusion, df = Degrees of freedom.

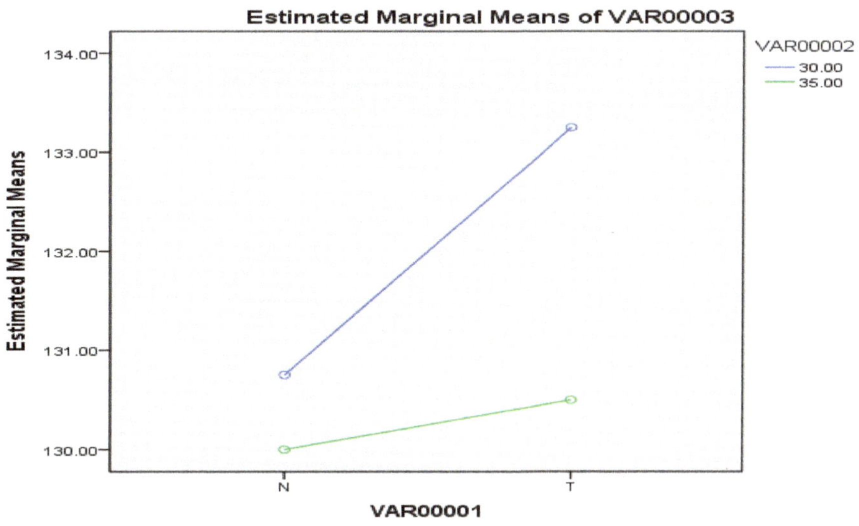

Figure 2. Profile plots of marginal means of weight gain. N = Untreated diets, T = Treated diets, VAR00001 = Dietary treatment, VAR00002 = Level of inclusion, VAR00003 = Dependent variable (Weight Gain).

Table 8. ANOVA tests of between-subjects effects.

Source	Type III sum of squares	df	Mean square	F	P/Sig.
DT	0.194	1	0.194	0.550	0.465
LI	0.137	1	0.137	0.387	0.539
DT * LI	0.047	1	0.047	0.134	0.717
Error	9.861	28	0.352		

Dependent Variable: Feed Conversion Ratio, (α = 0.05); TM = Dietary treatment, LI = Level of inclusion, df = Degrees of freedom.

and level of inclusion. The non significant level of interaction is demonstrated in the profile plots of the marginal means (Figure 3). This indicates similar utilization of nutrients by the experimental rabbits. The similar performance of rabbits observed in this study may be due to the similar proximate composition of the experimental diets.

The results of the nutrient digestibility of the rabbits are presented in Table 9. The feed dry matter, crude protein, crude fibre, and ash showed similar digestibility within the dietary treatments and the different levels of inclusion. The digestibility ranged from 77.25 to 78.55% for dry matter, 67.51 to 69.45% for crude protein, 49.80 to 50.45% for crude fibre, and 63.89 to 65.35% for ash. The similar nutrient digestibility recorded among the different group of the rabbits may be attributed to the similarity in the chemical composition of the diets. Report has shown that feed which vary relatively little in composition will show little variation in digestibility (McDonald et al., 1987 a, b). The values obtained in the present study are

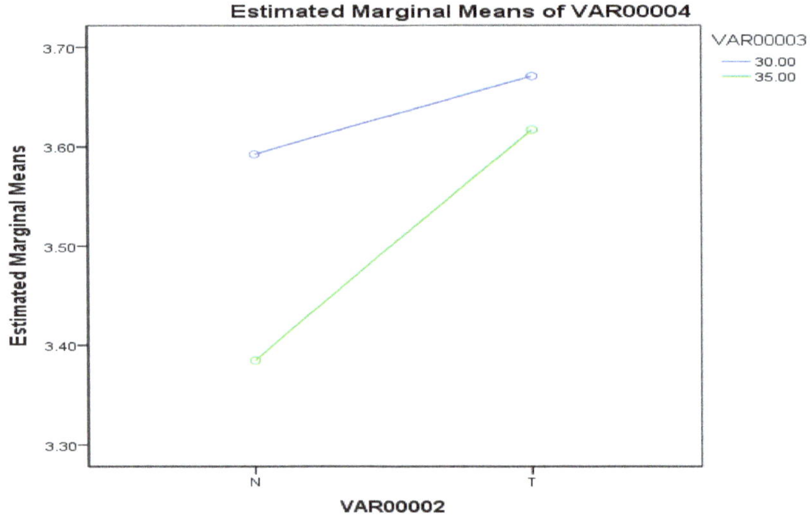

Figure 3. Profile plots of marginal means of feed conversion ratio. N = Untreated diets, T = Treated diets, VAR00002 = Dietary treatment, VAR00003 = Level of inclusion, VAR00004 = Dependent variable (Feed conversion ratio).

Table 9. Nutrient digestibility by rabbits fed different levels of untreated and urea treated rice milling waste based diets.

Parameter (%)	**D1**	**D2**	**D3**	**D4**
Dry matter	78.55	78.07	77.40	77.25
Crude protein	69.45	68.27	67.51	68.01
Crude fibre	50.03	51.15	49.80	50.45
Ash	65.35	64.27	63.89	64.02

comparable with those earlier reported for weaner rabbits fed with similar diets (Abonyi et al., 2012).

Conclusion

The result of this study indicated that the chemical composition of rice milling waste can be highly varied when treated with urea which resulted in modification of the chemical composition of the waste. The modification however, did not significantly affect the growth performance and nutrient digestibility of the experimental rabbits in this work. Thus, weaner rabbits can be successfully raised on a commercial chick mash diet blended with 35% rice milling waste with good initial quality treated or untreated without any adverse effect on growth performance and nutrient digestibility.

ACKNOWLEDGEMENT

The authors would like to thank the management of Federal College of Education (Technical) Asaba, for allowing the use of their metabolic cage for rearing the rabbits.

REFERENCES

Abonyi FO, Iyi EO, Machebe NS (2012). Effects of feeding sweet potato (*Ipomoea batatas*) leaves on growth performance and nutrient digestibility of rabbits. Afr. J. Biotechnol. 11(15):3709-3712.

Adeniji AA (2010). Effects of dietary grit inclusion on the utilization of rice rusk by pullet chicks. Trop. Subtrop. Agroecosyst. 12:175-180.

Aduku AO (2004). Animal nutrition in the tropics: feeds and feeding, pasture management, monogastric and ruminant nutrition. Zaria, Nigeria, pp. 31-43.

Amaefule KU, Iheukwumere FC, Lawal AS, Ezekwonna AA (2006). The effect of treated rice milling waste on performance, nutrient retention, carcass and organ characteristics of finisher broilers. Int. J. Poult. Sci. 5(1):51-55.

Amaefule KU, Nwogu RK, Ohazuluike N (2003). Influence of treatment of rice mill waste on its nutritional value for broilers. J. Sustain. Agric. Environ. 5:196-203.

Ambasankar K, Chandrasekan D (2002). Feeding value of rice waste for layers and its effect on egg quality. Indian Vet. J. 79:39-42.

Ambaye TD (2009). On-farm evaluation of urea treated rice straw and rice bran supplementation on feed intake, milk yield and composition of fogera cows, North Western Ethiopia. MSc. Thesis, Bahir Dar University, Bahir Dar, pp.6-10.

Amdt DC, Richardson CR, Sheirol LB (1980). Digestibility of chemically treated cotton plant by-product and effect on mineral balance, urine volume, and pH. J. Anim. Sci. 51(1):215-218.

AOAC (1980). Association of Official Analytical Chemists, Official methods of analysis. 15th ed., Washington DC.

Beagle EC (1978). Rice husk conversion to energy.FAO Consultant. Rome, Italy.

Belewu MA, Babalola FT (2009). Nutrient enrichment of waste agricultural residues after solid state fermentation using *Rhizopusoligosporus.* J. Appl. Biosci. 13:695-699.

Bhattacharya SC, Joe MA, Kandhekar ZP, Salam A, Shrestha RM (1999). Greenhouse-gas emission mitigation from the use of agricultural residues: the case of rice husk. Energy 24:43-59.

Chenost M (1995). Optimizing the use of poor quality roughage through treatments and supplementation in warm climate countries with particular emphasis on urea treatment.*First Electronic Conference on Tropical Feeds* FAO, Rome.

Crampton EW, Harris LE (1969). Applied animal nutrition. In:Ambasankar, K.and Chandrasekan, D (2002). Feeding value of rice waste for layers and its effect on egg quality. Indian Vet. J. 79:39-42.

Farell DJ (1994). Utilization of rice bran in diets for domestic and ducklings. Worlds Poult. Sci. J. 50:115-132.

Maikano A (2007). Utilization of rice offal in practical ration of broilers. Zoologist 5:1-7.

McDonald P, Edwards RA, Greenhalgh JFD (1987a). Animal nutrition Longman Group (FE) Ltd 4thed., pp. 24-25.

McDonald P, Edwards RA, Greenhalgh JFD (1987b). Animal nutrition Longman Group (FE) Ltd 4thed., pp. 24-25.

NAERLS and PCU (2004). Field situation assessment of 2004. In: Maikano A (2007). Utilization of rice offal in practical ration of broilers. *The* Zoologist. 5:1-7.

Oluokun JA (2005). Intake, digestion and nitrogen balance of diets blended with urea treated and untreated cowpea husk by growing rabbit. Afr. J. Biotechnol. 4(10):1203-1208.

Preston TR, Leng RA (1987). Matching ruminant production system with available resources in tropics and sub-tropics; Australia, Penambul Books, Armidale, Australia. P. 259.

Raharjo YO, Cheeke PR, Patton NM (1986). Growth and reproductive performance of rabbits on a moderately low crude protein diet with or without methionine or urea supplementation. J. Anim. Sci. 63:795-803.

Rexen F, Vestergaard KT (1976).The effect of digestibility of a new technique for alkali treatment of straw. Anim. Feed Technol. 1:73-83.

Saunders RM (1990). The properties of rice bran as a food stuff. Cereal Foods World 35:632-639.

Singh KS, Marwaha VK (1968). In Ambasankar, K. and Chandrasekan, D. (2002).Feeding value of rice waste for layers and its effect on egg quality. Indian Vet. J. 79:39-42.

Sundstol F, Coxworth EM (1984). Ammonia treatment. In: Sundstol,F. And Owen, E.(Eds). Straw and other fibrous by-products as feed. Elsevier, Amesterdam., pp. 196-247.

Taiwo AA, Adebowale EEA, Greenhalgh JFD, Akinsoyinu AO (1992). Effects of urea treatment on the chemical composition and degradation characteristics of some crop residues. Nig. J. Anim. Prod. 19:25-34.

Thipwimon C, Shabbir HG, Suthum P (2004).Environmental profile of power generation from rice husk in Thailand. *The Joint International Conference on"Sustainable Energy and Environment (SEE)".*

Yakubu B, Adegbola TA, Bogoro S, Yusuf HB (2007). Effect of urea treated and untreated rice offal on the performance of broilers: growth performance and economy of production. J. Sustain. Develop. Agric. Environ. 3:7-13.

Anthocyanins from mulberry (*Morus rubra*) fruits as potential natural colour additives in yoghurt

Robert Byamukama[1], Moses Andima[1,2], Angella Mbabazi[1] and Bernard T. Kiremire[1]

[1]Department of Chemistry, Makerere University, P. O. Box 7062, Kampala, Uganda.
[2]Department of Chemistry, Busitema University, P. O. Box 236, Tororo, Uganda.

Colouring potential of anthocyanins from whole fruit juice of mulberry (*Morus rubra*) was studied in yoghurt. Whole fruit juice from *M. rubra* rich in non-acylated anthocyanins was incorporated into plain yoghurt (100 g) at increasing concentration levels of the juice; 10, 20, 25, 30, 40 and 50 mg cyanidin-3-glucopyranoside equivalents (cy-3-glu eqv) and stored under refrigerated condition (< 8°C) for two weeks. Colour properties, pigment and colour stability and degradation kinetics were studied using a UV-Vis spectrophotometer (UV-1700 CE Shimadzu, Japan).Yoghurt coloured with mulberry anthocyanins between 25 to 40 mg cy-3-glu eqv concentration levels of anthocyanins produced a colour which was very much comparable to commercial brand strawberry yoghurt coloured with 20 mg FD & C red No. 3 in 100 g of yoghurt. Pigment and colour stabilities of the anthocyanins increased with increasing concentration of anthocyanins added to yoghurt. The tendency to polymerise decreased with increasing concentration of the pigments added to yoghurt. The degradation of the anthocyanins was fitted to first-order reaction kinetics. Moderate concentration levels (25 to 40 mg cy-3-glu eqv in 100 g of yoghurt) of mulberry anthocyanins were found to be ideal to colour yoghurt.

Key words: Mulberries, anthocyanins, cyanidin-3-glucopyranoside (cy-3-glu), natural colourants, colour stability, yoghurt.

INTRODUCTION

One of the vital constituents of some foods and beverages is colour. In addition to other functions (Mendi et al., 2000), colour plays a very important role in the acceptability of some foods by many consumers (Giusti and Wrolstad, 2003). In practice, most manufacturers tend to colour products which have dull colours and look unappealing to most consumers. Synthetic colourants have often been used in attempts to colour some foods and beverages (Delgados-Vargas and Paredes-López, 2003). However, the demand for foods with synthetic colours is diminishing drastically due to associated health problems and legislative action against some of them (Cevallos-Casals and Cisneros-Zevallos, 2004; Markakis, 1982).

Anthocyanins have been of great interest as alternatives to synthetic colourants due to their bright colours and associated health benefits (Cevallos-Casals and Cisneros-Zevallos, 2004; Fan et al., 2008). They are considered to be safe because they have been consumed for centuries in fruits, and vegetables without any health risks (Bridle and Timberlake, 1997). The application of anthocyanins as food colourants faces some limitations such as low stability to several processing, formulation and storage conditions. Despite the drawback, highly acylated anthocyanins possess high colour stability in some food matrices (Giusti and Wrolstad, 2003; Kirca et al., 2006). Nevertheless, Wallace and Giusti (2008) reported that whole fruit extracts containing non-acylated anthocyanins from *Berberis boliviana* L. showed improved colour and pigment stability when incorporated in yoghurt. This information led us to believe that whole fruit juice extracts from mulberries (*Morus rubra*) could serve as an appropriate colourant and nutraceautical in yoghurt.

Mulberries (*M. rubra*) are a rich source of anthocyanins, with fully ripe fruits yielding between 2.53 to 8.30 mg cy-3-glu/gfresh weight of anthocyanins (Özgen et al., 2009). Other studies reported the content of anthocyanins to vary from 0.59 mg/ g to 10.90 mg/ g fresh weight infrom different cultivars of mulberries (Liu et al., 2004).The fruits possess many health benefits such asantithrombotic, antioxidant, antimicrobial, anti-inflammatory, and neuroprotective properties(Aramwit et al., 2010). In recent years, mulberry anthocyanins have in many occasions been applied in baked foods, and confectioneries (Wu et al., 2011). In Turkey, they are used to make special important traditional products such as mulberry pekmez, mulberry pestil, and mulberry kome (Ercisli and Orchan, 2007). The anthocyanins are stable at low temperatures and in the dark (Aramwit et al., 2010). In this study, we decided to investigate whether anthocyanins from mulberries (*M. rubra*) could be used as potential colour additives in yoghurt since yoghurt has a low pH and it is stored under refrigerated conditions.

MATERIALS AND METHODS

Plant materials and yoghurt

Ripe fruits of mulberries (*M. rubra)* were collected from Masaka District in Uganda and stored in a laboratory freezer before use. Plain yoghurt (1.5% fat) at the point of adding colours and commercial brand strawberry flavoured yoghurt were kindly provided by fresh dairy factory (Sameer Agriculture and Livestock (U) ltd, Plot 49-53/55 fifth street, Industrial Area, Kampala, Uganda).

Chemicals, reagents and equipment

Analytical grade methanol, potassium chloride, sodium acetate, and potassium metabisulphite used in this study were purchased from BDH laboratory suppliers (U) ltd. A UV-Vis spectrophotometer (UV-1700 CE Shimadzu) was used for spectrophotometric measurements using 1 cm path length disposable cuvette tubes.

Preparation of whole-fruit juice concentrate

Frozen fruits of mulberry were allowed to defrost and the fruits were pasteurised immediately to denature enzymes such as polyphenol oxidase (PPO) which causes browning of anthocyanins (Wang and Xu, 2007). The fruits were then blended in a laboratory blender (Waring 700S 1 L, Lab depot, Inc.) and the juice was filtered through a clean piece of nylon clothe. The juice was centrifuged at 4000 rpm and the supernatant was concentrated in a vacuum rotary evaporator at 35°C to 15% brix. The concentrate was kept below 4°C in a freezer until analysis.

High-performance liquid chromatography (HPLC) analysis

To confirm the anthocyanins in *M. rubra,* high performance liquid chromatography (Shimadzu, Columbia, Md.,U.S.A) was performed on a Hypercil gold™ reverse phase 250 x 4.6 mm symmetry, 5 µm size C18 column (Thermo Fisher Scientific Inc.) equipped with a photo array detector (PAD). Separation of the anthocyanins was achieved using a linear gradient. Solvent A was $HCOOH:H_2O$; 1:19, Solvent B was $HCOOH:H_2O:MeOH$; 1:9:10. The elution profile was 0 to 23 min, 10 to 100% B, 0 to 5 min 10% B, 5 to 23 min, 100% B, 23 to 28 min, 100% B 28 to 29 min, 10% B at a flow rate of 0.75 ml min^{-1} and the injection volume was 15 µl. UV-Vis spectral data was obtained from 250 to 560 nm (Wallace and Giusti, 2008).

Colouring yoghurt with the juice extract

100 g portions of plain yoghurt at the point of adding colour were coloured with the whole fruit juice extract of mulberries at increasing concentration levels of the juice (10, 20, 25, 30, 40, and 50 mg Cy-3-glu eqv) in 100 g of yoghurt. The amount of the juice extract for each portion of yoghurt was chosen in order to study the effect of increasing pigment concentration on the colour of yoghurt and the stability of the pigment in yoghurt (Wallace and Giusti, 2008). Plain yoghurt was manually mixed gently and thoroughly with the juice concentrate to ensure uniformity and avoid syneresis of the set yoghurt. Yoghurt samples were prepared in triplicate at each colouring level. The coloured yoghurt samples were packed in McCartney bottles (50 ml, 144 bottles in total, 3 at each colouring level) and stored in a laboratory refrigerator (< 8°C)for two weeks during which, colour properties, pigment concentration and pH were measured at regular time intervals (Wallace and Giusti, 2008).

Extraction of anthocyanins from the yoghurt matrix

Anthocyanins were recovered from the coloured yoghurt matrix using the method adopted by (Wallace and Giusti, 2008). Coloured yoghurt samples (100 g) at each concentration level, were thoroughly mixed with methanol acidified with hydrochloric acid (99.9%, 0.1%, 30 ml) for 5 min and centrifuged at 4000 rpm for 15 min. The extraction procedure was carried out twice on each sample to ensure maximum recovery of anthocyanins from the yoghurt matrix. The supernatant in each case was carefully filtered and concentrated to a known volume at 35°C with a rotary evaporator. The concentrates were used for spectrophotometric analysis in triplicate. To determine the recovery rates, juice containing anthocyanins (20 mg cy-3-glu eqv) were added to plain yoghurt (100 g) and extracted similarly with acidified methanol. The recovery rate was expressed as a percentage of the recovered amount in the total initial amount added to yoghurt Equation (1). This procedure was repeated six times to determine the extraction efficiency.

$$Recovery\ rate = \frac{amount\ recovered}{Original\ amount\ added\ to\ yoghurt}\ x\ 100 \quad (1)$$

Determination of anthocyanin content

The content of total anthocyanins in the coloured samples

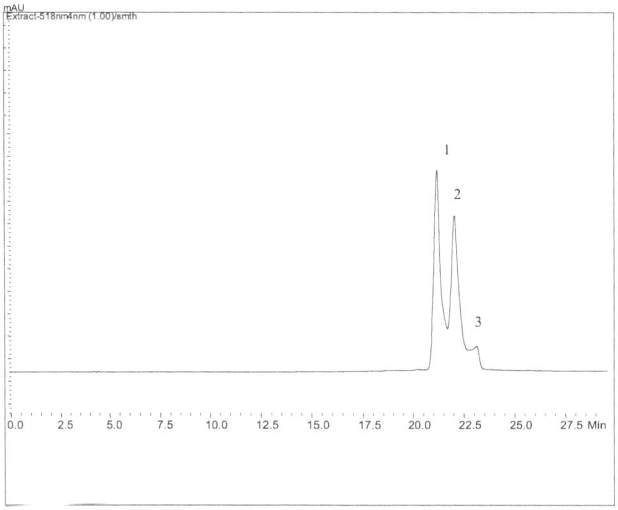

Figure 1. HPLC-PAD chromatogram of anthocyanins in M. rubra fruit juice extract recorded at 520 nm. Cyanidin-3-O-glucopyranoside (peak 1), cyanidin-3-O-rutinoside (peak 2), pelagornidin 3-O-glucopyranoside (peak 3).

remaining at any time was determined using the pH-differential method described by Giusti and Wrolstad (2001).

Evaluation of pigment and colour stability

To evaluate pigment and colour stability, the extent of degradation of mulberry anthocyanin pigments and colour in yoghurt was determined by calculating the percent change in absorbance using Equation (2) (Katsaboxakis et al., 1998).

$$\text{Pigment and colour stability} = \frac{A_t}{A_o} x100 \quad (2)$$

Where A_t is the absorbance of anthocyanins at any time *t*, A_0 is the initial absorbance of anthocyanins added to yoghurt at each concentration level. Any variations in colour were confirmed by visual inspection (Brenes et al., 2005; Torskangerpoll and Andersen, 2005). Visual inspection is the commonest method used according to Sameer Agriculture and livestock (U) ltd factory manual basing on the fact that consumers rely on the visual colour to product selection.

Measurement of polymeric colour

The extent of anthocyanin polymerisation in yoghurt matrix was determined using the bisulphite bleaching method (Giusti and Wrolstad, 2001).

Kinetic calculations and statistical analysis

In order to evaluate the degradation kinetics of mulberry anthocyanins in yoghurt, anthocyanin concentration was plotted against time and linear regression analysis was used to determine the adequacy of the anthocyanin degradation kinetic model using Minitab (Mtb14) software. The degradation constant (*k*) was determined from the 1st derivative of the curves plotted using Equation (4) and the half-lives ($t_{1/2}$) at each concentration level were calculated from Equation (5) (Kirca et al., 2006; Wallace and Giusti, 2008).

$$ln\frac{C_t}{C_o} = -kt \quad (4)$$

$$t_{1/2} = \frac{ln2}{k} \quad (5)$$

Where C_t is the concentration of anthocyanins at time *t*, C_o is the initial concentration of anthocyanins and *t* is the storage time (shelf life) and *k* is the degradation constant.

Graphs and statistical analysis

Summary statistics (mean and standard deviation), graphs and linear regression analysis were obtained using Microsoft® excel 2007 (Microsoft corp., 2013) as well as Minitab (Mtb16. Exe) software (Minitab® Inc.). Results for colour, pigment stability and degree of polymerisation were subjected to regression analysis and analysis of variance (ANOVA) with an alpha $P<0.05$ acceptance level (Kirca et al., 2006; Wallace and Giusti, 2008).

RESULTS AND DISCUSSION

Confirmation of the anthocyanins in the *M. rubra* extract

Two major anthocyanins (Cyanidin-3-*O*-glucopyranoside and cyanidin-3-*O*-rutinoside) and two minor anthocyanins (Pelargonidin 3-*O*-glucopyranoside and pelargonidin 3-*O*-rutinoside) have been identified to be present in mulberry fruits (Qin et al., 2010). In this study, the identities of these anthocyanins were confirmed by comparing their order of elution from a C-18 column (Figure 1). Cyanidin-

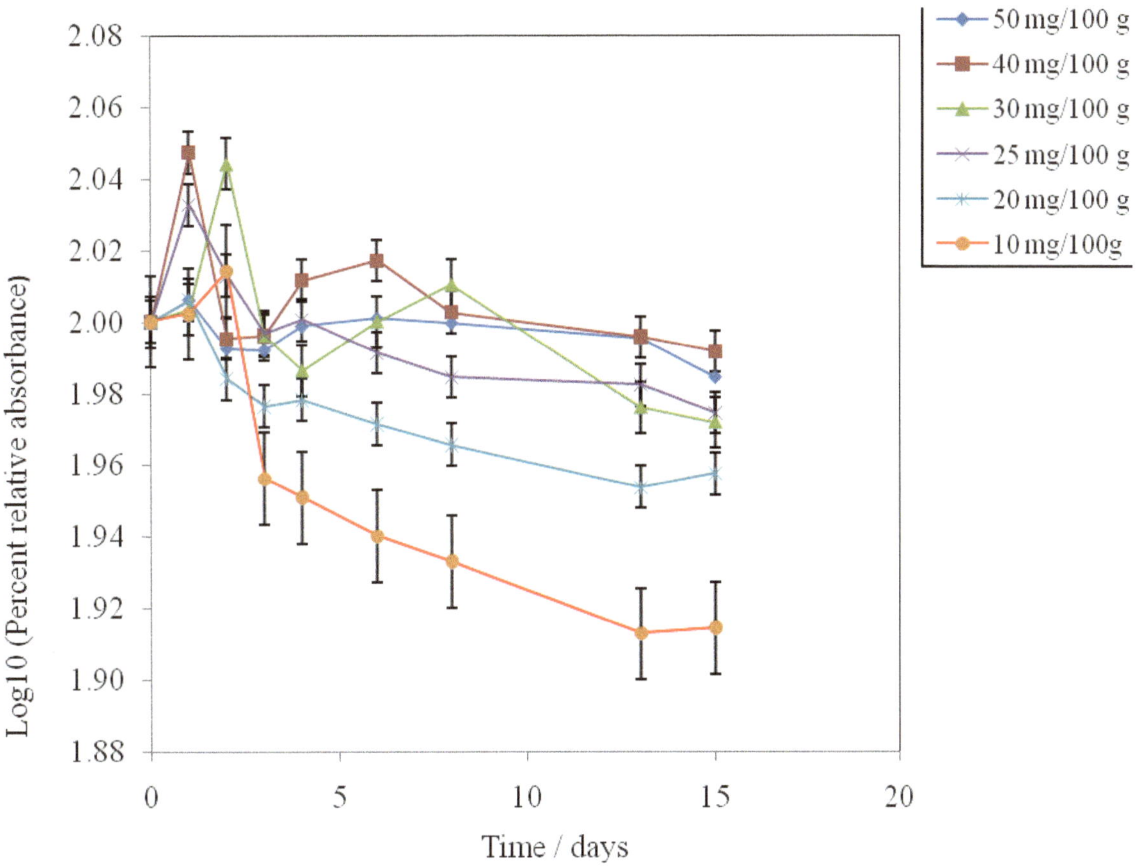

Figure 2. Variation of percent relative absorbance at all concentration levels of anthocyanins; 1020, 25, 30, 40 and 50 mg cyanidin-3-glucopyranoside equivalents in 100 g of yoghurtin 14 days.

3-*O*-glucopyranoside (peak 1) and cyanidin-3-*O*-rutinoside (peak 2) were the major anthocyanins with only one minor anthocyanin (peak 3) (Figure 1). The average content of monomeric anthocyanins in *M. rubra* was 7.30 mg/g fresh weight (STDEV = 0.01, n = 3) as determined by pH differential method (Giusti and Wrolstad, 2001). This concentration was within the same range as reported by Özgen et al. (2009).

Pigment stability

In this study, pigment stability was evaluated in terms of percent relative absorbance. The anthocyanin recovery rate was calculated to be 96.55% (n= 6, STDEV 1.04). Generally, percent relative absorbance increased in the first two days of the investigations and, thereafter, it decreased at all concentration levels (Figure 2). Stabilisation of anthocyanin molecules as they associate with constituents of the yoghurt matrix such as sugars, stabilizers and milk fats could lead to the increased stability in the first two days (Jing and Giusti, 2005; Wrolstad et al., 1990). However, at 30 and 40 mg cy-3-glu eqv of the juice extract in 100g of yoghurt, percent relative absorbance increased gradually between the fourth and eighth days before decreasing, up to the end of the second week (Figure 2).

Concentration levels of 30-50 mg cy-3-glu eqv of mulberry juice containing anthocyanins added to yoghurt showed increased pigment stability ($P > 0.05$) than at lower concentration levels (10 to 25 mg cy-3-glu eqv) ($P < 0.05$) (Figure 2). The results indicated that pigment stability increased with increasing concentration level of the pigments added to yoghurt. Nevertheless, concentration level of 40 mg cy-3-glu eqv in 100 g of yoghurt was the best ($P = 0.248$) because it gave the highest percent relative absorbance than at 30 and 50 mg concentration levels ($P = 0.079$ and 0.099) respectively.

Increased pigment stability at high concentration is attributed to anthocyanin stabilisation through intramolecular association between anthocyanins and intermolecular association with other flavonoids, amino acids, organic acids, nucleotides, polysaccharides or metals that could be present in the matrix (Castañeda-Ovando et al., 2009). These stabilisation mechanisms increase with increasing concentration. At lower concentration levels, intermolecular association

Table 1. Variation of λvis-max and absorbance with storage time at all anthocyanins concentration levels;10,20, 25, 30, 40 and 50 mg cyanidin-3-glucopyranoside equivalents in 100 g of yoghurt (n = 3, P<0.05).

Time (days)	10 mg cy-3-glu eqv			20 mg cy-3-glu eqv			25 mg cy-3-glu eqv			30 mg cy-3-glu eqv			40 mg cy-3-glu eqv			50 mg cy-3-glu eqv		
	λvis-max (nm)	Abs (Au)	Monomeric anthocyanin (cy-3-glu eqv	λvis-max (nm)	Abs (Au)	Monomeric anthocyanin (cy-3-glu eqv	λvis-max (nm)	Abs (Au)	monomeric anthocyanin(cy-3-glu eqv)	λvis-max (nm)	Abs (Au)	Monomeric anthocyanin (cy-3-glu eqv)	λvis-max (nm)	Abs (Au)	Monomeric anthocyanin (cy-3-glu eqv)	λvis-max (nm)	Abs (Au)	monomeric anthocyanin (cy-3-glu eqv)
0	510.8	0.240	10.02± 0.01	510.8	0.48	20.04± 0.00	510.8	0.60	25.06± 0.02	511.0	0.36	30.00± 0.01	511.0	0.48	40.01± 0.02	511.0	0.60	50.01±0.01
1	511.8	0.241	10.05± 2.3	511.8	0.49	20.32± 0.03	511.0	0.65	26.94± 0.22	511.0	0.36	30.23± 0.72	511.0	0.53	44.59± 0.23	511.0	0.61	50.72± 0.12
2	511.0	0.248	10.35± 0.11	511.0	0.46	19.32± 0.24	510.5	0.62	25.74± 0.19	511.0	0.40	33.20± 0.60	511.0	0.47	39.57± 0.08	511.0	0.59	49.16± 2.21
3	511.0	0.217	9.06± 0.07	511.0	0.45	18.98± 0.18	510.7	0.59	24.79± 0.13	511.0	0.36	29.72± 0.49	511.0	0.47	39.64± 1.65	511.0	0.59	49.09± 0.24
4	519.3	0.214	8.95± 3.0	521.0	0.46	19.06± 0.30	520.0	0.60	25.03± 0.32	511.5	0.35	29.06± 0.23	511.5	0.49	41.09± 0.13	511.5	0.60	49.83± 0.13
6	511.0	0.209	8.73± 0.17	508.0	0.45	18.76± 0.29	508.0	0.59	24.5± 0.44	511.0	0.36	30.00± 0.27	511.0	0.50	41.61± 0.50	511.0	0.60	50.11± 1.40
8	511.0	0.206	8.59± 0.10	507.8	0.44	18.51± 0.11	510.0	0.58	24.13± 0.28	511.0	0.37	30.72± 0.56	511.0	0.48	40.26± 0.59	511.0	0.60	49.96± 0.14
13	511.0	0.196	8.19± 0.09	512.5	0.43	18.02± 0.05	512.0	0.58	24.01± 0.20	511.0	0.34	28.39± 0.31	511.0	0.47	39.60± 0.03	511.0	0.59	49.47± 1.06
15	512.5	0.197	8.22± 0.15	511.0	0.44	18.18± 0.94	512.8	0.56	23.56± 0.48	511.0	0.34	28.11± 0.05	510.5	0.47	39.24± 1.05	510.5	0.58	48.27± 0.74

Monomeric anthocyanin recorded as mean±Stdev (n= 3, P <0.05).

predominates but at higher concentration levels, both intra and intermolecular associations are present (Gauche et al., 2010; Nielsen et al., 1993). In similar studies, Giusti and Wrolstad (2003) noted that increased anthocyanin concentration in food systems promoted higher colour stability. Likewise, pigment and colour stability of strawberry syrup increased greatly by increasing the concentration of anthocyanins in the syrup (Cavalcanti et al., 2011).

Despite the high stability of mulberry anthocyanins in this study, some amount of anthocyanins added to yoghurt was lost in the two weeks study at all concentration levels. For example, on the last day, the percent relative absorbencies of anthocyanins at each concentration level were 96.5, 98.1, 93.7, 94.3, 90.7 and 82.1% at 50, 40, 30, 25, 20 and 10 mg cy-3-glu eqv in 100 g of yoghurt respectively (Figure 2). Polymerisation of monomeric anthocyanins with other phenolic compounds and degradation by endogenous or exogenous enzymescan cause gradual loss of monomeric anthocyanins (Brownmiller et al., 2008; Ścibisz et al., 2012; Skrede et al., 2000). Different authors cited by Ścibisz et al. (2012) established that lactic acid culture which is added to yoghurt as a starter produces hydrogen peroxide which accelerates anthocyanin degradation or the lactic acid culture probably produces β-glucosidases which converts anthocyanins to their more unstable aglycone forms.It is clear that one or more of these enzymes or other factors could have caused some gradual degradation of anthocyanins added to yoghurt in this study.

Colour stability

In addition to using percent relative absorbance as a measure of pigment and colour stability, variations in colour hue of mulberry anthocyanins described as absorption spectra, were evaluated at the storage temperature (< 8°C), pH 3.4, at all concentration levels of anthocyanins that were added to yoghurt. On average, the visible λ_{max} for the pigments at all concentration levels was 511.7 nm over the 2 weeks period (Table 1). The absorbance values at visible maximum absorption wavelength decreased gradually with time at all concentration levels as the pigments stayed longer in yoghurt (Table 1). However, the decreases in absorbencies with time at each concentration level were small, suggesting high colour stability, which corresponds to high pigment stability.

An increase in the maximum absorption wavelength (bathochromic shift, $\Delta\lambda_{Vis-max}$ from 511 to 520 nm) and absorbance (hyperchromic effect, $\Delta\lambda_{Vis-max}$) were observed on the fourth day at 10-

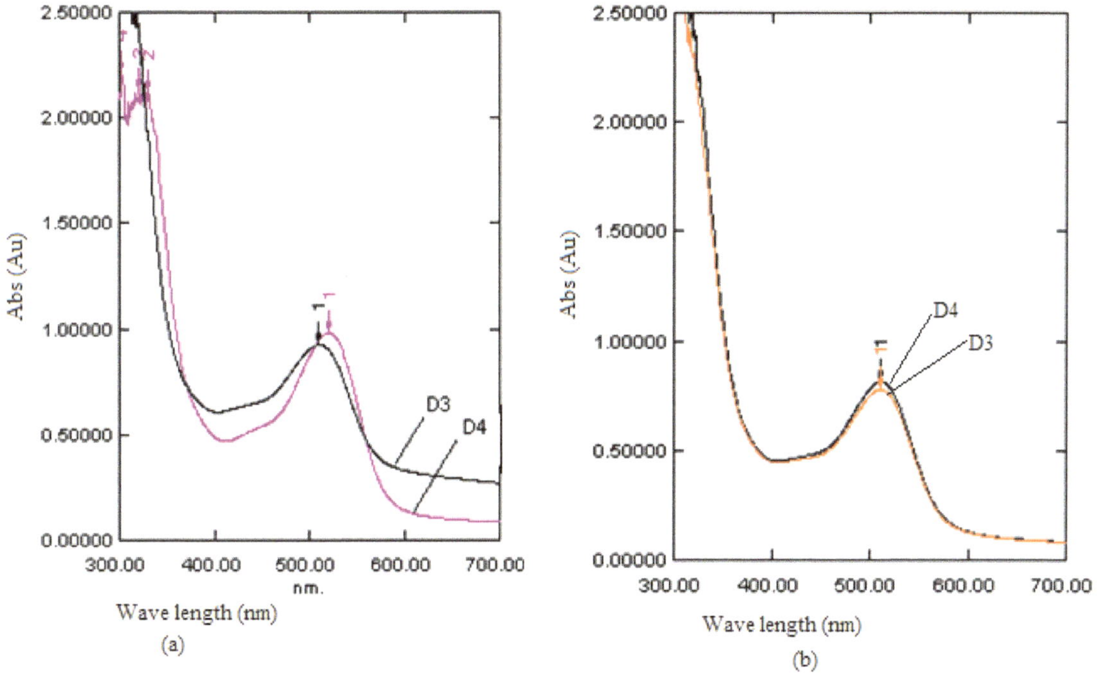

Figure 3. Comparison of absorption spectral pattern of mulberry anthocyanins; (a) 25mg cy-3-glu eqv/100 g and (b) 40 mg cy-3-glu eqv/100 g concentration levels on the third (D3) and fourth (D4) days (n = 3).

25 mg/100 g concentration levels (Table 1). The bathochromic effect was negligibly small between 30-50 mg/100 g concentration levels. For example at 25 mg cy-3-glu concentration level, there was a bathochromic shift from 511 to 521.0 nm between the third and fourth days whereas the bathochromic shift was only 0.5 nm from the third to the fourth day at 40 mg/100 g concentration level (Table 1, Figure 3).

The observed bathochromic and hyperchromic effects at 10 to 25 mg cy-3-glu concentration levels could be attributed to intermolecular association with other polyphenols and other flavonoids (Heredia et al., 1998). At lower concentration levels, intermolecular association of anthocyanins with polyphenols and other flavonoids are stronger than intramolecular associations leading to the observed bathochromic shift and hyperchromic effects (Gauche et al., 2010; Nielsen et al., 1993). Plant constituents such as cell wall and vacuolar components in the mulberry whole fruit juice could have associated with the anthocyanin pigments. Recent studies showed that interaction of plant constituents such as hydroxycinnamic acids in canned strawberries with nonacylated anthocyanins enhanced the colour stability of anthocyanins (Kammerer et al., 2007). Various components in the yoghurt matrix such as added sugar, fats, and stabilizers could have associated with anthocyanins to increase their colour stability (Jing and Giusti, 2005; Wrolstad et al., 1990). On the other hand, under refrigerated conditions, anthocyanins have reduced reaction rates, which could enhance the colour stability of the pigments added to yoghurt.

The variations in colour hue of the yoghurt coloured with mulberry anthocyanins were visually confirmed by comparing with commercial brand strawberry flavoured yoghurt (containing 20 mg FD & C red No. 3 in 100 g of yoghurt) and plain yoghurt as controls (Brenes et al., 2005; Torskangerpoll and Andersen, 2005). Between 25 to 40 mg cy-3-glu eqv in 100 g of yoghurt, the anthocyanins imparted a bright pink colour similar to that of commercial brand strawberry flavoured yoghurt. On the other hand, at 20 mg cy-3-glu eqv in 100 g of yoghurt, a dull/faint pink colour was imparted to yoghurt while at 10 mg cy-3-glu eqv in 100 g of yoghurt, a faded pink colour was imparted to the yoghurt. At 50 mg cy-3-glu eqv in 100 g of yoghurt, the pigments mixed with cream plain yoghurt, producing a colour close to red, which was rather unappealing. It is clear from these results, that concentration levels ranging from 25-40 mg cy-3-glu equivalents in100 g of yoghurt would be ideal for imparting an appealing colour to yoghurt.

Polymeric colour analysis

Polymeric colour was measured to determine the level of anthocyanin pigment polymerisation on addition to yoghurt matrix at all concentration levels. The results indicated that, there was a general increase in percent polymeric colour up to about the sixth day followed by a gradual decrease after the eighth day of storage at all

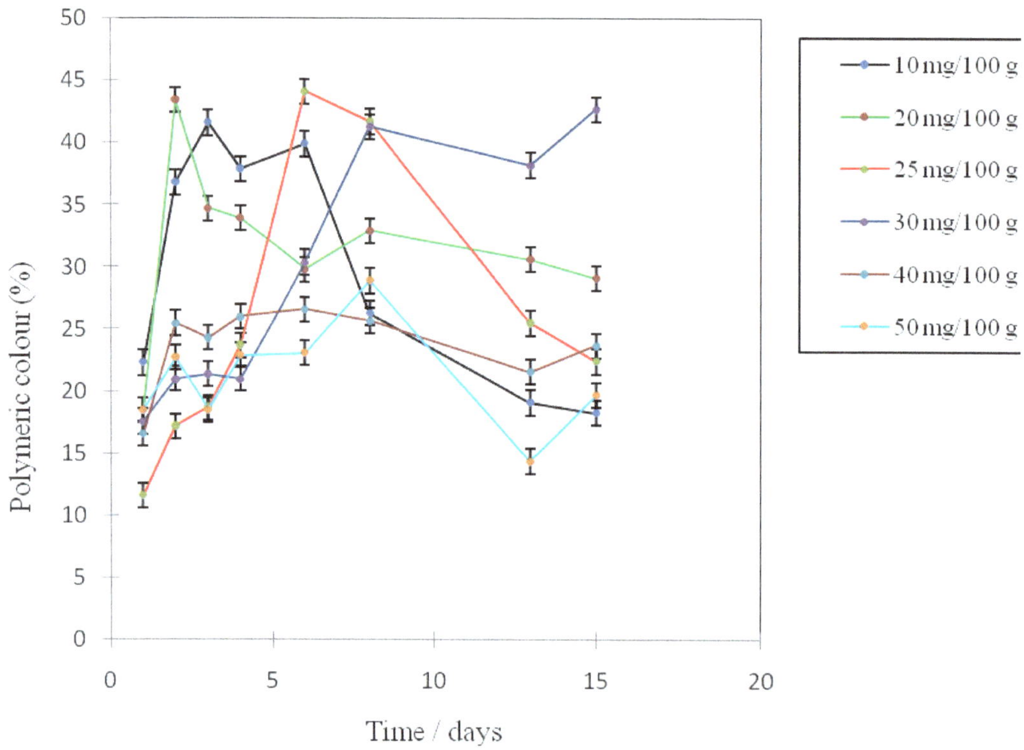

Figure 4. Variation of polymeric colour with time at all concentrations of anthocyanins;10, 20, 25, 30, 40 and 50 mg cyanidin-3-glucopyranoside equivalents in 100 g of yoghurt (n = 3).

concentration levels (Figure 4). The extent of anthocyanin polymerisation was higher at 10, 20, 25 and 30 mg Cy-3-glu eqv ($P < 0.05$) than at higher concentration levels (40 and 50 mg Cy-3-glu eqv) ($P > 0.05$) in 100 g of yoghurt (Figure 4). It was again observed that on the eighth day, concentration 40 mg had the lowest percentage of polymerisation hence more stable anthocyanins. This again supports the trend observed in pigment and colour stability at concentration 40 mg in Figure 2.

Anthocyanins that polymerize may absorb at the maximum wave length of absorption in the visible range (in this case at 511 to 520 nm). This means that they maintain a uniform and stable colour over time (Wallace and Giusti, 2008). However, large polymers tend to precipitate resulting in a decrease in polymeric colour. The increase in polymeric colour in the first six days was therefore due to increase in absorption by polymerised anthocyanins. As the pigments stayed longer in the yoghurt matrix, they could have formed high molecular weight polymers with other polyphenols which precipitated out leading to a drop in polymeric colour and consequently leading to decrease in colour stability.

Degradation kinetics

The degradation of anthocyanins from *M. rubra* in yoghurt matrix was fitted to first-order kinetics. The rates of degradation at lower concentration levels (10 to 25 mg) were higher than at higher concentration levels (30 to 50 mg) (Table 2) which further supports the variation in pigment and colour stability. The half-life of the pigments increased with increasing concentration level suggesting that the pigments were more stable at higher concentration levels than at lower concentration levels (Table 2).

Linear regression analysis was used to determine the adequacy of the degradation kinetic model at alpha (α) < 0.05 acceptance level which indicates that the concentration of the anthocyanins decreased linearly with time (Figure 5). For example, the degradation equation at 20 mg cy-3-glu eqv concentration level was $lnC = -0.00691t + 2.985$ ($P < 0.05$).

A good regression was obtained between concentration and time at 10-25 mg cy-3-glu eqv concentration levels ($P < 0.05$) (Table 2). However other factors in addition to concentration tend to influence the stability of the pigments added to yoghurt between 30-50 mg cy-glu eqv concentration levels ($P > 0.05$) (Table 2). Intermolecular and intramolecular associations are evident at higher concentration levels (Nielsen et al., 1993).Whole fruit juice of mulberries used in this study contains cell wall and vacuolar components which could have exerted protective effect on the pigments at higher concentration levels (Kammerer et al., 2007).

Table 2. Degradation kinetic and linear regression parameters.

Concentration levels (mg cy-3-glu in 100 g)	k (day^{-1})	$t_{1/2}$ (days)	Regression coefficients (R^2)(%)	P > F
10	1.48 x 10^{-2}	46.7	76.9	0.002
20	6.9 x 10^{-3}	100.3	80.2	0.001
25	6.18 x 10^{-3}	112.7	76.9	0.008
30	5.0 x 10^{-3}	138.6	37.6	0.079
40	3.0 x 10^{-3}	231.05	18.5	0.248
50	1.0 x 10^{-3}	693.1	34	0.099

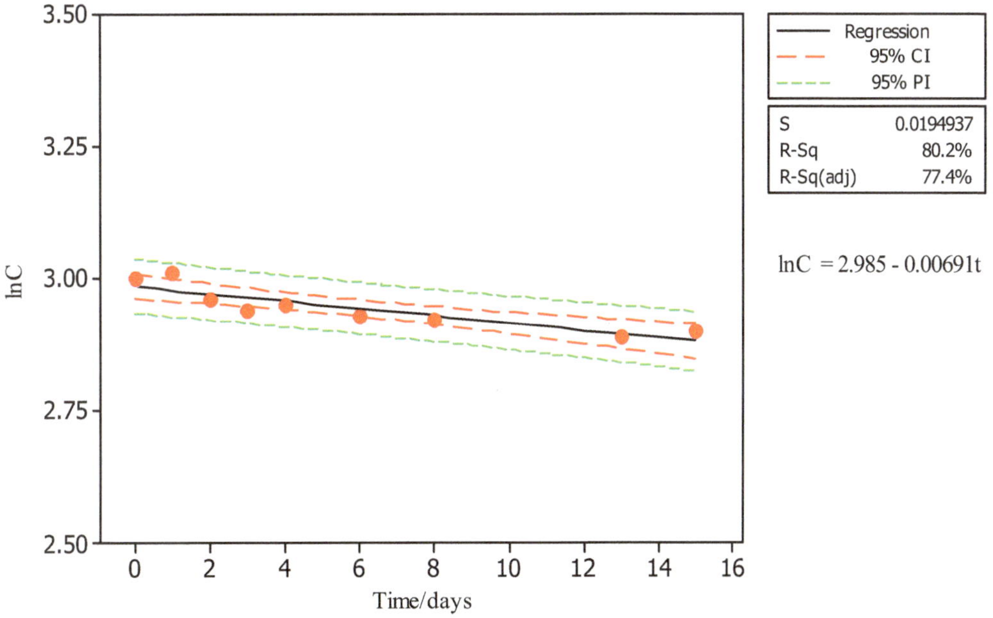

Figure 5. Linear regression predictions of degradation kinetic model at 20 mg cy-3-glu eqv in100 g of yoghurt concentration levels.

Conclusions

Addition of juice from whole fruit of mulberry (25-40 mg cy-3-eqv) rich in anthocyanins to plain yoghurt (100 g) achieved a good colour, matching with that of commercial brand strawberry flavoured yoghurt which was coloured using erythrosine (FD & C Red No. 3, 20 mg in 100 g of yoghurt). Colour and pigment stabilities of the anthocyanin pigments were very high between 25-40 mg cy-3-eqv in 100 g concentration levels. At the same time, the tendency of the pigments to polymerise was very low between 25-40 mg cy-3-glu concentration levels. The half-life for the pigments wasvery high compared to the shelf-life of commercial brand strawberry yoghurt indicating that, anthocyanins from mulberries are very stable when added to yoghurt. Moderate concentration levels (25-40 mg cy-3-glu eqv in 100 g of yoghurt) of mulberry anthocyanins would be ideal to colour yoghurt.

Conflict of Interest

The authors have not declared any conflict of interest.

ACKNOWLEDGEMENTS

The authors are grateful to the Uganda National Council for Higher Education (UNCHE) and Busitema University for financial support towards this study. Authors also appreciate the support provided by Fresh dairy factory, Sameer Agriculture and Livestock (U) ltd in providing the yoghurt samples used in this study.

REFERENCES

Aramwit P, Bang N, Srichana T (2010). The properties and stability of anthocyanins in mulberry fruits. Food Res. Int. 43:1093-1097.

http://dx.doi.org/10.1016/j.foodres.2010.01.022
Brenes CH, Del Pozo-Insfran D, Talcott ST (2005). Stability of copigmented anthocyanins and ascorbic acid in a grape juice model system. J. Agric. Food Chem. 53(1):49-56. http://dx.doi.org/10.1021/jf049857w
Bridle P, Timberlake CF (1997). Anthocyanins as natural food colors-selected aspects. Food Chem. 58(1):103-109. http://dx.doi.org/10.1016/S0308-8146(96)00222-1
Brownmiller C, Howard LR, Prior RL (2008). Processing and storage effects on monomeric anthocyanins, percent polymeric color, and antioxidant capacity of processed blueberry products. J. Food Sci. 73(5):72-79. http://dx.doi.org/10.1111/j.1750-3841.2008.00761.x
Castañeda-Ovando A, Pacheco-Hernández MDL, Páez-Hernández ME, Rodríguez JA, Galán-Vidal CA (2009). Chemical studies of anthocyanins: A review. Food Chem. 113(4):859-871. http://dx.doi.org/10.1016/j.foodchem.2008.09.001
Cavalcanti RN, Santos DT, Meireles MAA (2011). Non thermal stabilization mechanisms of anthocyanins in model and food systems-an overview. Food Res. Int. 44(2):499-509. http://dx.doi.org/10.1016/j.foodres.2010.12.007
Cevallos-Casals BA, Cisneros-Zevallos L (2004). Stability of anthocyanin-based aqueous extracts of Andean purple corn and red-fleshed sweet potato compared to synthetic and natural colorants. Food Chem. 86(1):69-77. http://dx.doi.org/10.1016/j.foodchem.2003.08.011
Delgados-Vargas F, Paredes-López O (2003). Pigments as colourants. In: Natural Colourants for Food and Nutraceutical Uses. Delgados-Vargas F, Paredes-López O. (Eds.). CRC Press: Boca Raton London New York Washington, D. C, Chapter 4
Ercisli S, Orchan E (2007). Chemical composition of white (*Morus alba*), red (*Morus rubra*) and black (*Morus nigra*) mulberry fruits. Food Chem.103:1380-1384. http://dx.doi.org/10.1016/j.foodchem.2006.10.054
Fan G, Han Y, Gu Z, Gu F (2008). Composition and colour stability of anthocyanins extracted from fermented purple sweet potato culture. LWT-Food Sci. Tech. 41(8):1412-1416.
Gauche C, Elisa da Silva M, Luiz MTB (2010). Effect pf pH on the copigmentation of anthocyanins from Cabernet Sauvignon grape extracts with organic acids. Sci. Agric. (Piracicaba, Braz.) 67(1):41-46.
Giusti MM, Wrolstad RE (2001). Characterization and Measurement of Anthocyanins by UV-Visible Spectroscopy. In: Current Protocols in Food Analytical Chemistry. RWE John Wiley & Sons, Inc. F1.2.1-F1.2.13
Giusti MM, Wrolstad RE (2003). Acylated anthocyanins from edible sources and their applications in food systems. Biochem. Eng. J.14(3): 217-225. http://dx.doi.org/10.1016/S1369-703X(02)00221-8
Heredia FJ, Francia-Aricha EM, Rivas-Gonzalo JC, Vicario IM, Santos-Buelga C (1998). Chromatic characterisation of anthocyanins from red grapes - I. pH effect. Food Chem.63:491-498. http://dx.doi.org/10.1016/S0308-8146(98)00051-X
Jing P, Giusti MM (2005). Characterisation of anthocyanin-rich waste from purple corncobs (Zea mays L.) and its application to color milk. J. of Agri. Food Chem. 53(22):8775-8781. http://dx.doi.org/10.1021/jf051247o
Kammerer D, Schillmoller S, Maier O, Schieber A, Carle R (2007). Colour stability of canned strawberries using black carrot and elderberry juice concentrates as natural colourants. Eur. Food Res. Tech. 224(6):667-679. http://dx.doi.org/10.1007/s00217-006-0356-3
Katsaboxakis K, Papanicolaou D, Melanitou M (1998). Stability of pigmented orange anthocyanins in model and real food system. Italian J. Food Sci.10:17-25.
Kirca A, Ozkan M, Cemerog˘lu B (2006). Stability of black carrot anthocyanins in various fruit juices and nectars. Food Chem. 97(4):598-605. http://dx.doi.org/10.1016/j.foodchem.2005.05.036
Liu X, Xiao G, Chen W, Xu Y, Wu J (2004). Quantification and purification of mulberry anthocyanins with macroporous resins. BioMed. Res. Int. 5:326-331.
Markakis P (1982). Stability of anthocyanins in foods. In: Anthocyanins as food colors. Markakis P, Academic Press: New York, pp. 163-180. http://dx.doi.org/10.1016/B978-0-12-472550-8.50010-X
Mendi SD, Peters B, Kamga TP (2000). The Effects of pH and Heat Treatment Processing on the Stability of Natural Food colours used in Dairy Products. The J. Food Tech. Africa. 2(5):59-61.
Nielsen SS, Marcy JE, Sadler GD (1993). Chemistry of aseptically processed foods. In: Principles of aseptic processing and packaging. Chambers JV, Nelson P E (Eds.). Food Processors Institute: Washington DC, pp. 87-111.
Özgen M, Serçe S, Kaya C (2009). Phytochemical and antioxidant properties of anthocyanin-rich *Morus nigra* and *Morus rubra* fruits. Scientia Hort.119(3):275-279. http://dx.doi.org/10.1016/j.scienta.2008.08.007
Qin C, Li Y, Niu W, Ding Y, Zhang R, Shang X (2010). Analysis and characterisation of Anthocyanins in Mulberry Fruit. Czech. J. Food Sci. 28(2):117-126.
Ścibisz I, Ziarno M, Mitek M, Zaręba D (2012). Effect of probiotic cultures on the stability of anthocyanins in blueberry yoghurts. LWT-Food Sci. Tech. 49:208-212.
Skrede G, Wrostad RE, Durst RW (2000). Changes in Anthocyanins and polyphenolics During Juice Processing of Highbush Blueberries (*Vaccinium corymbosum* L.). J. Food Sci. 65(2):357-364. http://dx.doi.org/10.1111/j.1365-2621.2000.tb16007.x
Torskangerpoll K, Andersen ØM (2005). Colour stability of anthocyanins in aqueous solutions at various pH values. Food Chem. 89(3):427-440. http://dx.doi.org/10.1016/j.foodchem.2004.03.002
Wallace TC, Giusti MM (2008). Determination of Color, Pigment, and Phenolic Stability in Yogurt Systems Colored with Nonacylated Anthocyanins from *Berberis boliviana* L. as Compared to Other Natural/Synthetic Colorants. J. Food Sci.73(4): C241-C248. http://dx.doi.org/10.1111/j.1750-3841.2008.00706.x
Wang WD, Xu SY (2007). Degradation kinetics of anthocyanins in blackberry juice and concentrate. J. Food Eng.82(3): 271-275. http://dx.doi.org/10.1016/j.jfoodeng.2007.01.018
Wrolstad RE, Skrede G, Ernersen G (1990). Influence of sugar on anthocyanin pigment stability in frozen strawberries. J. Food Sci.55(4): 1064-1065,1072. http://dx.doi.org/10.1111/j.1365-2621.1990.tb01598.x
Wu X, Liang L, Zou Y, Zhao T, Zhao J, Li F, Yang L (2011). Aqueous two-phase extraction, identification and antioxidant activity of anthocyanins from mulberry (*Morus atropurpurea* Roxb.). Food Chem.129(2): 443-453. http://dx.doi.org/10.1016/j.foodchem.2011.04.097

30

Thermodynamics of the adsorption of palmitate and laurate soaps onto some metal ore surfaces in aqueous media

Millicent U. Ibezim-Ezeani and Alphonso C. I. Anusiem

Department of Pure and Industrial Chemistry, University of Port Harcourt, P. M. B. 5323, Choba, Port Harcourt, Nigeria.

The thermodynamics of the adsorption of sodium-palmitate and sodium-laurate onto galena, hematite and cassiterite surfaces in aqueous media has been studied. The adsorption capacities of the soaps (adsorbates) were determined and the Langmuir, Freundlich and Temkin isotherm models were used to fit the experimental data. The Langmuir isotherm model was the best to describe the adsorption process with the highest correlation coefficient. Evaluation of the complete monolayer adsorption capacity at 29°C using the Langmuir equation showed 0.0506 mmol/g for galena, 0.0494 mmol/g for hematite, 0.0310 mmol/g for cassiterite (for sodium-palmitate adsorption) and 0.0802 mmol/g for galena, 0.0778 mmol/g for hematite, 0.0467 mmol/g for cassiterite (for sodium-laurate adsorption). These values are in the order: galena > hematite > cassiterite. The change in Gibbs free energy, ΔG, enthalpy, ΔH and entropy, ΔS of adsorption were computed and the values indicate that the adsorption processes are spontaneous, exothermic and favoured at low temperature suggesting physisorption as the predominant process.

Key word: Thermodynamics, adsorption, soap, collector reagents, langmuir isotherm.

INTRODUCTION

Thermodynamic studies of the adsorption of palmitate and laurate soaps (adsorbates) onto some metal ores (adsorbents) in aqueous phase were investigated as a prelude to determine the possible use of these soaps as collector reagents in metal ore beneficiation. Molecules of soap possess hydrophobic and hydrophilic parts which can be adsorbed selectively on the surface of the desired mineral particles to reduce wetting by water. The balance of dynamic forces between the concentration of adsorbate in bulk solution with that at the interface for a particular system leads to adsorption equilibrium that can be described in terms of adsorption isotherm. Such experimental isotherms are useful for a number of applications such as the predictive modeling procedures for analysis and design of adsorption systems (Upendra and Manas, 2006). The energy changes associated with the adsorption processes as molecules migrate from the bulk phase to the adsorptions site and the extent of adsorption depends on some physical parameters such as temperature, pressure, concentration in the bulk phase, and the surface area of the adsorbent, as well as on the nature of the adsorbate and the adsorbent. Various studies have justified the validity of evaluating the effect of these thermodynamic parameters in adsorption process (Arivoli et al., 2008; Abiola and Oforka, 2004; Abdel-aal and Morad, 2001).

In the adsorption of copper, nickel, cobalt and manganese by kaolinite from aqueous solution, Yavuz et al. (2003) observed a small but positive ΔG at low temperature which became negative at high temperature suggesting that the process is spontaneous at such high temperature. Similar results were obtained by Vinod and Anirudhan (2001) for the sorption of tannic acid on zirconium pillared clay which showed the process to be spontaneous, exothermic and the restriction of adsorbate molecules at the surface of the adsorbent. Gaikwad (2004), Gomma and Wahdan (1994) observed similar trend in their studies. Horsfall and Spiff (2005)

*Corresponding author. E-mail: millibejmj@yahoo.com.

Abbreviations: SSA, Specific surface area; **EGME,** ethylene-glycol-monethyl ether.

investigated the effects of temperature on the sorption of Pb^{2+} and Cd^{2+} from aqueous solution by caladium bicolour biomass and obtained results which showed that the adsorption process was physical in nature. The purpose of this study is to evaluate the necessary thermodynamic parameters of adsorption process which will be useful in describing the rates, mechanisms and spontaneity or non- spontaneity of the adsorption process. These thermodynamic parameters will then be exploited in the ultimate beneficiation of local ores using sodium-palmitate and sodium-laurate soaps as collector reagents in the froth flotation process which is mainly dependent on the selective provision of hydrophobic surface on the component of the ores.

MATERIALS AND METHODS

The metal ores used in this investigation are galena (obtained from Enyigba-Abakiliki, Nigeria) contained 59.3% lead, hematite (obtained from Itakpe, Nigeria) contained 76.9% iron and cassiterite (obtained from Jos, Nigeria) contained 58.7% tin. The samples were crushed in the laboratory jaw and roll crushers. Gravimetric method of jigging and tabling was employed in removing siliceous materials, while magnetic separation method was used in separating magnetic materials from the ore sample. Sieve analyses were performed using the British standard sieve plates to obtain samples of galena 65 micron, hematite 70 micron and cassiterite 65 micron. Mineralogical analysis of the ore samples was performed with volumetric and spectrophotometric (Buck scientific atomic absorption spectrophotometer model 205A) methods of analyses. Solutions of reagents were prepared with distilled deionised water, sodium hydroxide and hydrochloric acid solutions were used for pH adjustments. BDH chemical reagents analar grade (of not less than 98% purity) were used in this study. Determination of specific surface area (SSA) was done by the ethylene-glycol-monethyl ether (EGME) method (Cerator and Lutenegger, 2002). Soap preparation was performed by saponification reaction (Tooley, 1976).

Determination of thermodynamic parameters

0.5 g sample galena was added to 40 mL of 0.2 mmol/L soap solution in a conical flask. The capped flask was agitated in a mechanical shaker (120 oscillations per min.) for 1 h to attain equilibrium adsorption. The mixture was centrifuged, decanted and filtered to obtain a clear filtrate. The equilibrium conductance of the clear filtrate was measured and the equilibrium soap concentration was determined from the calibration curve. The adsorption capacity (q_e) in mol/g, was determined after the experiment was repeated at different concentrations and temperatures. The experiments were repeated using hematite and cassiterite.

Data evaluation

The concentration of soap samples were obtained from the calibration plots of the metal soaps based on Kohlrausch law equation in Equation.1 (Ibezim-Ezeani and Anusiem, 2009; Atkins, 1998).

$$\Lambda_m = \Lambda^o{}_m - KC^{\frac{1}{2}} \qquad (1)$$

Where Λm is the molar conductivity, Λ°m is the limiting molar conductivity, C is the concentration of the solution and K is a constant. The adsorption capacity (q_e) at equilibrium in mol/g, was determined using the mass balance equation (Ibezim-Ezeani and Anusiem, 2010) expressed as Equation 2.

$$q_e = \frac{V}{m}(C_o - C_e) \qquad (2)$$

Where C_e is the soap concentration in solution (mol/L) at equilibrium, C_o is the initial soap concentration in solution, V is volume of initial soap solution used (L), and m is mass of adsorbent used (g). The adsorption capacities were used in the isotherm equation plots. Linearized form of some adsorption isotherm models (Vikrant and Pant, 2006; Ho et al., 2005, 2001; Abiola et al., 2004; Cordero et al., 2004; Horsfall et al., 2004; Onyango et al., 2004; Haghseresht and Lu, 1998; Muhammad et al., 1998) used for data analyses are presented in Equation 3. (Langmuir), Equation 4 (Freundlich) and Equation 5 (Temkin).

$$\frac{C_e}{q_e} = \frac{1}{X_m b} + \frac{C_e}{X_m} \qquad (3)$$

$$\ln q_e = \ln K_f + \frac{1}{n}\ln C_e \qquad (4)$$

$$q_e = \frac{RT}{b_T}\ln K_T + \frac{RT}{b_T}\ln C_e \qquad (5)$$

The constants b, K_f and K_T are the Langmuir, Freundlich and Temkin isotherm constants, respectively. X_m is the maximum adsorption capacity upon complete coverage of the adsorption sites. 1/n is a measure of the surface heterogeneity, with values between 0 and 1, b_T is related to the adsorption intensity, R is the gas constant and T is temperature in Kelvin. Thermodynamic parameter ΔH, ΔG and ΔS are calculated from the variation of Langmuir equation constant, b (L/mol) with change in temperature, T, using equations 6 to 9 (Kim et al., 2005; Krishnan and Auirudhan, 2003).

$$\ln b = \ln b' - \frac{\Delta H}{RT} \qquad (6)$$

$$\ln b = -\frac{\Delta G}{RT} \qquad (7)$$

$$\ln b' = \frac{\Delta S}{R} \qquad (8)$$

$$\Delta S = \frac{\Delta H - \Delta G}{T} \qquad (9)$$

The plot of ln b against $^1/_T$ is linear and ΔH, ΔS and ΔG can be computed from the slope and intercept of such plots. The coefficient of determination, R^2 measurements were used in this study to ascertain the reliability of this data and best fitting models (Chase and Brown, 1992).

RESULTS AND DISCUSSION

Specific surface area of adsorbents

Calculations from experimental results and sample

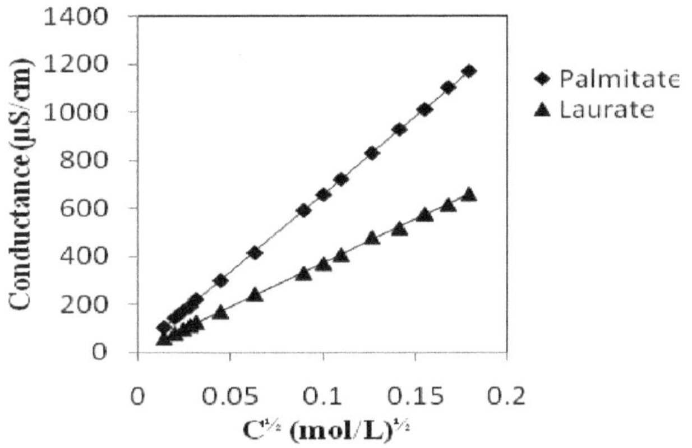

Figure 1. Plot of conductance versus soap concentration.

Table 1. Comparison of coefficient of determination (R^2) For the Isotherms.

Metal ore	Sodium-palmitate adsorption			Sodium-laurate adsorption		
	Langmuir	Freundlich	Temkin	Langmuir	Freundlich	Temkin
Galena	0.9989	0.9552	0.9952	0.9803	0.9471	0.9796
Hematite	0.9971	0.9498	0.9925	0.9977	0.9642	0.9961
Cassiterite	0.9968	0.9067	0.9590	0.9855	0.8582	0.9144

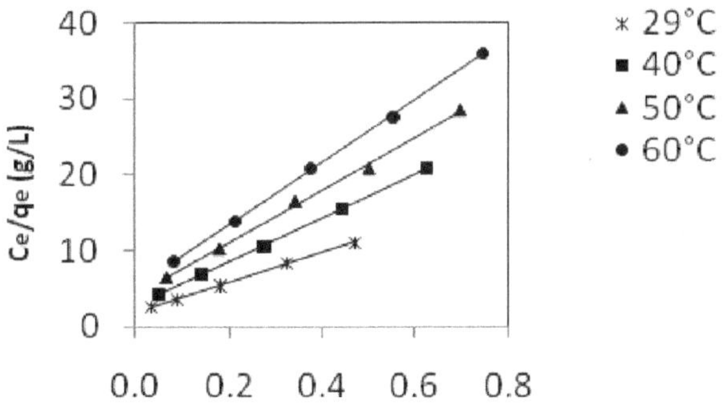

Figure 2. Langmuir isotherms for sodium-palmitate adsorption onto galena at different temperatures and concentrations.

analysis show that the specific surface area of galena is 128.7m^2/g, hematite is 118.5m^2/g and cassiterite is 88.1m^2/g. These values indicate that the surface area is in the order: galena > hematite > cassiterite. It is expected that if the specific surface area is the primary factor for adsorption, that the adsorption capacity will be greatest for galena. Ho (2006), Horsfall et al. (2004), Horsfall and Abia (2003) reported increased adsorption capacity with increasing surface area in their various studies.

Calibration of conductometer

Plots of the conductance of sodium-palmitate / sodium-laurate against the square root of initial soap concentrations are linear (Figure 1). This was used for subsequent determination of concentration of the soaps in the various experiments.

Adsorption isotherm

The Langmuir, Freundlich and Temkin isotherm models were considered as possible models to identify the isotherm that best describes the equilibrium adsorption data. The comparison of the coefficient of determination, R^2, for the three isotherm models is presented in Table 1. From our results, the Langmuir isotherm model with the highest R^2 values best described the adsorption process. This model assumes the formation of monolayer of adsorbate molecules on the adsorbent surface with all adsorption sites being energetically equivalent and that there is no migration or interaction between adsorbed ions on the surface area (Adamson, 1976; Langmuir, 1918). The Langmuir isotherm model was therefore used to analyze our equilibrium data at various temperatures and concentrations and presented in Figures 2 to 4 for sodium-palmitate and Figures 5 to 7 for sodium-laurate. The Langmuir constants Xm (the maximum adsorption capacity upon complete coverage of available adsorption sites), and b (related to energy / intensity of adsorption) for the sodium-palmitate and sodium-laurate adsorption are presented in Tables 2 and 3, respectively. The change in enthalpy, ΔH, Gibbs free energy, ΔG, and entropy, ΔS of adsorption were therefore calculated from

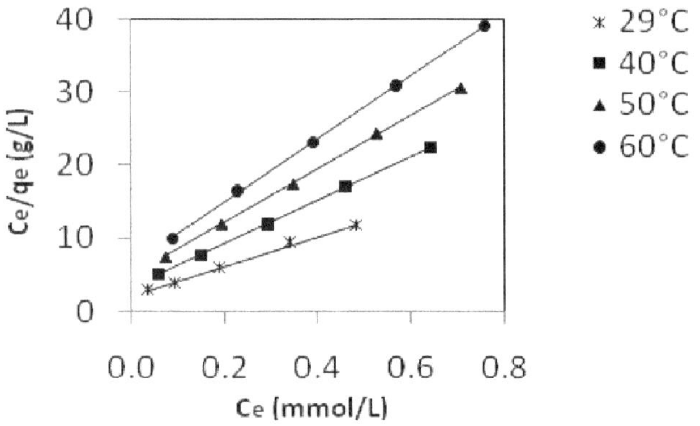

Figure 3. Langmuir isotherms for sodium-palmitate adsorption onto hematite at different temperatures and concentrations.

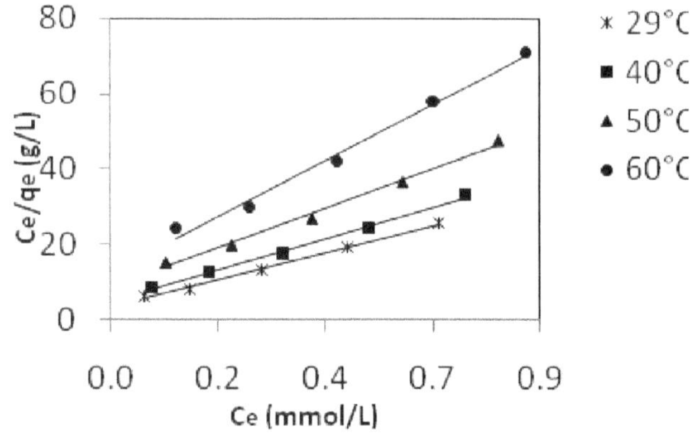

Figure 4. Langmuir isotherms for sodium-palmitate adsorption onto cassiterite at different temperatures and concentrations.

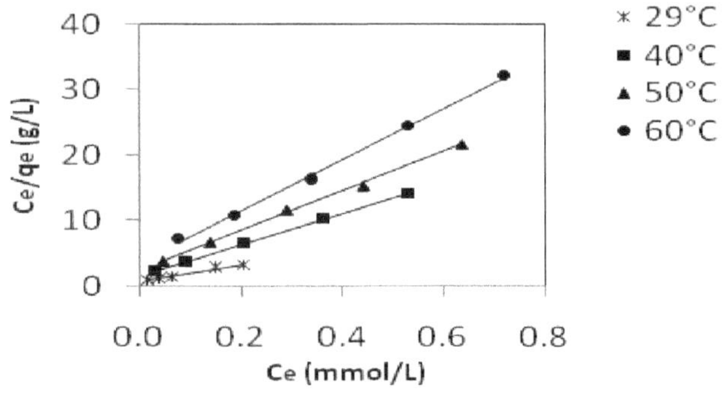

Figure 5. Langmuir isotherms for sodium-laurate adsorption onto galena at different temperatures and concentrations.

the variations of the Langmuir constant, b (L/mol) with change in temperature (Figures 8 and 9 for the sodium-palmitate and sodium-laurate adsorption, respectively) by plotting ln b versus $^{1}/_{T}$. The values of ΔG, ΔH and ΔS are shown in Table 4.

The negative values of ΔG at 29°C (approximately -23.00 kJ/mol for sodium-palmitate adsorption and -24.00 kJ/mol for sodium-laurate adsorption) indicate that the

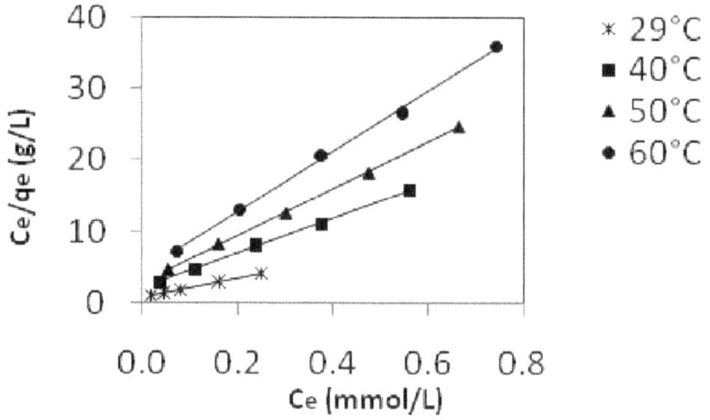

Figure 6. Langmuir isotherms for sodium-laurate adsorption onto hematite at different temperatures and concentrations.

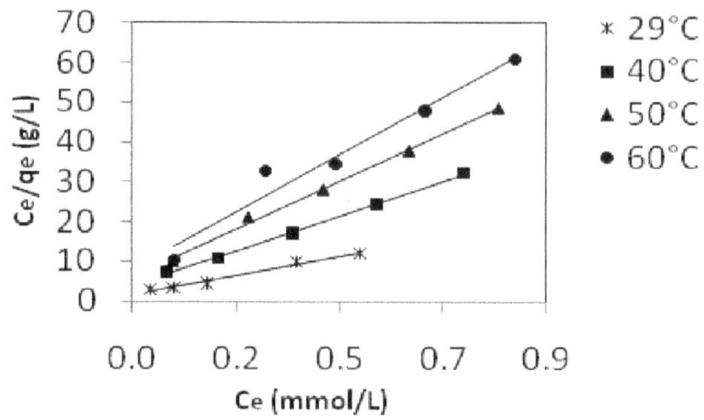

Figure 7. Langmuir isotherms for sodium-laurate adsorption onto cassiterite at different temperatures and concentrations.

adsorption process is to a large extent spontaneous and feasible. The feasibility of the process decreases with increase in temperature judging from the decrease in the values of ΔG with increase in temperature suggesting some structural changes in the adsorbate-adsorbent interaction. Generally, values of ΔG up to -20 kJ/mol are consistent with electrostatic interaction between adsorbate molecules and the adsorbent surface (physical adsorption) while ΔG values more negative than -40kJ/mol involve charge sharing or transfer from the adsorbate molecules to the adsorbent surface to form a coordinate type of bond which indicates chemical adsorption (Horsfall et al., 2006; Ebenso, 2004). The negative values of ΔH indicate the exothermic nature of the adsorption process which is favoured at low temperature. It is of note that enthalpy change due to chemical adsorption (>40 kJ/mol) is considerably larger than that due to physical adsorption. The ΔG and ΔH values obtained in this study (Table 4) indicate physisorption as the predominant adsorption process.

The positive values of ΔS suggest freedom of adsorbate molecules on the adsorbent surfaces confirming predominant physisorption process. In the adsorption processes studied, the values of ΔH and ΔS are higher for the sodium-palmitate when compared to

Table 2. Langmuir constants at different temperatures for sodium-palmitate adsorption.

Temp. (°C)	Galena		Hematite		Cassiterite	
	b (L/mol)	Xm (mmol/g)	b (L/mol)	Xm (mmol/g)	b (L/mol)	Xm (mmol/g)
29	10596.02	0.0506	9674.12	0.0494	8920.61	0.0310
40	10167.38	0.0348	8960.44	0.0336	7772.48	0.0263
50	8494.74	0.0288	7599.19	0.0274	5457.30	0.0210
60	7994.50	0.0242	6763.35	0.0231	5211.02	0.0149

Table 3. Langmuir constants at different temperatures for sodium-laurate adsorption.

Temp. (°C)	Galena		Hematite		Cassiterite	
	b (L/mol)	Xm (mmol/g)	b (L/mol)	Xm (mmol/g)	b (L/mol)	Xm (mmol/g)
29	15746.75	0.0802	12977.58	0.0778	12103.03	0.0467
40	14544.73	0.0423	11480.06	0.0409	9788.57	0.0258
50	11889.72	0.0332	11107.64	0.0306	8690.64	0.0188
60	13039.59	0.0249	9595.36	0.0237	7826.38	0.0156

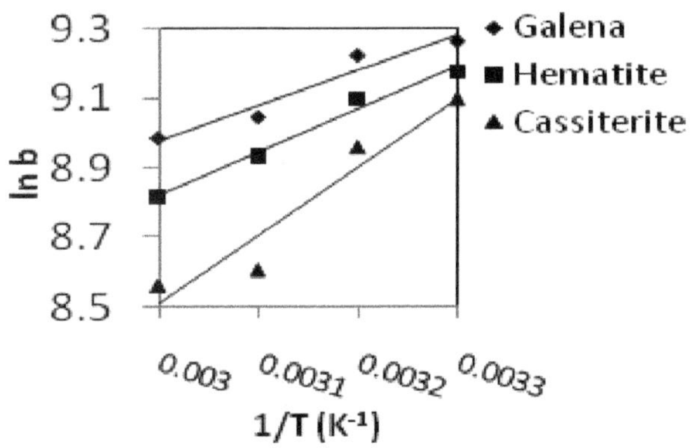

Figure 8. Plot of natural logarithm of langmuir constant at different temperature for sodium-palmitate adsorption.

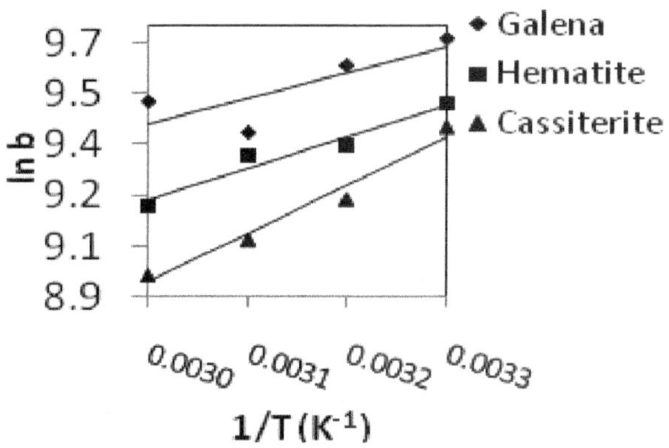

Figure 9. Plot of natural logarithm of langmuir constant at different temperature for sodium-laurate adsorption.

that of sodium-laurate. This observation could be due to differences in the molecular structure of the two compounds which are similar with respect to the position of the carboxylate group, but with different numbers of the carbon atoms (the palmitate soap has 16 carbon atoms, while the laurate soap has 12 carbon atoms). This difference in the length of carbon chain could induce a steric effect. The longer chain of the palmitate soap / its bulkier nature may hinder the proper orientation of the soap and thereby decrease the ease of its migration to the active sites which could lead to coverage of fewer adsorption sites, while the laurate soap bearing smaller carbon-chain may diffuse faster in the aqueous medium and thereby enhance higher surface coverage. Also in all the experimental cases, the values of ΔH and ΔS for the adsorbents are in the order: galena > hematite > cassiterite. This trend is consistent with the values of their specific surface area, confirming the importance of surface area in the evalution of the dynamics and feasibility of the process for a given application.

Conclusion

This study has demonstrated that the sodium-palmitate and sodium-laurate adsorption onto galena, hematite and cassiterite in aqueous phase is feasible. Results obtained show that our data fits the Langmuir isotherm model suggesting the formation of monolayer with energetically equivalent adsorption sites and with no lateral interaction between the adsorbed molecules on the adsorbent surface. Thermodynamic data analysis revealed that the adsorption process is spontaneous, exothermic, physical in nature and related to the specific surface area of the adsorbent. Our results suggest that the locally produced soaps can be exploited as collector reagents in designing an appropriate froth flotation system for harnessing galena, hematite and cassiterite from their ores.

Table 4. Thermodynamic parameters for the adsorption processes.

Adsorbent	ΔH (kJ/mol)	ΔG at 29^{0}C(kJ/mol)	ΔS (J/K/mol)
Galena (sodium-palmitate)	-8.52	-23.34	49.08
Hematite (sodium-palmitate)	-10.30	-23.12	42.46
Cassiterite (sodium-palmitate)	-16.35	-22.90	21.70
Galena (sodium-laurate)	-6.38	-24.23	59.11
Hematite (sodium-laurate)	-7.81	-23.80	52.95
Cassiterite (sodium-laurate)	-11.86	-23.57	38.76

REFERENCES

Abdel-aal MS, Morad MS (2001). Inhibiting effects of some quinolines and organic phosphonium compounds on corrosion of mild steel in 3M HCl solution and their adsorption characteristics. Br. Corrosion J., 36: 253-260.

Abiola OK, Oforka NC (2004). Adsorption of (4-amino-2-methyl-5-pyrimidinyl methylthio) acetic acid on mild steel from hydrochloric acid solution (HCl) – part 1. Material Chem. Phys., 83: 315-322.

Abiola OK, Oforka NC, Ebenso EE (2004). A potential corrosion inhibitor for acid corrosion of mild steel. Bull. Electrochem., 20(9): 409-413.

Adamson AW (1976). Physical Chemistry of Surface. John Wiley and Sons, New York, pp. 56-89.

Arivoli S, Hema M, Barathiraja C (2008). Comparative study on metal ions adsorption on a low cost carbonaceous adsorbent kinetic equilibrium and mechanistic studies. Iranian J. Environ. Health, Sci. Eng., 5: 1-10.

Atkins PW (1998). Physical Chemistry, 6th ed., Oxford University Press, New York, pp. 775-777.

Cerato AB, Lutenegger AJ (2002). Determination of surface area of fine-grained soils by the ethylene glycol monoethyl ether (EGME) method. Geotechnical Testing J., 25 (3): 1-7.

Chase W, Brown F (1992). General Statistics, 2nd ed., John Wiley and Sons, New York, pp. 610-630.

Cordero B, Lodeiro P, Herrero R, Sastre de Vicente ME (2004). Biosorption of cadmium by *Fucus Spiralis*, Environ. Chem., 1: 180-187.

Ebenso EE (2004). Effects of methyl red and halide ions on the corrosion of aluminum in H_2SO_4. Part 2. Bull. Electrochem., 20: 551-559.

Gaikwad RW (2004). Removal of Cd (II) from aqueous solution by activated charcoal derived from coconut shell. Electro. J. Environ. Agric. Food Chem., 3: 1-13.

Gomma GK, Wahdan MH (1994). Effect of temperature on the acidic dissolution of copper in the presence of amino acids. Mater. Chem. Phys., 39: 142-148.

Haghseresht, Lu GO (1998). Adsorption Characteristics of phenolic compounds onto coal-reject-derived adsorbents. Energy Fuels, 12: 1100-1107.

Ho YS (2006). Second-order kinetic model for the sorption of cadmium onto tree fem: A comparison of linear and non-linear methods. Water Res., 40: 119-125.

Ho YS, Chiu WT, Chung-Chi W (2005). Regression analysis for the sorption isotherms of basic dyes on sugarcane. Dust. Bioresour. Technol., 96: 1285-1291.

Ho YS, Ngjcy, McKay G (2001). Removal of lead (II) from effluents by sorption of peat using second.

Horsfall M, Abia AA (2003). Sorption of cadmium (II) and zinc (II) ions from aqueous solutions by cassava waste biomass (*Manihot Esculenta Vanz*). Water Res., 37: 4913-4923.

Horsfall M, Abia AA, Spiff AI (2006). Kinetic studies on the adsorption of Cd^{2+}, Cu^{2+} and Zn^{2+} ions from aqueous solutions by cassava (*Manihot Sculenta Vanz*) tuber bark waste. Bioresour. Technol., 97: 283-291.

Horsfall M, Spiff AI (2005). Equilibrium sorption study of Al^{3+}, Co^{2+} and Ag^{2+} in aqueous solutions by fluted pumpkin (*telfairia occidentalis hook f*) waste biomass. Acta Chim. Slov., 52: 174-181.

Horsfall M, Spiff AI, Abia AA (2004). Studies on the influence of mercaptoacetic acid (MAA) modification of cassava (*Manihot Esculenta Vanz*) waste biomass on the adsorption of Cu^{2+} and Cd^{2+} from aqueous solution. Bull. Korean Chem. Soc., 25(7): 969-976.

Ibezim-Ezeani MU, Anusiem ACI (2009). Studies on the adsorption of palmitate and laurate soaps onto some metal ore surfaces in aqueous media. Scientia Africana, 8(2): 31-41.

Ibezim-Ezeani MU, Anusiem ACI (2010). Kinetic studies of adsorption of palmitate and laurate soaps onto some metal ores in aqueous media. Int. J. Phys. Sci., 5(1): 62-67.

Kim TY, Sun-Kuu P, Sung-tony C, Kim HB, Kang Y, Kim SD, Kim SJ (2005). Adsorption of heavy metals by brewery biomass. Korean J. Chem. Eng., 22(1): 91-98.

Krishnan KA, Anirudhan TS (2003). Removal of cadmium (II) from aqueous solutions by steam-activated sulphurised carbon prepared from sugar-cane baggase pith: Kinetics and equilibrium studies. Water SA., 29(2): 147-156.

Langmuir I (1918). The adsorption of gases on plane surfaces of glass, mica and platinum. J. Am. Chem. Soc., 40(9): 1361-1403.

Muhammad N, Jeremy P, Smith MD, Wheatleg AD (1998). Adsorption of heavy metals in slow sand filters. 24th WEDC Conference Islamadbad, Pakistan, pp. 346-349.

Onyango MS, Kojma Y, Aoyi O, Bernardo EC, Matsuda H (2004). Adsorpton equilibrium modeling and solution chemistry dependence of fluoride removal from water by trivalent-cation-exchanged zeolite F-9. J. Colloid Interface Sci., 279: 341-350.

Tooley P (1976). Experiments in Applied chemistry, John Murray, London, p. 29.

Upendra K, Manas B (2006). Sorption of cadmium from aqueous solution using pretreated rice husk. Bioresour. Technol., 97: 104-109.

Vikrant S, Pant KK (2006). Removal of chromium from industrial waste by using eucalyptus bark. Bioresour. Technol., 97: 15-20.

Vinod VP, Anirudhan TS (2001). Sorption of tannic acid on zirconium pillared clay. J. Chem. Technol. Biotechnol., 77: 92-101.

Yavuz O, Yalcin A, Fuat G (2003). Removal of copper, nickel, cobalt and manganese from aqueous solution by kaolinite. Water Res., 37: 948-952.

Colour of starch-iodine complex as index of retrogradability of starch pastes

Louis M. Nwokocha[1] and Gabriel B. Ogunmola[2]

[1]Department of Chemistry, University of Ibadan, Ibadan, Nigeria.
[2]Institute of Genetic Chemistry and Laboratory Medicine, Bodija, Ibadan, Nigeria.

Starch retrogradation is mainly due to the amylose fraction of starch. Amylose quantification is usually done by iodine staining with which it forms a blue colour complex while starch retrogradability can be monitored by freeze-thaw stability measurements. In this work, spectroscopic determination of the blue value and visual monitoring of starch–iodine colour complex were used to study starch retrogradation. The results obtained were compared with that from freeze-thaw stability measurements. Native cassava starch, its carboxymethylated and cyanoethylated derivatives of different degrees of substitution (D.S) were used in the study. From the results, increase in starch derivatization reduced amylose ability to bind iodine, decreased the blue value with resultant decay in the blue colour of the starch-iodine complex. The blue black colour of the starch-iodine complex was lost in carboxymethyl starch at D.S of 0.145 and cyanoethyl starch at D.S of 0.141; at these degrees of substitution, the helical structure of amylose was no longer maintained by the starch molecules and retrogradation eliminated. Freeze-thaw stability study showed clear pastes with no evidence of retrogradation over 10 freeze-thaw cycles for derivatized starches above these degrees of substitutions. This showed a good agreement with that obtained from the colour of the starch-iodine complex. Hence derivatized starch products, which showed absence of blue black colour of starch-iodine complex would be freeze-thaw stable over a long period of cold storage. The disappearance of the blue colour of starch-iodine complex thus becomes an index of retrogradability and freeze-thaw stability of starch pastes on cold storage.

Key words: Starch pastes, retrogradability, blue value, starch-iodine colour.

INTRODUCTION

When pasted starch products are subjected to cold storage the starch macromolecules in the starch dispersions tend to re-associate and exude water, a process referred to as syneresis (Thomas and Atwell, 1999; Balagopolan et al., 1998). This re-association and subsequent crystallization from the dispersions (retrogradation) is a major problem limiting starch utilization. Retrogradation problem can minimized by derivatization of the starch hydroxyls through introducing groups which retard the side-by-side association of starch molecules thereby stabilizing the aqueous dispersion of the paste (Thomas and Atwell, 1999; Balagopolan et al., 1998; Yeh and Yeh, 1993). When starch is derivatized, there is a change in the molecular organization within the granules and in starch physicochemical properties (Ogunmola and Nwokocha, 2002); and since starch retrogradation is a phenomenon primarily associated with amylose molecules (Collison, 1968) (amylopectin retrogrades very slowly), the effective changes must have been those associated with amylose units. Starch reacts with iodine to form a blue colour (Thomas and Atwell, 1999). Studies on starch fractions have shown that the blue colour complex observed with iodine is due to the colour formation of amylose units (Radley, 1968). Amylopectin reacts with iodine to form a violet-red colour. The ability to form blue colour complex with iodine is gradually destroyed with increasing derivatization until the blue colour disappears (Radley, 1968). Similarly with increasing degree of derivatization, retrogradation is found to decrease until a certain degree of derivatization when retrogradation is completely removed. Since the blue colour of starch-iodine complex and retrogradation in starch pastes are primarily associated with amylose, it is possible to use the disappearance of the blue colour to monitor the retrogradability of aqueous starch pastes and thus provide an assay method of retrogradability in starch products evaluation.

MATERIALS AND METHODS

Preparation of cassava starch derivatives

Cassava starch used in this study was isolated in Chemistry Department, University of Ibadan, Nigeria. All reagents used including iodine crystals, potassium iodide, acetic acid, ethanol and sodium hydroxide are products of Sigma-Aldrich Chemical Co., St Louis, MO, USA) and of reagent grade. Carboxymethylation (Reaction 1) was achieved by treating aqueous starch dispersions with different concentrations of monosodium chloroacetate in presence of sodium hydroxide as the alkalizing agent. In similar manner, cyanoethylation (Reaction 2) was effected by treating aqueous starch dispersions with acrylonitrile using sodium hydroxide as the alkalizing agent. The detailed procedures are reported elsewhere (Ogunmola and Nwokocha, 2002; Nwokocha and Ogunmola, 2008). Derivatization can occur at any of the three hydroxyl groups of the glucopyranosyl unit.

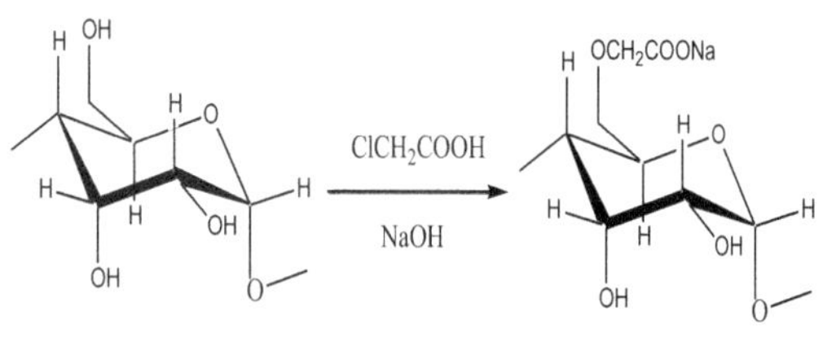

Reaction 1

Starch Sodium carboxymethyl starch

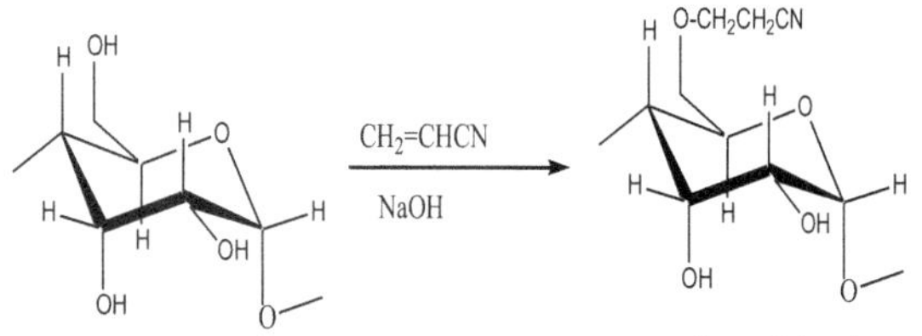

Reaction 2

Starch Cyanoethylstarch

Measurement of freeze-thaw stability of native, carboxymethyl and cyanoethyl starches

The freeze-thaw stability of the starch samples was determined on 5% starch pastes. 0.5 g dry starch samples were weighed into centrifuge tubes and 10 ml distilled water added and sealed. The samples were shaken to disperse the starch and then pasted in a Clifton water bath at 95°C for 30 min with occasional shaking throughout the heating regime. The samples were removed, cooled and subjected to alternate freezing (freezer temperature) and thawing (25°C) for 18 and 3 h, respectively and centrifuged at 5000 rpm for 15 min and the amount of water separated after each freeze-thaw cycle measured. The freeze-thaw stability was determined as percentage of water exuded per weight of paste.

Determination of the blue value of native, carboxymethyl and cyanoethyl starch-iodine complexes

0.1 g dry starch sample was weighed into a boiling tube, 1 ml ethanol (95%) was added followed by 9 ml of 1 M NaOH solution and heated in a boiling water bath for 10 min to solubilize the starch. The starch solution was cooled and quantitatively transferred into a 100 ml standard volumetric flask and the volume made up to 100 ml mark with distilled water. 2.5 ml of starch solution was taken into 50 ml standard flask; 0.5 ml of 1 M acetic acid was added followed by 1 ml of stock iodine (0.2 g I_2 / 2.0 g KI/ 100 ml) and the solution made up to the 50 ml mark with distilled water. The resulting colour was left for 20 min to fully develop before the absorbance reading was monitored at 620 nm with a Perkin-Elmer Lambda 3B double beam UV/ visible spectrophotometer. Iodine solution of same concentration as above but without starch sample was used in the reference cell. The colour of the starch- iodine complex was monitored visually. The blue value was calculated according to the method of Gilbert and Spragg (1964) using the formula:

$$Blue\ value = \frac{Absorbance\ at\ 620\,nm\ X\ 4}{Concentration\,(mg/dl)}$$

RESULTS AND DISCUSSION

Freeze-thaw stability of carboxymethyl and cyanoethyl starches

Tables 1 and 2 show the freeze-thaw stability of carboxymethyl and cyanoethyl starches respectively. Both modified starches exhibited unusual stability even at low degrees of substitution in comparison with native starch. In carboxymethyl starch, stability to syneresis was observed at D.S of 0.064 and above. At lower levels of substitution (D.S< 0.064), syneresis was observed from the 5th freeze-thaw cycle in D.S of 0.028 and 0.061 with maximum exudates achieved decreasing with increasing D.S. In cyanoethyl starch, maximum exudates of 11.50% were obtained after 10 freeze-thaw cycles in D.S of 0.038. However, D.S of 0.100 and higher levels were stable to freeze-thaw cycles. Native cassava starch (D.S. 0.0) could not survive the first freeze-thaw cycle and attained maximum exudates of 31.30% after seven freeze-thaw cycles. Thus derivatization introduced branching groups onto the amylose chains which

Table 1. Freeze–thaw stability of native and carboxymethyl starches at 5% (w/v) concentration.

No. freeze-thaw cycles (days)	0	1	2	3	4	5	6	7	8	9	10
D.S					Percentage water separated						
0.0	0.7	22.10	5.60	18.60	26.70	30.80	31.30	31.30	31.30	31.30	31.30
0.028	-	-	-	-	-	21.79	24.75	24.75	29.74	30.00	30.00
0.061	-	-	-	-	-	0.028	6.32	20.10	23.07	23.07	23.07
0.064	-	-	-	-	-	-	-	-	-	-	-
0.076	-	-	-	-	-	-	-	-	-	-	-
0.111	-	-	-	-	-	-	-	-	-	-	-
0.131	-	-	-	-	-	-	-	-	-	-	-
0.134	-	-	-	-	-	-	-	-	-	-	-
0.187	-	-	-	-	-	-	-	-	-	-	-
0.194	-	-	-	-	-	-	-	-	-	-	-
0.197	-	-	-	-	-	-	-	-	-	-	-
0.280	-	-	-	-	-	-	-	-	-	-	-

- clear paste with no water separation, D.S of native starch = 0.0.

Table 2. Freeze–thaw stability of native and cyanoethyl starches at 5% (w/v) concentration.

No. freeze-thaw cycles (days)	0	1	2	3	4	5	6	7	8	9	10
D.S					Percentage water separated						
0.0	0.7	5.60	18.60	22.10	26.70	30.80	31.30	31.30	31.30	31.30	31.30
0.038	-	-	6.90	8.14	8.56	10.15	10.51	10.51	11.03	11.08	11.50
0.100	-	-	-	-	-	-	-	-	-	-	-
0.141	-	-	-	-	-	-	-	-	-	-	-
0.168	-	-	-	-	-	-	-	-	-	-	-
0.199	-	-	-	-	-	-	-	-	-	-	-
0.255	-	-	-	-	-	-	-	-	-	-	-
0.303	-	-	-	-	-	-	-	-	-	-	-
0.368	-	-	-	-	-	-	-	-	-	-	-
0.603	-	-	-	-	-	-	-	-	-	-	-
0.644	-	-	-	-	-	-	-	-	-	-	-
0.699	-	-	-	-	-	-	-	-	-	-	-
0.963	-	-	-	-	-	-	-	-	-	-	-

- clear paste with no water separation, D.S of native starch = 0.0.

hindered reassociation of amylose and reduced their ability to retrograde.

Relationship between colour of starch-iodine complex and degree of substitution

From Tables 3 and 4, a general decrease in blue value was observed with increase in D.S. This implied a gradual derivatization of the amylose as the D.S increased. Also the helical structure of amylose in solution responsible for complex formation with iodine was gradually destroyed with increase in D.S. It was observed that the decrease in blue value did not vary uniformly with increase in D.S. This implied both amylose and amylopectin were randomly derivatized. The blue colour of the starch-iodine complex was gradually destroyed as D.S increased until a green colour was observed. In carboxymethyl starch, the transition to green colour occurred at a D.S of 0.145. Higher D.S gave various shades of green colour. At D.S of 0.145, the helical structure of amylose was destroyed to the extent that retrogradation was completely removed. Hence, D.S of 0.145 and above should produce pastes that are stable to cold storage. In cyanoethyl starch, the transition to green colour occurred at a D.S of 0.141. Higher D.S gave

Table 3. Relationship between the starch-iodine colour and degree of carboxymethylation.

Degree of carboxymethylation	Blue value	Starch-iodine colour
0.0	0.375±0.0005	Blue black
0.014	0.374±0.0007	Blue black
0.039	0.370±0.0010	Blue black
0.055	0.342±0.0015	Blue black
0.116	0.290±0.0004	Blue black
0.128	0.300±0.0011	Blue black
0.145	0.251±0.0013	Deep green
0.185	0.170±0.0006	Green
0.205	0.208±0.0004	Green
0.212	0.185±0.0004	Green
0.269	0.186±0.0002	Green
0.296	0.159±0.0004	Green

Table 4. Relationship between the starch-iodine colour complex and degree of cyanoethylation.

Degree of cyanoethylation	Blue value	Starch-iodine colour
0.0	0.375±0.0005	Blue black
0.038	0.268±0.0009	Blue black
0.100	0.266±0.0010	Blue black
0.141	0.266±0.0006	Green-blue
0.169	0.246±0.0005	Green-blue
0.199	0.114±0.0015	Green-blue
0.255	0.032±0.0006	Light green
0.274	0.081±0.0013	Light green
0.303	0.079±0.0013	Light green
0.368	0.214±0.0010	Green-blue
0.484	0.171±0.0012	Green
0.591	0.194±0.0008	Green-blue
0.603	0.075±0.0006	Light green
0.639	0.181±0.0005	Green-blue
0.644	0.149±0.0005	Green-blue
0.699	0.164±0.0010	Green-blue
0.963	0.127±0.0015	Green

various shades of green colour which in some cases had bluish tint. This has been attributed to crosslinking and surface derivatization which tends to occur when starch is cyanoethylated under high acrylonitrile concentration (Hefrieter, 1986). Due to surface derivatization and crosslinking some long segments amylose inside the granules were not accessible to the reagents and hence remained long enough to form blue complex with iodine, this was responsible for the blue tint observed at high cyanoethylation degree.

Comparison of results from freeze-thaw stability and colour of starch-iodine complex showed that stability was reached at slightly different points. The difference arose from the fact that each method monitored retrogradation by different parameters. Retrogradation by freeze-thaw stability measures re-association between adjacent starch molecules while retrogradation by colour of starch-iodine complex measures the extent of complexation of the amylose with iodine. At D.S less than 0.064 in carboxymethyl and 0.100 in cyanoethyl starches, re-association was significant enough to result in the crystallization of the starch molecules. At these degrees of substitution the amylose chains were long enough to enter into helical structure with iodine to give the blue colour complex. At higher D.S the associative forces were very weak to result in starch crystallization and at

D.S of 0.145 in carboxymethyl and 0.141 in cyanoethyl starches, the helical structure was no longer maintained resulting in destruction of the blue colour. The method based on complex formation of starch with iodine has the advantage of rapidity because the result can be obtained in less than an hour unlike the freeze-thaw stability method that requires at least a day to get result.

Conclusion

The stabilization of starch paste by derivatization was related to the colour of starch-iodine complex. The blue colour of starch-iodine complex decayed with increase in degree of substitution (D.S) of the derivatized starch until it completely disappeared. This corresponded to when freeze-thaw stability was achieved without any noticeable sign of retrogradation. The disappearance of the blue colour of starch-iodine complex thus becomes an index of retrogradability and freeze-thaw stability of starch on cold storage.

Conflict of Interests

The author(s) have not declared any conflict of interests.

ACKNOWLEDGEMENT

This study was supported by University of Ibadan Research Grant 94/95.

REFERENCES

Balagopolan C, Padmaja G, Nanda SK, Moorthy SN (1998). Soil Stability. In: Cassava in Food, Feed and Industry, CRC Press Inc. Boca Raton, Florida. P. 127.

Collison R (1968). Retrogradation of Starch. In: Starch and its Derivatives, 4th Edition, J.A. Radley (ed.), Chapman and Hall, London. pp. 194-200. http://dx.doi.org/10.1002/star.19680201105

Gilbert GA, Spragg SP (1964). Iodometric Determination of Amylose. In: Methods in Carbohydrate Chemistry. R. L. Whistler (ed.), Academic Press, Orlando, Florida. pp. 168-169.

Hefrieter BT (1986). Miscellaneous Modifications: Cyano.ethyl Starch, In: Modified Starches: Properties and Uses. O.B Wurzburg (ed.), CRC Press Inc., Boca Raton, Florida. pp.188-189.

Nwokocha LM, Ogunmola GB (2008). Carboxymethylation of starch in different solvents and solvent-water mixtures. Optimization of reaction conditions. J. Appl. Sci. 8(8):1581-1585. http://dx.doi.org/10.3923/jas.2008.1581.1585

Ogunmola GB, Nwokocha LM (2002). Cyanoethylation of starch in varied solvents. J. Chem. Soc. Niger. 27:150-157.

Radley JA (1968). In: Starch and Its Derivatives. 4th Edition, Chapman and Hall Inc, London. P. 38.

Thomas DJ, Atwell WA (1999). Light microscopy; Retrogradation, In: Starches. Eagan Press Handbook Series, Eagan Press. St. Paul Minnesota, USA. pp. 13-15, 28-29.

Yeh AI, Yeh SL (1993). Some characteristics of hydroxypropylated and crosslinked rice starch. Cereal Chem. 70(5):596-600.

Extraction of silica gel from *Sorghum bicolour* (L.) moench bagasse ash

Mupa M.[1], Hungwe C. B. [1], Witzleben S. [2], Mahamadi C.[1] and Muchanyereyi N.[1]

[1]Department of Chemistry, Bindura University of Science Education, P. Bag 1020, Bindura, Zimbabwe.
[2]Bonn-Rhein-Sieg University of Applied Sciences, Inorganic Chemistry and Materials Analysis, von Liebig Str. 20, 53359 Rheinbach, Germany.

Sweet sorghum (*Sorghum bicolor* (L.) moench), a crop that is grown by subsistence farmers in Zimbabwe was used to extract silica gel in order to assess its possible use as a raw material for the production of silica-based products. The gel was prepared from sodium silicate extracted from sweet sorghum bagasse ash by sodium hydroxide leaching. Results show that maximum yield can be obtained at pH 5 and with 3 M sodium concentration. The silica gel prepared at optimum pH 5 had a bulk density of 0.5626 g/cm^3 and anestimated porosity of 71.87%. Silica gel aged over 10 h had improved moisture adsorption properties. X-ray fluorescence (XRF) determinations show that the silica content in the ash is 40.1%. Characterization of sweet sorghum ash and silica gels produced at pH 5, 7 and 8.5 by Fourier Transform Infrared spectroscopy gave absorption bands similar to those reported by other researchers.Transmission electron micrographs show that silica prepared under optimum conditions is amorphous and consisted of irregular particles. Sweet sorghum proved to be a potential low cost raw material for the production of silica gel.

Key words: Silica gel, sweet sorghum, extraction, ash, bagasse, X-ray powder diffraction (XRD).

INTRODUCTION

Silica gel is a non-toxic and inert inorganic polymer composed of amorphous silicon dioxide. It is characterized by a large surface area of up to 800 m^2/g. This property and its ability to adsorb a number of compounds has been exploited in a number of ways especially in catalysis and sorbent technology. Silica has also been used as molecular sieves in petroleum refining, air separation and nuclear waste management, as stationary phase material in liquid chromatography and as reinforcement for rubber and plastic (Devi and Dhanalakshmi, 2012; Tongjian et al., 2009).

Silica gel is prepared commercially through acid precipitation from sodium silicate solution. The sodium silicate is prepared through high temperature reaction of quartz with soda ash. The main disadvantage of this approach is that it is energy intensive requiring temperatures of up to 1300°C and the silica product may contain heavy metal impurities making it unsuitable for certain applications where purity is crucial. Spherical silica gel particles can also be prepared through hydrolysis of Stoiber reagents such as tetraethoxysilane and tetramethoxysilane (Sing et al., 2011). These reagents have been reported to be hazardous and are generally very expensive making them unsuitable for large scale production (Dorcheh and Abbasi, 2008).

A number of researches have shown that silica gel can be extracted from plant waste (Worathanakul et al., 2009; Alayande et al., 2012). Extraction from rice husk ashes has been widely reported (Prasad and Pandey, 2012). Research has so far focused on the preparation of silica gels of various morphology e.g. nano particles, meso, macro and microporous particles (Noushad et al., 2012; Thuadaji and Nuntiya, 2008). Another potential silica source is the sugarcane bagasse ash. Ugheoke and Mamat (2012) reported a new approach of preparing ordered and disordered silica gel particles from bagasse ash. A cation exchange method was used to extract silica from sugarcane.

Sorghum bicolor (L.) moench, better known as sweet sorghum is widely grown in sub-Saharan Africa as a subsistence crop under various climatic conditions. A number of researchers have investigated potential applications of sweet sorghum and these include bio-fuels and bio-ethanol production(Almodares and Hadi, 2009; Srinivasa et al., 2009; Nadir et al., 2009). Sweet sorghum can also be a potential raw material for the production of syrups and jaggery (Nimbkar et al., 2006; Gnansounou et al., 2005). The by-product bagasse can be used for firing plants such as in ethanol production. Just like sugarcane and other plants in the grass family, *sorghum bicolor* (L.) *moench*is a silicon accumulator.Venkata et. al., 2012 found that sweet sorghum has a silica content of 5.48%. When produced in large quantities, sweet sorghum can therefore be a potential source of silica.

This research paper reports a new biobased approach of extracting silica from sweet sorghum bagasse ash. Sweet sorghum is a drought resistant crop that grows under various climatic conditions and growing it does not require special skills. Sweet sorghum bagasse ash is renewable since is derived from plant source.

METHODS

Materials and reagents

Sweet sorghum was harvested from Hippo Valley High School Garden near Chiredzi, in south eastern Zimbabwe. Chemical reagents used in this research were of analytical grade and these were sodium hydroxide (NaOH), hydrochloric acid (HCl) and ammonia (NH_3) obtained from Associated Chemical Enterprises (PTY) Ltd, Republic of South Africa.

Sampling and sample preparation

Mature sweet sorghum (SS) stalks were cut into smaller pieces and dried overnight at 110°C. The dried samples were ignited in a muffle furnace for three hours. The ash content of the whole plant was determined atignition temperatures of 200, 400, 600 and 800°C.

Preparation of silica gel

Silica gel extraction was adapted from methods reported in the literature with minor changes (Gnansounou et al., 2005; Venkata et al., 2012). Five grams of sweet sorghum bagasse ash was leached for 30 min in 100 ml of 1 M HCl solution to remove heavy metals. The leached ash was washed with copious amounts of distilled water. This was followed by refluxing the ash suspended in 100 ml of 3 M NaOH for three hours. The resultant solution which predominantly contained sodium silicate was filtered off to remove any suspended solids. The clear solution was titrated with 3 M HCl until an optimum pH 5 whereby then silica gel would have started precipitating. The gel was allowed to age for 10 h. Silica gel was separated by centrifugation and washed until the aqueous part reacted negative to a chloride test. The experiments were repeated for pH 7 and 9 and aging times of 10, 18 and 48.

Characterization of silica gels

The bulk density and porosity were determined using reported methods by Kalapathy et al. (2002) and Adam et al. (2011) and calculated as follows:

$$\rho = \frac{M}{V} \quad (1)$$

Where ρ is the bulk density, M and V the mass and volume of silica gel respectively.

The porosity of the gel was estimated using the following equation:

$$p = (1 - \frac{\rho_{Si}}{\rho_s}) \times 100 \quad (2)$$

Where ρ_{Si} is the bulk density of silica gel and ρ_s the specific density of amorphous silica assumed to be 2.0 g/cm^3 (Adam et al., 2011).

FTIR spectra of samples were recorded on a Thermo-Fisher Nicolet 600 FTIR instrument.The samples were prepared as KBr pellets. The composition of the ash and purity of extracted silica was determined using an XRF-spectrometer, Oxford X-Supreme 8000 with a tungsten X-ray tube and powder sample holder.The results are shown in Table 1. A BrukerD2 phase XRD instruments equipped with Cu-Kα, powder sample holder, powder sample holder and LynxEye detector in the range 2 theta 10° to 65° was used to record diffractograms of the silica samples and the results are shown in Figure 5. SEM images were recorded on a Jeol JSM 6510 instrument. Samples were prepared on a powder sample holder.

RESULTS AND DISCUSSION

The effect of temperature on ash content was investigated for the temperature range 200 to 800°C. At temperatures above 600°C, no significant changes in ash content were observed and 600°C was hence selected as the optimum ignition temperature. The effect of ignition temperature on ash content is shown on Figure 1. The average ash content for the whole plant was found to be 5.6%. A similar value 5.48% was reported by other researchers (Alayande et al., 2012).

Results of XRF analysis on the ash and silica prepared under optimum conditions are shown in Table 1. The ash

Table 1. Percentagecomposition of sweet sorghum bagasse ash and extracted silica.

Component	Ash	Extracted silica gel
SiO_2	40.16	96.36
K_2O	26.46	0.922
Na_2O	-	2.146
Al_2O_3	15.23	0.467
CaO	6.91	-
P_2O_5	5.92	-
SO_3	0.55	-
Fe_2O_3	0.41	-
Cl^-	4.25	-
MnO	0.03	-
TiO_2	0.03	-

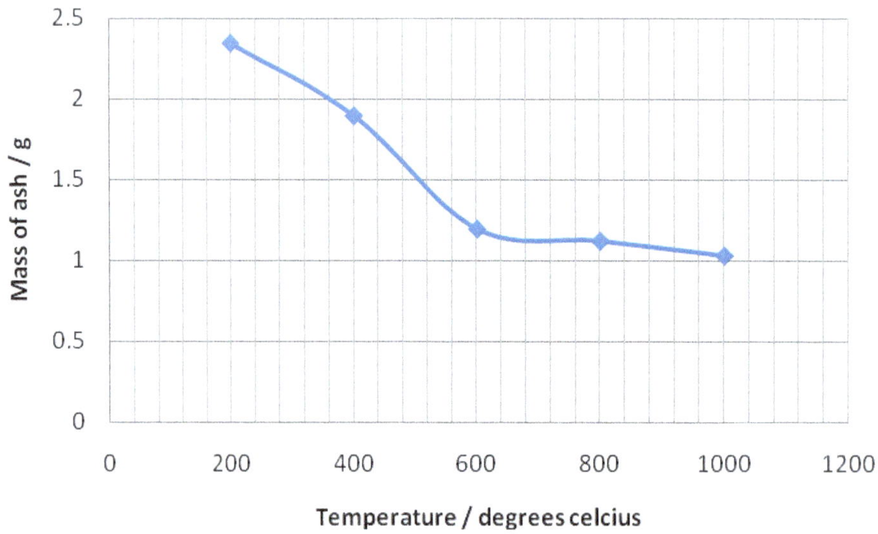

Figure 1. Effect of ignition temperature on ash content.

was found to have an average silica content of 40.16%. This shows that silica can commercially be extracted from sweet sorghum bagasse ash in processes where it is produced in large quantities as a waste product and adds value to the crop that is a potential raw material for bio-ethanol and syrups.

An XRF spectrum of sweet sorghum ash is shown in Figure 2. A small peak at 1.74 keV is typical of Si-k_α and hence confirms the presence of Si in the ash. The spectrum shows that the ash contains fewer heavy metals or very low concentrations of heavy metals that can be leached out prior to extraction.

Silica gel yield was dependent upon the pH and decreased from pH 5 to 8.5. At pH outside this range no meaningful precipitation was observed. The silica gel prepared at optimum pH 5 had a bulk density of 0.5626 $g{\cdot}cm^3$and an estimated porosity of 71.87%.

Water adsorption properties of silica gel prepared at pH 5 but at different aging times were determined. Silica gel that was aged for 10 hours had better moisture adsorption properties than any gel aged over longer periods of time.The results of selected aging times are shown in Table 2. It can be assumed that the decrease in adsorption capacities is due to a decrease in surface area for gels aged over longer periods.

FTIR spectra of sweet sorghum bagasse ash and dried silica gel samples prepared at different pH are shown in Figure 3. The strong absorption bands at 1044 cm^{-1} was observed in all the samples and was expectedly much stronger in silica gel samples than the ash. The absorption band is associated with asymmetric vibrations of the siloxane, Si-O-Si. The spectra also show absorption bands at about 800 cm^{-1} which corresponds to a stretching vibration of Si-O-Si. The weak absorption

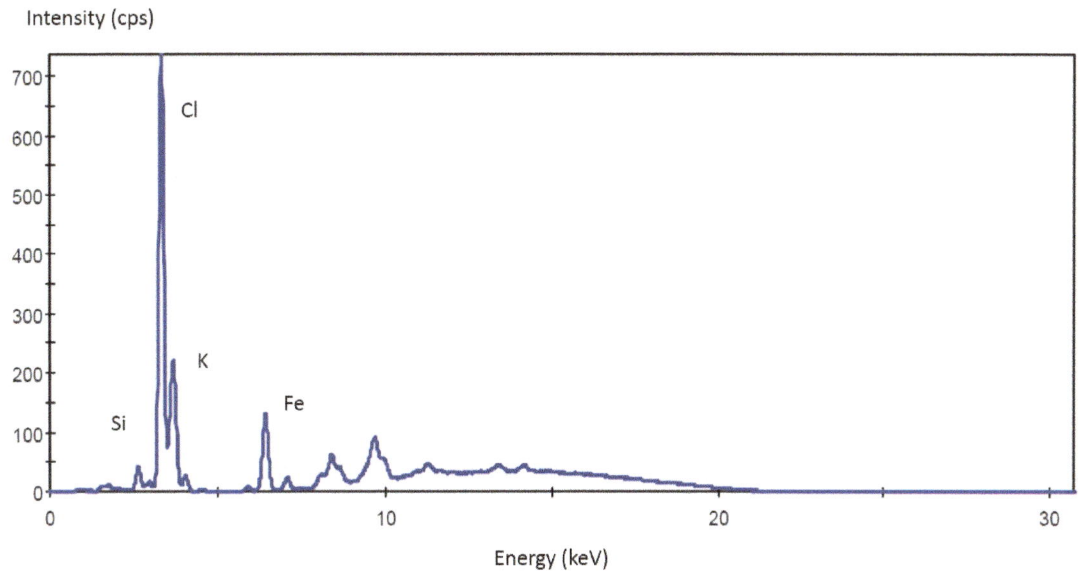

Figure 2. XRF spectrum of sweet sorghum ash.

Table 2. Effect of aging time on moisture absorption for silica gel prepared at pH 5.

Aging time (hours)	Maximum moisture absorption (%)
10	167.0
18	126.7
48	122.3

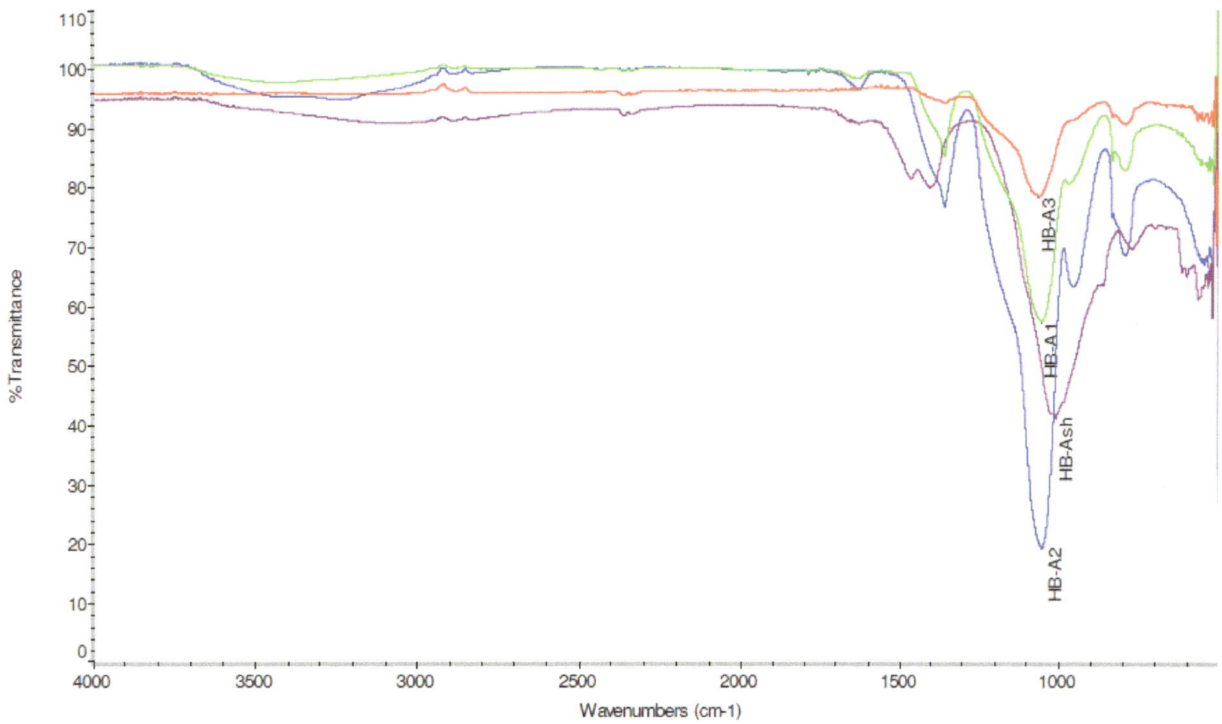

Figure 3. FTIR Spectra of sweet sorghum ash, silica gels prepared at pH 5 (HB-A1), pH 7 (HB-A2) and pH 9 (HB-A3).

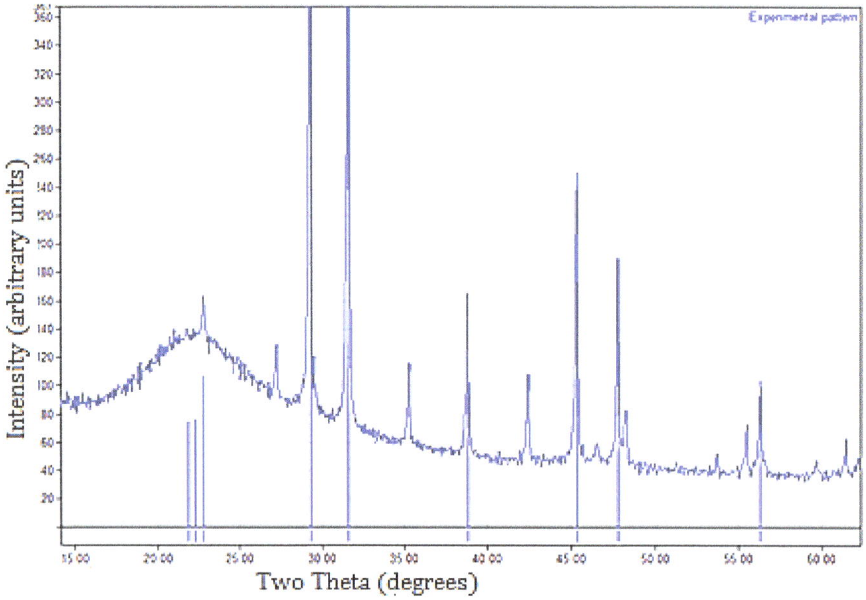

Figure 4. XRD diffractogram of sweet sorghum silica.

Figure 5. Micrographs of sweet sorghum silica at (A)100x, (B) 1000x and (C) 4500x magnification.

band at about 480 cm^{-1} is due to bending vibration of Si-O-Si. These Si-O-Si absorption bands are typical of silica gels and have also been observed by various researchers (Kalapathy et al., 2002; Essien et al., 2012; Nayak and Bera, 2009; Uzma et al., 2008).

An XRD diffractogram of sweet sorghum silica is shown in Figure 4. This shows that silica gel particles werehighly amorphous and the other phase was to sodium salts trapped in the gel. Figure. 5 shows micrographs of sweet sorghum silica at magnification of 100x, 1000x and

4500x. The silica gel consist of irregular particles of non-uniform size ranging from 5 μm to more than 100 μm.

Conclusion

Sweet sorghum is an important agricultural crop from which a number of products can be derived. The main advantage of this crop is its ability to grow under different climatic conditions. This research was able to demonstrate that sweet sorghum bagasse ash, a waste product of heat energy generation in the production of bio-ethanol can be used for extraction of silica gel for various applications.

Conflict of Interest

The authors have not declared any conflict of interest.

ACKNOWLEDGEMENT

The researchers would like to thank the Research Borad of the Bindura University of Science Education for providing funds under the Research Grant No 123/2013 and the International Office, Bonn-Rhein-Sieg University of Applied Sciences for having facilitated my research visit to the institution.

REFERENCES

Adam F, Chew TS, Andas J (2011). A simple template-free sol-gel synthesis of spherical nanosilica from agricultural biomass. J. Sol-Gel Sci. Technol. 59:580-583.

Alayande OS, Dare OE, Ayinde WB, Bamigbose J, Ayedun PA, Osinkolu GA (2012). Development of Ordered and Disordered Macroporous Silica from Bagasse Ash. Afr. J. Pure Appl. Chem. 6(1):10–14.

Almodares A, Hadi MR (2009). Production of Bio-ethanol from Sweet Sorghum: A review. Afr. J. Agric. Res. 4(9):772-780.

Devi PR, Dhanalakshmi KG (2012). Application of Mesoporous silica nanomaterial: An Overview. Int. J. Adv. Life Sci. 4:1-9.

Dorcheh AS, Abbasi MH (2008). Silica aerogel: Synthesis, properties and characterization. J. Mater. Process. Technol.199:10-26.

Essien ER, Olaniyi OA, Adams LA, Shaibu RO (2012). Sol-gel derived porous silica: Economic Synthesis and Characterization. J. Minerals Mater. Charact. Eng.11:976–981.

Gnansounou E, Dauriat A, Wyman CE (2005). Refining sweet sorghum to ethanol and sugar: Economic Trade-offs in the context of North China. Bioresource Technol. 96(6):985–1002.

Kalapathy U, Proctor A, Schultz J (2002). An improved method for the production of silica from rice hull ash. Bioresource Technol. 85:285-289.

Nadir N, Miel M, Karim MI, Yunus RM (2009). Comparison of sweet sorghum and cassava for ethanol production by using *Saccharomyces cerevisiae*. J. Appl. Sci. 9(17):3068-3073.

Nayak JP, Bera J (2009). Preparation of Silica aerogel by ambient pressure drying process using rice husk ash as raw material. Trns. Ind. Ceram. Soc. 68(2):1-4.

Nimbkar N, Akande NMJM, Rajavanshi AK (2006). Syrup production from sweet Sorghum. National Agricultural Research Institute (NARI).

Noushad M, Rahman IA, Huein A, Mohamad D, Ismail AR (2012). A Simple Method for obtaining Spherical Nanosilica from Rice Husk, Int. J. Adv. Sci. Eng. Infor. Technol. 2(2):28–30.

Prasad R, Pandey M (2012). Rice Husk Ash as a renewable source for the production of value added silica gel and its application: An Overview. Bull. Chem. Reaction Eng. Catalysis. 7(1):1-25.

Sing LP, Agrawal SK, Bhattacharyya SK, Sharma U, Ahalawat S (2011). Preparation of silica nanoparticles and its beneficiation role in cementitious materials. Nanomater. Nanotechnol.1:44-51.

Srinivasa RP, Rao SS, Seetharama N, Umakath AV, Reddy S, Reddy BVS, Gowda CLL (2009). Sweet Sorghum for biofuels and strategies for its improvement. Information Bulletin No. 77, Patancheru 502 324, Andra Pradeshi, India: International Crops Research Institute for the Semi-Arid Tropics, 80 pp, ISBN: 978-92-9066-518-2.

Thuadaij N, Nuntiya A (2008). Synthesis and characterization of Nanosilica from Rice Husk Ash Prepared by Precipitation Method, CMU. J. Nat. Sci. Nanotechnol. 7(1):59–65.

Tongjian Z, Yun X, Jiang W, Xin F (2009). Effect of application Rates of Assistant on the Properties of Nano SiO_2/NR Composites. J. Wuhan Univ. Technol. Mater. Scie. Ed. 24(3):387–392.

Ugheoke IB, Mamat O (2012). A critical assessment and new research directions of rice husk silica processing methods and properties. Maejo Int. J. Sci. Technol. 6(03):430–448.

Uzma KH, Bangi A, Venkateswara R, Parvathy RA (2008). A new method of sodium silicate-based hydrophobic silica aerogels via ambient-pressure drying. Sci. Technol. Adv. Mater. 9:035006(10 pp).

Venkata SCh, Ramana YR, Nagalashmi D, Jagadeeswara RS (2012). Evaluation of Sweet Sorghum (*Sorghum bicolour* (L.) moench) bagasse by Chemical, In sacco and *In vivo* Techniques in Graded Murrah Buffallo. Bulls. J. Veterinary Adv. 2(8):418–423.

Worathanakul P, Payubnop W, Muangpet A (2009). Characterization for post-treatment effect of Bagasse Ash for Silica extraction. World Acad. Sci. Eng. Technol. 32:360–262.

Efficiency enhancement in P3HT:PCBM blends using Squarylium III dye

M. Tembo[1], M. O. Munyati[2], S. Hatwaambo[1] and M. Maaza[3]

[1]Department of Physics, School of Natural Sciences, University of Zambia, P. O. Box 32379, Lusaka, 10101 Zambia.
[2]Department of Chemistry, School of Natural Sciences, University of Zambia, P. O. Box 32379, Lusaka, 10101 Zambia.
[3]Nanosciences Laboratories, Materials Research Department, iThemba LABS, P. O. Box 722, Somerset West 7129, Western Cape, South Africa.

Nano-size thin films comprising poly (3-hexylthiophene) (P3HT) and a fullerene derivative [6, 6] phenyl-C_{61}-butyric acid methyl ester (PCBM) incorporating squarylium dye III (SQ3) are reported. The materials prepared were evaluated for their optical, electrical and photo-conversion efficiency. Active layer materials comprising a blend of P3HT:SQ3:PCBM were deposited by spin-coating to produce thin films measuring 100 nm and subsequently annealed at 140°C for 10 min. The films were characterized by UV-Vis-NIR spectroscopy for their optical properties, atomic force microscopy for surface morphology and film thickness, and electrical properties. Optical measurements for blends incorporating different amounts of dye showed increased photo-absorbance with increasing dye concentration. The combined contribution of SQ3 and thermal annealing resulted in increased power conversion efficiency (η) of pristine P3HT:PCBM solar cells from 1.9 to 3.9 %. The dye in the active layer improved photo-absorption by enhanced light harvesting while thermal treatment improved the nanoscale morphology leading to better metal-film interface contact and broadening of the absorption wavelength range.

Key words: Polymer solar cell (PSC), squarylium dye III (SQ3), bulk hetero-junction (BHJ), power conversion efficiency (PCE), poly (3-hexylthiophene (P3HT), [6, 6]-phenyl C_{60} butyric acid methyl ester (PCBM).

INTRODUCTION

Energy is essential to modern society. Its global demand has continued to increase due to increasing population and rapid economic development in Africa. Depletion in fossil fuel resources together with the demand for more environmentally friendly energy sources has given impetus to research on 'green' energy alternatives.

Conversion of solar energy via photo-voltaic systems has emerged as one of the most promising of the green energy alternatives that can be deployed on a large scale. The cost of generating power using the current solar cells is more expensive than commercial power generation (Miyake et al., 2010). Conventional solar cells currently are made of silicon which is associated with high manufacturing cost due to processes that requires high temperature and vacuum. Additionally, dissemination has been hampered by high installation costs. Polymer solar cells offer a tantalizing possibility of lower cost devices. One of the most promising polymer solar cell (PSCs) in terms of efficiency and long term stability, is the system based on regio-regular poly(3-hexylthiophene) (rrP3HT) as the electron donor and the fullerene derivative [6,6]-phenyl-C61-butyric acid methyl ester (PCBM) as the electron acceptor material. To attain high efficiencies in PSCs, there is need to understand the fundamental electronic interactions between the polymeric donors and the fullerene acceptors as well as the complex interplay of device architecture, morphology and processing. Additionally, a sound understanding of the fundamental electronic processes involved is important.

The state of the art in the field of PSCs is currently represented by bulk hetero-junction (BHJ) solar cells based on P3HT and PCBM (Yu et al., 1995; Hoppe, 2006; Kim et al., 2006; Ma et al., 2005; Padinger et al., 2003; Reyes-Reyes et al., 2005). Low photo-conversion efficiencies of polymer solar cells compared to their inorganic counter-parts have remained a major drawback towards commercialization. Photo-conversion is typically below 9%. The limitation towards achieving high efficiencies can in-part be attributed to their narrow absorption band which lies in the 450 to 750 nm region. P3HT has a large band gap of about 1.9 eV with a limiting wavelength of harvested light of 650 nm. Almost half of the sun's power spectrum lies outside the absorption range of many organic systems (Bird et al., 1986; Diffey, 1991). Hence, thicker layers in the order of 300 nm would be required to enhance photo-absorption. However, use of layers of this thickness would lead to recombination of charges. As a compromise, polymer layers are typically 200 nm (Stratakis and Kymakis, 2013).

Organic dyes have been used to increase light harvesting capacity in P3HT:PCBM blend (Ameri et al., 2013). Squaraine dyes have been reported as promising small molecules that can be incorporated into polymer solar cells to improve light harvesting capacity owing to their intense absorption in the long wavelength range (Chen et al., 2014). Power conversion efficiencies of 5.5% have been obtained in ternary P3HT:PCBM systems incorporating a squaraine dye following thermal annealing (Rao et al., 2014). They observed enhanced absorption in the range 650 to 780 nm relative to P3HT:PCBM blend. Huang et al. (2013) demonstrated experimentally that in addition to extending absorption in the solar spectrum, small molecules, such squaraine dyes (SQ), improved excit on harvesting via Förster resonance energy transfer (FRET). Inclusion of SQ provides two mechanisms for charge transfer following excitation. The first one involves excitation of P3HT and charge transfer to PCBM while the second is charge transfer involving FRET of the P3HT excitation to SQ and subsequent dissociation. In this paper we report the improvement in power conversion efficiency of pristine P3HT:PCBM through the combined contribution of squarylium III dye (SQ3) and thermal annealing. Ternary P3HT:SQ3:PCBM systems showed an increase in absorption maximum and a red shift in the absorption band. The combined effect of dye inclusion and thermal annealing was to increase power conversion efficiency from 1.9 to 3.9%.

EXPERIMENTAL PROCEDURES

Materials and solutions

Regio-regular P3HT, PCBM and SQ3 were used without further purification. P3HT:PCBM blends containing SQ3 concentrations of 4.76, 9.09, 13.04 and 16.67% were prepared to investigate the effect of dye loading. The blend solutions were vigorously stirred for more than 24 h at room temperature under nitrogen atmosphere in a glove box to maximize mixing.

Film and device fabrication

For optical absorption measurements, the P3HT:PCBM and P3HT:SQ3:PCBM blend films (with different concentrations of the dye) were prepared by spin-coating the solutions onto cleaned micro-glass substrates. Thin film preparation and thermal annealing conditions were kept the same as those of the corresponding devices for accurate comparison. For solar cell fabrication, ITO-glass (15 Ω/ϒ) substrates were sequentially cleaned in an ultrasonic bath using acetone and methanol, rinsed with de-ionized water then finally dried. PEDOT:PSS was spin-coated onto the ITO layer in air and dried using a digitally controlled hot plate at 100°C for10 minutes. The PEDOT:PSS layer increased the electrode work function and its electrical connection with the organic active layer. The P3HT:PCBM and P3HT:SQ3:PCBM blend solutions were spin-coated (1500 rpm) on top of the insoluble PEDOT:PSS layer. Thin films were then heated at 100°C for 10 min to removing any residual solvent. A 100 nm thick aluminum (Al) electrode was thermally deposited onto the active layer using a vacuum deposition system at a pressure of about 3 x 10^{-4} Pa through a shadow mask to obtain cells with an active area of 1cm^2. Devices were thermally annealed at 140°C for 4 min. The complete organic solar cells, with a configuration shown in Figure 1, were stored in the dark under suitable pressure until measurement.

Measurements

Films prepared were characterized for their optical properties using a Lambda 90 UV-Vis-NIR spectrophotometer before and after annealing. Morphological characteristics were established by atomic force microscopy. The power conversion efficiency (PCE) was calculated from the Current-Voltage (I-V) characteristics under

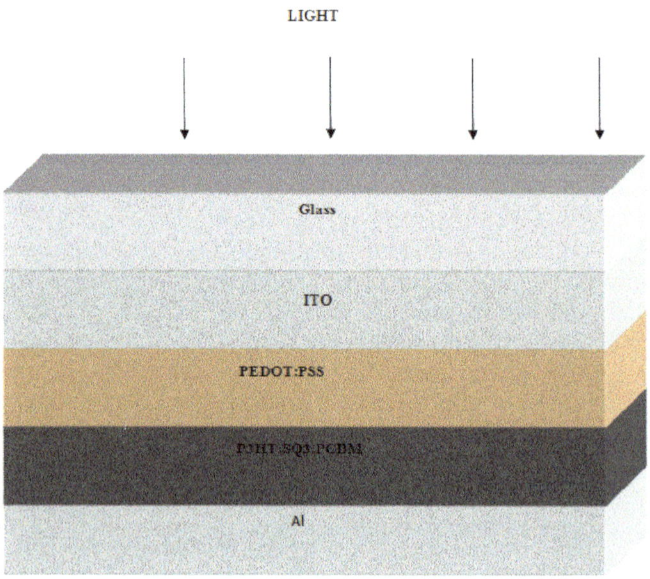

Figure 1. Polymer solar cell configuration showing the different layers of materials.

air mass 1.5 global (1.5 G) solar simulated light irradiation of100 mW/cm^2. Current-voltage characteristics were measured using a KEITHLEY 4200 at room temperature in nitrogen atmosphere.

RESULTS AND DISCUSSION

Absorption characteristics of P3HT:SQ3:PCBM blends

The chemical structures of SQ3 are shown in Figure 2. Absorption spectra of the active layer materials (P3HT, PCBM and SQ3 dye) are shown in Figure 3. It is shown in this figure that PCBM absorbs mainly below the wavelength region of 500 nm. The P3HT spectrum shows an absorption peak at 511nm and absorption edge extending to 630 nm. P3HT also has vibronic shoulders at 547 and 607 nm and these are associated with π-π* stacking in P3HT and indicative of its crystallinity. SQ3 shows an absorption band extending in the longer wavelength region extending to 650 nm with an absorption peak at 627 nm.

The contribution of SQ3 in light harvesting was investigated by incorporating SQ3 in quantities of 4.76, 9.09, 13.04 and 16.67% in P3HT:PCBM solar cell active layer are shown in Figure 4. Inclusion of SQ3 improves the absorption profile of the P3HT:SQ3:PCBM blend with an increase in absorption intensity and widening the absorption range from the visible to near – infra red. This can be attributed to the combined effect of the P3HT absorption (400 to 700 nm) and SQ3 (550 to 575 nm) spectrum. The shape of the shoulders at about 547 and 607 nm remain relatively the same as SQ3 increased.

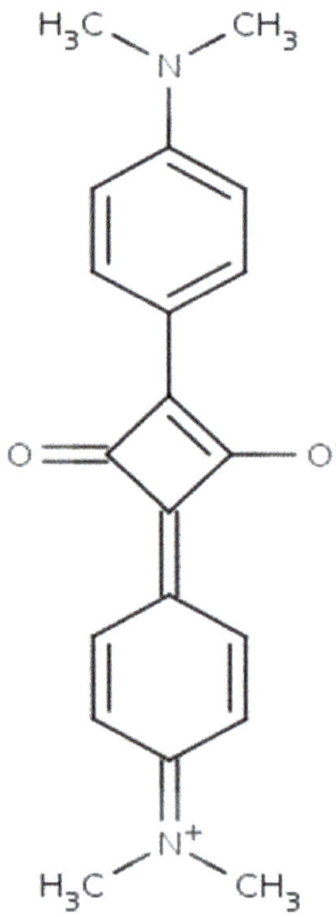

Figure 2. Structure of SQ3 dye .

This indicates that there was no disruption to P3HT crystallinity (Aziz et al., 2014). Increasing the dye content in the P3HT:PCBM blend increases light harvesting through the increase in absorbance and broadening of

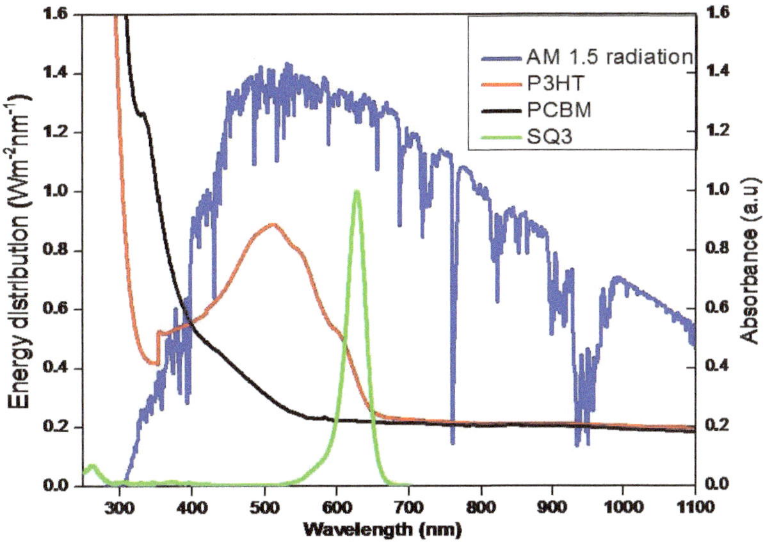

Figure 3. Absorption spectra for spin coated P3HT, PCBM and SQ3 dye thin films compared with the standard AM1.5 solar spectrum.

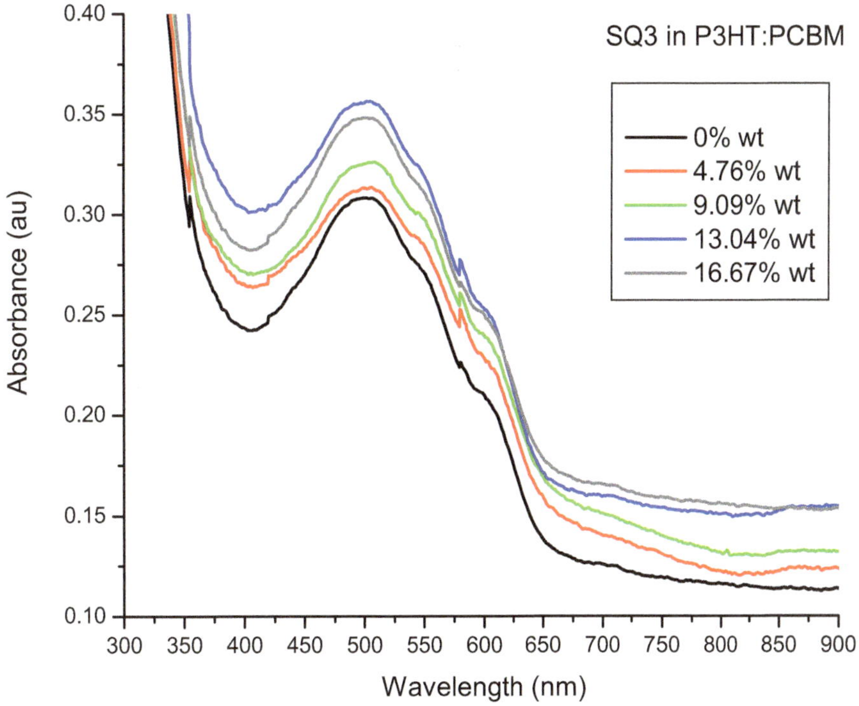

Figure 4. Absorbance spectra of pristine P3HT:SQ3:PCBMwith varying dye loading.

the absorption range. Additionally, the peak absorbance increased from 0.31 to 0.36 a.u representing about 16% increment. Since the absorption at about 600 nm is associated with P3HT crystallinity, an increase would suggest the formation of a more ordered morphology promoted by SQ3 molecules. Thus, the total number of photons absorbed by the solar cell active layer increases when SQ3 dye is added to the P3HT:PCBM blend. The increase is due to a synergistic effect in the absorption characteristics of P3HT and SQ3. However, there is a decrease in absorbance when the concentration of SQ3 was 16.67%. The decrease may be attributed to the

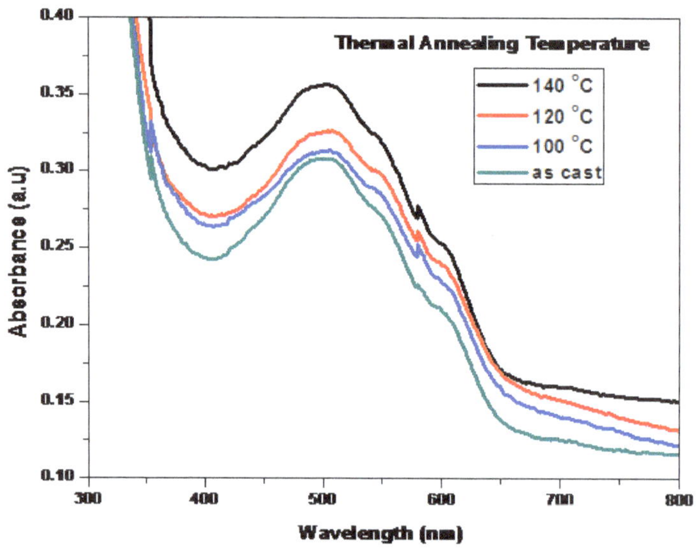

Figure 5. UV-Vis absorbance spectra of 100 nm P3HT:PCBM blend films spin coated at 1400 rpm treated in the following ways: as cast (green), thermally annealed at 100°C (blue), 120 °C (red) and 140°C.

disturbed interpenetrating network between P3HT and PCBM with the relative decrease in the P3HT:PCBM content with the high concentration of SQ3, which hinders the charge transportation (An et al., 2014; Cha et al., 2013; Rao et al., 2014).

Effect of annealing

UV-Vis absorption spectroscopy was used to study the effect of thermal annealing on thin films of P3HT:PCBM spin coated on glass substrates. Thermal annealing optimized the morphology of P3HT:PCBM thin films and thus improved photo-absorbance as shown in Figure 5. The untreated film has peak absorbance of 0.30 a.u. and shoulders at 501, 545 and 597 nm. Thermal annealing at temperatures of 100, 120 and 140°C increased the peak absorbance to 0.31, 0.32 and 0.36 a.u respectively. The increase in absorption intensity is associated with higher crystallinity of P3HT following reordering in annealed samples. Additionally, films treated at 140°C show a red shift of peak absorption wavelength from 501 to 510 nm while the shoulders shift to 549 and 604 nm respectively. The red shift is ascribed to increased inter-chain interaction among the P3HT chains arising from annealing. This results in lowering of the band gap and increase in of the optical π-π^* transition, an indication of formation of crystallites.

Atomic Force Microscope (AFM) studies

AFM was used to evaluate the effect of adding SQ3 and thermal annealing on P3HT: PCBM blends. The images are shown in Figure 6. The results show that as cast films without SQ3 had a surface roughness of 0.38 nm while samples annealed at 140°C had a roughness of 0.75 nm. A significant increase in roughness was observed on addition of SQ3 weight of 13.04%. However, annealing at140°C of P3HT:SQ3:PCBM showed a slight decrease in roughness to 0.8 nm. An increase in roughness is associated with increased order in the P3HT phase (Aziz et al., 2014). Thus, the slight reduction suggests reduced crystallinity and disturbance in the morphology at a dye load of 6 mg and annealing temperature of 140°C. Huang et al. (2013) showed that SQ molecules was prefer to reside at the interface between P3HT and PCBM increasing phase separation in the polymer blend. Decrease in roughness maybe a result of loss of the favoured morphology following annealing at 140°C.

Electrical properties

The solar cell electrical characteristics are determined by measuring the current density to voltage (I-V) characteristics, both in dark and under illumination. The open circuit voltage (V_{oc}), short circuit current (I_{sc}), fill factor (FF) and maximum power point (P_{max}) were deduced from the I-V curve. The power conversion efficiency is subsequently determined using these data and is tabulated in Table 3.

Current -voltage (I-V) characterization

The polymer solar cell device structure is shown in Table 2 and their I-V curves in Figure 7. The active layer

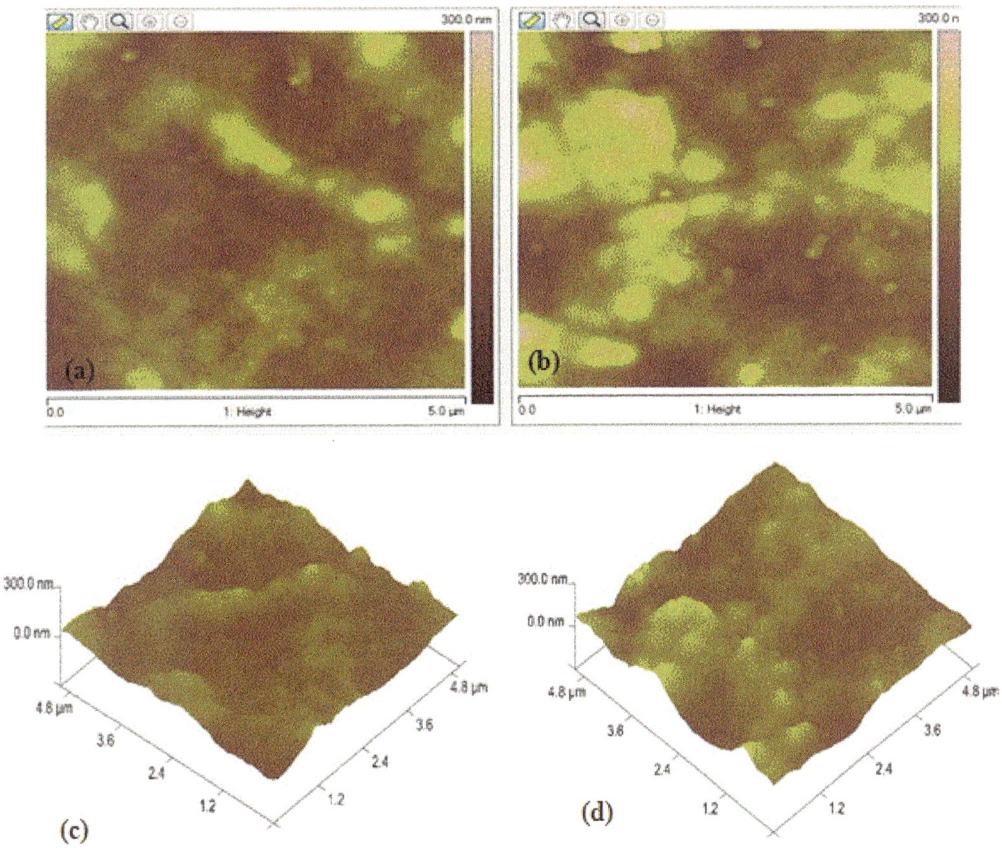

Figure 6. Two dimensional AFM images of P3HT:PCBM (a) and PCBM:SQ3:PCBM (b), 3 Dimension AFM images of P3HT:PCBM (c) and PCBM:SQ3:PCBM (d).

Table 1. Weight percentage of SQ3 in P3HT:PCBM.

SQ3 mass (mg)	PCBM mass (mg)	P3HT mass (mg)	SQ3 weight (%)
2	20	20	4.76
4	20	20	9.09
6	20	20	13.04
8	20	20	16.67

Table 2. Polymer solar cells device structure.

Solar Cell	Device Structure	Illumination	Annealing
A	ITO/PEDOT:PSS/P3HT:SQ3:PCBM/Al	NO	NO
B	ITO/PEDOT:PSS/P3HT:PCBM/Al	YES	NO
C	ITO/PEDOT:PSS/P3HT:SQ3:PCBM/Al	YES	NO
D	ITO/PEDOT:PSS/P3HT:PCBM/Al	YES	YES
E	ITO/PEDOT:PSS/P3HT:SQ3:PCBM/Al	YES	YES

consisted of an active layer of 13.04% weight SQ3 inP3HT:PCBM. Thermal annealing was done at a temperature of 140°C as the optimum temperature. The electrical parameters determined from Figure 7 of the devices and are summarized in Table 3. The control device comprising P3HT:PCBM without SQ3 with no

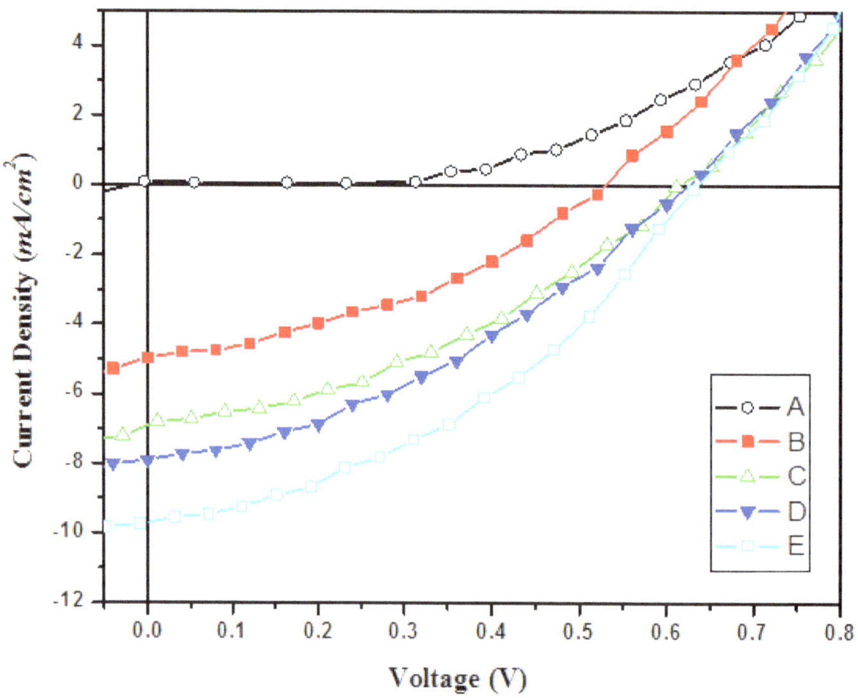

Figure 7. Current versus voltage (I-V) curves obtained for solar cells under AM 1.5 solar spectrum simulations (light) at irradiation intensity of 100 mW.cm^{-2}.

Table 3. Electrical characteristics of P3HT:SQ3:PCBM solar cells.

Solar Cell	V_{oc} (V)	I_{sc} (mA)	I_{max} (mA)	P_{max}	FF	η (%)
B	0.53	5.78	3.39	1.08	0.35	1.3
C	0.61	6.88	4.29	1.59	0.38	2.5
D	0.62	7.88	5.04	1.81	0.37	2.8
E	0.64	9.68	6.86	2.41	0.40	3.9

thermal treatment had a I_{sc} value of 5.7 mA, V_{oc} of 0.32 V and FF of 0.55. The power conversion efficiency (PCE) was established to be 1.3%. The inclusion of SQ3 combined with thermal annealing resulted in a device with 3.9% PCE. The I_{sc} increased from 5.78 mA for the control to 9.68 mA for the annealed ternary blend. The V_{oc} and FF increased from 0.53 to 0.64 mA respectively. Increase in I_{sc} is due to the increase in photo generated excitons arising from the broadened absorption range to the NIR. The V_{oc} is related to the energy difference between the donor HOMO and acceptor LUMO, a change is expected if exciton separation occurs at the donor-acceptor interface. It has been reported by Huang et al. (2013) that the highest occupied molecular orbital (HOMO) of SQ is located in-between the HOMOs of P3HT and PCBM, similarly for the lowest unoccupied molecular orbital (LUMO) levels. Consequently the initial excitation leads to two possible routes for current generation: (i) photo induced charge transfer between P3HT and PCBM, (ii) FRET excitation of P3HT to SQ3 followed by charge dissociation. Thus, increase in PCE from 1.3 to 3.9% can be attributed to the synergistic effect arising from the change in absorption profile to near infra red (NIR) and establishment of an addition charge transfer route in the system. This represents a two-fold increase in PCE.

Conclusion

In this work, we have investigated the effect of adding SQ3 dye to P3HT:PCBM forming a ternary system. The light harvesting capacity was increased by broadening the absorption band to NIR. A red shift was observed with vibronic shoulders shifting from 501 and 547 nm to 510 and 600 nm respectively. The increase in absorption intensity of these shoulders suggest an increase in P3HT crystallinity. Furthermore, inclusion of SQ3 resulted in broadening of the absorption spectrum. A similar effect was observed as a result of thermal treatment of the ternary system at 140°C.The combined contribution of SQ3 and thermal annealing resulted in increased power conversion efficiency (η) of pristine P3HT:PCBM solar cell from 1.9 to 3.9%. Increase in PCE is attributed to the extension of the absorption range thus improving light harvesting capacity and introduction of an additional charge transfer mechanism through SQ3.

Conflict of Interest

The authors have not declared any conflict of interest.

ACKNOWLEDGMENTS

The authors would like to express gratitude for the financial support from the International Science Program (ISP). Support from Nanoscience Africa Network (NANOAFNAC) and Materials Science and Solar Energy Network for Eastern and Southern Africa (MSSEESA) is duly acknowledged. Special mention should also be made to iThemba LABS, Physics Department of University of Western Cape in South Africa and Physics Department of the University of Dar-es-salaam in Tanzania.

REFERENCES

An Q, Zhang FJ, Li L, Wang J, Zhang J, Zhou L, TangW (2014).Improved efficiency of bulk heterojunction polymer solar cells by doping low-bandgap small molecule. ASC Appl. Interfaces. 6:6537-6544.

Ameri T, Khoram P, Min J, Brabec CJ (2013). Organic Solar Cells: A Review. Adv. Mater. 25:4245-4266.

Aziz F, Ismail FA, Soga T (2014). Effect of solvent annealing on crystallanity of spray coated ternary blend films prepared using low boiling point solvent. Chem. Eng. Process. 79:48-55.

Bird RE, Riordan C (1986).Simple solar spectral model for direct and diffuse irradiance on horizontal and tilted planes at the earth's surface for cloudless atmospheres. J. Climate Appl. Meteorol. 25:87-97.

Cha H, Chung DS, Bae SY, Lee M, An TK, Hwang J, Kim KH, Kim YH, Choi DH, Park CE (2013). Complementary Absorbing Star-Shaped Small Molecules for the Preparation of Ternary Cascade Energy Structures in Organic Photovoltaic Cells. Adv. Funct. Mater. 23(12): 1556-1565.

Chen G, Sasabe H, Wang XF, Hong Z, Kido J (2014). A squaraine dye as molecular snsitizer for increasing light harvesting in polymer solar cells. Synthetic Metals. 192:10-14.

Diffey BL (1991). Solar ultraviolet radiation effects on biological systems. Rev. Phys. Medicine Biol. 36(3):299-328.

Hoppe H, Sariciftci NS (2006).Morphology of polymer/fullerene bulk heterojunction solar cells. J. Mater. Chem. 16:48–58.

Huang JS, Goh T, Li X, Sfeir MY, Bielinsk EA (2013). Improving performance of P3HT/PCBM with squarianedye.In Z. H. Kafafi (Ed.), Organic Photovoltaics XIV, 8830.

Kim JY, Kim SH, Lee H, Lee K, Ma W, Gong X, Heeger AJ (2006). New architecture for high-efficiency polymer photovoltaic cells using solution - based titanium oxide as an optical spacer. Adv. Mater. 18: 572–576.

Ma W, Yang C, Gong X, Lee K, Heeger AJ (2005). Thermally stable, efficient polymer solar cells with nanoscale control of the interpenetrating network morphology. Adv. Funct. Mater.15:1617–1622.

Miyake K, Uetani Y, Seike T, Kato T, Oya K (2010). Development of Next Generation Organic Solar Cell. Sumitomo Chemical Co. Ltd.

Padinger F, Rittberger RS, Sariciftci NS (2003). Effects of postproduction treatment on plastic solar cells. Adv. Funct. Mater. 13: 85-88.

Rao AB, Kumar SM, Sivakumar G, Singh SP, Bhanuprakash K (2014). Effect of incorporation of squaraine dye on the photovoltaic response of buli heterojunction solar cells based on P3HT:PC70BM blend. ACS Sustainable Chem. Eng. 2:1743-1751.

Reyes-Reyes K, Kim D, Carroll L (2005). "High-efficiency photovoltaic devices based on annealed poly(3-hexylthiophene and 1-(3-methoxycarbonyl)-propyl-1-phenyl-(6,6) C61 blends." Appl. Phys. Lett.87, 083506.

Stratakis E, Kymakis E (2013). Nanoparticle based Plasmonic photovoltaic devices. Materials Today 16(4):133–146.

Yu G, Gao J, Hummelen JC, Wudl F, Heeger AJ (1995). Polymer photovoltaic cells: Enhanced efficiencies via a network of internal donor–acceptor heterojunctions. Science. 270:1789–1791.

Effect of solvent type on the yields and mineral compositions of the leaf extracts of *Moringa oleifera* L.

Adeoye, M. Dasola[1], Lawal, A.Tunbosun[2], Azeez, L. Adeyemi[1] and Olayiwola, O. Abidemi[3]

[1]Industrial and Environmental Chemistry Unit, Chemical Sciences Department, Fountain University, Osogbo Osun State, Nigeria.
[2]Industrial and Environmental Chemistry Unit, Chemical Sciences Department, Al-Hikmah University, Ilorin Kwara State, Nigeria.
[3] Chemical Sciences Department. Oduduwa University,Ipetumodu, Ile-Ife, Osun State, Nigeria.

The present study was conducted for the leaves extract of *Moringa oleifera* L (*Moringeaeae*) in different solvents: Fermented maize marsh (Omidun /FMM), hot water (HW), cold water (CW), palm-wine (PW), ethanol, dichloromethane (DCM) and n-hexane to determine the effects of solvent properties on the extraction yields and the compositions of dietary minerals (k, Ca, Mg, Zn, Cu, and Fe) for *M. oleifera.* The observed yields follow the trend: PW > CW > FMM > ethanol >HW > DCM> n-hexane. Fermented maize marsh extracts recorded the highest mean compositions for Ca and Mg (560.50 ± 34.5 and 485.75 ± 67.32); palm-wine extracts recorded the highest compositions of Fe and Cu (7.42 ± 1.22 and 0.50 ± 0.01), while cold water and hot water extracts recorded the highest compositions of K (2045.14 ± 265.02) and Zn (2.16 ± 0.11) respectively. The lowest concentrations of these trace metals which range from 0.06 ± 0.01 to 6.47 ± 0.35 was recorded in n-hexane extracts. Statistical analysis of the mineral constituents using two sample t-tests studies at p < 0.05 also revealed great significant differences between the means of the mineral elements in the extracts except in few cases. The high yields of these extracts in PW, CW and FMM indicate that these solvents have great affinity and similar solubility properties for the different compositions of *M. oleifera*.

Key words: *Moringa oleifera,* fermented maize marsh, solvatochromic, extraction, toxicological.

INTRODUCTION

Green plants provide most of the world's molecular oxygen and are the basis of a good number of the earth's ecologies, especially on land. Plants enrich our lives as flowers and ornaments, and generally contain high amounts of essential nutrients, vitamins, minerals and fatty acids. These are important in animal feed and human nutrient for osmotic adjustment and to activate enzymes, hormones and other organic molecules that enhance the growth, function and maintenance of life processes by living organisms (Aslam et al., 2005). Minerals cannot be synthesized by animals and must be provided from plants or mineral-rich water (Mosha et al., 1995; Al-Kharusi et al., 2009).

Moringa oleifera Lam [syn. *M. ptreygosperma* Gaertn.]

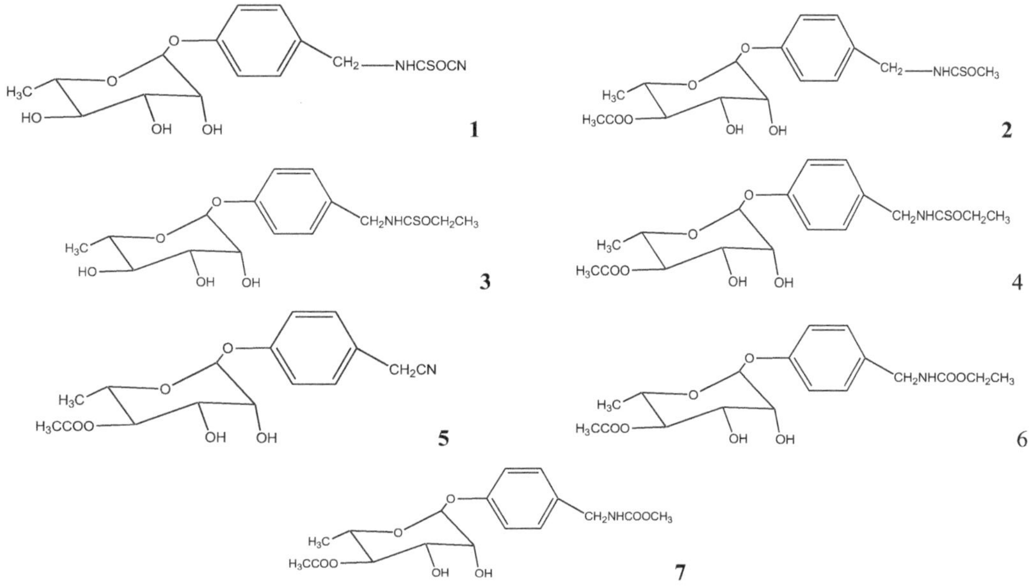

Figure 1. Structures of some selected phytochemicals from moringa (1) niazidin (2) niazicin (3) niazimcin (4) niaziminin (5), niazirinin (6) niazinin (7) niazicinin (Reubena, 1995).

plants, widely distributed and naturalized species of a monogeneric family Moringaceae is commonly called ben oil tree but locally known as *Zogeli* among the Hausa speaking people of Nigeria (Toba et al., 2010). It has been reported to be a rich source of protein, vitamins, β - carotene, aminoacids, various phenolics and essential minerals like calcium (Ca) and potassium (K) (Toba et al., 2010; Makkar and Becker 1996). It has also been reported to be a good source of natural antioxidants such as ascorbic acid, flavonoids, phenolics and carotenoids which enhance the shelf-life of fat containing foods. Its oil and micronutrients have also been acclaimed to contain antitumour, antiepileptic, antidiuretic, anti-inflammatory and venomous bite characters (Hsu, 2006), while its aqueous leaf extracts has been reported to regulate thyroid hormone and can be used to treat hyperthyroidism since it exhibits an antioxidant effect (Kumar and Pari, 2003). The biodiesel derived from *M. oleifera* oil by alkali-catalyzed trans-esterification with methanol is reported to be an acceptable substitute for petrol and diesel. Its cetane number was found to be the highest reported for a biodiesel fuel with much better oxidative stability (Mahajan et al., 2007). The seeds also serve as one of the cheap and best coagulants for water treatment (Kalogo et al., 2000).

The minerals and nutritional values of *M. oleifera* and their bioavailability have been a subject of great studies and there have been reported considerable variations among these constituents in both the leaves and seeds of this plant. These depend on factors like genetic background, location, environment and cultivation methods (Aslam et al., 2005; Brisibe et al., 2009; Toba et al., 2010).

Cold water, hot water, palm wine or fermented maize mash (omidun) has been used traditionally as solvent for herbal extraction in Nigeria. The leaves, barks or roots of some herbal plants are either cooked or soaked in these solvents to treat certain diseases. However, there are limited reports on the influence of these solvents on the mineral and bioactive compositions of these herbal extract. Most of the bioactive components contained in plant matrices are medium-sized molecules which due to the presence of aromatic delocalized π-electrons in it, the molecules are highly polarizable. Their high polarizability makes the molecules liable to a variety of specific interactions with various solvents, e.g., protonation, hydrogen bonding, and specific solvation (Andri et al., 2009). Thus, for toxicological reasons, drug and medicine manufacturers are increasingly required to minimize the number of solvents employed in pharmaceutical processes (Kolar et al., 2002). The presence of a solvent in the extract may also affect the kinetics of crystallization and the morphology of the crystals, which is an important factor in determining the product quality (Andri et al., 2009).

To date, twenty three active compounds had been isolated and elucidated from *M. Oleifera* leaves and these had been claimed to depend on their species, sources and ages (Rubeena, 1995; Toba et al., 2010). These compounds are novel glucosides with hypothensive properties; some of these are niaziminin, niazinin, niazicinin, niazimicin, niazirin, niazirinin, niazicin and niazimin in Figure 1 (Rubeena, 1995). It is therefore pertinent in this work to confirm the presence/ retention of these compounds in the extracts of *Moringa* leaves

Figure 2. Flow diagram for the preparation of ogi /omidun (FAO, 1999).

ecotype in Osogbo, a South Western Nigeria town through their functional groups and to study the effect of solvent on its mineral compositions.

MATERIALS AND METHODS

Solvents

Fermented maize mash (*omidun)*, prepared by soaking and fermentation of maize grains in cold water for 28 to 72 h, wet milled, sieved and left to sediment for 48 h (Figure 2) were collected from traditional processors of ogi in Osogbo, Osun State, Nigeria. This solvent was from ogi prepared from white maize grains (FAO, 1999; Abdulrahaman and Kolawole, 2006). The *omidun* samples were collected aseptically in sterile screw cap bottles. Fresh palm-wine was bought in undiluted form from the local palm-wine taper while other reagents/chemicals (dichloromethane, ethanol, n- hexane) used were analytical grade products of British Drug House (BDH). The reagents were redistilled twice and their boiling points were used as index of purities. The standards used were prepared using distilled water.

Collections and identification of plant materials

Fresh leaves of *Moringa* were collected from *M. oleifera* plants located in Fountain University compound, Osogbo in Osun State southern/western Nigeria between January and February, 2013. The plant materials had earlier been taxonomically identified by Dr. Awodoyin, R.O. of Botany Herbarium, University of Ibadan. The leaves were air- dried until a constant weight was obtained (one week). They were pulverized to powdery form using a well-cleaned 240 v 4 L blender machine. The powdered samples were then stored in airtight containers, properly kept and labeled for further analysis. Soil samples of the land on which *M. oleifera* is grown were also taken within 10 cm depth. And, its pH determination was done at ratio of 1:1 in distilled water (Anjorin et al., 2010).

Determination of moisture and ash contents

2 g fresh samples of *M. oleifera* leaves (in duplicate) were weighed. The moisture contents in the samples were determined using the method of the Association of Official Analytical Chemists (AOAC, 1990). This was done by weighing the leaves in the crucible and drying in oven at 105°C until a constant weight was obtained. Determination of the ash content was done by ashing at 550°C for about 3 hrs (Ogbe and John, 2011). The ash resulted from the combustion was then allowed to cool to lowest temperature and weighed.

Preparation of *M. oleifera* extracts

The extracts of *M. oleifera* in different solvents were prepared by soaking 8 g of dried powdered samples each in 100 ml of distilled coldwater, distilled hot water, omidun, palmwine, dichloromethane, n- hexane, ethanol respectively for 24 h in an airtight reagent bottles. These were then filtered. The filtrates were concentrated in rotary vacuum evaporator to yield an oily golden yellowish mass. The concentrates of all extracts mentioned above were finally air dried to constant weight and the total yield of each sample were determined. 2 g of each of the concentrates were treated with 20 ml of concentrated Trioxonitrate (V) acid (HNO_3) heated to reflux on an electric hot plate at 70 to 90°C until digestion was completed (Anjorin et al., 2010). Side by side, 20 ml of HNO_3 was also added to empty beakers which served as blank. The contents of the beaker were allowed to cool, filtered through whatman No. 2 filter paper into volumetric flasks and made up to volume of 100 ml with distilled water before the Atomic Absorption Spectroscopy analysis was done. These were carried out in triplicates.

Determination of the compositions of dietary minerals in the *M. oleifera* extracts

Chemical analysis

Macro elements (K, Ca, Mg) and trace elements (Zn, Cu, Fe) concentrations in the *M. oleifera* extracts in different solvents were estimated on Atomic Absorption Spectrophotometer (AAS) (Buck Scientific Modelling 2010/2011 V.G.T made in England) at the Agronomy Department of the University of Ibadan. Infra-red analysis of the functional groups in the *M. oleifera* extract and UV-Visible spectroscopy analysis of the concentrates in the wavelength range of 190 to 800 nm in different solvents were also determined.

Statistical analyses

The analyses in this study were carried out in triplicates. For each, a mean value and standard deviation were calculated. Statistical analysis of the results obtained involved the use of Aspin-Welch unequal variance two sample t-tests.

RESULTS AND DISCUSSION

Characterization of the extracts of *M. oleifera* leaves

The extracts of *M. Oleifera* leaves were obtained in amorphous form as golden to dark brown colour in different solvents. The moisture contents, the extracts yields, the mineral composition and the results of the spectra analysis (UV and IR) showing the effect of

Table 1. Yields of *M. oleifera* extracts in different solvents.

Solvent	Yield (g)
Hot water	1.13 ± 0.02
Cold water	2.0 ± 0.01
n-hexane	0.2 ± 0.02
D.C.M	1.0 ± 0.11
Ethanol	1.2 ± 0.03
F.M.M	1.9 ± 0. 01
Palm-wine	6.77 ± 0.12

F.M.M., Fermented maize mash; DCM, dichloromethane.

Table 2. Mineral composition of *M. oleifera* leaves extracts in different solvents.

Extracts	Ca (mg/L)	Mg (mg/L)	K (mg/L)	Fe(mg/L)	Cu (mg/L)	Zn (mg/L)
Cold water	484.33±75.77^{a}	292.82±16.09^{a}	2045.14±265.02^{a}	1.38±0.04^{a}	0.08±0.01^{a}	0.64±0.12^{a}
Hot water	512.18±17.18^{b}	197.25±26.37^{b}	1775.99±98.66^{b}	2.78±0.17^{a}	0.06±0.01^{b}	2.16±0.11^{b}
F.M.M	560.50±34.5^{b}	485.75±67.32^{c}	1875.15±183.53^{b}	3.47±0.07^{b}	0.11±0.02^{a}	1.14±0.05^{c}
D.C.M	20.40±1.43^{c}	12.71±3.02^{d}	33.93±3.77^{c}	6.63±0.33^{c}	0.09±0.02^{a}	0.29±0.02^{d}
n-hexane	4.22±0.28^{d}	1.56±0.09^{e}	6.47±0.35^{d}	0.70±0.01^{d}	0.06±0.01^{b}	0.14±0.03^{e}
Ethanol	5.19±0.04^{d}	54.75±4.12^{f}	625.21±28.27^{e}	5.89±1.38^{c}	0.13±0.05^{a}	0.33±0.02^{d}
Palm-wine	152.50±24.47^{e}	212.50±35.85^{b}	216.50±12.56^{f}	7.42±1.22^{c}	0.50±0.01^{b}	0.44±0.03^{f}

D.C.M, 1, 2-dichloromethane, F.M.M, fermented maize mash (Omidun). Data are expressed as Mean ±SD. Means with different superscripts in a column are significantly different at $p < 0.05$.

Table 3. Summary of the Infra-red (IR) analysis showing the different functional groups in the extracts of *M. oleifera* leaves.

Functional groups	(υ_{max}) cm^{-1}						
	Cold water	Hot water	F.M.M	Palm- wine	Ethanol	D.C.M	n-hexane
$\upsilon_{\text{O-Hstr}}$	3396	3425	3430	3404s	3457.82	3453	3445.71
$\upsilon_{\text{OH/ N-Hstr}}$	-	-	-	3373	-	-	-
$\upsilon_{\text{C-H str}}$ in CH_2	2976.95	2980.2	-	2978.34	-	-	2927.75
$\upsilon_{\text{C=C str (aromatic ring)}}$	1648.66	1643.09	1637.86	1643.94	1639.21	1638.36	1638.54
$\upsilon_{\text{C=O str (carbamate)}}$	-	-	-	-	-	-	1726.42
$\upsilon_{\text{-C=O amide}}$	-	-	-	-	-	-	1589.63
$\upsilon_{\text{C-O thioamide}}$	1400	1398		1399.50	-	-	1374.23
$\upsilon_{\text{C-OH stretch/def}}$	1047	1048.30	1090	1048.86s	1048.57	-	-
υ Subst bzc	662.26	668.33	609.32	662.26	601.00	-	

F.M.M, Fermented maize mash; D.C.M, dichloromethane

different functional groups and the effect of extractive solvents in the spectra shifts of *M. oleifera* leaves extracts are as presented in Tables 1 to 4 and in Figures 3 to 15. Moisture contents: 10.22% ± 0.11. Ash values: 7. 93% ± 0.32

pH Values and soil samples

The soil on which the plants grow is loamy soil while its pH was slightly acidic (pH 6.55). This was in line with the findings of Anjorin et al. (2010). The moisture contents (10.22% ± 0.11) and total ash values (7. 93% ± 0.32) for *M. oleifera* leaves obtained in this study were slightly higher than those reported by Moyo et al. (2011) and Ogbe and John (2011) for *M. oleifera* leaves. The variations in the reported values of these parameters may be due to differences in the ages of the trees or their different stages of maturity and possibly due to difference in agro-climatic conditions (Ogbe and John, 2011).

Table 4. UV-Visible Absorption Bands of different solvents extracts of *M. oleifera.*

Extracts	Band I		Band II		Band III		Band IV	
	(λ_{max})nm	Absorbance	(λ_{max})nm	Absorbance	(λ_{max})nm	Absorbance	(λ_{max})nm	Absorbance
Cold water	300	2.304	-	-	-	-	-	-
Hot water	295	1.565	-	-	-	-	-	-
F.M.M	300	2.081	-	-	-	-	-	-
Palm wine	300	2.018	-	-	-	-	-	-
Ethanol	295	1.553	410	0.452	-	-	665	0.116
D.C.M	295	1.331	410	0.724	-	-	665	0.250
n-Hexane	-	-	415	0.234	435	0.126	470	0.143

F.M.M., Fermented maize mash; D.C.M., dichloromethane.

Figure 3. IR Spectrum of n-hexane extract of *M. oleifera.*

Extracts yields

Plant matrices including leaf, stem and root contain various solute molecules with more than one functional group. Therefore it is difficult to predict the solubility and interactions of the solute in a particular solvent (Andri et al., 2009). Polarity was defined by Snyder as the relative ability of a molecule to engage in strong interactions of all kinds and is the sum of all possible interactions (Andri et al., 2009). Thus, the concept of polarity can be used as alternative way of considering solubility.

The total yields of *M. oleifera* in the solvents of choice (dichloromethane, n-hexane, Fermented Maize Mash, Ethanol and water) as shown in Table 2 indicated that palm-wine and non polar (n- hexane) extracts had the highest and lowest yields respectively. All the extracts obtained from the solvents used were observed as dark brown or golden brown semi-solids. The high yields of these extracts in palm-wine and fermented maize marsh (Omidun) indicated that these solvents has high affinity and similar solubility properties for the different compositions of *M. oleifera* because they contain hydroxyl group which is hydrophyllic showing that the molecules in these extracts dissolves better in these

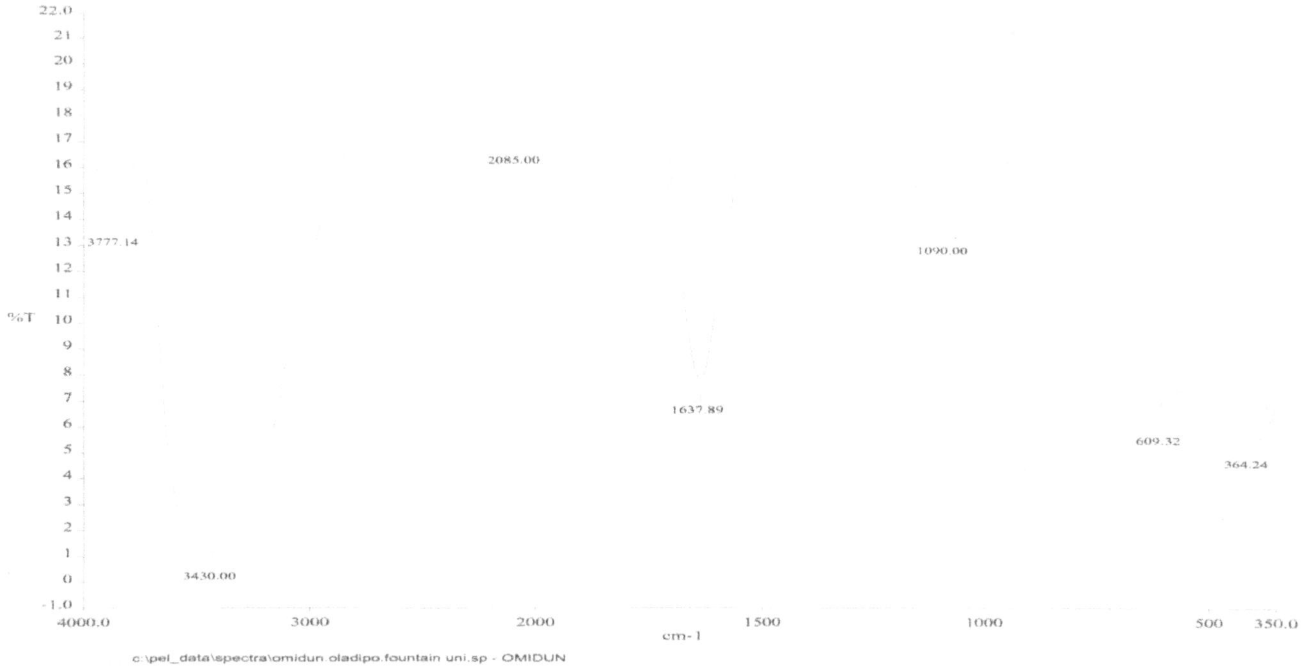

Figure 4. IR Spectrum of Fermented maize mash extracts of *M. oleifera.*

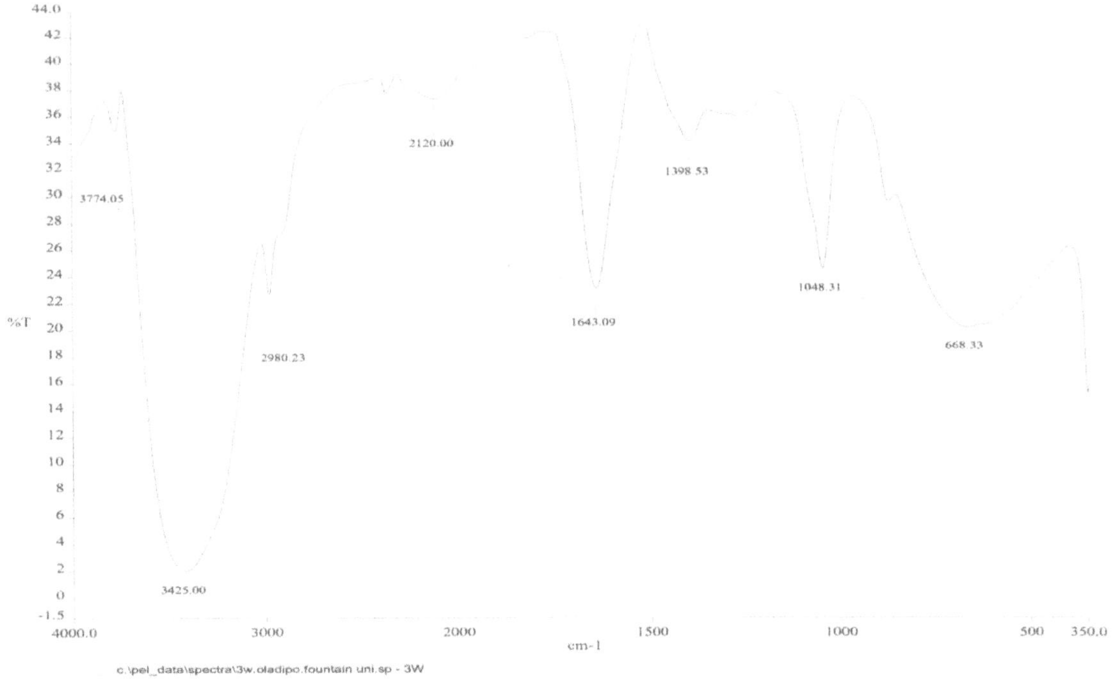

Figure 5. IR Spectrum of Hot water extracts of *M. oleifera.*

solvents. The extract yields of the plants decreased with decreasing solvent polarity. Low polar solvents such as dichloromethane and n-hexane were not able to significantly extract the *M. oleifera*. However, there is also requirement from environment considerations and toxicological implication to reduce the quantity of organic solvents used in particular chlorinated solvents (including dichloromethane) in the extraction of consumable items

Figure 6. IR Spectrum of cold water extracts of *M. oleifera.*

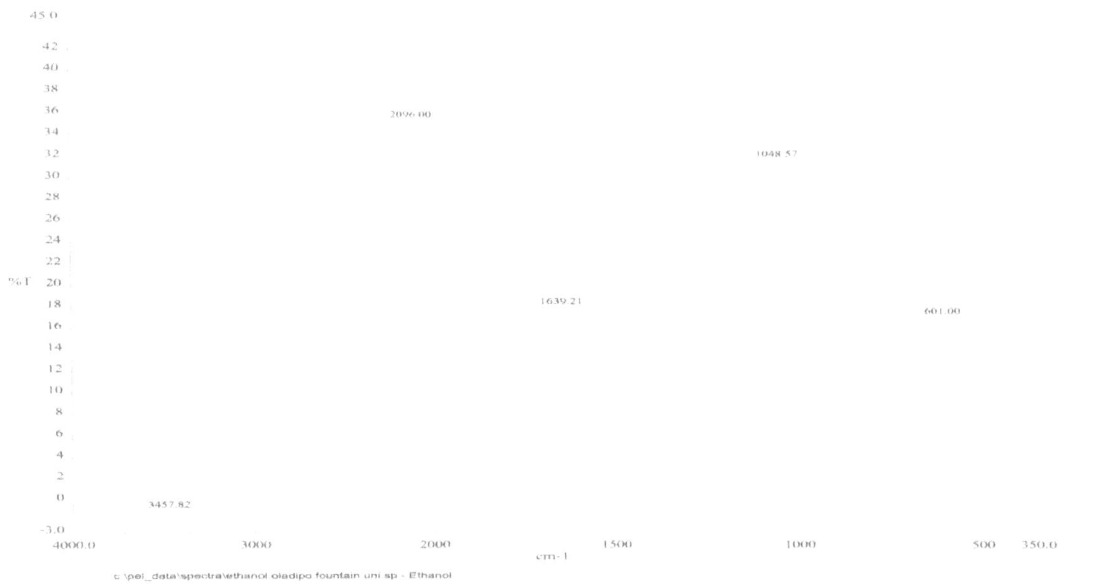

Figure 7. IR Spectrum of ethanol extracts of *M. oleifera.*

(Barwick, 1997).

Distribution of minerals in *M. oleifera* extracts

The level of the important dietary mineral constituents (k, Ca, Mg, Zn, Cu and Fe) that are determined in this study in different solvent extracts of *M. oleifera* are as summarized in Table 2. In all the extracts and for the macronutrients analysed, concentrations of K was the highest while that of Mg was the lowest. For the micronutrients analysed, Fe concentrations was the highest while Cu was the least. This report supported the previous findings of Moyo et al. (2011) that *Moringa*

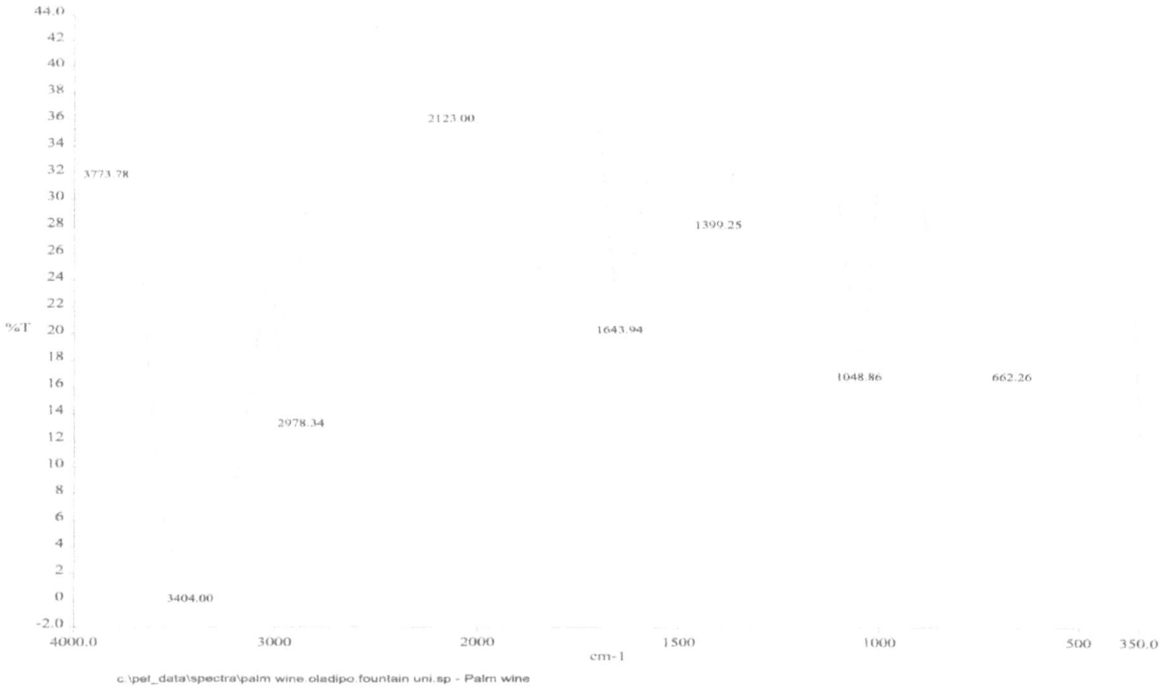

Figure 8. IR Spectrum of alm wine extracts of *M. oleifera.*

seeds and leaves could be a potential source of Fe supplement for humans and livestock.

Fermented maize marsh extracts recorded the highest mean compositions for Ca and Mg (560.50 ± 34.5 and 485.75 ± 67.32); palm-wine extracts recorded the highest compositions of Fe and Cu (7.42 ± 1.22 and 0.50 ± 0.01) while cold water and hot water extracts recorded the highest compositions of K (2045.14 ± 265.02) and Zn (2.16 ± 0.11) respectively. The lowest concentrations of these trace metals which range from 0.06 ± 0.01 to 6.47 ± 0.35 was recorded in n-hexane extracts. Statistical analysis of the mineral constituents using two sample t-tests studies at $p < 0.05$ also revealed great significant differences between the means of the mineral element in the extracts except for Ca contents in hot water and FMM extracts, as well as, in n-hexane and ethanol extracts pair; K concentrations in hot water and FMM extracts; Fe concentrations in cold water and hot water, in addition to, dichloromethane and ethanol extracts pair; Cu in cold water, FMM and dichloromethane; and Zn concentrations in ethanol and dichloromethane. This indicates that solvents polarity and types has great influence on the mineral and bioactive compositions of *M. oleifera* extracts.

The high values of Ca and K in Omidun extracts may be as a result of the high nutritional and therapeutic values of *ogi.* Earlier reports have also shown the presence of microorganisms and their attendant minerals, vitamins and soluble nitrogenous compounds present in ogi which cannot also be ruled out in omidun. This might be an added advantage for the highest yields of these elements in the omidun extracts. This indicated that the extraction process is a function of the solvents compositions.

The similarity in some of the mineral contents of the extracts obtained with cold water, hot water and omidun also attests to the high affinity and similar solubility properties of these polar solvents for *M. oleifera.*

Ultra violet/visible spectroscopy (UV/Visible) analysis

From the ultra violet/ visible spectral (Figures 9 to 15) and the summary of the ultra violet/ visible spectra analysis, that is, the wavelength maximum absorption (λ_{max}) and their corresponding absorbance for the leaves extract of *M. oleifera* in different solvents media at room temperature in Table 4, two prominent bands were observed for the extracts in the organic solvents used: Polar protic (ethanol), polar aprotic (dichloromethane), and non polar (n-hexane) while three bands were observed in the aqueous solvents (cold water, boiled water, FMM and palm-wine). The highest absorption bands of these extracts appear at 295 nm/300 nm while the lowest were obtained at 210 to 285 nm in all the solvents except in non polar n-hexane. This non appearance of band 1 in the spectra of the extracts in organic solvent could be due the non transparency of this band to these solvent.

Band 1(295 nm) in cold water is shifted to longer wavelength, in both palm-wine and FMM and the intensities as observed also increased in this solvent.

Figure 9. UV-Visible spectrum of n-hexane extracts of *M. oleifera*.

This shows that solvents have great influence (through interaction e.g., protonation, hydrogen bonding, and specific solvation effect) on the molecules of the compounds (Kolar et al., 2002). The absorbance at 295 nm was retained for the extracts in dichloromethane (DCM) and ethanol but with lower intensity (hypochromic shifts), Band 2 (410 nm) was also retained in ethanol but with increase in intensity of absorption. Increase in solvents polarity has little or no effects on the absorption maximum (λ_{max}). The band which occurred at 295 nm in DCM, boiled water and ethanol were shifted to longer wavelength, that is, 300 nm (Batochromic shift) in Omidun and cold water.

Infrared spectroscopy (IR)

Two or more compounds are one if their spectra agree in position and relative intensities

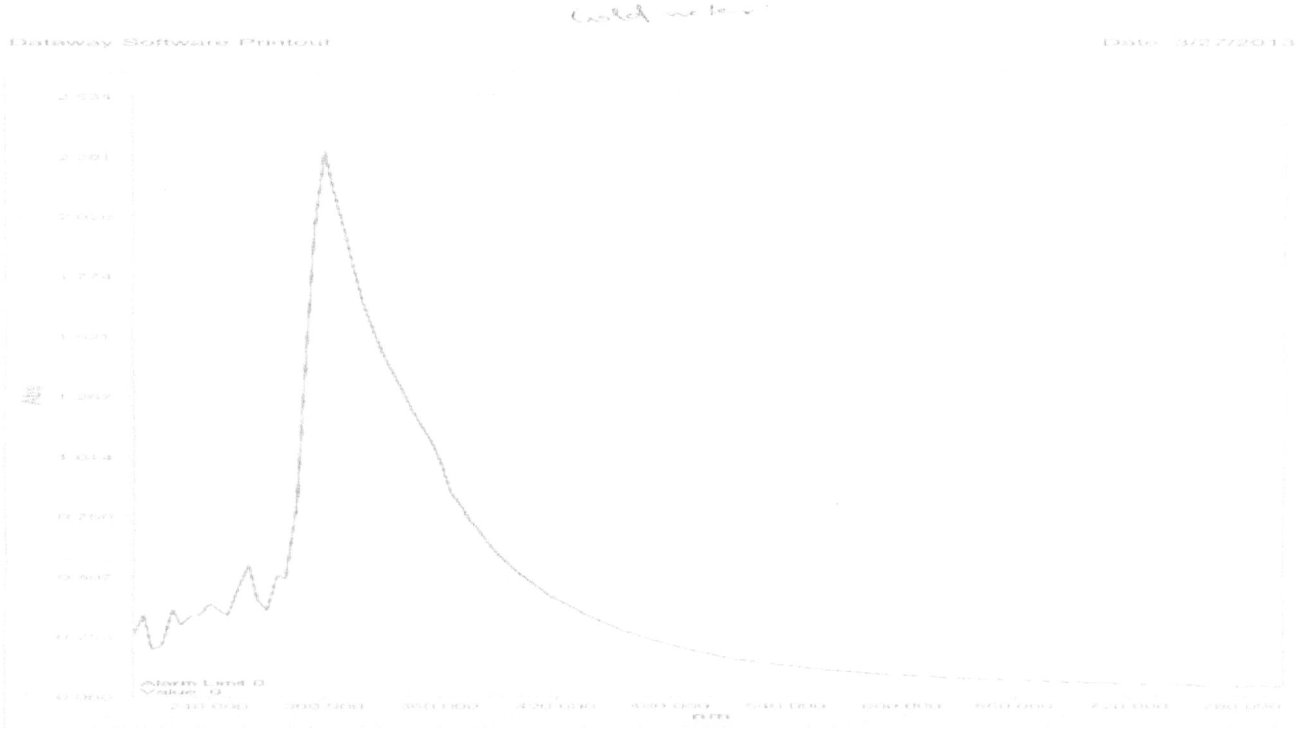

Figure 10. UV-Visible spectrum of cold water extracts of *M. oleifera*.

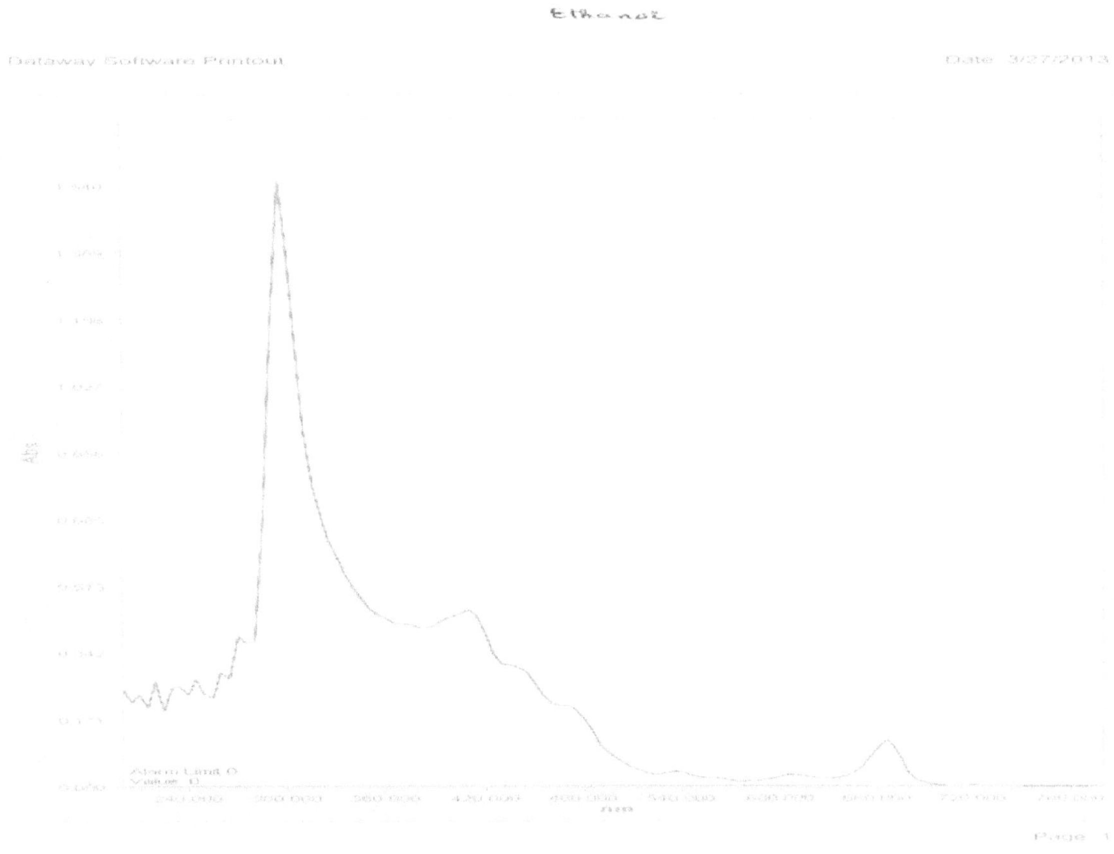

Figure 11. UV-Visible spectrum of ethanol extracts of *M. oleifera*.

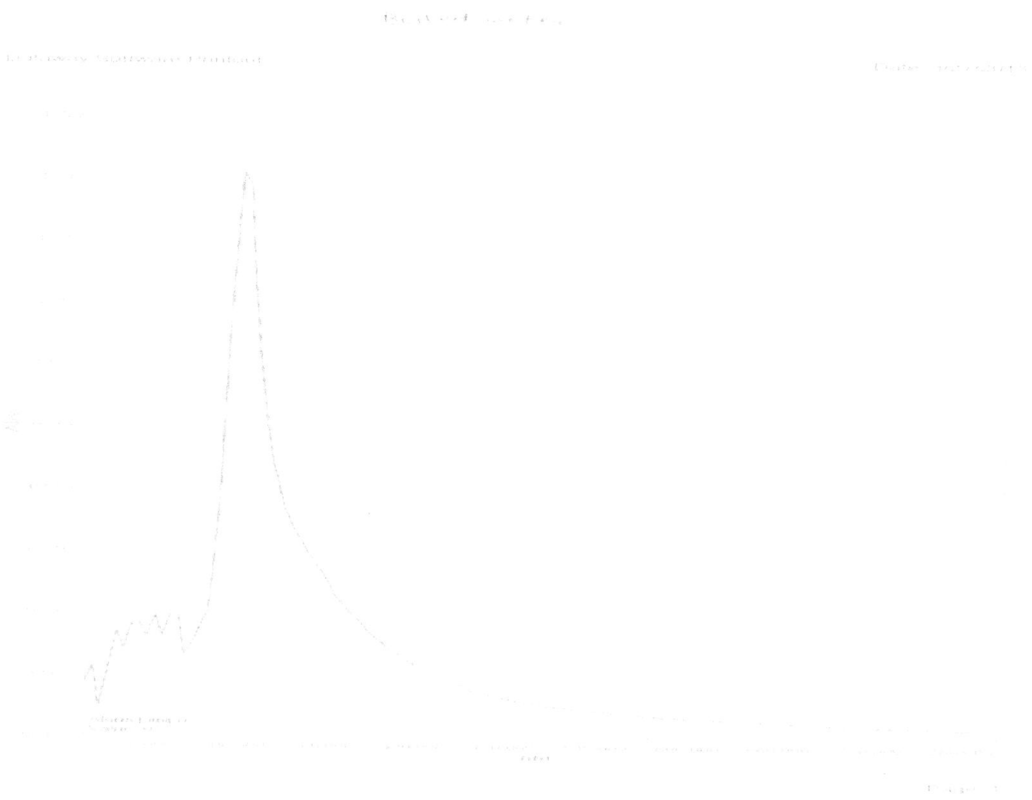

Figure 12. UV-Visible spectrum of Hot water extracts of *M. oleifera.*

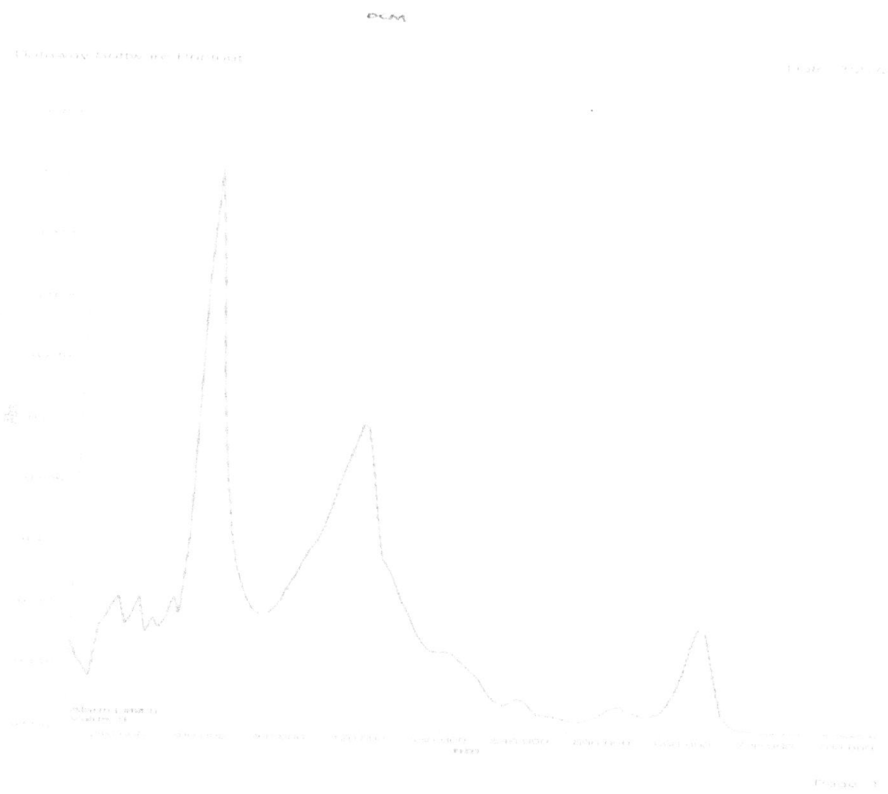

Figure 13. UV-Visible spectrum of Dichloromethane extracts of *M. oleifera.*

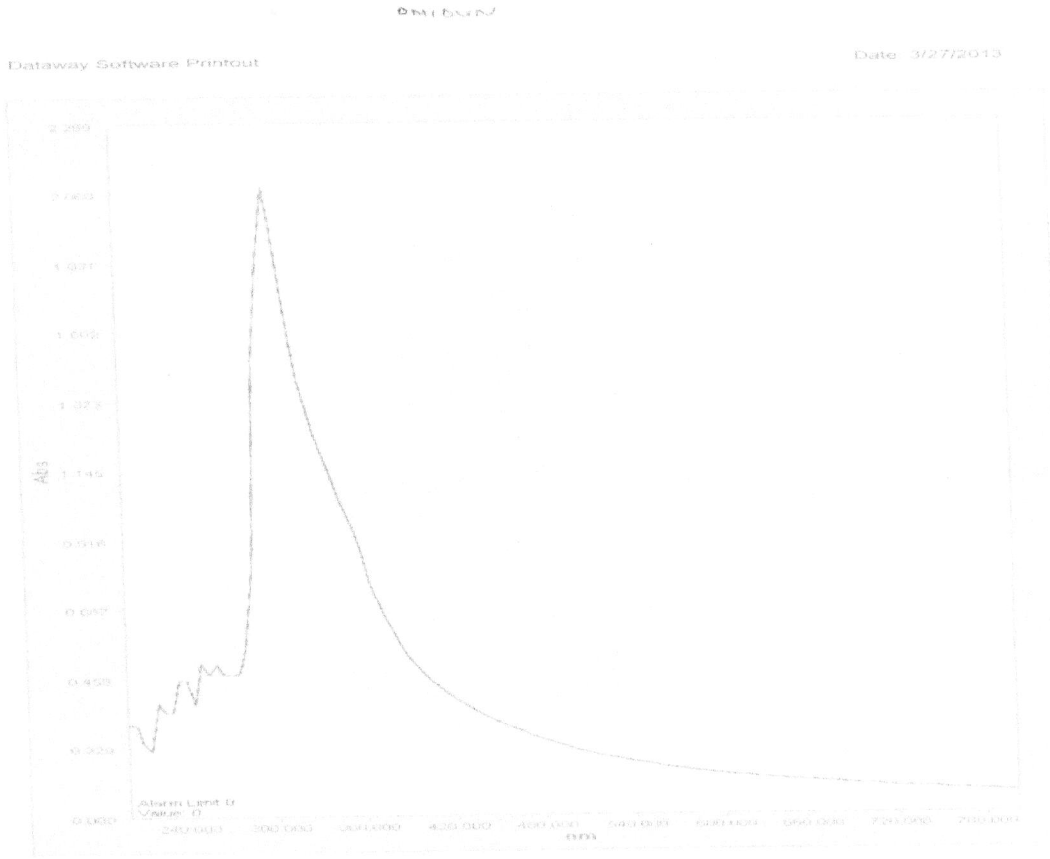

Figure 14. UV-visible spectrum of fermented maize marsh extracts of *M. oleifera.*

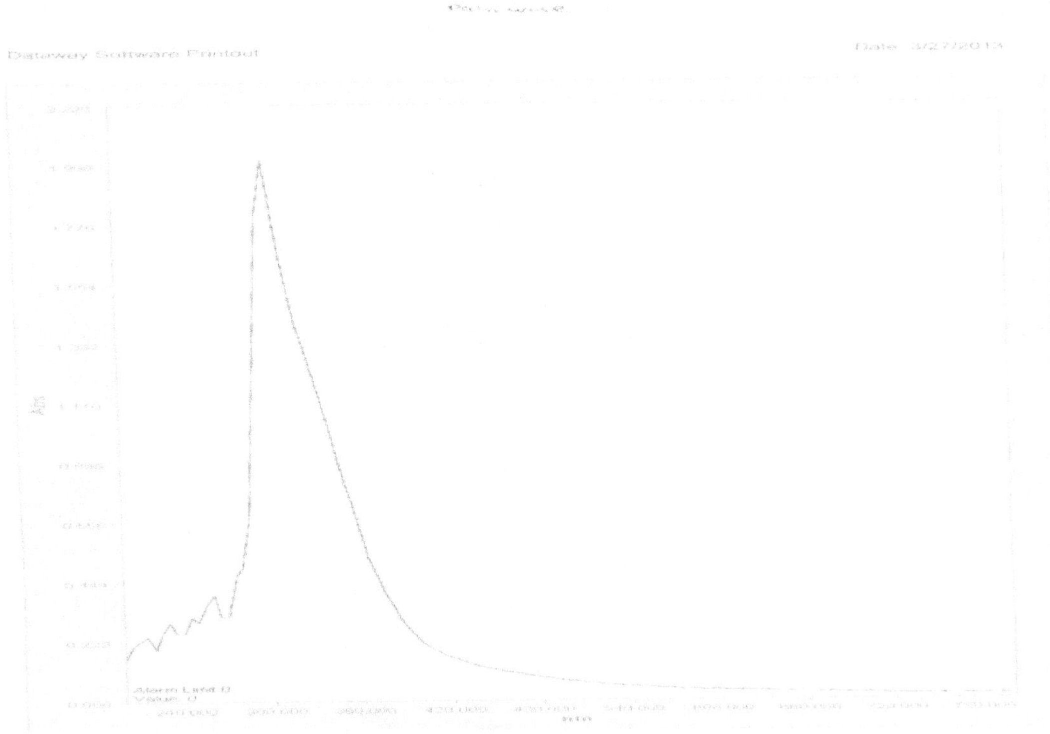

Figure 15. UV-Visible spectrum of palm-wine extracts of *M. oleifera*.

of the bands. Thus, the functional groups of the active compounds (Figure 1) in the studied *M. oleifera* extracts and the effect of different solvents on them were confirmed by (IR) analysis. Figures 3 to 8 and Table 3 show the characteristic absorption bands exhibited by the IR spectrum of the studied leave extracts in different solvents. These observed functional groups are comparable with those obtained from the proposed structures of the isolated compounds in Figure 1 with slight variations due to changes in solvent (Rubeena, 1995).

Conclusion

This study shows that solvents polarities and types have great influence on the various chemical compositions of *M. oleifera* plant extracts due to its various interactions with them. Palm wine gave the highest yields of the extracts and highest compositions of micronutrients (Fe and Cu). Fermented maize marsh gave highest composition of the macronutrients (Ca and Mg) and micronutrients (Zn) components. Statistical analysis of the mineral constituents using two sample t-tests studies at $p<0.05$ generally reveals great significant differences in the attendants studied mineral compositions in the extracts except in few cases. The Moringa from this source was found to be a good source of important minerals and this further confirmed that this plant is a viable food supplement feeds additives for dietary minerals in animal and human beings.

Conflict of Interest

The authors have not declared any conflict of interest.

REFERENCES

Abdulrahaman AA, Kolawole OM (2006). Traditional preparations and uses of maize in Nigeria. Ethnobot. Leaflets 10:219-227.

Al-Kharusi LM, Elmardi MO, Ali A, Al-Said FAJ, Abdelbasit KM, Al-Rawahi S (2009). Effect of mineral and organic fertilizers on the chemical characteristics and quality of date fruits. Int. J. Agric. Biol. 11:290–296.

Andri CK, Masitah H, Harcharan S (2009). Effects of solvent properties on the Soxhlet extraction of diterpenoid lactones from Andrographis paniculata leaves Sc. Asia. 35:306-309.

Anjorin TS, Ikokoh P, Okolo S (2010). Mineral composition of Moringa oleifera leaves, pods and seeds from two regions in Abuja, Nigeria. Int. J. Agric. Biol. 12:431-434.

AOAC (1990). Official methods of analysis, Association of Official Analytical Chemists, Washington, D.C., USA. 15th Edition. pp. 807-928.

Aslam M, Anwar F, Nadeem R, Rashid U, Kazi TG, Nadeem M (2005). Mineral composition of *Moringa oleifera* leaves and pods from different regions of Punjab, Pakistan. Asian J. Plant Sci. 4:417-421. http://dx.doi.org/10.3923/ajps.2005.417.421

Barwick VJ (1997). Strategies for solvent selection - a literature review. Trends Anal. Chem. 16:293-309. http://dx.doi.org/10.1016/S0165-9936(97)00039-3

Brisibe EA, Umoren UE, Brisibe F, Magalhaes PM, Ferreira JFS, Luthria D, Wu X, Prior RL (2009). Nutritional characterization and antioxidant capacity of different tissues of Artemisia annua L. Food Chem. 115:1240-1246. http://dx.doi.org/10.1016/j.foodchem.2009.01.033

FAO corporate document repository (1999). Fermented cereals. A global perspective. Chapter 2: Cereal fermentations in African countries. pp. 1-2.

Hsu R (2006). Moringa Oleifera Medicinal and Economic Uses. International course on Economic Botany. National herbarium, Leiden, The Netherlands.

Kalogo Y, Rosillon F, Hammes F, Verstraete W (2000). Effect of a water extract of Moringa oleifera seeds on the hydrolytic microbial species diversity of a UASB reactor treating domestic wastewater. Lett. Appl. Microbiol. 31:259-264. http://dx.doi.org/10.1046/j.1365-2672.2000.00814.x

Kolar P, Shen JW, Tsuboi A, Ishikawa T (2002). Solvent selection for pharmaceuticals. Fluid Phase Equil. 194(7):771-82. http://dx.doi.org/10.1016/S0378-3812(01)00716-6

Kumar A, Pari L (2003). Antioxidant action of Moringa oleifera Lam. (drumstick) against antitubercular drugs induced lipid peroxidation in rats. J. Med. Food. 6:255-259. http://dx.doi.org/10.1089/10966200360716670

Mahajan SG, Mali RG, Mehta AA (2007). Effect of Moringa oleifera Lam. seed extract on toluene diisocyanate –induced immune-mediated inflammatory responses in rats. J. Immunotoxicol. 4:85-96. http://dx.doi.org/10.1080/15476910701337472

Makkar HPS, Becker K (1996). Nutritional value and anti-nutritional components of whole and ethanol extracted Moringa oleifera leaves. Animal Feed Sci. Tech. 63:211-228. http://dx.doi.org/10.1016/S0377-8401(96)01023-1

Mosha TC, Pace RD, Adeyeye S, Mtebe K, Laswai H (1995). Proximate composition and mineral contents of selected Tanzanian vegetables and effect of traditional processing on the retention of ascorbic acid, riboflavin in and thiamine. Plant Food Hum. Nutri. 48:235–245. http://dx.doi.org/10.1007/BF01088445

Moyo B, Patrick J, Masika AH, Voster M (2011). Nutritional characterization of Moringa (*Moringa oleifera* Lam.) leaves Afri. J. Biotech. 10(60):12925-12933.

Ogbe AO, John PA (2011). Proximate study, Mineral and Anti-nutrient composition of *Moringa oleifera* leaves harvested from Lafia, Nigeria: Potential benefits in poultry nutrition and health. J. Microbiol. Biotech. Food Sci. 12:1(3):296-308.

Rubeena S (1995). Studies in the chemical constituents of M. Oleifera L and preparation of potential biologically significant derivatives of 8-Hydroxyquinoline. H.E.J. Research Institute of Chemistry University of Karachi, Pakistan Ph.D thesis.

Toba SA, Pius I, Simon O (2010). Mineral Composition of Moringa oleifera Leaves, Pods and Seeds from Two Regions in Abuja, Nigeria. Int. J. Agric. Biol. ISSN Print: 1560–8530; ISSN Online:1814–959609–398/SBC/2010/12–3–431–434 http://www.fspublishers.org

Permissions

All chapters in this book were first published in AJPAC, by Academic Journals; hereby published with permission under the Creative Commons Attribution License or equivalent. Every chapter published in this book has been scrutinized by our experts. Their significance has been extensively debated. The topics covered herein carry significant findings which will fuel the growth of the discipline. They may even be implemented as practical applications or may be referred to as a beginning point for another development.

The contributors of this book come from diverse backgrounds, making this book a truly international effort. This book will bring forth new frontiers with its revolutionizing research information and detailed analysis of the nascent developments around the world.

We would like to thank all the contributing authors for lending their expertise to make the book truly unique. They have played a crucial role in the development of this book. Without their invaluable contributions this book wouldn't have been possible. They have made vital efforts to compile up to date information on the varied aspects of this subject to make this book a valuable addition to the collection of many professionals and students.

This book was conceptualized with the vision of imparting up-to-date information and advanced data in this field. To ensure the same, a matchless editorial board was set up. Every individual on the board went through rigorous rounds of assessment to prove their worth. After which they invested a large part of their time researching and compiling the most relevant data for our readers.

The editorial board has been involved in producing this book since its inception. They have spent rigorous hours researching and exploring the diverse topics which have resulted in the successful publishing of this book. They have passed on their knowledge of decades through this book. To expedite this challenging task, the publisher supported the team at every step. A small team of assistant editors was also appointed to further simplify the editing procedure and attain best results for the readers.

Apart from the editorial board, the designing team has also invested a significant amount of their time in understanding the subject and creating the most relevant covers. They scrutinized every image to scout for the most suitable representation of the subject and create an appropriate cover for the book.

The publishing team has been an ardent support to the editorial, designing and production team. Their endless efforts to recruit the best for this project, has resulted in the accomplishment of this book. They are a veteran in the field of academics and their pool of knowledge is as vast as their experience in printing. Their expertise and guidance has proved useful at every step. Their uncompromising quality standards have made this book an exceptional effort. Their encouragement from time to time has been an inspiration for everyone.

The publisher and the editorial board hope that this book will prove to be a valuable piece of knowledge for researchers, students, practitioners and scholars across the globe.

List of Contributors

Sherifat A. Aboaba
Department of Chemistry, University of Ibadan, Ibadan, Nigeria

Yeye, Emmanuel
Department of Chemistry, University of Ibadan, Ibadan, Nigeria

S. Oluwagbemiga Alayande
Centre for Energy Research and Development, Obafemi Awolowo University, Ile-Ife, Osun State, Nigeria
Department of Chemistry, Federal University of Agriculture, Abeokuta, Ogun State, Nigeria

E. Olugbenga Dare
Department of Chemistry, Federal University of Agriculture, Abeokuta, Ogun State, Nigeria

W. Babatunde Ayinde
Department of Chemistry, Federal University of Agriculture, Abeokuta, Ogun State, Nigeria

Janet Bamigbose
Department of Chemistry, Federal University of Agriculture, Abeokuta, Ogun State, Nigeria

P. A. Ayedun
Department of Mechanical Engineering, Federal University of Agriculture, Abeokuta, Ogun State, Nigeria

G. Ayodeji Osinkolu
Centre for Energy Research and Development, Obafemi Awolowo University, Ile-Ife, Osun State, Nigeria

J. C. Nnaji
National Institute for Freshwater Fisheries Research, New Bussa, Nigeria

A. Uzairu
Department of Chemistry, Ahmadu Bello University, Zaria, Nigeria

C. E.Gimba
Department of Chemistry, Ahmadu Bello University, Zaria, Nigeri

J. A. Kagbu
Department of Chemistry, Ahmadu Bello University, Zaria, Nigeria

M. Djebbar
Faculty of Chemistry, University of Technology, BP 1505, 31045 El-Mnouar, Oran, Algérie

F. Djafri
Faculty of Chemistry, University of Technology, BP 1505, 31045 El-Mnouar, Oran, Algérie.
University of Technology, Mascara, Algeria

M. Bouchekara
Faculty of Chemistry, University of Technology, BP 1505, 31045 El-Mnouar, Oran, Algérie

A. Djafri
University of Technology, Mascara, Algeria

A. I. Ajai
Department of Chemistry, Federal University of Technology, Minna Nigeria

M. Bankole
Department of Chemistry, Federal University of Technology, Minna Nigeria

J. O. Jacob
Department of Chemistry, Federal University of Technology, Minna Nigeria

U. A. Audu
Department of Chemistry, Federal University of Technology, Minna Nigeria

M. I. Yusufu
Department of Food Science and Technology, Federal Polytechnic, Idah, Nigeria

C. C. Ariahu
Department of Food Science and Technology, Federal Polytechnic, Idah, Nigeria

B. D. Igbabul
Department of Food Science and Technology, University of Agriculture, Makurdi, Nigeria

Ali Moghimi
Department of Chemistry, Varamin (Pishva) Branch Islamic Azad University, Varamin, Iran

Nilgül Özmen
Kahramanmaraş Sütçü İmam University, Faculty of Forestry, Kahramanmaraş, Turkey

Nihat Sami Çetin
Kahramanmaraş Sütçü İmam University, Faculty of Forestry, Kahramanmaraş, Turkey

Arun K. Singh
Indian Institute of Petroleum, Dehradun 248005, India

Satish Kumar
Indian Institute of Petroleum, Dehradun 248005, India

B. A. Akinwande
Department of Food Science and Engineering, Ladoke Akintola University of Technology, Ogbomoso, Oyo State, Nigeria

I. O. Adeoye
Department of Pure and Applied Chemistry, Ladoke Akintola University of Technology, Ogbomoso, Oyo State, Nigeria

I. B. Anjorin
Department of Food Science and Engineering, Ladoke Akintola University of Technology, Ogbomoso, Oyo State, Nigeria

S. T. Ubwa
Department of Chemistry, Benue State University, Makurdi - Nigeria

K. Asemave
Department of Chemistry, Benue State University, Makurdi - Nigeria

G. O. Igbum
Department of Chemistry, Benue State University, Makurdi - Nigeria

A. J. Hotton
National Agency for Food and Drug Administration and Control P. M. B 1018 Jalingo, Taraba State, Nigeria

J. T. Barminas
Department of Chemistry, Federal University of Technology, P. M. B. 2076, Yola, Adamawa State, Nigeria

S. A. Osemeahon
Department of Chemistry, Federal University of Technology, P. M. B. 2076, Yola, Adamawa State, Nigeria

W Ivwurie
Department of Chemistry, Federal University of Petroleum Resources, Effurun, Delta State, Nigeria

C. E Snape
Nottingham Fuel and Energy Centre, the University of Nottingham, University Park, Nottingham, United Kingdom

C Sun
Nottingham Fuel and Energy Centre, the University of Nottingham, University Park, Nottingham, United Kingdom.

Abid A. Lone
Division Of Post Harvest Technology, Sher-e-Kashmir University of Agricultural Sciences and Technology-Shalimar, Kashmir, India

Shaiq A. Ganai
Division Of Post Harvest Technology, Sher-e-Kashmir University of Agricultural Sciences and Technology-Shalimar, Kashmir, India

Rayees A. Ahanger
Division Of Post Harvest Technology, Sher-e-Kashmir University of Agricultural Sciences and Technology-Shalimar, Kashmir, India

Hilal A. Bhat
Division Of Post Harvest Technology, Sher-e-Kashmir University of Agricultural Sciences and Technology-Shalimar, Kashmir, India

Tauseef A. Bhat
Division Of Post Harvest Technology, Sher-e-Kashmir University of Agricultural Sciences and Technology-Shalimar, Kashmir, India

Imtiyaz A. Wani
Division Of Post Harvest Technology, Sher-e-Kashmir University of Agricultural Sciences and Technology-Shalimar, Kashmir, India

Rashid Mahmood
Division of Science and Technology, University of Education, Township Lahore-54000, Pakistan

Afia Naseer
Division of Science and Technology, University of Education, Township Lahore-54000, Pakistan

C. E. Snape
Nottingham Fuel and Energy Centre, the University of Nottingham, University Park, Nottingham, United Kingdom

Khaldoun Al-Sou'od
Department of Chemistry, Faculty of Science, Al al-Bayt University, P. O. Box 130040, Mafraq 25113, Jordan

Md. Ahsan Habib
Department of Chemistry, Faculty of Science, University of Dhaka, Dhaka 1000, Bangladesh

Iqbal Mohmmad Ibrahim Ismail
Department of Chemistry, Faculty of Science, King Abdulaziz University Jeddah 21589, Saudi Arabia

Abu Jafar Mahmood
Department of Chemistry, Faculty of Science, University of Dhaka, Dhaka 1000, Bangladesh

Md. Rafique Ullah
Department of Chemistry, Bangladesh University of Engineering and Technology, Dhaka 1000, Bangladesh

Sameh H. Othman
Nuclear Research Center, Atomic Energy Authority, Cairo, Egypt

N. Belghiti
Laboratoire de Recherche «Chimie-Biologie appliquées à l'environnement», Faculté des Sciences, Université Moulay Ismail Meknès, Marocco

M. Bennani
Laboratoire de Recherche «Chimie-Biologie appliquées à l'environnement», Faculté des Sciences, Université Moulay Ismail Meknès, Marocco

M. Hamidi
URMM/UCTA, Faculté des Sciences et Techniques d'Errachidia, Université Moulay Ismaïl, Maroc

S. M. Bouzzine
URMM/UCTA, Faculté des Sciences et Techniques d'Errachidia, Université Moulay Ismaïl, Maroc

M. Bouachrine
ESTM, Université Moulay Ismail, Meknes, Marocco

Jian-yu Shang
College of Environmental and Energy Engineering, Beijing University of Technology, Beijing City, Chaoyang, 100124, China

Zhong-liang Liu
College of Environmental and Energy Engineering, Beijing University of Technology, Beijing City, Chaoyang, 100124, China

Chun-bo Wang
School of Energy and Power Engineering, North China Electric Power University, Hebei Province, Baoding 071003, China

Acaye Ongwech
Department of Chemistry Gulu University, P. O. Box 166, Gulu, Uganda

George William Nyakairu
Department of Chemistry, School of Physical Sciences, College of Natural Sciences, Makerere University, P. O. Box 7062, Kampala, Uganda

Jolocam Mbabazi
Department of Chemistry, School of Physical Sciences, College of Natural Sciences, Makerere University, P. O. Box 7062, Kampala, Uganda

Justus Kwetegyeka
Department of Chemistry, Kyambogo University, P. O. Box 1, Kyambogo, Uganda

Margaret Masette
Food Bioscience Research Centre - National Agriculture Research Organisation Kawanda, P. O. Box 7852, Kampala Uganda

F. M. Alabi
Raw Materials Research and Development Council PMB 232, Garki Abuja, Nigeria

M. O. Omojola
Raw Materials Research and Development Council PMB 232, Garki Abuja, Nigeria

N.Oragwa Leonard
Chemistry Department, University of Abuja - Nigeria

O.Olajide Olutayo
Chemistry Advanced Laboratory, Sheda Science and Technology Complex, Abuja- Nigeria

O. Efiom Otu
Chemistry Department, University of Abuja - Nigeria

K. Okwute Simon
Chemistry Department, University of Abuja - Nigeria

S. Isah
Bells University of Technology, Nigeria

A. A. Oshodi
Bells University of Technology, Nigeria

V. N. Atasie
Bells University of Technology, Nigeria

Simon Terver Ubwa
Department of Chemistry, Benue State University, P. M. B 102119, Makurdi, Nigeria

James Abah
Department of Chemistry, Benue State University, P. M. B 102119, Makurdi, Nigeria

Barnabas Atsinafe Oshido
Department of Chemistry, Benue State University, P. M. B 102119, Makurdi, Nigeria

Esther Otokpa
Department of Chemistry, Benue State University, P. M. B 102119, Makurdi, Nigeria

Robert Byamukama
Department of Chemistry, Makerere University, P. O. Box 7062, Kampala, Uganda

Moses Andima
Department of Chemistry, Makerere University, P. O. Box 7062, Kampala, Uganda

Angella Mbabazi
Department of Chemistry, Makerere University, P. O. Box 7062, Kampala, Uganda
Department of Chemistry, Busitema University, P. O. Box 236, Tororo, Uganda

Bernard T. Kiremire
Department of Chemistry, Makerere University, P. O. Box 7062, Kampala, Uganda

Millicent U. Ibezim-Ezeani
Department of Pure and Industrial Chemistry, University of Port Harcourt, P. M. B. 5323, Choba, Port Harcourt, Nigeria

Alphonso C. I. Anusiem
Department of Pure and Industrial Chemistry, University of Port Harcourt, P. M. B. 5323, Choba, Port Harcourt, Nigeria

Louis M. Nwokocha
Department of Chemistry, University of Ibadan, Ibadan, Nigeria

Gabriel B. Ogunmola
Institute of Genetic Chemistry and Laboratory Medicine, Bodija, Ibadan, Nigeria

M. Mupa
Department of Chemistry, Bindura University of Science Education, P. Bag 1020, Bindura, Zimbabwe

C. B. Hungwe
Department of Chemistry, Bindura University of Science Education, P. Bag 1020, Bindura, Zimbabwe

S. Witzleben
Bonn-Rhein-Sieg University of Applied Sciences, Inorganic Chemistry and Materials Analysis, von Liebig Str. 20, 53359 Rheinbach, Germany

C. Mahamadi
Department of Chemistry, Bindura University of Science Education, P. Bag 1020, Bindura, Zimbabwe

N. Muchanyereyi
Department of Chemistry, Bindura University of Science Education, P. Bag 1020, Bindura, Zimbabwe

M. Tembo
Department of Physics, School of Natural Sciences, University of Zambia, P. O. Box 32379, Lusaka, 10101 Zambia

M. O. Munyati
Department of Chemistry, School of Natural Sciences, University of Zambia, P. O. Box 32379, Lusaka, 10101 Zambia

S. Hatwaambo
Department of Physics, School of Natural Sciences, University of Zambia, P. O. Box 32379, Lusaka, 10101 Zambia

M. Maaza
Nanosciences Laboratories, Materials Research Department, iThemba LABS, P. O. Box 722, Somerset West 7129, Western Cape, South Africa

Adeoye, M. Dasola
Industrial and Environmental Chemistry Unit, Chemical Sciences Department, Fountain University, Osogbo Osun State, Nigeria

A.Tunbosun Lawal
Industrial and Environmental Chemistry Unit, Chemical Sciences Department, Al-Hikmah University, Ilorin Kwara State, Nigeria

L. Adeyemi Azeez
Industrial and Environmental Chemistry Unit, Chemical Sciences Department, Fountain University, Osogbo Osun State, Nigeria

O. Abidemi Olayiwola
Chemical Sciences Department. Oduduwa University,Ipetumodu, Ile-Ife, Osun State, Nigeria